# Foundations in Signal Processing, Communications and Networking

Volume 11

More information about this series at http://www.springer.com/series/7603

Rudolf Ahlswede

# Transmitting and Gaining Data

## Rudolf Ahlswede's Lectures on Information Theory 2

*Edited by*

Alexander Ahlswede
Ingo Althöfer
Christian Deppe
Ulrich Tamm

Springer

*Author*
Rudolf Ahlswede (1938–2010)
Faculty of Mathematics
University of Bielefeld
Bielefeld
Germany

*Editors*
Alexander Ahlswede
Bielefeld
Germany

Ingo Althöfer
Friedrich-Schiller University
Jena
Germany

Christian Deppe
University of Bielefeld
Bielefeld
Germany

Ulrich Tamm
Bielefeld University of Applied Sciences
Bielefeld
Germany

ISSN 1863-8538 ISSN 1863-8546 (electronic)
ISBN 978-3-319-35656-3 ISBN 978-3-319-12523-7 (eBook)
DOI 10.1007/978-3-319-12523-7

Mathematics Subject Classification (2010): 94, 94A, 68P, 68Q

Springer Cham Heidelberg New York Dordrecht London



Printed on acid-free paper

Springer is part of Springer Science+Business Media (www.springer.com)

# Preface[1]

Classical information processing consists of the main tasks of **gaining** knowledge, **storage**, **transmission**, and **hiding** data.

The first named task is the prime goal of Statistics and for the next two Shannon presented an impressive mathematical theory, called Information Theory, which he based on probabilistic models.

The basics in this theory are concepts of codes—lossless and lossy—with small error probabilities in spite of noise in the transmission, which is modeled by channels.

Another way to deal with noise is based on a combinatorial concept of error correcting codes, pioneered by Hamming. This leads to another way to look at Information Theory, which instead of being looked at by its tasks can also be classified by its mathematical structures and methods: primarily **probabilistic** versus **combinatorial**.

Finally, Shannon also laid the foundations of a theory concerning hiding data, called Cryptology. Its task is in a sense dual to transmission and we therefore prefer to view it as a subfield of Information Theory.

Viewed by mathematical structures there is again already in Shannon's work a **probabilistic** and a combinatorial or **complexity**-theoretical model.

The lectures are suitable for graduate students in Mathematics, and also in Theoretical Computer Science, Physics, and Electrical Engineering after some preparations in basic Mathematics.

The lectures can be selected for courses or supplements of courses in many ways.

Rudolf Ahlswede

[1] This is the original preface written by Rudolf Ahlswede for the first 1,000 pages of his lectures. This volume consists of the second third of these pages.

# Contents

# Words and Introduction of the Editors

The editors feel somewhat uneasy. Rudolf Ahlswede's work in Information Theory will be covered in 10 volumes. We have to write a preface for each of them. Can you imagine how Rudolf Ahlswede would have done this? On one of his pragmatic days he would have given simple order: "Ok, let's copy the preface of Volume 1 and include it in each of the other nine volumes." But then, someone from Springer-Verlag would have complained: "This is not the way we expect things to run. Please write proper prefaces." A way out of the dilemma might be the following: There will be a new preface by the editors in each volume, but not very long ones. They will contain some comments on the content, and each one also one of the many anecdotes from Rudolf Ahlswede's long academic life.

So, we start with a short introduction to the second volume. Whereas the first volume contained introductory lectures on data compression, universal coding, channel coding, information inequalities, etc., which Rudolf Ahlswede regularly presented in a first lecture on Information Theory, the contents of the subsequent volumes are not so clearly assigned, say, to Lectures on Information Theory II, III, and so on. In further lectures, Rudolf Ahlswede rather decided to present selected chapters on Information Theory which varied over the years, since these chapters were usually closely related to his current research. In this way, many lecture notes, especially on channel coding, identification, quantum information theory, and related combinatorics, were collected over the years.

Besides, he decided to include further material, which he wanted to have in his book, also in the form of lecture notes. The advantage is that in contrast to a presentation in a standard textbook, the chapters may be rather independent of each other. A change in notation, which occurs over the lecture notes, then can make sense depending on the various aspects of the concept. For instance, in some lecture notes, the mutual information is denoted as $I(X, Y)$ depending on the input and output variables of the channel, in other places $I(P, W)$ is used more emphasizing the underlying probability distributions $P$ for $X$ and $PW$ for $Y$ when it seemed appropriate to him. The reader should be aware of such changes. However, this also has as a consequence that some "lecture notes" may contain a little more stuff and more advanced material than suitable for one lecture. As already pointed out in our editors' words to the first volume, Rudolf Ahlswede several times planned to (and was asked to) publish some parts of these lecture notes. However, finally he decided to wait in order to further shape the material according to his "General Theory of Information Transfer." This means that the references of some chapters which he worked on a long time ago may be not up to date. For this volume this hopefully will not be a big problem, since here Rudolf Ahlswede presents in some detail very classical stuff from the early days of information theory including developments in the USA, Russia, Hungary, and (which probably nobody else can describe any more) in the German school around his supervisor Konrad Jacobs.

In Chaps. 3 and 5, Rudolf Ahlswede made the approach to a rigorous justification of the foundations of information theory. It is known that Shannon expounded his ideas, the severity of which is not covered in the most general case, the task available to mathematicians.

In the subsequent anecdote from 1991 (told by Ingo Althöfer), the trouble with another delay of a publication is described: The Faculty of Mathematics at Bielefeld University had a very successful "Sonderforschungs-Bereich" on "Discrete Mathematics." Rudolf Ahlswede was a central figure, involved with two projects: one on Information Theory, the other one on Combinatorics. The SFB started in 1989 and lasted for 12 years. In Spring and Summer 1991, the application for the first prolongation had to be written. All project leaders were busy with their new descriptions. Only Ahlswede was confident that he would not have to do paper work: "In my original project descriptions (from 1988) there are so many important open problems left that we can simply continue to work on them." SFB's director, Prof. Walter Deuber, gave him a warning: "Rudi, they expect us—and also you—to write new descriptions. Please, do so.""Sorry Walter, but that is not necessary. My projects are fine." Deuber knew how things run and people functioned and said "ok," shrugging his shoulders. One Friday morning in July 1991, I was working in my office. Prof. Ahlswede knocked at the (open) door: "Herr Althöfer, can you please come for a moment. It seems we have ein kleines Problem." In Ahlswede's office, Deuber was sitting on the sofa and grinning: "We just got a call from the headquarter of Deutsche Forschungs-Gemeinschaft (DFG). They looked through our application for the SFB prolongation and found that the Ahlswede project descriptions are unchanged from 1988. They urge us to provide new descriptions until Tuesday." On Tuesday and Wednesday the inspection of the SFB by an international commission of experts would take place. Now Prof. Ahlswede took over: "Sehen Sie, Herr Althöfer. You have to help me in writing the new application." Boom! Just in those days I had some complicated private affair, and now this… We arranged a special date. "Herr Althöfer, please start to think now—Prof. Deuber will give you helpful formal instructions—and come to my home on Saturday, at 2 p.m." It became a long session in Stapenhorststr. 150, and in the remaining hours of the weekend we wrote draft versions for the application. For me it was the chance to include also own research topics—some of them from my beloved "theory of game tree search." On Monday we made the final touch, with some TeX help by Prof. Deuber's assistant Dr. Wolfgang Thumser, and got 60 pages ready just in time. On Tuesday morning each referee found on his place a blue rapid letter-file with this last-minute work. It was a success...

The comment to this book was written by Gerhard Kramer. Rudolf Ahlswede met him in 1996 at the Winter School on Coding and Information Theory in Mölle, Sweden. Gerhard was at this time a Ph.D. student of James L. Massey (1934–2013). Christian Deppe (one of the editors of this book) was also at this meeting. He was a Ph.D. student of Rudolf Ahlswede. On the day the Ph.D. students gave talks about their first results (most of them gave the first talk outside their universities), Christian Deppe cited in his talk one of the results of Kamil Zigangirov. It was a capacity function, which was presented by Christian as a graph. During the talk, Kamil Zigangirov stood up and said that the figure was wrong. Instead of the second part of the function, the first part of the function had to be linear. During the talk the problem could not be solved, because Christian thought his figure was correct. During the break, Gerhard noticed that Christian used the y-axis for the rate, but Kamil used in his original paper the x-axis for the rate. Thus Gerhard could solve the discussion between them. After the break, Kamil stood up and apologized for his comment. Now Gerhard Kramer was one of the leading experts in Information Theory. He was President of the IEEE Information Theory Society (at the time we asked him for his comments to this volume) and organized many Conferences. Even now he is still very diplomatic and good at solving conflicts. He is one of the people who fought for the Information Theory Society and tried to keep the memory of the fathers of Information Theory.

Our thanks go to Regine Hollmann, Carsten Petersen, and Christian Wischmann for helping us in typing, typesetting, and proofreading. Furthermore, our thanks go to Bernhard Balkenhol who combined the first approximately 2,000 pages of lecture scripts in different styles (amstex, latex, etc.) to one big lecture script. He can be seen as one of the pioneers of Ahlswede's lecture notes.

At this place, we would like to remember Vladimir Balakirski who passed away in August 2013. We were shocked by this sad news while we were working on this second volume. Vladimir was a member of Rudolf Ahlswede's research group in Bielefeld from 1995 to 1999. During this time he recorded Ahlswede's lectures on tape and also transferred several of them to TEX, which will be included in the forthcoming volumes. Vladimir was an enthusiastic researcher, and an open and friendly person with a good sense of humor. He will be missed by the editors.

Alexander Ahlswede
Ingo Althöfer
Christian Deppe
Ulrich Tamm

# Part I
# Transmitting Data

# Chapter 1
# Special Channels

## 1.1 Lecture on the Weak Capacity of Averaged Channels

Coding theorem and weak converse of the coding theorem are proved for averaged semicontinuous stationary channels and for almost periodic discrete channels, whose phases are statistically known. Explicit formulas for the capacities are given. The strong converses of the coding theorems do not hold.

### *1.1.1 Introduction*

Averaged channels were introduced by Jacobs [29]. In this class of channels he found the first example of a channel for which the strong converse of the coding theorem does not hold, but the coding theorem and the weak converse of the coding theorem hold (unpublished). The proof uses Strassen's version of Shannon's method of random codes. The disadvantage of this method is that the channel capacity cannot be computed. In [44] Wolfowitz gave simpler examples of channels without strong capacity and he gave the reason for which the strong converse of the coding theorem fails to hold, but he does not show if these channels have a weak capacity. In Sect. 1.1.3, we give a simple example of a channel without strong capacity, but with weak capacity. The proof of the coding theorem and its weak converse uses classical results for simultaneous channels [17, 43]. The main idea consists in a comparison of codes for an averaged channel with its corresponding simultaneous channel. This leads to a method which is representative for the more complicated proof in Sects. 1.1.3–1.1.6, where we deal with coding theorem and weak converses for averaged channels under different assumptions on the time structure (stationary, almost periodic, nonstationary) and the output alphabet (finite, nonfinite) of the channel. *It is remarkable that we need the strong converse for simultaneous channels for the proof of weak converses for averaged channels.* This demonstrates the importance of the concept of a strong converse—which is due to Wolfowitz [42, 43]—even to coding

 
A. Ahlswede et al. (eds.), *Transmitting and Gaining Data*,
Foundations in Signal Processing, Communications and Networking 11,
DOI 10.1007/978-3-319-12523-7_1

theorists who are only interested in weak converses. The result can be formulated roughly as follows:

The weak capacity of the averaged discrete stationary channel equals the strong capacity of the corresponding simultaneous channel (Theorem 2).

The weak capacity of the averaged semi-continuous stationary channel is in general greater than the weak capacity of the corresponding simultaneous channel (Theorem 3).

The weak capacity of an almost periodic discrete channel, whose phase is statistically known, is greater than or equal to the strong capacity of the corresponding simultaneous channel. Equality holds if each phase has positive probability (Corollary of Theorem 4).

In all these cases, we average with respect to a discrete PD. In Sect. 1.1.3, we prove the coding theorem and the weak converse for stationary discrete averaged channels, where the average is taken with respect to a general PD. The proof is based on the proof for Theorem 3 and several new ideas.

### *1.1.2 Definitions*

Let $\mathcal{X}_t = \mathcal{X} = \{1, \dots, a\}$ for $t = 1, 2, \dots$ and $(\mathcal{Y}_t, \mathcal{L}_t) = (\mathcal{Y}, \mathcal{L})$ for $t = 1, 2, \dots$ where $\mathcal{Y}$ is an arbitrary set and $\mathcal{L}$ is a $\sigma$-algebra of subsets in $\mathcal{Y}$. Furthermore let $S = \{s, \dots\}$ be a nonempty (index-) set and let $F_t(\cdot|1|s), \dots, F_t(\cdot|a|s)$ be PDs on $(\mathcal{X}_t, \mathcal{L}_t), t \in N, s \in S$. For each $x^n = (x_1, \dots, x_n), \mathcal{X}^n = \prod_{t=1}^n \mathcal{X}_t$ we define a PD on

$$\left(\mathcal{Y}^n = \prod_{t=1}^n \mathcal{Y}_t, \mathcal{L}^n = \prod_{t=1}^n \mathcal{L}_t\right) \quad \text{by} \quad F^n(\cdot|x^n|s) = \prod_{t=1}^n F_t(\cdot|x_t|s).$$

The sequence of kernels $(F^n(\cdot|\cdot|s))_{n=1}^{\infty}$ forms a semi-continuous (in general nonstationary) channel without memory. [In case $\mathcal{Y}_t = \mathcal{X}_t$ is finite, the kernels $F_t(\cdot|\cdot|s)$ are given by stochastic matrices $w_t(k|i|s) = F_t(\{k\}|i|s), i \in \mathcal{X}, k \in \mathcal{Y}$. We speak then of a discrete channel without memory.] Thus we have assigned to each $s \in S$ a semi-continuous channel. If we are interested in the simultaneous behavior of all these channels, then we call this indexed set of channels a simultaneous channel (semi-continuous, without memory). Common properties of the individual channels are assigned to the simultaneous channel: stationarity, almost periodicity, etc., as follows. The set $\{F^n(\cdot|\cdot|s) : s \in S\}$ designated by $S^n$ is called simultaneous channel $S^n$ in the discrete time interval $[1, n]$.

A code $(n, N, \lambda)$ for the simultaneous channel $S^n$ is a series of pairs

$$\{(u_1, D_1), \dots, (u_n, D_n)\},$$

where $u_i \in \mathcal{X}^n$, $D_i \in \mathcal{L}^n$ for $i = 1, \dots, N$, $D_i \cap D_j = \emptyset$ for $i \neq j$ and $F^n(D_i|u_i|s) \geq 1 - \lambda, 0 < \lambda < 1$ for $i = 1, 2, \dots, N$, $s \in S$ (sometimes instead of the code $(n, N, \lambda)$ we speak of a $\lambda$-code). $N$ is the length of the code $(n, N, \lambda)$ $N(n, \lambda)$ is the maximal length of a $\lambda$-code in $[1, n]$.

Let $\mathcal{A}^n = (A_1, \dots, A_b)$ be a partition of $\mathcal{Y}^n$ in finitely many elements of $\mathcal{L}^n$ and let $\mathcal{B}$ be the set of all such finite partitions. $\mathcal{P}^n$ is the set of all PDs on $(\mathcal{Y}^n, \mathcal{L}^n)$, where $\mathcal{L}^n$ is the $\sigma$-algebra of all subsets of $\mathcal{Y}^n$.

$$I(p^n, \mathcal{A}^n, s) = \sum_{i=1,\dots,b} \sum_{x^n \in \mathcal{X}^n} p^n(x^n) F^n(D_i|x^n|s) \log \frac{F^n(D_i|x^n|s)}{\sum_{y^n \in \mathcal{Y}^n} p^n(y^n) F^n(D_i|y^n|s)}$$

is the mutual information for $p^n, \mathcal{A}^n, F^n(\cdot|\cdot|s)$.

Kesten [32] has shown by proving the coding theorem and the weak converse that

$$C_1 = \lim_{n\to\infty} \frac{1}{n} \sup_{p^n \in \mathcal{P}^n} \sup_{\mathcal{A}^n \in \mathcal{B}^n} \inf_{s \in S} I(p^n, \mathcal{A}^n, s)$$

is the weak capacity of the stationary semi-continuous memoryless channel.

$$\begin{aligned} C &= \max_{p_1} \inf_{s \in S} I(p_1, F(\cdot|\cdot|s)) \\ &= \max_{p_1} \inf_{s \in S} \sum p_1(x_1) \int_{\mathcal{Y}_1} \log \frac{dF(\cdot|x_1|s)}{\sum_{y_1 \in \mathcal{Y}_1} p_1(y_1)\, dF(\cdot|y_1|s)}\, dF(\cdot|x_1|s) \end{aligned}$$

is the strong capacity of the stationary discrete memoryless simultaneous channel. This was proved by Wolfowitz [43]. Kemperman gave an example of a semi-continuous simultaneous channel with $C_1 < C$ (published in [32]). However, $C$ is the weak capacity of the averaged semi-continuous channel, defined by:

$$P^n(\cdot|x^n) = \sum_{s \in S} q_s F^n(\cdot|x^n|s),$$

where $q$ is a PD on a countable index set $S$ and $(F^n(\cdot|\cdot|s))_{n=1}^\infty$ are semi-continuous stationary simultaneous channels without memory (Theorem 3).

### 1.1.3 A Channel Without Strong Capacity

Given $\mathcal{X} = \mathcal{Y} = [1, \dots, a]$ and the stochastic matrices $w(\cdot|\cdot|1)$ and $w(\cdot|\cdot|2)$ with $a$ rows and columns. For $s = 1, 2$, we define the discrete memoryless channel (DMC) $(P^n(\cdot|\cdot|s))_{n=1}^\infty$ by

$$P^n(y^n|x^n|s) = \prod_{t=1}^n w(y_t|x_t|s) \quad \text{for all } x^n \in \mathcal{X}^n, y^n \in \mathcal{Y}^n, \quad n = 1, 2, \dots$$

and the averaged channel (AC) $(P^n(\cdot|\cdot))_{n=1}^{\infty}$ by

$$P^n(y^n|x^n) = \frac{1}{2}P^n(y^n|x^n|1) + \frac{1}{2}P^n(y^n|x^n|2), \qquad x^n \in \mathcal{X}^n, y^n \in \mathcal{Y}^n, \quad n = 1, 2, \dots.$$

**Theorem 1** (Coding theorem and weak converse) *Let*

$$C = \max_{p} \inf_{s=1,2} I(p, w(\cdot|\cdot|s)) = \textit{strong capacity of the SC } (S^n)_{n=1}^{\infty}.$$

*For the maximal code length $N(n,\lambda)$ of the AC $(P^n)_{n=1}^{\infty}$ the following estimates hold:*

*(i) Given $0 < \lambda < 1$, $\delta > 0$ then there exists an $n_0 = n_0(\lambda, \delta)$ such that*

$$N(n, \lambda) > e^{(C+\delta)n} \quad \textit{for } n \geq n_0.$$

*(ii) Given $\delta > 0$ then there exists a $\lambda$ and an $n_0 = n_0(\lambda, \delta)$ such that*

$$N(n, \lambda) > e^{(C+\delta)n} \quad \textit{for } n \geq n_0.$$

*Proof* (i) A $\lambda$-code for the simultaneous channel

$$S^n = \{P^n(\cdot|\cdot|s) : s = 1, 2\} \quad \text{is a } \lambda\text{-code for } P^n(\cdot|\cdot).$$

(ii) Choose $\lambda < \frac{1}{2}$. For a $\lambda$-code $\{(u_i, D_i) : i = 1, \dots, N\}$ of $P^n$ we have

$$P^n(D_i|u_i) \geq 1 - \lambda, \quad i = 1, \dots, N$$

and therefore

$$P^n(D_i|u_i|s) \geq 1 - \lambda - \frac{1}{2} = 1 - \lambda' > 0, \quad s = 1, 2; \quad i = 1, \dots, N.$$

A $\lambda$-code for $P^n$ is a $\lambda'$-code for $S^n$, if $\lambda < \frac{1}{2}$.

By the strong converse for $S^n$ we have $N(n, \lambda) \leq e^{Cn + k(\lambda')\sqrt{n}}$, where $k(\lambda')$ is a known function [42]. This proves (ii). □

*Remark* The strong converse of the coding theorem gives, if it holds, an estimate of the following type:

Given $0 < \lambda < 1$, $\delta > 0$, then there exists an $n_0(\lambda, \delta)$ such that for $n \geq n_0$

$$N(n, \lambda) < e^{(C+\delta)n}$$

holds, where $C$ is the maximal constant for which estimation (i) holds. (For nonstationary channels, we have instead of $C$ a capacity function $C^n$.)

Choose $w(\cdot|\cdot|s)$ such that $(P^n(\cdot|\cdot|1))_{n=1}^{\infty}$ has capacity 0 and $(P^n(\cdot|\cdot|2))_{n=1}^{\infty}$ has capacity $C_2 > 0$. Then a fortiori $C = 0$. A $\lambda$-code for $\frac{1}{2}P^n(\cdot|\cdot|2)$ is a $\lambda$-code $P^n$. Choose $\lambda > \frac{1}{2}$ and let $N_2(n, \lambda)$ be the maximal code length for $P^n(\cdot|\cdot|2)$. Thus we have $N(n, \lambda) \geq N_2\left(n, \frac{1+\lambda}{2}\right) > e^{C_2 - k'(\lambda)\sqrt{n}}$ for all $n$. Thus the strong converse does not hold.

*Remark* A weaker form of the strong converse can be introduced; namely, there exists an $\alpha > 0$, such that for all $\lambda$ with $0 < \lambda < \alpha$, $\delta > 0$ there exists an $n_0(\lambda, \delta)$ with the property:

$$N(n, \lambda) < e^{(C+\delta)n}; \quad n \geq n_0.$$

For the channel we discussed above this estimate holds with $\alpha = \frac{1}{2}$. However, for the channels considered in Sect. 1.1.4 even this weaker estimate does not hold.

*Remark* For several stationary channels (e.g., discrete channel without memory, discrete channel with finite memory) $\lim_{n\to\infty} I(p^\infty, F^n)$ exists for each stationary PD $p^\infty$ on

$$(\mathcal{X}^\infty, \mathcal{L}^\infty) = \left(\prod_{t=1}^{\infty} \mathcal{X}_t, \prod_{t=1}^{\infty} \mathcal{L}_t\right) \quad \text{and} \quad C_{\text{stat.}} \stackrel{\text{def}}{=} \sup_{p^\infty} \lim_{n\to\infty} \frac{1}{n} I(p^\infty, F^n)$$

turns out to be the (weak or strong) channel capacity $C$.

The proofs of the coding theorem work mostly with ergodic stationary PD. In the first step it is shown that the joint source-channel distribution is ergodic under suitable conditions on the channel. Then McMillan's theorem gives that the information function $\frac{1}{n} i(p^\infty, F^n)$ converges to

$$\lim_{n\to\infty} \frac{1}{n} I(p^\infty, F^n) \quad (L^1 \text{ or with probability one}).$$

In the last step an application of Feinstein's maximal code theorem leads to the coding theorem. The channel defined under the first remark has the remarkable property that

$$C < \max_{\substack{p\times p\times\cdots\times p \\ \text{independent} \\ \text{sources}}} \lim_{n\to\infty} \frac{1}{n} I(p \times \cdots \times p, P^n) \tag{1.1.1}$$

$$\leq \max_{\substack{p^\infty \text{ ergodic} \\ \text{stat.}}} \lim_{n\to\infty} I(p^\infty, R^n) \stackrel{\text{def}}{=} C_{\text{erg.}} \leq \max_{p^\infty \text{ stat.}} \lim_{n\to\infty} \frac{1}{n} I(p^\infty, P^n) \stackrel{\text{def}}{=} C_{\text{stat.}}$$

*Therefore, the usual method of proving a coding theorem is not applicable.*

Some authors already speak of channel capacity if $C_{\text{stat.}}$ exists—without having proved the coding theorem and a converse (1.1.1) shows that this is not permissible.

Let us now prove (1.1.1). It is known, that the mean entropy $\bar{H}$ of a stationary PD on the product measure-space has the linearity property:

Given a PD $q_1, \dots, q_k$ and stationary PD $p_1, \dots, p_k$ on $(\mathcal{X}^\infty, \mathcal{L}^\infty)$, then

$$\bar{H}\left(\sum_{i=1}^{k} q_i p_i\right) = \sum_{i=1}^{k} q_i \bar{H}(P_i). \tag{1.1.2}$$

We now choose $p^n = p \times \cdots \times p$ for $n = 1, 2, \dots$ and define

$$p'^n(y^n|s) = \sum_{x^n \in \mathcal{X}^n} p^n(x^n) P^n(y^n|x^n|s) \quad s = 1, 2, \dots; y^n \in \mathcal{Y}^n$$

$$\tilde{p}(y^n|x^n|s) = p^n(x^n) P^n(y^n|x^n|s) \quad s = 1, 2, \dots, x^n \in \mathcal{X}^n, y^n \in \mathcal{Y}^n$$

$$p'^n(y^n) = \sum_{s=1|}^{2} \frac{1}{2} p'^n(y^n|s)$$

$$\tilde{p}^n(y^n, x^n) = \sum_{s=1}^{2} \frac{1}{2} \tilde{p}^n(y^n, x^n|s).$$

From $I(p^n, P^n) = H(p^n) + H(p'^n - H(\tilde{p}^n)$ and the above definitions, we conclude that

$$\lim_{n\to\infty} \frac{1}{n} I(p^n, P^n) = \lim_{n\to\infty} \frac{1}{n} \sum_{s=1}^{2} \frac{1}{2} \left[ H(p^n) + H(p'^n(\cdot|s)) - H(\tilde{p}^n(\cdot|\cdot|s)) \right].$$

Using (1.1.2) and the independence we get

$$\lim_{n\to\infty} \frac{1}{n} I(p^n, P^n) = \sum_{s=1}^{2} \frac{1}{2} \left[ H(p_1) + H(p'_1(\cdot|s)) - H(\tilde{p}_1(\cdot|\cdot|s)) \right]$$

$$= \sum_{s=1}^{2} \frac{1}{2} R(p_1, w(\cdot|\cdot|s))$$

and therefore

$$\max_{p^n = p\times p\times\cdots\times p} \lim_{n\to\infty} \frac{1}{n} I(p^n, P^n) = \max_p \frac{1}{2} I(p, w(\cdot|\cdot|2)) = \frac{1}{2} C_2 > C.$$

In [21] Hu Guo Ding proved that the coding theorem and the weak converse hold if and only if the channel is "information stable" (Def. given in [21, 22], and in Sect. 5.6). Information stability, however, is difficult to verify for our channel and it is even more difficult to find a formula for the channel capacity by this method.

### 1.1.4 The Weak Capacity of an Averaged Discrete Channel

Given a set of stochastic matrices

$$\left\{(w(ijs))_{\substack{i=1,\dots,a\\ j=1,\dots,a}} : s \in S = \{1, 2, \dots\}, \mathcal{X} = \mathcal{Y} = \{1, \dots, a\}\right\}$$

and a PD $q$ on $S$. We can assume w.l.o.g. that $q_s > 0$ for all $s \in S$. Otherwise, we would take instead of $S$ a subset $S'$ with $q'_s > 0$, $s' \in S'$, and $\sum_{s' \in S'} q'_s = 1$.

We define the individual discrete channel by

$$P^n(y^n|x^n|s) = \prod_{t=1}^{n} w(y_t|x_t|s), \quad y^n \in \mathcal{Y}^n, x^n \in \mathcal{X}^n, s \in S, \quad n = 1, 2, \dots;$$

the averaged discrete channel by

$$P^n(y^n|x^n) = \sum_{s \in S} q_s P^n(y^n|x^n|s)$$

and the simultaneous channel by

$$S^n = \{P^n(\cdot|\cdot|s) : s \in S\}, \quad n = 1, 2, \dots.$$

We need the following:

**Lemma 1** *If $f_n$ is a decreasing sequence of continuous, convex, nonnegative, functions defined on a simplex in $R^n$, then $\lim_{n\to\infty} f_n = f$ is continuous and the convergence is uniform.*

*Sketch of the Proof.* It is clear that the limit exists. $f$ is convex and could have discontinuities only on the extreme points, but this does not happen. From Dini's theorem, we conclude uniform convergence.

**Lemma 2** *(i) $f_k(p) = \inf_{s=1,\dots,k} I(p, F(\cdot|\cdot|s))$ converges uniformly in $p$ to $\inf_{s\in S} I(p, F(\cdot|\cdot|s))$.*

*(ii)*

$$\lim_{k\to\infty} C_k = \lim_{k\to\infty} \max_p \inf_{s=1,\dots,k} I(p, F(\cdot|\cdot|s)) = \max_p \inf_{s\in S} I(p, F(\cdot|\cdot|s)).$$

*Proof* The set of all PD on $\{1, \dots, a\}$ forms a simplex. $I(p, F(\cdot|\cdot|s))$ as a function of $p$ is continuous, convex, and nonnegative. $f_k(p)$ satisfies the hypotheses of Lemma 1. This gives us (i) and as an easy consequence (ii) (cf. [3]). □

We can now formulate:

**Theorem 2** (Coding theorem and weak converse for the DAC) *Let* $C = \max_p \inf_{s \in S} I(p, w(\cdot | \cdot |s))$. *Then the following estimates hold:*

*(i) Given* $0 < \lambda < 1$, $\delta > 0$, *then there exists an* $n_0 = n_0(\lambda, \delta)$ *such that*

$$N(n, \lambda) > e^{(C-\delta)n} \quad \textit{for } n \geq n_0.$$

*(ii) Given* $\delta > 0$, *there exists a* $\lambda$ *and an* $n_0 = n_0(\lambda, \delta)$ *such that*

$$N(n, \lambda) > e^{(C+\delta)n} \quad \textit{for } n \geq n_0.$$

*Proof* (i) A $\lambda$-code for the simultaneous channel $S^n$ is a $\lambda$-code for $P^n$. The statement follows from the coding theorem for simultaneous stationary channels without memory [43].

(ii) Define

$$\varepsilon_k = \sum_{\kappa=k+1} q_\kappa, \quad \eta_k = \inf_{\kappa=1,\ldots,k} q_\kappa > 0.$$

For the given $\delta > 0$ choose $k$ such that $|C_k - C| \leq \delta/2$ then choose

$$\lambda = \eta_k/2 \quad \text{and} \quad \lambda' = \varepsilon_k + \tfrac{1}{2}\eta_k. \tag{1.1.3}$$

A $\lambda$-code $\{(u_i, D_i) : i = 1, \ldots, N\}$ for $P^n$ is a $\lambda'$-code for $\sum_{s=1}^k q_s P^n(\cdot | \cdot |s)$, since

$$\sum_{s=1}^{k} q_s P^n(D_i|u_i|s) \geq P^n(D_i|u_i) - \varepsilon_k \geq 1 - \left(\frac{1}{2}\eta_k + \varepsilon_k\right). \tag{1.1.4}$$

But

$$\sum_{s=1}^{k} q_s P^n(D_i|u_i|s) - q_s' P^n(D_i|u_i|s) \leq 1 - \varepsilon_k - \eta_k \quad \text{for} \quad s' = 1, \ldots, k. \tag{1.1.5}$$

From (1.1.4) and (1.1.5) we have

$$q_s P^n(D_i|u_i|s) \geq \frac{1}{2}\eta_k \quad \text{for} \quad s = 1, \ldots, k, \quad i = 1, \ldots, N$$

and therefore

$$P^n(D_i|u_i|s) \geq \frac{1}{2}\eta_k \quad \text{for} \quad s = 1, \ldots, k, \quad i = 1, \ldots, N.$$

Now we apply the *strong* converse of the coding theorem for simultaneous discrete channels without memory and conclude that

$$N(n, \lambda) < e^{nC_k} + k(\eta_k)\sqrt{n} \quad \text{for all } n. \tag{1.1.6}$$

Statement (ii) follows from (1.1.3) and (1.1.6). □

*Remark*

1. The proof uses essentially the strong converse for simultaneous discrete channels without memory. Fano's lemma would lead to the estimate

$$N(n, \lambda) < e^{(nC_k+1)\eta_k^{-1}}, \quad \text{but} \lim_{k\to\infty} \frac{C_k}{\eta_k} = \infty.$$

2. From the proof of the theorem it follows that the weak capacity remains unchanged if we average with respect to $q^*$, where $q^*$ is equivalent to $q$.

### 1.1.5 The Weak Capacity of an Averaged Semi-continuous Channel

We return to the semi-continuous case as described in Sect. 1.1.2.

**Theorem 3** *Let $C = \max_p \inf_{s\in S} I(p, w(\cdot|\cdot|s))$. Then the following estimates hold:*

*(i) Given $0 < \lambda < 1$, $\delta > 0$, then there exists an $n_0 = n_0(\lambda, \delta)$ such that*

$$N(n, \lambda) > e^{(C-\delta)n} \quad \textit{for } n \geq n_0.$$

*(ii) Given $\delta > 0$, there exists a $\lambda$ and an $n_0 = n_0(\lambda, \delta)$ such that*

$$N(n, \lambda) > e^{(C+\delta)n} \quad \textit{for } n \geq n_0.$$

*Proof* (i) Given $0 < \lambda < 1$, $\delta > 0$. Choose $k$ such that $\varepsilon_k < \lambda$ and define $\lambda_k := \frac{\lambda-\varepsilon_k}{1-\varepsilon_k} < 1$. A $\lambda_k$-code for ${}_kS^n = \{P^n(\cdot|\cdot|s) : s = 1, \dots, k\}$ is a $\lambda$-code for $P^n$. The coding theorem for ${}_kS^n$ gives

$$e^{C_k n - k(\lambda_k)\sqrt{n}} \leq N_k(n, \lambda_k) \leq N(n, \lambda)$$

and therefore $N(n, \lambda) > e^{n(C-\delta)}$ for $n$ sufficiently large.

(ii) If we use Lemma 2 and the strong converse of the coding theorem for ${}_kS^n$, the proof of Theorem 2 (ii) carries over verbatim. □

*Remark* An example of Kemperman (published in [32]) shows that there are semi-continuous channels with $C > C_1$. $C_1$ is the (weak) capacity of $(S^n)_{n=1}^{\infty}$; $C$ is the weak capacity of $(P^n)_{n=1}^{\infty}$. Therefore we can give the intuitive interpretation: The statistical knowledge of the individual channels which govern the transmission increases the weak capacity.

### 1.1.6 Nonstationary Averaged Channels

Given stochastic $a \times a$-matrices $w_t(\cdot|\cdot|s)$, $s \in S$, $t = 1, 2, \dots$, we define the nonstationary simultaneous channel (cf. [2])

$$(S^n)_{n=1}^{\infty} = \{P^n(\cdot|\cdot|s) : P^n(y^n|x^n|s) = \prod_{t=1}^{n} w_t(y_t|x_t|s), x^n \in \mathcal{X}^n, y^n \in \mathcal{Y}^n, s \in S\} \quad n = 1, 2, \dots.$$

Let $q = (q_1, q_2, \dots)$ be a discrete PD on the countable set $S$ and $q_s > 0$ for all $s \in S$.
Define

$$C_k(n) = \max_{p^n} \inf_{s=1,\dots,k} I(p^n, P^n(\cdot|\cdot|s)) \quad \text{and} \quad C(n) = \inf_{k=1,2,\dots} C_k(n).$$

**Theorem 4** *If for the nonstationary averaged channel* $(P^n)_{n=1}^{\infty}$ *the condition: For each* $\delta_1 > 0$ *there exists a* $k$ *and an* $n_1(\delta_1, k)$ *such that*

$$|C(n) - C_k(n)| < \delta_1 n \quad \text{for } n \geq n_1(\delta_1, k) \tag{1.1.7}$$

*holds, then we have the estimates*

*(i) Given* $0 < \lambda < 1$, $\delta > 0$, *then there exists an* $n_0 = n_0(\lambda, \delta)$ *such that*

$$N(n, \lambda) > e^{C(n)-\delta n} \quad \text{for } n \geq n_0.$$

*(ii) Given* $\delta > 0$, *there exists a* $\lambda$ *and an* $n_0 = n_0(\lambda, \delta)$ *such that*

$$N(n, \lambda) < e^{C(n)+\delta n} \quad \text{for } n \geq n_0.$$

(cf. [2]).

*Proof* (i) Given $0 < \lambda < 1$, $\delta > 0$. Choose $k$ such that $\lambda_k := \frac{\lambda}{1-\varepsilon_k} < 1$. A $\lambda_k$-code for ${}_kS^n = \{P^n(\cdot|\cdot|s) : s = 1, \dots, k\}$ is a $\lambda$-code for $P^n$. The coding theorem for ${}_kS^n$ gives

$$e^{C_k(n)-\delta n} \leq N_k(n, \lambda_k) \leq N(n, \lambda)$$

for $n$ sufficiently large (Satz 2 in [2], Chap. 3).

(ii) Using (1.1.7) and the strong converse of the coding theorem for ${}_kS^n$ (Satz 3 in [2], Chap. 3) the proof of Theorem 2 (ii) carries over verbatim. □

**Example** Almost periodic discrete averaged channels. Let $(w_t(\cdot\cdot))_{t=1}^{\infty}$ be an almost periodic sequence of stochastic matrices (cf. [2], Chap. 2), then we can define the simultaneous almost periodic channel

$$(S^n)_{n=1}^{\infty} = \{P^n(\cdot|\cdot|s) : P^n(y^n|x^n|s)$$
$$= \prod_{t=1}^{n} w_{t+s}(y_t|x_t), x^n \in \mathcal{X}^n, y^n \in \mathcal{Y}^n, \qquad s = 0, 1, \ldots\} \quad n = 1, 2, \ldots.$$

and the averaged almost periodic channel

$$(P^n(\cdot\cdot)_{n=1}^{\infty} = \left(\sum_{s=0} q_s P^n(\cdot|\cdot|s)\right) \quad n = 1, 2, \ldots.$$

From Theorem 4, we conclude the

**Corollary 1** *For the almost periodic averaged channel $(P^n)_{n=1}^{\infty}$, the coding theorem and the weak converse hold with*

$$C = \lim_{n\to\infty} \frac{1}{n} \max_{(p_1,\ldots,p_n)=p^n} \inf_{s=0,1,\ldots} I(p^n|P^n(\cdot|\cdot|s)) = \lim_{n\to\infty} \frac{1}{n} C^n.$$

*Proof* We have to show that (1.1.7) is satisfied. But this follows from the almost periodicity and the norm-continuity of $I(p, w)$ in $w$ as is shown in [2], Chap. 2, p. 2. The Bedingung 1, there is exactly the same as (1.1.7). It follows from the definition of $C^n$ that $\frac{1}{n}C^n$ is monotone nondecreasing and $\frac{1}{n}C^n \leq \log a$; hence, $\lim_{n\to\infty} \frac{1}{n}C^n$. The capacity function is constant. □

*Remark*

1. $C$ can be greater than

$$\max_p \inf_{t=1,2,\ldots} I(p, w_t(\cdot\cdot)).$$

**Example** Choose two stochastic matrices $w(\cdot|\cdot|1)$, $w(\cdot|\cdot|2)$ with

$$I(p, w(\cdot|\cdot|1)) \equiv 0 \equiv C_1 \quad \text{and} \quad \max_p I(p, w(\cdot|\cdot|2)) = C_2 > 0.$$

Define

$$w_{2s}(\cdot|\cdot) = w(\cdot|\cdot|2)$$
$$w_{2s-1}(\cdot|\cdot) = w(\cdot|\cdot|1) \quad \text{for} \quad s = 1, 2, \ldots.$$

Then $(w_t(\cdot\cdot)_{t=1}^{\infty}$ is a periodic sequence of stochastic matrices. The corresponding periodic channel $(S^n)_{n=1}^{\infty}$ has the capacity

$$C = \frac{1}{2}C_2 > \max_p \inf_{t=1,2,\dots} I(p, w_t(\cdot\cdot)) = 0.$$

2. The corollary says: if we know the phase of an almost periodic channel statistically and each phase has positive probability, then the (weak) capacity of this channel equals the (strong) capacity for $(S^n)_{n=1}^{\infty}$.

   [Coding theorem and strong converse for $(S^n)_{n=1}^{\infty}$ were proved in [2].] The statistical knowledge of the phase increases the maximal code length in such a way that instead of the strong converse only a weak converse holds.

   If $q_s$ is not positive for all $s$, then the capacity of the averaged channel can of course be greater than $C$.

   **Example** Choose $w(\cdot|\cdot|1)$, $w(\cdot|\cdot|2)$ such that

$$I(p, w(\cdot|\cdot|1)) \quad \text{and} \quad I(p, w(\cdot|\cdot|2))$$

   have their maximum for different arguments $p_1, p_2$ and

$$I(p_1, w(\cdot|\cdot|1)) > I(p_1, w(\cdot|\cdot|2))$$
$$I(p_2, w(\cdot|\cdot|2)) > I(p_2, w(\cdot|\cdot|1)).$$

   For $q_1 = 1$ the averaged channel is the periodic channel $(w_t(\cdot\cdot))_{t=1}^{\infty}$ and has a capacity greater than the capacity of the corresponding simultaneous channel.

### 1.1.7 Averages of Channels with Respect to General Probability Distributions

Until now, we considered averages with discrete probability distributions. What happens, if we take averages with respect to nondiscrete PDs, for instance, the Lebesque measure on [0, 1]?

I. Let us look at a discrete averaged stationary channel with $S = [0, 1]$, $q =$ Lebesque measure on $S$.

Define

$$P^n(\cdot|\cdot) = \int_{[0,1]} P^n(\cdot|\cdot|s)q(ds).$$

Of course, $w(i|j|s)$ has to be measurable in $s$ for $i, j = 1, \dots, a$. In this case, our method from Sect. 1.1.4 is not applicable as can be seen by the following example:

$$w(i|j|s) = 1 \quad i = 1, \ldots, a, s \in (0, 1]$$
$$w(i|j|0) = 1/a \quad i, j = 1, \ldots, a.$$

Then $(P^n)_{n=1}^{\infty}$ has strong capacity $\log a$, because $N(n, \lambda) = e^{n \log a}$; but $(S^n)_{n=1}^{\infty}$ has capacity 0, because $w(\cdot|\cdot|0)$ has capacity 0. We have to give another approach.

II. Let $(S, \mathcal{C}, q)$ be a normed measure space, $S$ is index set as usual. Divide $[0, 1]$ in disjoint intervals of length $\beta$. $s_1$ and $s_2$ are $\beta$-equivalent, if $w(i|j|s_1)$ and $w(i|j|s_2)$ are in the same interval for all $i, j = 1, \ldots, a$. This equivalence relation leads to a partition of $S$ in at most $(1/\beta)^{a^2}$ measurable sets

$$_lS, \quad l = 1, \ldots, L(\beta) \leq 1/\beta^{a^2}$$

and therefore to a partition $S^n$ in the sets

$$_lS^n = \{P^n(\cdot|\cdot|s) : s \in_l S\}, \quad l = 1, \ldots, L(\beta).$$

For $0 < \alpha < 1$ define

$$C(\alpha, \beta) = \max_p \sup_{\substack{l_1, \ldots, l_k \leq L(\beta) \\ q\left(\bigcup_{i=1}^k {}_{l_i}S\right) \geq 1-\alpha}} \inf_{s \in \bigcup_{i=1}^k {}_{l_i}S} I(p, w(\cdot|\cdot|s)) \tag{1.1.8}$$

Instead of max sup we can write max max, because we vary over a finite set of index constellations.

Furthermore, we define

$$C(\alpha) = \max_p \sup_{\{S' : s' \subset S, q(S') \geq 1-\alpha\}} \inf_{s \in S'} I(p, w(\cdot|\cdot|s)) \quad \text{and} \quad C = \lim_{\alpha > 0} C(\alpha).$$

It follows from the definition that

$$C(\alpha, \beta) \leq C(\alpha) \quad \text{for all } \beta. \tag{1.1.9}$$

**Theorem 5** *For the general stationary discrete channel*

$$(P^n(\cdot|\cdot))_{n=1}^{\infty} = \left( \int_S P^n(\cdot|\cdot|s) q(ds) \right)_{n=1}^{\infty},$$

*the following estimates hold with* $C = \inf_{\alpha > 0} C(\alpha)$*:*

*(i) Given* $0 < \lambda < 1$, $\delta > 0$, *then there exists an* $n_0 = n_0(\lambda, \delta)$ *such that*

$$N(n, \lambda) > e^{Cn - \delta n} \quad \textit{for } n \geq n_0.$$

*(ii) Given $\delta > 0$, there exists a $\lambda > 0$ and an $n_0 = n_0(\lambda, \delta)$ such that*

$$N(n, \lambda) > e^{Cn+\delta n} \quad \textit{for} \quad n \geq n_0.$$

*Proof* (i) Given $\lambda, \delta > 0$, choose $\alpha < \lambda$ and $S'$ such that $q(S') \geq 1 - \alpha$ and

$$\left| \max_p \inf_{s \in S'} I(p, w(\cdot | \cdot |s)) - C(\alpha) \right| \leq \delta/2.$$

Define $\lambda' := \frac{\lambda-\alpha}{1-\alpha}$. A $\lambda'$-code for $S'^n$ is a $\lambda$-code for $P^n$, because $(1-\lambda')(1-\alpha) = 1-(1-\alpha)\lambda' - \alpha = 1 - \lambda$. Hence

$$N(n, \lambda) \geq N'(n, \lambda') \geq e^{(C(\alpha)-(\delta/2))n - K(\lambda')\sqrt{n}} \geq e^{(C-\delta)n}$$

for $n$ sufficiently large.

(ii) First of all choose $\alpha$ such that $|C(\alpha) - C| \leq \delta/2$. Let ${}_{l_i}S, \dots, {}_{l_k}S$ be a family of sets such that the maximum is attained in (1.1.8), then $q\left(\bigcup_{i=1}^{k} {}_{l_i}S\right)$ has to be greater than $1 - \alpha$. We define $1 - \varepsilon(\alpha, \beta) = q({}_{l_i}S \cup \dots \cup {}_{l_k}S)$.

$\varepsilon(\alpha, \beta)$ is by definition smaller than or equal to $\alpha$. Define now

$$\eta(\alpha, \beta) = \inf_{i=1,\dots,k} q({}_{l_i}S) > 0$$

and choose

$$\lambda = \frac{\eta(\alpha, \beta)}{2}, \quad \lambda' = \varepsilon(\alpha, \beta) + \frac{1}{2}\eta(\alpha, \beta).$$

Then a $\lambda$-code for $P^n$ is a $\lambda'$-code for

$$\int_{{}_{l_1}S \cup \dots \cup {}_{l_k}S} P^n(\cdot | \cdot |s) q(ds),$$

since

$$\int_{{}_{l_1}S \cup \dots \cup {}_{l_k}S} P^n(D_i|u_i|s) \geq P^n(D_i|u_i) - \varepsilon(\alpha, \beta) \geq 1 - \left(\frac{1}{2}\eta(\alpha, \beta) + \varepsilon(\alpha, \beta)\right). \tag{1.1.10}$$

But from (1.1.10) and the definitions given above, it follows that

$$\int_{{}_{l_1}S \cup \dots \cup {}_{l_k}S} P^n(D_i|u_i|s) - \int_{{}_{l_j}S} P^n(D_i|u_i|s) q(ds)$$
$$\leq 1 - \varepsilon(\alpha, \beta) - \eta(\alpha, \beta) \quad \text{for} \quad i = 1, \dots, N; \quad j = 1, \dots, k. \tag{1.1.11}$$

From (1.1.10), (1.1.11), we have

$$\int_{l_j S} P^n(D_i|u_i|s)q(ds) \geq \frac{1}{2}\eta(\alpha,\beta) \quad \text{for} \qquad j = 1, \dots, k; \quad i = 1, \dots, N. \tag{1.1.12}$$

We need the trivial

**Lemma 3** (Combinatorial lemma) *Let $B_i$, $i = 1, \dots, I$ be measurable sets with $q(B_i) \geq \gamma$, $i = 1, \dots, I$. If we define $m$ as the maximal number of sets $B_i$ with a common element, then the estimate $m \geq \gamma$ holds.*

Denote by $D_{ji}$ the set

$$\left\{ s : P^n(D_i|u_j|s) \geq \frac{1}{4}\eta(\alpha,\beta), s \in_{l_i} S \right\} \qquad j = 1, \dots, N; \quad i = 1, \dots, k.$$

It follows from (1.1.12) that

$$q(D_{ji}) \geq \frac{1}{4}\eta(\alpha,\beta) \quad j = 1, \dots, N; \quad i = 1, \dots, k.$$

The sets $D_{11}, D_{21}, \dots, D_{N1}$ satisfy the hypothesis of Lemma 3. Hence, there exists an element $s_1 \in_{l_i} S$ which is contained in at least $\frac{1}{4}\eta(\alpha,\beta)N$ of these sets. That means there exists a subcode of length $\frac{1}{4}\eta(\alpha,\beta)N$ of the code

$$\{(u_i, D_i) : i = 1, \dots, N\}$$

such that

$$P^n(\bar{D}_{i_1}|\bar{u}_{i_1}|s_1) \geq \frac{1}{4}\eta(\alpha,\beta) \quad \text{for} \qquad i_1 = 1, \dots, \left[\frac{1}{4}\eta(\alpha,\beta)N\right].$$

Apply now the same arguments to $D_{12}, \dots, D_{N2}$. Thus we find a subcode of our subcode which is now a simultaneous code for $P^n(\cdot|\cdot|s_1)$ and $P^n(\cdot|\cdot|s_2)$ of length greater than $\frac{1}{4}\eta(\alpha,\beta) \cdot \frac{1}{4}\eta(\alpha,\beta) \cdot N$. Proceeding in the same way, we have after $k \leq L(\beta)$ steps a subcode of length $N^*$ greater than $\left(\frac{1}{4}\eta(\alpha,\beta)\right)^{L(\beta)} \cdot N$ with

$$P^n(D_j^*|u_j^*|s_i) \geq \frac{1}{4}\eta(\alpha,\beta), \qquad j = 1, \dots, N^*; \quad i = 1, \dots, k.$$

From the strong converse of the coding theorem for simultaneous channels and the norm continuity of $R(p, w)$ in $w$ uniformly in $p$, we conclude:

$$N\left(\frac{1}{4}\eta(\alpha,\beta)\right)^{L(\beta)} \leq N^* \leq \exp C(\alpha,\beta)n + f(\beta)n + K(\eta)\sqrt{n} \text{ where } \lim_{\beta\to 0} f(\beta) = 0.$$

Using (1.1.10) we have

$$N(n, \lambda) \leq expC(\alpha)n + f(\beta)n + K(\eta)\sqrt{n} + L(\beta)\log \frac{1}{4}\eta(\alpha, \beta).$$

Choose now $\beta$ such that $f(\beta) \leq \delta/4$ and use $|C(\alpha) - C| \leq \delta/2$. Then we have $N(n, \lambda) \leq e^{Cn+\delta n}$ for $n$ sufficiently large. That proves (ii).

*Remark*

1. Theorem 5 can be extended to the semi-continuous case, if

$$\begin{aligned} C(\alpha) &= \max_p \sup_{\{S':q(S')\geq 1-\alpha\}} \inf_{s\in S'} R(p, F(\cdot|\cdot|s)) \\ &= \lim_{n\to\infty} \frac{1}{n} \max_{p^n} \sup_{\{S':q(S')\geq 1-\alpha\}} \sup_{\mathcal{D}^n} \inf_{s\in S'} I(p^n, \mathcal{D}^n, P^n(\cdot|\cdot|s)) \end{aligned}$$

for all $\alpha > 0$. Part (i) follows then from the coding theorem for simultaneous semi-continuous channels [32]. For the proof of part (ii), we use that for an arbitrary set of channel kernels $\{F(\cdot|\cdot|s) : s \in S\}$ the corresponding set of information functions $\{I(p, F(\cdot|\cdot|s)) : s \in S\}$ is totally bounded in the norm of uniform convergence. (This is a consequence of Hilfssatz 1 in [2], Chap. 1, Sect. 4.) Hence, we can find for given $\beta > 0$ a family of sets $\{_jS^* : j = 1, \dots, L(\beta)\}$ such that for $s_1, s_2 \in_j S^*$

$$\sup_p |I(p, F(\cdot|\cdot|s_1)) - I(p, F, (\cdot|\cdot|s_2))| \leq \beta \quad j = 1, \dots, L(\beta).$$

We redefine the $_jS$, which we used in the proof of Theorem 5, as follows:

$$_jS :=_j S^*$$

then, the proof of part (ii) carries over to the semicontinuous case.

2. The extension of Theorem 5 to the nonstationary case seems to be difficult. It could be of interest for the "arbitrarily varying channel"-problem [33].

## 1.2 Lecture on Further Results for Averaged Channels Including C(λ)-Capacity, Side Information, Effect of Memory, and a Related, Not Familiar Optimistic Channel

In [3] (see Sect. 1.1), we proved a coding theorem and the weak converse of the coding theorem for averaged channels under different assumptions on the time structure and the output alphabet of the channel, and we gave explicit formulas for the weak capacity. The strong converse of the coding theorem does not hold;

therefore, it is of interest to know the $\lambda$-capacity $C_1(\lambda) = \lim_{n\to\infty} \frac{1}{n} \log N(n, \lambda)$ [44]. The paper [1] gives the (seemingly) first serious investigation of this quantity. Its continuation in [12] is the subject of Sect. 1.4. We give here upper and lower bounds for $\limsup_{n\to\infty} \frac{1}{n} \log N(n, \lambda)$ and $\liminf_{n\to\infty} \frac{1}{n} \log N(n, \lambda)$, (Sect. 1.2.1.2).

In the Sect. 1.2.1.5, we proved a coding theorem and the weak converse for stationary semi-continuous averaged channels, where the average is taken with respect to a general probability distribution. (We obtained this result in [3] (Sect. 1.1.3, Remark 2), by a different method only under an additional restriction.) The new method (Lemma 6) applies to averaged channels with side information (cf. [40, 43]).

In 1.2.2, we introduce another compound channel: the sender can choose for the transmission of a code word the individual channel over which he wants to transmit.

## 1.2.1 *Averaged Channels Where Either the Sender or the Receiver Knows the Individual Channel Which Governs the Transmission*

### 1.2.1.1 Introduction and Definitions

Simultaneous discrete memoryless channels, where the individual channel which governs the transmission is known to the sender or receiver were discussed for the first time by Wolfowitz [43, 45]. Kesten gave an extension to the semi-continuous case [32]. Compound channels where the channel probability function (CPF) for each *letter* is stochastically determined were introduced earlier by Shannon [40]. Wolfowitz proved strong converses [45]. In Sect. 1.1, we proved a coding theorem and its weak converse for averaged channels under different assumptions on the time structure (stationary, almost periodic, nonstationary) and the output alphabet (finite, infinite) of the channel. We introduce now averaged channels with side information. First, let us repeat the definition of a general averaged channel.

Let $\mathcal{X}_t = \{1, \dots, a\}$ for $t = 1, 2, \dots$ and $(\mathcal{Y}_t, \mathcal{L}_t) = (\mathcal{Y}, \mathcal{L})$ for $t = 1, 2, \dots$ where $\mathcal{Y}$ is an arbitrary set and $\mathcal{L}$ is a $\sigma$-algebra of subsets in $\mathcal{Y}$.

Furthermore, let $S = \{s, \dots\}$ be a nonempty (index) set, $(S, \mathcal{M}, q)$ a normed measure space and let $F_t(\cdot|1|s), \dots, F_t(\cdot|a|s)$ be probability distributions (PD) on $(\mathcal{Y}, \mathcal{L}), t \in N, s \in S$.

For each $x^n = (x_1, \dots, x_n) \in \mathcal{X}^n = \prod_{t=1}^n \mathcal{X}_t$, we define a PD on $\big(\mathcal{Y}^n = \prod_{t=1}^n \mathcal{Y}_t$, $\mathcal{L}^n = \prod_{t=1}^n \mathcal{L}_t\big)$ by $F^n(\cdot|x^n|s) = \prod_{t=1}^n F_t(\cdot|x_t|s)$.

The sequence of kernels $(F^n(\cdot| \cdot |s))_{n=1}^\infty$ forms a semi-continuous (in general nonstationary) channel without memory. [In case $\mathcal{Y}_t = \mathcal{Y}$ is finite, the kernels $F_t(\cdot| \cdot |s)$ are given by stochastic matrices $w_t(k|i|s) = F_t(\{k\}|i|s), i \in \mathcal{X}, k \in \mathcal{Y}$. We speak then of a discrete channel without memory.] Thus we have assigned to each $s \in S$ a semi-continuous channel. If we are interested in the simultaneous behavior of all these channels, then we call this indexed set of channels a simultaneous channel (semi-continuous without memory). The set $\{F^n(\cdot|\cdot|s) : s \in S\}$ designed by $S_n$ is called a simultaneous channel in the discrete time interval $[1, n]$. (cf. [2, 32, 43])

If $F^n(D|x^n|s)$ is a measurable function on $(S, \mathcal{M}, q)$ for each $D \in \mathcal{L}^n$, $x^n \mathcal{X}^n$, then we can define an averaged channel by

$$P^n(D|x^n) = \int_S F^n(D|x^n|s)\, \mathrm{d}q(s)$$

for $A \in \mathcal{L}^n, x^n \in \mathcal{X}^n, n = 1, 2, \ldots$.

A more intuitive description of this channel can be given as follows: at the beginning of the transmission of each word of length $n$ an independent random experiment is performed according to $(S, \mathcal{M}, q)$ with probability $q(s)$ that the outcome of the experiment be $s \in S$. If $s$ is the outcome of the experiment the word (of length $n$) is transmitted according to $F^n(\cdot|\cdot|s)$.

The definition of a code depends on the knowledge of the channel $F^n(\cdot|\cdot|s)$ by the sender and or receiver. If neither knows the channel over which the message is transmitted, a $(n, N, \lambda)$ code for the compound channel is defined a a set $\{(u_1, D_1), \ldots, (u_N, D_N)\}$, where $u_i \in \mathcal{X}^n$, $D_i \in \mathcal{L}^n$ for $i = 1, \ldots, N$, $D_i \cap D_j = \emptyset$ for $i \neq j$, such that

$$\int_S F^n(D_i|u_i|s)\, \mathrm{d}q(s) \geq 1 - \lambda \quad i = 1, \ldots, N \tag{1.2.1}$$

The $u_i$ and $D_i$ do not depend on $s$. (cf. [3], Sect. 1.1)

The next subsection is concerned with existence problems of the $\lambda$-capacity of this channel. (cf. [44])

If only the sender knows the channel of transmission, the $u_i$'s but not the $D_i$'s may depend on $s$. A $(n, N, \lambda)$ code $\{(u_1(s), D_1), \ldots, (u_N(s), D_N)\}$ must now satisfy

$$\int_S F^n(D_i|u_i(s)|s)\, \mathrm{d}q(s) \geq 1 - \lambda \quad i = 1, \ldots, N \tag{1.2.2}$$

If only the receiver knows the channel, the $D_i$ but not the $u_i$ may depend on $s$ and $D_i$ in (1.2.1) is replaced by $D_i(s)$. Finally, if both the sender and the receiver know the channel, $u_i$ and $D_i$ may depend on $s$.

We put $N_1(n, \lambda) =$ maximal $N$ for which an $(n, N, \lambda)$ code exists if neither sender nor receiver knows the channel over which the message is transmitted.

$N_i(n, \lambda)$ for $i = 2, 3, 4$ are the maximal $N$ for which an $(n, N, \lambda)$ code exists, respectively, if the sender only $(i = 2)$, the receiver only $(i = 3)$, and the sender and receiver $(i = 4)$ know the channel.

We designate the different compound channels by $\mathcal{C}_1, \mathcal{C}_2, \mathcal{C}_3, \mathcal{C}_4$. In the fourth and fifth subsection, we prove a coding theorem and the weak converse for $\mathcal{C}_i$, $i = 1, \ldots, 4$, in the stationary case. (For results in the nonstationary case cf. [3] (Sect. 1), [29])

In the third subsection, we show that in general memory need not "increase capacity." (cf. [46])

#### 1.2.1.2 A Remark on the $\lambda$-Capacity of $\mathcal{C}_1$ $\lambda$

According to Wolfowitz [44] $\lim_{n\to\infty} \frac{1}{n} \log N_1(n, \lambda)$ is, if the limit exists, the $\lambda$-capacity $C_1(\lambda)$ of the channel $\mathcal{C}_1$. It is known [3, 29, 44] (see Sect. 1.1) that $C_1(\lambda)$ cannot be constant, because the strong converse of the coding theorem does not hold in general. We need the

**Lemma 4** *If*

$$\frac{1}{N}\sum_{i=1}^{N} P^n(D_i|u_i) \geq 1-\lambda \ \textit{for} \quad i = 1, 2, \ldots, N \textit{ and } 1 > \gamma, \beta > 0$$

*such that* $\gamma\beta > \lambda$, *then*

$$q\left\{s|\frac{1}{N}\sum_{i=1}^{N} F^n(D_i|u_i|s) \geq 1-\gamma\right\} \geq 1-\beta$$

*Proof* Assume $q\left\{s|\frac{1}{N}\sum_{i=1}^{N} F^n(D_i|u_i|s) \geq 1-\gamma\right\} < 1-\beta$, then we have, if we write $f(s)$ instead of $\frac{1}{N}\sum_{i=1}^{N} F^n(D_i|u_i|s)$:

$$\begin{aligned}
\int_S f(s)\,dq(s) &= \int_{f<1-\gamma} f(s)\,dq(s) + \int_{f\geq 1-\gamma} f(s)dq(s) \\
&\leq (1-\gamma)q\{s|f<1-\gamma\} + q\{s|f\geq 1-\gamma\}, \quad \text{since } f(s)\leq 1 \\
&= (1-\gamma)(1-q\{s|f\geq 1-\gamma\}) + q\{s|f\geq 1-\gamma\} \\
&= 1-\gamma+\gamma q\{s|f\geq 1-\gamma\} \\
&\leq 1-\gamma+\gamma(1-\beta) = 1-\gamma\beta < 1-\lambda,
\end{aligned}$$

in contradiction to the proposition.

This proves the lemma. □

For a PD $p$ on $\mathcal{X}$ let $I(p, F(\cdot|\cdot|s))$ be the mutual information of the channel $F(\cdot|\cdot|s)$.

**Theorem 6** *Let $\mathcal{Y}$ be finite.*

*(i)*

$$\begin{aligned}
{}_+C_1(\lambda) &:= \inf_{\gamma\beta>\lambda} \frac{1}{1-\gamma} \sup_p \sup_{\{S':q(S')\geq 1-\beta\}} \inf_{s\in S'} I(p, F(\cdot|\cdot|s)) \\
&\geq \lim_{n\to\infty} \sup \frac{1}{n} \log N_1(n, \lambda)
\end{aligned}$$

*(ii)*

$$
\begin{aligned}
{}_{-}C_1(\lambda) &:= \sup_{0<\varepsilon<1} \sup_{p} \sup_{S':q(S')>1-\varepsilon\lambda} \inf_{s\in S'} I(p, F(\cdot|\cdot|s)) \\
&\leq \lim_{n\to\infty} \inf \frac{1}{n} \log N_1(n,\lambda)
\end{aligned}
$$

*Proof* Choose $\varepsilon$, $S'$ with $q(S') > 1 - \varepsilon\lambda$ such that

$$
\left| \sup_{p} \inf_{s\in S'} I(p, F(\cdot|\cdot|s)) - {}_{-}C_1(\lambda) \right| \leq \frac{\delta}{2},
$$

then $\eta$ such that $(1-\varepsilon\lambda)(1-\eta) = 1-\lambda$.

An $\eta$-code for the simultaneous channel $S^n = \{F^n(\cdot|\cdot|s) : s \in S'\}$ is a $\lambda$-code for $P^n$.

(ii) follows now from the coding theorem for $S^n$.

A $\lambda$-code for $P^n$ is an averaged $\lambda$-code for $P^n$:

$$
\frac{1}{N}\sum_{i=1}^{N} P^n(D_i|u_i) \geq 1-\lambda.
$$

From Lemma 4 follows that for every pair $(\gamma, \beta)$ with $0 < \gamma, \beta < 1$, $\gamma\beta > \lambda$ there exists a subset $S''$ of $S$ with $q(S'') \geq 1-\beta$ such that

$$
\frac{1}{N}\sum_{i=1}^{N} F^n(D_i|u_i|s) \geq 1-\gamma \quad s \in S''. \tag{1.2.3}
$$

Applying Fano's lemma for averaged errors, we get

$$
\frac{1}{n}\log N_1(n,\lambda) \leq \frac{1}{1-\gamma} \sup_{p} \inf_{s\in S''} I(p, F(\cdot|\cdot|s))
$$

and furthermore

$$
\frac{1}{n}\log N_1(n,\lambda) \leq_{+} C_1(\lambda),
$$

this proves (i). □

It is an easy consequence of the definition for ${}_{+}C_1(\lambda)$ and ${}_{-}C_1(\lambda)$ that ${}_{+}C_1(\lambda)$, ${}_{-}C_1(\lambda)$ are monotone increasing in $\lambda$, that $\lim_{\lambda\to 0+} C_1(\lambda)$, $\lim_{\lambda\to 0}\, {}_{-}C_1(\lambda)$ exist and that $\lim_{\lambda\to 0+} C_1(\lambda) = \lim_{\lambda\to 0-} C_1(\lambda)$. Let us denote this limit by $C_1$. Then we have as consequence of Theorem 6.

**Corollary 2** *$C_1$ is the weak capacity for $\mathcal{C}_1$.*

For Theorem 5 in Sect. 1.1 we gave a different proof.

*Remark* For an individual channel it is unessential whether we work with a $\lambda$-code or with an averaged $\lambda$-code, however, it makes a difference for simultaneous channels. If we use averaged $\lambda$-codes for simultaneous discrete memoryless channels, then the strong converse of the coding theorem does not hold.

**Example**

$$\mathcal{X} = \mathcal{Y} = \{1, 2, \dots, 5\}, \quad S = \{1, 2\}$$

$$(w(j|i|1))_{i,j=1,\dots,5} = \begin{pmatrix} 1 & 0 & 0 & 0 & 0 \\ 0 & 1 & 0 & 0 & 0 \\ 0 & 0 & 1 & 0 & 0 \\ 0 & 0 & 1 & 0 & 0 \\ 0 & 0 & 1 & 0 & 0 \end{pmatrix}$$

$$(w(j|i|2))_{i,j=1,\dots,5} = \begin{pmatrix} 0 & 0 & 1 & 0 & 0 \\ 0 & 0 & 1 & 0 & 0 \\ 0 & 0 & 1 & 0 & 0 \\ 0 & 0 & 0 & 1 & 0 \\ 0 & 0 & 0 & 0 & 1 \end{pmatrix}$$

The capacity of the simultaneous channel is given by

$$\max_p \min_{s=1,2} I(p, w(\cdot|\cdot|s)) = \max_p \min_{s=1,2} \sum_{i,j=1}^{5} p_i w(j|i|s) \log \frac{w(j|i|s)}{\sum_{k=1}^{5} w(j|k|s)}$$

$$I(p, w(\cdot|\cdot|1)) = (p_3 + p_4 + p_5) \log \frac{1}{p_3 + p_4 + p_5} + p_1 \log \frac{1}{p_1} + p_2 \log \frac{1}{p_2}$$

$$I(p, w(\cdot|\cdot|2)) = (p_1 + p_2 + p_3) \log \frac{1}{p_1 + p_2 + p_3} + p_4 \log \frac{1}{p_4} + p_5 \log \frac{1}{p_5}$$

$$\max_p I(p, w(\cdot|\cdot|1)) = \log 3 \tag{1.2.4}$$

The maximum is attained for $p = \left(\frac{1}{3}, \frac{1}{3}, p_3, p_4, p_5\right)$ and no other PDs.

$$\max_p I(p, w(\cdot|\cdot|2)) = \log 3, \tag{1.2.5}$$

the maximum is attained for $p = \left(p_1, p_2, p_3, \frac{1}{3}, \frac{1}{3}\right)$ and no other PDs.

From (1.2.4) and (1.2.5), it follows that

$$\max \min_{s=1,2} I(p, w(\cdot|\cdot|s)) < \log 3 \tag{1.2.6}$$

Consider the sets

$$V^n = \left\{x^n | x^n = (x_1, \dots, x_n) \in \mathcal{X}^n, x_t \in \{3, 4, 5\}\right\}$$
$$W^n = \left\{x^n | x^n = (x_1, \dots, x_n) \in \mathcal{X}^n, x_t \in \{1, 2, 3\}\right\}.$$

Define the code $\{(\overline{u}_i, D_i) : \overline{u}_i \in V^n \cap W^n, D_i = \{y^n : y_t = t\text{th component of } \overline{u}_i\}\}$.

The length of this code is

$$N = 2 \cdot 3^n - 1 > 3^n.$$

For $\lambda \geq 1/2$ the code is an averaged simultaneous $\lambda$-code:

$$\frac{1}{N}\sum_{i=1}^{N} F^n(D_i|\overline{u}_i|s) \geq \frac{1}{2} \geq 1-\lambda, \quad s = 1, 2.$$

If we denote the maximal length of an averaged simultaneous $\lambda$-code in

$$N_a(n, \lambda) > 3^n = e^{\log 3 \cdot n}.$$

However, it follows from Fano's lemma that

$$\max_p \min_{s=1,2} I(p, w(\cdot|\cdot|s)) < \log 3$$

is the weak capacity for our simultaneous channel with averaged error. The strong converse of the coding theorem does not hold. In special cases, we can give a sharper estimate than that given by Theorem 6.

**Example** Given $\mathcal{X} = \mathcal{Y} = \{1, \dots, a\}$ and the stochastic matrices $w(\cdot|\cdot|1)$, $w(\cdot|\cdot|2)$ with $a$ rows and $a$ columns. For $s = 1, 2$, we define the discrete memoryless channel $(P^n(\cdot|\cdot|s))_{n=1}^{\infty}$ by $P^n(y^n|x^n|s) = \prod_{t=1}^{n} w(y_t|x_t|s)$ for all $x^n \in \mathcal{X}^n$, $y^n \in \mathcal{Y}^n$, $n = 1, 2, \dots$ and the averaged channel $\mathcal{C}_1$ by

$$P^n(y^n|x^n|s) = \frac{1}{2}P^n(y^n|x^n|1) + \frac{1}{2}P^n(y^n|x^n|2), \quad x^n \in \mathcal{X}^n, \quad y^n \in \mathcal{Y}^n, \quad n = 1, 2, \dots.$$

We get

$$C_1(\lambda) = \max_{s=1,2} \max_p I(p, w(\cdot|\cdot|s)) = \overline{\overline{C}} \text{ for } \lambda > \frac{1}{2} \tag{1.2.7}$$

and

$$C_1(\lambda) = \max_p \inf_{s=1,2} I(p, w(\cdot|\cdot|s)) = \overline{\overline{C}} \quad \text{for } \lambda < \frac{1}{2} \tag{1.2.8}$$

and

$$\overline{\overline{C}} \geq \limsup_n \frac{1}{n} \log N\left(n, \frac{1}{2}\right) \geq \liminf_{n\to\infty} \frac{1}{n} \log N\left(n, \frac{1}{2}\right) \geq \overline{C} \qquad \text{for } \lambda = \frac{1}{2}. \tag{1.2.9}$$

[$C_1(\lambda) \geq \max_{s=1,2} \max_p I(p, w(\cdot|\cdot|s)) = \overline{\overline{C}}$ for $\lambda > \frac{1}{2}$ follows from the coding theorem for an individual channel. It remains to show that $C_1(\lambda) \leq \overline{\overline{C}}$: a $\lambda$-code $\{(u_i, D_i) : i = 1, \dots, N\}$ for $C_1$ has the property that either

$$P^n(D_i|u_i|1) \geq 1 - \lambda \quad \text{or} \quad P^n(D_i|u_i|2) \geq 1 - \lambda \;\; i = 1, \dots, N. \tag{1.2.10}$$

Therefore:

$$N(n, \lambda) \leq 2e^{\overline{\overline{C}}n + k(\lambda)\sqrt{n}} \quad \text{and} \quad C_1(\lambda) \leq \overline{\overline{C}}.$$

(1.2.8) is trivial. (1.2.9) is a consequence of (1.2.7) and (1.2.8).]

It is possible that $C_1(\lambda)$ exists for $\lambda = \frac{1}{2}$ and is unequal tp $\overline{\overline{C}}$ and to $\overline{C}$. Choose, for example, $w(\cdot|\cdot|1)$ such that $0 < C_0 < \max_p I(p, w(\cdot|\cdot|1))$ ([39]) and

$$w(1|i|2) = 1 \quad \text{for} \quad i = 1, \dots, a$$
$$w(j|i|2) = 0 \quad \text{for} \quad j \neq 1, i = 1, \dots, a.$$

Then we get

$$\overline{\overline{C}} > C_0 = \lim_{n\to\infty} \frac{1}{n} \log N\left(n, \frac{1}{2}\right) > \overline{C} = 0.$$

In general, we have

$$\overline{\overline{C}} \geq \lim_{n\to\infty} \sup \log N\left(n, \frac{1}{2}\right) \geq \liminf_{n\to\infty} \log N\left(n, \frac{1}{2}\right) > \max\left(\overline{C}, \overline{\overline{C}}_0\right),$$

where $\overline{\overline{C}}_0$ is the maximum of the zero error capacities of $w(\cdot|\cdot|1)$, $w(\cdot|\cdot|2)$.

A formula for $C\left(\frac{1}{2}\right)$ would imply a formula for $C_0$, which is unknown [39]. But even the existence of $\lim_{n\to\infty} \frac{1}{n} \log N\left(n, \frac{1}{2}\right)$ is not obvious. This seems to be a difficult problem. It is easy to construct channels for which $C(\lambda)$ has countable many jumps but for which the weak capacity $C = \lim_{\lambda\to 0} C(\lambda)$ exists, as was shown in [3]. Probably, there exist even channels for which $C(\lambda)$ does not exist for all $\lambda$ but for which $C$ still exists.

*Remark* In case $S$ is finite, we can give a sharper estimate than (i) in Theorem 6:

$$\inf_{0<\eta<1} \sup_{p} \sup_{S':q(S')\geq 1-\frac{\lambda}{\eta}} \inf_{s\in S'} I(p, F(\cdot|\cdot|s)) \geq \limsup_{n\to\infty} \frac{1}{n} \log N_1(n, \lambda)$$

This can be proved by extending the argument used under (1.2.10) for $S = \{1, 2\}$ to the general finite case.

#### 1.2.1.3 A Remark on the Paper of Wolfowitz: "Memory Increases Capacity"

Given a stochastic matrix $w(i|j)_{\substack{i=1,\dots a\\ j=1,\dots,a}}$, we define the channel 0 without memory:

$$P^n(y^n|x^n) = \prod_{t=1}^{n} w(y_t|x_t), \quad x^n \in \mathcal{X}^n, \qquad y^n \in \mathcal{Y}^n, \quad n = 1, 2, \dots.$$

Let $(P^n(\cdot|\cdot|M))_{n=1}^{\infty}$ be any channel with the property

$$P_t(y_t|x_t|M) = w(y_t|x_t), \qquad x_t \in \mathcal{X}_t, \qquad y_t \in \mathcal{Y}_t, \quad t = 1, 2, \dots.$$

Thus, the two channels are directly comparable. Wolfowitz [46] proved: suppose that in channel $M$ the power of the memory ([45], Chap. 6.7) between blocks of letters separated by $d$ letters approaches zero as $d \to \infty$, uniformly in the blocks, then the capacity of channel $M$ is not less than that of channel 0.

It follows from Dobrushin's inequality [22] that

$$C_M = \limsup_{n\to\infty} \frac{1}{n} \sup_{p^n} I(p^n, P^n(\cdot|\cdot|M)) \geq C_0.$$

Wolfowitz's result holds, iff $C_M$ is capacity. But $C$ need not be the capacity of channel $M$. [cf. Sect. 1.1.3, Remark 3.]

Averaged channels are channels with memory. They give us examples of channels, where memory decreases capacity.

**Example** Let $w(\cdot|\cdot|1)$, $w(\cdot|\cdot|2)$ be stochastic matrices with

$$\max_p I(p, w(\cdot|\cdot|2)) = 0$$
$$\max_p I(p, w(\cdot|\cdot|1)) > 0.$$

Let $w(\cdot|\cdot) = q_1 w(\cdot|\cdot|1) + q_2 w(\cdot|\cdot|2)$.

For $q_1$ sufficiently near to 1:

$$\max_p I(p, w) > 0,$$

but the weak capacity of

$$\left(P^n(\cdot|\cdot|M)\right)_{n=1}^{\infty} = \left(\sum_{s=1}^{2} q_s P^n(\cdot|\cdot|s)\right)_{n=1}^{\infty}$$

is $\max_p \inf_{s=1,2} I(p, w(\cdot|\cdot|s)) = 0$.

In general, we have

$$C_M \geq \max_p \sum_{s=1,2} q_s I(p, w(\cdot|\cdot|s)) \quad \text{[cf. Sect. 1.1.3, Remark 3]}$$

$$C_0 = \max_p I(p, q_1 w(\cdot|\cdot|1) + q_2 w(\cdot|\cdot|2))$$

the weak capacity of $M$ equals $\max_p \inf_{s=1,2} I(p, w(\cdot|\cdot|s))$.

If $\max_p \inf_{s=1,2} I(p, w(\cdot|\cdot|s)) < C_0$, then the memory decreases capacity.

If $\max_p \inf_{s=1,2} I(p, w(\cdot|\cdot|s)) \geq C_0$, then memory increases capacity.

**Example**

$$w(\cdot|\cdot|1) = \begin{pmatrix} 1 & 0 \\ 0 & 1 \end{pmatrix}$$

$$w(\cdot|\cdot|2) = \begin{pmatrix} 0 & 1 \\ 1 & 0 \end{pmatrix}$$

$$q_1 = q_2 = \frac{1}{2}$$

$$w(\cdot|\cdot) = \begin{pmatrix} \frac{1}{2} & \frac{1}{2} \\ \frac{1}{2} & \frac{1}{2} \end{pmatrix}$$

$$\max_p \inf_{s=1,2} I(p, w(\cdot|\cdot|s)) = \log 2$$

$$C_0 = 0$$

Even if the strong converse holds for channel $M$, the capacity need not be greater than $C_0$. (For the definition of the general discrete channel see [45], Chap. 5)

**Example**

$$\mathcal{X} = \mathcal{Y} = \{1, 2\}$$

$$P^n(111\ldots1|111\ldots1) = 1$$

$$P^n(000\ldots0|000\ldots0) = 1$$

$$P^n(y^n|x^n) = \frac{1}{2^n} \text{ iff } x^n \begin{matrix} \neq (1,1,1\dots 1) \\ \neq (0,0,0\dots 0) \end{matrix} \quad n = 1, 2, \dots, x^n \in \mathcal{X}^n, y^n \in \mathcal{Y}^n.$$

$$w(\cdot|\cdot) = \begin{pmatrix} 1 & 0 \\ 0 & 1 \end{pmatrix}$$

The strong capacity of $(P^n(\cdot|\cdot))_{n=1}^{\infty}$ is 0 and $C_0 = \log 2$.

#### 1.2.1.4 Discrete Averaged Channels Where Either the Sender or the Receiver Knows the CPF

Assume $\mathcal{Y}_t = \{1, \dots, a\}$, $t = 1, 2, \dots$; $S = \{1, 2, \dots\}$, $q = (q_1, \dots)$, $q_i > 0$ PD on $S$.

The discrete averaged channel is defined by

$$(P^n(D|x^n))_{n=1}^{\infty} = \left( \sum_{s=1}^{\infty} q_s F^n(D|x^n|s) \right)_{n=1}^{\infty} \qquad A \subset \mathcal{Y}^n, x^n \in \mathcal{X}^n$$

**Lemma 5** *For $q = (q_1, \dots)$ define*

$$\eta_k = \inf_{\kappa=1,\dots,k} q_\kappa > 0.$$

*If $\{(u_1(s), D'_1(s)) \dots (u_N(s), D'_N(s))\}$ is a set of pairs, where $u_i(s) \in \mathcal{X}^n$, $D'_i(s) \in \mathcal{L}_n$ for $i = 1, \dots, N$, $s \in S$, $D'_i(s) \cap D'_j(s) = \emptyset$ for $i \neq j$, $s \in S$ and furthermore*

$$\sum_{s \in S} q_s P^n(D'_i(s)|u_i(s)|a) \geq 1 - \frac{\eta_k}{2} \tag{1.2.11}$$

*then*

$$P^n(D'_i|u_i(s)|s) \geq \frac{1}{2}\eta_k \quad \text{for } s = 1, \dots, k; \quad i = 1, \dots, N. \tag{1.2.12}$$

*Proof* Define $\varepsilon_k = \sum_{\kappa=k+1}^{\infty} q_\kappa$.

From (1.2.11), we conclude

$$\sum_{s=1}^{k} q_s P^n(D'_i(s)|u_i(s)|s) \geq 1 - \varepsilon_k - \frac{1}{2}\eta_k,$$

since

$$\sum_{s=1}^{k} q_s P^n(D_i'(s)|u_i(s)|s) \geq \sum_{s=1}^{\infty} q_s P^n(D_i'(s)|u_i(s)|s) - \varepsilon_k$$
$$\geq 1 - \left(\frac{1}{2}\eta_k + \varepsilon_k\right). \tag{1.2.13}$$

But

$$\sum_{s=1}^{k} q_s P^n(D_i'|u_i(s)|s) - q_{s'} P^n(D_i'(s')|u_i(s')|s') \leq 1 - \varepsilon_k - \eta_k \quad \text{for } s' = 1, \ldots, k. \tag{1.2.14}$$

From (1.2.13), (1.2.14), we have

$$q_s P^n(D_i'|u_i(s)|s) \geq \frac{1}{2}\eta_k \quad \text{for} \quad s = 1, \ldots, k; \quad i = 1, \ldots, N$$

and therefore

$$P^n(D_i'|u_i(s)|s) > \frac{1}{2}\eta_k \quad \text{for} \quad s = 1, \ldots, k; \quad i = 1, \ldots, N$$

(cf. Sect. 1.1, proof of Theorem 2) □

*Remark* The proof goes through verbatim for the semi-continuous case. Averages with respect to general PD can be treated in the same way as in Sect. 1.1.3. However, in the fifth subsection we give a different proof, which covers all these cases.

**Theorem 7** (Coding theorem and weak converse for $\mathcal{C}_3$) *Let* $C_3 = \max_p \inf_{s \in S} I(p, w(\cdot|\cdot|s))$.

*Then the following estimates hold:*

*(i) Given* $0 < \lambda < 1$, $\delta > 0$, *then there exists an* $n_0 = n_0(\lambda, \delta)$ *such that* $N_3(n, \lambda) > e^{(C_3 - \delta)n}$ *for* $n \geq n_0$.

*(ii) Given* $\delta > 0$, *then there exists a* $\lambda$ *and an* $n_0 = n_0(\lambda, \delta)$ *such that* $N_3(n, \lambda) < e^{(C_3 + \delta)n}$ *for* $n \geq n_0$.

*Proof* (i) A $\lambda$-code for the simultaneous channel $\{P^n(\cdot|\cdot|s) : s \in S\}$ is a $\lambda$-code for $\mathcal{C}_3$.

(ii) For the given $\delta > 0$ choose $k$ such that $|C_k - C| \leq \frac{\delta}{2}$, then choose $\lambda = \frac{\eta_k}{2}$. It follows from Lemma 5, that for a $\lambda$-code $\{(u_i, D_i'(s)) : i = 1, \ldots, N; s \in S\}$ for $\mathcal{C}_3$

$$P^n(D_i'|u_i|s) \geq \frac{1}{2}\eta_k \quad s = 1, \ldots, k; \quad i = 1, \ldots, N$$

holds.

Statement (ii) follows now by usual arguments. ([3], Theorem 4.5.2) □

**Theorem 8** (Coding theorem and weak converse for $\mathcal{C}_2$ and $\mathcal{C}_4$) *Let* $C_2 = C_4 = \inf_{s\in S} \max_p I(p, w(\cdot|\cdot|s))$.

*Then the following estimates hold:*

*(i) Given* $0 < \lambda < 1$, $\delta > 0$, *then there exists an* $n_0 = n_0(\lambda, \delta)$ *such that* $N_j(n, \lambda) > e^{(C_j-\delta)n}$ *for* $n \geq n_0$, $j = 2, 4$.

*(ii) Given* $\delta > 0$, *then there exists a* $\lambda$ *and an* $n_0 = n_0(\lambda, \delta)$ *such that* $N_j(n, \lambda) < e^{(C_j+\delta)n}$ *for* $n \geq n_0$, $j = 2, 4$.

*Proof* (i) follows from Theorem 4.5.3 in [45].

(ii) By the same arguments as in the proof for Theorem 7 (ii) we get

$$P^n(D_i'(s)|u_i(s)|s) \geq \frac{1}{2}\eta_k \quad s = 1, \ldots, k; \quad i = 1, \ldots, N.$$

Applying the strong converse of the coding theorem for individual channels we get (ii). □

*Remark* The strong converse of the coding theorem does not hold for $\mathcal{C}_2$, $\mathcal{C}_3$, and $\mathcal{C}_4$:

**Example** (see Sect. 1.1.3, Remark 1) Choose $w(\cdot|\cdot|s)$, $s = 1, 2$ such that $(P^n(\cdot|\cdot|1))_{n=1}^{\infty}$ has capacity 0 and $(P^n(\cdot|\cdot|2))_{n=1}^{\infty}$ has capacity $C(2) > 0$. Then a fortiori $C_2 = C_3 = C_4 = 0$.

A $\lambda$-code for $\frac{1}{2}P^n(\cdot|\cdot|2)$ is a $\lambda$-code for $P^n$. For $\lambda > \frac{1}{2}$, we get

$$N_i(n, \lambda) > e^{C(2)n-k(\lambda)\sqrt{n}} \quad \text{for } i = 2, 3, 3.$$

That $\mathcal{C}_3$ does not have a strong capacity was earlier shown by Wolfowitz ([45], Chap. 7.7).

#### 1.2.1.5 The General Case

We return to the case, where the individual channels $(F^n(\cdot|\cdot|s))_{n=1}^{\infty}$ are semi-continuous and $q$ is a general PD (as described in the first subsection).

**Lemma 6** *If*

$$\frac{1}{N}\sum_{i=1}^{N}\int_S P^n(D_i(s)|u_i(s)|s)dq(s) \geq 1 - \lambda \quad \textit{for } i = 1, \ldots, N \textit{ and } 1 > \gamma, \beta > 0$$

*such that* $\gamma\beta > 0$ *then*

$$q\left\{s|\frac{1}{N}\sum_{i=1}^{N}(D_i(s)|u_i(s)|s) \geq 1 - \lambda\right\} \geq 1 - \beta.$$

*Proof* Define $f^*(s) = \frac{1}{N}\sum_{i=1}^{N} P^n(D_i(s)|u_i(s)|s)$, then the proof of Lemma 4 turns over verbatim. □

**Theorem 9** (Coding theorem and weak converse for $C_1$) *Let*

$$C_1 = \inf_{\alpha>0} \sup_{\{S'\subset S: q(S')\geq 1-\alpha\}} \lim_{n\to\infty} \frac{1}{n} \sup_{\mathcal{D}^n\in\mathcal{Z}^n} \sup_{p^n} \inf_{s\in S'} I(p^n|\mathcal{D}^n|F^n(\cdot|\cdot|s)),$$

*where $\mathcal{D}^n = (D_1,\ldots,D_b)$ is a partition of $\mathcal{Y}^n$ in finitely many elements of $\mathcal{L}^n$ and, $\mathcal{Z}^n$ is the set of all such finite partitions, and*

$$I(p^n|\mathcal{D}^n|F^n(\cdot|\cdot|s) = \sum_{i=1,\ldots,b} \sum_{x^n\in\mathcal{X}^n} p^n(x^n)F^n(D_i|x^n|s) \log \frac{F^n(D_i|x^n|s)}{\sum_{y^n\in\mathcal{Y}^n} p^n(y^n)F^n(D_i|y^n|s)}$$

*then the following estimates holds:*

*(i) Given $0<\lambda<1, \delta>0$, then there exists an $n_0 = n_0(\lambda,\delta)$, such that $N_1(n,\lambda) > e^{(C_1-\delta)n}$ for $n \geq n_0$.*
*(ii) Given $\delta > 0$, then there exists a $\lambda$ and an $n_0 = n_0(\lambda,\delta)$, such that $N_1(n,\lambda) < e^{(C_1-\delta)n}$ for $n \geq n_0$.*

*Proof* (i) Define

$$C_1(\alpha) = \sup_{\{S'\subset S: q(S')\geq 1-\alpha\}} \lim_{n\to\infty} \sup_{\mathcal{D}^n} \sup_{p^n} \inf_{s\in S'} I(p^n|\mathcal{D}^n|F^n(\cdot|\cdot|s)).$$

Given $\lambda, \delta > 0$, choose $\alpha < \lambda$ and $S'$ such that $q(S') \geq 1-\alpha$ and

$$\left| \lim_{n\to\infty} \frac{1}{n} \sup_{\mathcal{D}^n} \sup_{p^n} \inf_{s\in S'} I(p|\mathcal{D}^n|F^n(\cdot|\cdot|s)) - C_1(\alpha) \right| \leq \frac{\delta}{2}.$$

Define $\lambda' = \frac{\lambda-\alpha}{1-\alpha}$. A $\lambda'$-code for the semi-continuous simultaneous channel $\{F^n(\cdot|\cdot|s) : s\in S'\}$ is a $\lambda$-code for $C_1$, because $(1-\lambda')(1-\alpha) = 1-(1-\alpha)\lambda'-\alpha = 1-\lambda$.

The coding theorem for semi-continuous simultaneous channels ([32], Theorem 1) gives us

$$N_1(n,\lambda) \geq e^{\left(C_1(\alpha)-\frac{\delta}{2}\right)n - k(\lambda')\sqrt{n}} \geq e^{(C_1-\delta)n}$$

for $n$ sufficiently large.

(ii) Choose $\alpha$ such that

$$|C_1(\alpha) - C_1| \leq \frac{\delta}{3}, \quad \text{then } \beta \text{ such that } \log a\left(\frac{1}{1-\beta} - 1\right) \leq \frac{\delta}{3} \tag{1.2.15}$$

and finally $\lambda$ such that $\alpha\beta > \lambda$.

By Lemma 6 there exists a set $S$ with $q(S) \geq 1 - \alpha$ and

$$\frac{1}{N}\sum_{i=1}^{N} F^n(D_i|u_i|s) \geq 1 - \beta \quad \text{for} \quad s \in S.$$

From Fano's lemma we obtain

$$N_1(n, \lambda) \leq e^{\frac{C_1(\alpha)}{1-\beta}n}.$$

If we use (1.2.15) we get statement (ii). □

**Theorem 10** (Coding theorem and weak converse for $C_3$) *Let*

$$C_3 = \inf_{\alpha>0} \sup_{p} \sup_{\{S' \subset S : q(S') \geq 1-\alpha\}} \inf_{s \in S'} I(p, F^n(\cdot | \cdot |s)),$$

*then the following estimates holds:*

*(i) Given $0 < \lambda < 1, \delta > 0$, then there exists an $n_0 = n_0(\lambda, \delta)$, such that $N_3(n, \lambda) > e^{(C_3-\delta)n}$ for $n \geq n_0$.*
*(ii) Given $\delta > 0$, then there exists a $\lambda$ and an $n_0 = n_0(\lambda, \delta)$, such that $N_3(n, \lambda) < e^{(C_3-\delta)n}$ for $n \geq n_0$.*

*Proof* (i) Define

$$C_3(\alpha) = \sup_{p} \sup_{\{S' \subset S : q(S') \geq 1-\alpha\}} \inf_{s \in S'} I(p, F^n(\cdot | \cdot |s)).$$

Given $\lambda, \delta > 0$, choose $\alpha > \lambda$ and $S'$ such that $q(S') \geq 1 - \alpha$ and

$$\left| \sup_{p} \inf_{s \in S'} I(p, F^n(\cdot | \cdot |s)) - C_3(\alpha) \right| \leq \frac{\delta}{2}.$$

Define $\lambda' = \frac{\lambda-\alpha}{1-\alpha}$. A $\lambda'$-code for the compound channel (with receiver knowledge) $S'^n(R) = \{F^n(\cdot | \cdot |s) : s \in S'\}$ is a $\lambda$-code for $C_3$, because $(1-\lambda')(1-\alpha) = 1-\lambda$.

The coding theorem for $S'^n(R)$ ([32], Theorem 4) gives us

$$N_3(n, \lambda) \geq N_R(n, \lambda') \geq e^{\left(C_3(\alpha) - \frac{\delta}{2}\right)n - k(\lambda')\sqrt{n}} \geq e^{(C_3-\delta)n}$$

for $n$ sufficiently large.

(ii) Choose $\alpha$ such that

$$|C_3(\alpha) - C_3| \leq \frac{\delta}{3}, \quad \text{then } \beta \text{ such that } \log a \left( \frac{1}{1-\beta} - 1 \right) \leq \frac{\delta}{3} \tag{1.2.16}$$

and finally $\lambda$ such that $\alpha\beta > \lambda$.

By Lemma 6 there exists a set $S$ with $q(S) \geq 1 - \alpha$ and

$$\frac{1}{N}\sum_{i=1}^{N} F^n(D_i(s)|u_i|s) \geq 1 - \beta \qquad \text{for } s \in S.$$

From Fano's lemma we obtain

$$N_3(n, \lambda) \leq e^{\frac{C_3(\alpha)}{1-\beta}n}.$$

If we use (1.2.16) we get statement (ii). □

**Theorem 11** (Coding theorem and weak converse for $\mathcal{C}_2$ *and* $\mathcal{C}_4$) *Let*

$$C_2 = C_4 = \inf_{\alpha>0} \sup_{\{S' \subset S : q(S') \geq 1-\alpha\}} \lim_{n\to\infty} \frac{1}{n} \sup_{\mathcal{D}^n} \inf_{s \in S'} \sup_{p^n} I(p, \mathcal{D}^n, F^n(\cdot|\cdot|s))$$

*(the existence of the limit was shown in* [32]*), then the following estimates holds:*

*(i) Given $0 < \lambda < 1, \delta > 0$, then there exists an $n_0 = n_0(\lambda, \delta)$, such that $N_j(n, \lambda) > e^{(C_j - \delta)n}$ for $n \geq n_0$, $j = 1, 2$.*
*(ii) Given $\delta > 0$, then there exists a $\lambda$ and an $n_0 = n_0(\lambda, \delta)$, such that $N_j(n, \lambda) < e^{(C_j - \delta)n}$ for $n \geq n_0$, $j = 1, 2$.*

*Proof* (i) It follows from the definition of $\mathcal{C}_2$ and $\mathcal{C}_4$ that

$$N_2(n, \lambda) \geq N_4(n, \lambda) \tag{1.2.17}$$

therefore it is enough to prove (i) for $\mathcal{C}_2$.

Define

$$C_2(\alpha) = \sup_{\{S' \subset S : q(S') \geq 1-\alpha\}} \lim_{n\to\infty} \frac{1}{n} \sup_{\mathcal{D}^n} \inf_{s \in S'} \sup_{p^n} I(p|\mathcal{D}^n|F^n(\cdot|\cdot|s)).$$

Given $\lambda, \delta > 0$, choose $\alpha > \lambda$ and $S'$ such that $q(S') \geq 1 - \alpha$ and

$$\left| \lim_{n\to\infty} \frac{1}{n} \sup_{\mathcal{D}^n} \inf_{s \in S'} \sup_{p^n} I(p|\mathcal{D}^n|F^n(\cdot|\cdot|s)) - C_2(\alpha) \right| \leq \frac{\delta}{2}.$$

Define $\lambda' = \frac{\lambda-\alpha}{1-\alpha}$. A $\lambda'$-code for the simultaneous channel with sender knowledge $S'^n(S) = \{F^n(\cdot|\cdot|s) : s \in S'\}$ is a $\lambda$-code for $\mathcal{C}_2$.

The coding theorem for $S'^n(S)$ ([32], Theorem 3) gives us

$$N_2(n, \lambda) \geq N_S(n, \lambda') \geq e^{\left(C_2(\alpha) - \frac{\delta}{2}\right)n}$$

for $n$ sufficiently large and therefore

$$N - 2(n, \lambda) \geq e^{(C_2 - \delta)n}$$

for $n$ sufficiently large.

(ii) Choose $\alpha$ such that

$$|C_4(\alpha) - C_4| \leq \frac{\delta}{3}, \quad \text{then } \beta \text{ such that } \log a \left( \frac{1}{1-\beta} - 1 \right) \leq \frac{\delta}{3} \tag{1.2.18}$$

and finally $\lambda$ such that $\alpha\beta > \lambda$.

By Lemma 6 there exists a set $S$ with $q(S) \geq 1 - \alpha$ and

$$\frac{1}{N} \sum_{i=1}^{N} F^n(D_i(s)|u_i(s)|s) \geq 1 - \beta \qquad \text{for } s \in S.$$

From Fano's lemma and (1.2.17) we get

$$N_2(n, \lambda) \leq N_4(n, \lambda) \leq e^{\frac{C_4(\alpha)}{1-\beta} n}.$$

If we use (1.2.18) we get statement (ii). □

### 1.2.2 Another Channel: The Optimistic Channel

Let $\mathcal{S}$ be an arbitrary (index-)set and to each $s \in \mathcal{S}$ assigned a semi-continuous nonstationary channel without memory, $(F^n(\cdot|\cdot|s))_{n=1}^{\infty}$.

In the theory of simultaneous channels [1, 43], one uses the following definition of a $\lambda$-code: a $\lambda$-code $(N, n, \lambda)$ is a set of pairs $\{(u_i|D_i) : i = 1, \dots, N\}$ with $u_i \in \mathcal{X}^n$, $D_i \in \mathcal{L}^n$ for $i = 1, \dots, N$. $D_i \cap D_j = \emptyset$ for $i \neq j$ and with

$$\inf_{s \in \mathcal{S}} F^n(D_i|u_i|s) \geq 1 - \lambda. \tag{1.2.19}$$

This describes the situation, where neither the sender nor the receiver knows the individual channel which governs the transmission of a code word $u_i$.

If the sender can choose for the transmission of a message the channel over which he wants to transmit, then we have in the code definition (1.2.19) to exchange by

$$\sup_{s \in \mathcal{S}} P^n(D_i|u_i(s)|s) \geq 1 - \lambda. \tag{1.2.20}$$

We denote the described optimistic channel by $\mathcal{OC}$.

### 1.2.2.1 The Discrete Memoryless Case

Assume that the channels $(F^n(\cdot|\cdot|s))_{n=1}^{\infty}$ are discrete, stationary, and memoryless.

**Theorem 12** (Coding theorem and strong converse for $\mathcal{A}$) *Let* $C = \sup_{s\in\mathcal{S}} \max_p I(p, w(\cdot|\cdot|s))$. *Given* $\delta > 0, 0 < \lambda < 1$, *then there exists an* $n_0$ *such that for* $n \geq n_0$

*(i)* $N(n, \lambda) > e^{(C-\delta)n}$
*(ii)* $N(n, \lambda) < e^{(C+\delta)n}$

*Proof* Part (i) follows from the coding theorem for an individual channel.
For the proof of part (ii) we need the

**Lemma 7** ([45], 4.2.2. p. 35, [2]) *Let b be greater than 0. There exists a null-sequence of positive real numbers* $\{a_n\}_{n=1}^{\infty}$ *with the property:*
*let* $n \in N$, $D \subset \mathcal{Y}^n$, $x^n \in \mathcal{S}^n$, $s, s^* \in \mathcal{S}$ *such that* $F^n(D|x^n|s) > b$ *and*

$$|w(i|j|s^*) - w(i|j|s)| \leq \frac{a}{2\sqrt{n}}, \quad 1 \leq i, j \leq a$$

*then*

$$\left| \frac{F^n(A|x^n|s^*)}{F^n(D|x^n|s)} - 1 \right| < a_n.$$

$\mathcal{S}$ can be written as a finite union of disjoint sets $\mathcal{S}_r$, $r = 1, \dots, \left(\frac{2\sqrt{n}}{a}\right)^{a^2} = R$, such that

$$|w(j|i|s) - w(j|i|s^*)| \leq a \cdot 2^{-\sqrt{n}} \quad i, j = 1, \dots, a \quad \text{for } s, s^* \in \mathcal{S}_r.$$

Let now $\{(u_i, D_i) : i = 1, \dots, N\}$ be a code $(N, n, \lambda)$ of maximal length.

To every $u_i$ corresponds an individual channel $P^n(\cdot|\cdot|s_i)$ and therefore a matrix $w(\cdot|\cdot|s_i)$. Let $\{(u_{i_r}, D_{i_r}) : i_r = 1, \dots, N_r\}$ be the subcode which corresponds to $\mathcal{S}_r$. It follows from Lemma 7 that for a $\lambda' > \lambda$ there exists an $n_0$ such that for $n \geq n_0$

$$P^n(A_{i_r}|u_{i_r}|s) \geq 1 - \lambda' \quad \text{for all} \quad s \in \mathcal{S}_r, \quad r = 1, \dots, R.$$

The strong converse of the coding theorem for an individual channel gives us

$$N(n, \lambda) = \sum_{r=1}^{R} N_r(n, \lambda) \leq R e^{Cn + k(\lambda')\sqrt{n}} \leq e^{(C+\delta)n}$$

for $n$ sufficiently large. □

*Remark* In the semi-continuous case we have in general no compactness property which leads to Lemma 7. The coding problem for channel $\mathcal{O}C$ is then equivalent to the coding problem of an individual channel with the input alphabet $\mathcal{X} = \bigcup_{s\in\mathcal{S}} \mathcal{X}_s$, $\mathcal{X}_s = \{(1, s), \dots, (a, s)\}$ for $s \in \mathcal{S}$, where the code words are restricted to have as components elements from one set $\mathcal{X}_s$.

### 1.2.2.2 The CPF Varies from Letter to Letter

Given a set of kernels $\{F(\cdot|\cdot|s) : s \in \mathcal{S}\}$. For every $n$-tuple $\mathcal{S}^n = (s_1, \dots, s_n)$, $s_i \in \mathcal{S}$, we can define the product kernel

$$F^n(\cdot|\cdot|s^n) = \prod_{t=1}^{n} F(\cdot|\cdot|s_t).$$

Consider now the class

$$\mathcal{C}^n = \{F^n(\cdot|\cdot|s^n) : s^n = (s_1, \dots, s_n), s_i \in \mathcal{S}\}.$$

A simultaneous $\lambda$-code for $\mathcal{C}^n$ is a set of pairs $\{(u_i, D_i) : i = 1, \dots, N\}$, $u_i \in \mathcal{X}^n$, $D_i \in \mathcal{L}^n$, $D_i \cap D_j = \emptyset$ for $i \neq j$ with $F^n(D_i|u_i|s^n) \geq 1-\lambda$ for all $s^n = (s_1, \dots, s_n)$, $i = 1, \dots, N$.

It is, in general, an unsolved problem to estimate the maximal length of such a $\lambda$-code.

If we replace the code definition (1.2.20) by

$$F^n(D_i|u_i(s^n)|s^n) > 1 - \lambda \quad \text{for} \quad i = 1, \dots, N \text{ and some } s^n \in \mathcal{Y}^n, \tag{1.2.21}$$

then this leads to a new channel $\mathcal{O}^*C$, which can be viewed as a memoryless channel $W$ with input alphabet $\mathcal{X} \times \mathcal{S}$, output alphabet $(\mathcal{Y}, \mathcal{L})$, $(x^n, s^n)$ : $((x_1, s_1), \dots, (x_n, s_n))$, and

$$F^n(A|(x^n, s^n)) = F^n(A|x^n|s^n) \quad \text{for}$$
$$A \in \mathcal{L}^n, \quad (x^n, s^n) \in (\mathcal{X} \times \mathcal{S}) \times \dots \times (\mathcal{X} \times \mathcal{S}) \quad n = 1, 2, \dots.$$

A $\lambda$-code for channel $\mathcal{O}^*C$ is a $\lambda$-code for $(F^n)_{n=1}^{\infty}$ and vice versa. We have reduced the coding problem for $\mathcal{O}^*C$ to a known situation: $(F^n)_{i=1}^{\infty}$ is a stationary memoryless channel with general input and output alphabets and can be treated by the methods of Chap. 6.

## 1.3 Lecture on The Structure of Capacity Functions for Compound Channels

### 1.3.1 Definitions and Introduction of the Capacity Functions $C(\overline{\lambda}), C(\lambda_R), C(\overline{\lambda}_R)$

Let $\mathcal{X} = \{1, \ldots, a\}$ and $\mathcal{Y} = \{1, \ldots, b\}$ be, respectively, the input and output alphabets which will be used for transmission over a channel (or a system of channels). Any sequence of $n$ letters $x^n = (x_1, \ldots, x_n) \in \prod_1^n \mathcal{X}$ is called a transmitted or sent $n$-sequence, any sequence $y^n = (y_1, \ldots, y_n) \in \prod_1^n \mathcal{Y}$ is called a received $n$-sequence.

Let $\mathcal{S} = \{1, \ldots, k\}$, and

$$\mathcal{C} = \{w(\cdot|\cdot|s) : s \in \mathcal{S}\},$$

where each $w(\cdot|\cdot|s)$ is an $(a \times b)$ stochastic matrix, also called a channel probability function (CPF). For each $x^n = (x_1, \ldots, x_n) \in \mathcal{X}^n = \prod_1^n \mathcal{X}$, we define a probability distribution (PD) on $\mathcal{Y}^n = \prod_1^n \mathcal{Y}$ by $W^n(y_n|x^n|s) = \prod_{t=1}^n w(y_t|x_t|s)$, $y^n \in \mathcal{Y}^n$. $P^n(y^n|x^n|s)$ is the probability that, when the $n$-sequence $x^n$ is sent, the (chance) sequence received is $y^n$. The sequence $(P^n(\cdot|\cdot|s))_{n=1}^{\infty}$ describes a discrete channel without memory (DMC).

Thus we have assigned to each $s \in \mathcal{S}$ a DMC. We call the system of channels

$$\mathcal{C}^* = \{(P^n(\cdot|\cdot|s))_{n=1}^{\infty} : s \in \mathcal{S}\}$$

a *compound* (or simultaneous) channel (cf. [45]), if the transmission is governed as follows: each $n$-sequence $x^n$ is transmitted according to some channel in $\mathcal{C}^*$ and the channel may vary *arbitrarily* in $\mathcal{C}^*$ from one such $n$-sequence to another.

We define a code $(n, N, \lambda)$ for the compound channel as a system

$$\{(u_i, D_i) : u_i \in \mathcal{X}^n, D_i \subset \mathcal{Y}^n, D_i \cap D_j = \emptyset \quad \text{for} \quad i \neq j, \quad i = 1, \ldots, N\}$$

which satisfies

$$P^n(D_i|u_i|s) \geq 1 - \lambda, i = 1, \ldots, N; s \in \mathcal{S}.$$

As usual the entropy of a probability vector $\pi = (\pi_1, \ldots, \pi_t)$ is defined to be $H(\pi) = -\sum_{i=1}^t \pi_i \log_2 \pi_i$. Denote the rate for the (row) probability vector $\pi$ on $\mathcal{X}$ and CPF $w(\cdot|\cdot|s)$ by $R(\pi, s) = H(\pi'(s)) - \sum_i \pi_i H(w(\cdot|i|s))$, where $\pi'(s) = \pi \cdot w(\cdot|\cdot|s)$. Let $N(n, \lambda)$ be the maximal length of an $(n, N, \lambda)$ code for $\mathcal{C}^*$. It is an easy consequence of Theorem 1 in [43], that

$$\lim \frac{1}{n} \log N(n, \lambda) = C \tag{1.3.1}$$

where $C$ is a constant, independent of $\lambda$, given by

$$C = \max_{\pi} \inf_{s \in \mathcal{S}} R(\pi, s).$$

(1.3.1) means that the coding theorem and strong converse of the coding theorem hold. $C$ is called the capacity.

A code $(n, N, \overline{\lambda})$ with average error $\overline{\lambda}$ is a system

$$\{(u_i, D_i) : u_i \in \mathcal{X}^n, D_i \subset \mathcal{Y}^n, D_i \cap \mathcal{D}_j = \emptyset \quad \text{for} \quad i \neq j, \quad i = 1, \ldots, N\}$$

which satisfies

$$\frac{1}{N} \sum_{i=1}^{N} P^n(D_i|u_i|s) \geq 1 - \overline{\lambda}, \quad s \in \mathcal{S}.$$

Let $N(n, \overline{\lambda})$ be the maximal length of an $(n, N, \overline{\lambda})$ code for $\mathcal{C}^*$. It was proved in [17], that

$$\inf_{\overline{\lambda} > 0} \lim_{n \to \infty} \frac{1}{n} \log N(n, \overline{\lambda}) = 0.$$

(The coding theorem and weak converse for average error.)

When $|\mathcal{S}| = 1$ it is immaterial whether we use maximal or average error (cf. [45], Chap. 3.1, Lemma 3.11). This has led to the belief—widespread among engineers—that this is true even for more complex channel systems. However, already for compound channels with $|\mathcal{S}| = 2$ one has to distinguish carefully between these errors, as was shown in [1] (see the first example of Sect. 1.2). In fact,

$$\overrightarrow{\underline{\lim}}_{n \to \infty} \frac{1}{n} \log N(n, \overline{\lambda})$$

is in general greater than $C$. This means that, when we use average errors for codes for $\mathcal{C}^*$, we can achieve longer code lengths. The following questions are therefore of interest:

1. For which $\overline{\lambda}$ does $\lim_{n \to \infty} \frac{1}{n} \log N(n, \overline{\lambda})$ exist?
2. What can we say about the capacity function $C(\overline{\lambda})$, where

$$C(\overline{\lambda}) = \lim_{n \to \infty} \frac{1}{n} \log N(n, \overline{\lambda})$$

whenever the latter exists?

3. When $C(\overline{\lambda}) > C$, which encoding procedure gives the longest codes?

We shall also study channel $\mathcal{C}^*$ under randomized encoding. A random code $(n, N, \lambda_R)$ is a system of pairs

$$\{(p_i, D_i) : p_i \text{ PD on } \mathcal{X}^n, D_i \text{ disjoint}, \quad i = 1, \ldots, N\}$$

which satisfy

$$\sum_{x^n \in \mathcal{X}^n} p_i(x^n) P^n(D_i|x^n|s) \geq 1 - \lambda_R \quad i = 1, \dots, N. \tag{1.3.2}$$

If we allow average error instead of maximal error, we have to replace (1.3.2) by

$$\frac{1}{N} \sum_{i=1}^{N} \sum_{x^n \in \mathcal{X}^n} p_i(x^n) P^n(D_i|x^n|s) \geq 1 - \overline{\lambda}_R \tag{1.3.3}$$

in order to define a random (randomized) $(n, N, \overline{\lambda}_R)$ code.

The use of a random code is as follows: A set of messages $N = \{1, \dots, N\}$ is given in advance. If message $i$ is to be sent the sender performs a random experiment according to $p_i$, and the outcome of the experiment is sent. The receiver, after receiving the $n$-sequence $y^n \in D_j$, decides that message $j$ was intended. [This code concept was described in [11] under 2.1.]

Questions of interest to us are:

1. For which values of $\lambda_R, \overline{\lambda}_R$ does $\lim_{n\to\infty} \frac{1}{n} \log N(n, \lambda_R)$, respectively, $\lim_{n\to\infty} \frac{1}{n} \log N(n, \overline{\lambda}_R)$, exist?
2. What is the structure of the *capacity functions*

$$C(\overline{\lambda}_R) = \lim_{n\to\infty} \frac{1}{n} \log N(n, \overline{\lambda}_R)$$

and

$$C(\overline{\lambda}_R) = \lim_{n\to\infty} \frac{1}{n} \log N(n, \lambda_R)$$

where these are well-defined?

All our results will be obtained under the restriction that $\mathcal{C}$ contains only finitely many, say $k$, CPF's.

A word about notations. The functions $C(\overline{\lambda})$, $C(\lambda_R)$, and $C(\overline{\lambda}_R)$ are distinguished only by their arguments; these will always appear explicitly. The result is that all our results have to be interpreted with this understanding. For example, one of our theorems says that

$$C(\lambda_R) = C(\overline{\lambda}) = C(\overline{\lambda}_R)$$

under certain conditions when $\lambda_R = \overline{\lambda} = \overline{\lambda}_R$. Taken literally, this is a trivial statement. In the light of our notation it means that the three functions coincide for certain values of the argument. This notation will result below in no confusion or ambiguity, and has the advantages of suggestiveness and typographical simplicity.

Throughout this lecture $\lambda$, $\overline{\lambda}$, $\lambda_R$, and $\overline{\lambda}_R$ take values only in the open interval $(0, 1)$. This assumption avoids the trivial and will not be stated again.

### 1.3.2 Auxiliary Results

**Lemma 8** *Let $\mathcal{S} = \{1, \dots, d\}$ and let $\{(u_i, D_i) : i = 1, \dots, N\}$ be a code with $\inf_{s \in \mathcal{S}} \frac{1}{n} \sum_{i=1}^{N} P^n(D_i|u_i|s) \geq 1 - \overline{\lambda}$. There exist sequences $\{u_{i_v} : v = 1, \dots, N_1\} \subset \{u_i : i = 1, \dots, N\}$ such that*

$$P^n(D_{i_v}|u_{i_v}|s) \geq 1 - (\overline{\lambda} + \varepsilon)d \quad \textit{for } v = 1, \dots, N_1 = \left[\frac{\varepsilon}{1+\varepsilon} N\right] \quad \textit{and for} \quad s = 1, \dots, d.$$

*Proof* Define the probability distribution $P^*$ on $\{1, \dots, N\}$ by $P^* = \frac{1}{N}$ for $i = 1, \dots, N$. Define the random variables $\{X_s : s = 1, \dots, d\}$ by $X_s(i) = 1 - P(D_i|u_i|s)$ for $i = 1, \dots, N$. Thus $X_s(i) \geq 0$ and

$$\mathbb{E}X_s = 1 - \frac{1}{N} \sum_{i=1}^{N} P(D_i|u_i|s) \leq \overline{\lambda}.$$

Hence

$$P^*\{X_s \leq d \cdot \mathbb{E}X_s \text{ for } s = 1, \dots, d\} \leq P^*\{X_s \leq d(\overline{\lambda} + \varepsilon) \quad \text{for} \quad s = 1, \dots, d\}.$$

Define

$$B^* = \{X_s \leq d(\overline{\lambda} + \varepsilon) \quad \text{for} \quad s = 1, \dots, d\}$$

and

$$B_s = \{X_s > d(\overline{\lambda} + \varepsilon)\}, \quad s = 1, \dots, d.$$

Then

$$P^*(B_{\mathcal{S}}) \leq \frac{\mathbb{E}X_s}{d(\overline{\lambda} + \varepsilon)} \leq \frac{\overline{\lambda}}{d(\overline{\lambda} + \varepsilon)}.$$

Hence

$$P^*\left(\bigcup_{s=1}^{d} B_s\right) \leq \frac{\overline{\lambda}}{\overline{\lambda} + \varepsilon}$$

and therefore

$$P^*(B^*) \geq 1 - \frac{\overline{\lambda}}{\overline{\lambda} + \varepsilon} = \frac{\varepsilon}{\overline{\lambda} + \varepsilon}.$$

By the definition of $P^*$

$$|B^*| \geq N \cdot \frac{\varepsilon}{\overline{\lambda} + \varepsilon} \geq N \cdot \frac{\varepsilon}{1 + \varepsilon} \geq \left[N \frac{\varepsilon}{1 + \varepsilon}\right].$$

The elements of $B^*$ are the desired sequences. This proves Lemma 8. □

In Lemmas 9 and 10 only we let $|\mathcal{S}| = 1$ and $(P^n(\cdot|\cdot))_{n=1}^{\infty}$ be the only element of $\mathcal{C}^*$. We then have:

**Lemma 9** (Shannon, Lemma 3.1.1 in [45]) *Let* $\{(u_i|D_i) : i = 1, \dots, N\}$ *be a code for* $P^n(\cdot|\cdot)$ *with average error* $\overline{\lambda}$, *then there exists a sub-code of length* $N_1 = \frac{N\varepsilon}{\overline{\lambda}+\varepsilon}$ *with maximal error* $\overline{\lambda} + \varepsilon$.

*Proof* Denote $|\{u_i|P^n(D_i|u_i) < 1 - \overline{\lambda} - \varepsilon\}|$ by Z, then $Z(1 - \overline{\lambda} - \varepsilon) + (N - Z) \geq N(1 - \overline{\lambda})$ and therefore $N_1 = N - Z \geq \frac{\varepsilon}{\overline{\lambda}+\varepsilon} N$. □

**Lemma 10** *Given a random code* $\{(p_i, D_i) : i = 1, \dots, N\}$ *for* $P^n(\cdot|\cdot)$ *with average error* $\overline{\lambda}$, *we can construct a nonrandom code of the same length N with average error* $\leq \overline{\lambda}$.

(As a consequence of Lemma 10, for given length $N$ the average error is minimized by a nonrandom code. Obviously, the maximal length of a code of average error $\overline{\lambda}$ increases with increasing $\overline{\lambda}$. Hence, for given average error, a nonrandom code is at least as long as any random code.)

*Proof* Let $\{(p_i, D_i) : i = 1, \dots, N\}$ be a random code with $\frac{1}{N}\sum_{i=1}^{N}\sum_{x^n\in\mathcal{X}^n} p_i(x^n) P^n(D_i|x^n) = 1 - \overline{\lambda}$. The contribution of message $i$ to $N(1 - \overline{\lambda})$ is clearly $\sum_{x^n\in\mathcal{X}^n} p_i(x^n)P^n(D_i|x^n)$. Suppose now that $P^n(D_i|x^n_{(1)}) \geq P^n(D_i|x^n_{(2)}) \geq \cdots \geq P^n(D_i|x^n_{(a^n)})$. Instead of using $\{x^n_{(1)}, \dots, x^n_{(a^n)}\}$ with the probabilities $\{p_i(x^n_{(1)}), \dots, p_i(x^n_{(a^n)}\}$ for message $i$, now use $x^n_{(1)}$ with probability 1, and keep $D_i$ as the decoding set which corresponds to message $i$. The contribution of message $i$ to $N(1 - \overline{\lambda})$ is now replaced by the larger quantity $P^n(D_i|x^n_{(1)})$. Using the same procedure for all $i$ one achieves a nonrandom code $\{(u_i|D_i) : i = 1, \dots, N\}$ with average error $\leq \overline{\lambda}$. □

(One can improve on the code even more by keeping the $u_i$ of the new code, and replacing the $D_i$ by the maximum-likelihood sets $B_i$.)

#### 1.3.2.1 Averaged Channels

Let $\mathcal{S} = \{1, \dots, d\}$, and let $g = (g_1, \dots, g_d)$ be a probability vector on $\mathcal{S}$. The sequence

$$(P^n(\cdot|\cdot))_{n=1}^{\infty} = \left(\sum_{s=1}^{d} g_s P^n(\cdot|\cdot|s)\right)_{n=1}^{\infty}$$

is called an *averaged* channel. Let $N_a(n, \lambda)$ be the maximal length of any code $(n, N, \lambda)$ for this channel. Denote $\lim_{n\to\infty}\frac{1}{n}\log N_a(n, \lambda)$ by $C_a(\lambda)$ for those $\lambda$ for which the limit exists.

Theorem 13 and the second remark of Lecture 2 imply that

$$C_a(\lambda) = \max_{\{\mathcal{S}': \mathcal{S}'\subset\mathcal{S},\, g(\mathcal{S}')>1-\lambda} \max_{\pi} \inf_{s\in\mathcal{S}'} R(\pi, s)$$

at least for $\lambda \notin \{\sum_{i \in \mathcal{S}'} g_i : \mathcal{S}' \subset \mathcal{S}\}$. Furthermore, as a consequence of Lemma 9 we have

$$C_a(\lambda) = C_a(\overline{\lambda}) \quad \text{for } \lambda = \overline{\lambda} \notin \left\{ \sum_{i \in \mathcal{S}'} g_i : \mathcal{S}' \subset \mathcal{S} \right\}.$$

Also, as a consequence of Lemma 10 we have

$$C_a(\overline{\lambda}_R) = C_a(\overline{\lambda}).$$

Obviously, $C_a(\overline{\lambda}_R) \geq C_a(\lambda_R) \geq C_a(\lambda)$ and therefore

$$C_a(\overline{\lambda}_R) = C_a(\lambda_R) = C_a(\overline{\lambda}) = C_a(\lambda) \quad \text{for } \lambda = \overline{\lambda} \notin \left\{ \sum_{i \in \mathcal{S}'} g_i : \mathcal{S}' \subset \mathcal{S} \right\}.$$

#### 1.3.2.2 Compound Channels (CC) with Side Information

CC with side information were introduced in [43]. If the sender knows the CPF in $\mathcal{C}$ which governs the transmission of a message to be sent, an $(n, N, \lambda)$ code is defined as a system

$$\left\{ (u_i(s), D_i) : u_i(s) \in \mathcal{X}^n, D_i \subset \mathcal{Y}^n, D_i \text{ disjoint}, \quad i = 1, \dots, N; s \in \mathcal{S} \right\}$$

which satisfies $P^n(D_i | u_i(s) | s) \geq 1 - \lambda$ for $i = 1, \dots, N$; $s \in \mathcal{S}$.

The capacity is then given by $\inf_{s \in \mathcal{S}} \max_\pi R(\pi, s)$ (Theorem 2 of [43]).

We will need a slightly more general theorem. In the situation just described, the sender knows precisely the channel which actually governs the transmission of any word; in other words, he has *complete* knowledge. We shall say that the sender has the partial knowledge

$$K = \{(\mathcal{S}_1, \dots, \mathcal{S}_h) : \mathcal{S}_i \subset \mathcal{S}, \quad i = 1, \dots, h\},$$

if the sender knows only that the governing channel has an index which belongs to a set of $K$, the set itself being known to him.

**Lemma 11** *The capacity of the compound channel $\mathcal{C}^*$ with the sender's partial knowledge $K = (\mathcal{S}_1, \dots, \mathcal{S}_h)$ equals*

$$\inf_{i=1,\dots,h} \max_\pi \inf_{s \in \mathcal{S}_i} R(\pi, s).$$

### 1.3.3 The Structure of $C(\overline{\lambda})$

The determination of $C(\overline{\lambda})$ at its points of discontinuity seems to be difficult, and it is even undecided whether $\lim_{n\to\infty} \frac{1}{n} \log N(n, \overline{\lambda})$ exists at these points. (Compare also [1, 44]). The determination of $C(\overline{\lambda})$ becomes more and more complicated as $|\mathcal{S}|$ increases, and it seems to us that a simple recursion formula does not exist. However, the following results help clarify the structure of $C(\overline{\lambda})$.

**Theorem 13** *Given $\mathcal{C} = \{w(\cdot|\cdot|s) : s = 1, \dots, k\}$, then $C(\overline{\lambda})$ is well-defined except perhaps for finitely many points $\lambda_1, \dots, \lambda_{K^*(k)}$, and for every $\overline{\lambda} \neq \lambda_i$, $i = 1, \dots, K^*(k)$, $C(\overline{\lambda})$ equals an expression*

$$C_{\ell r \dots} = \max_{\pi} \inf_{s=\ell, r \dots} R(\pi, s). \tag{1.3.4}$$

The points $\lambda_i$ belong to a finite set $D^*$ which is characterized in Theorem 14.

*Proof* Since $0 \leq \log N(n, \overline{\lambda}) \leq n \log a$, $C^+(\overline{\lambda}) = \overline{\lim}_{n\to\infty} \frac{1}{n} \log N(n, \overline{\lambda})$ and $C^-(\overline{\lambda}) = \underline{\lim}_{n\to\infty} \frac{1}{n} \log N(n, \lambda)$ are well-defined for all $\overline{\lambda}$. Let $\{(u_i, D_i) : i = 1, \dots, N\}$ be a $(n, N, \overline{\lambda})$-code for $\mathcal{C}_n^*$ of maximal length. For every $\varepsilon > 0$ define

$$G_{\ell r \dots}(\varepsilon) = \{u_i | P^n(D_i|u_i|s) > \varepsilon \quad \text{for } s = \ell, r, \dots \text{ and for no other index}\} \tag{1.3.5}$$

and

$$G_0(\varepsilon) = \{u_i | P^n(D_i|u_i|s) \leq \varepsilon \quad \text{for all } s \in \mathcal{S}\}.$$

The $G$'s form a partition of the code into disjoint sub-codes. Applying Lemma 9 with $\varepsilon$ sufficiently small for any one value of $s$, say $s = 1$, we obtain that $|G_0(\varepsilon)|$ is bounded by a fixed multiple of $N(n, \overline{\lambda})$. Since $N(n, \overline{\lambda})$ grows exponentially, we can, and do, omit $G_0(\varepsilon)$ from our code without any essential loss, provided $\varepsilon$ is sufficiently small.

Define $\alpha_{\ell r \dots}(n, \varepsilon) = \frac{|G_{\ell r \dots}(\varepsilon)|}{N(n, \overline{\lambda})}$. Let $n_1, n_2, \dots$ be a subsequence of the integers such that

$$\overline{\lim_{t\to\infty}} \frac{1}{n_t} \log N(n_t, \overline{\lambda}) = C^+(\overline{\lambda}). \tag{1.3.6}$$

We can now define

$$\alpha_{\ell r \dots}(\varepsilon) = \overline{\lim_{t\to\infty}} \alpha_{\ell r \dots}(n_t, \varepsilon). \tag{1.3.7}$$

Let

$$L(\varepsilon) = \{(\ell, r, \dots) : \alpha_{\ell r \dots}(\varepsilon) > 0\}.$$

If $(\ell, r, \dots) \in L(\varepsilon)$ then, as a consequence of the strong converse for compound channel (Theorem 4.4.1 of [45]), $C^+(\overline{\lambda}) \leq C_{\ell r \dots}$, and therefore

$$C^+(\overline{\lambda}) \leq \inf\{C_{\ell r \dots} : (\ell, r, \dots) \in L(\varepsilon)\}. \tag{1.3.8}$$

Since $\varepsilon$ was arbitrary,

$$C^+(\overline{\lambda}) \leq \lim_{\varepsilon \to 0} \inf\{C_{\ell r \dots} : (\ell, r, \dots) \in L(\varepsilon)\}. \tag{1.3.9}$$

Define

$$f_t(s) = |\{u_i | P_{n_t}(D_i | u_i | s) > \varepsilon\}|$$

for $s = 1, \dots, k$. Hence $f_t(s) + (N - f_t(s))\varepsilon \geq N(1 - \overline{\lambda})$ and consequently

$$f_t(s) \geq N\left(\frac{1 - \overline{\lambda} - \varepsilon}{1 - \varepsilon}\right) \quad s = 1, \dots, k$$

On the other hand,

$$\frac{f_t(s)}{N} = \sum_{(\ell, r, \dots)} \alpha_{\ell r \dots}(n_t, \varepsilon) \geq \frac{1 - \overline{\lambda} - \varepsilon}{1 - \varepsilon}, \qquad s = 1, \dots, k; \quad s \in \{\ell, r \dots\}. \tag{1.3.10}$$

Clearly, for $\eta > 0$ there exists a $n_0(\eta)$ such that, for $n_t \geq n_0(\eta)$, $\alpha_{\ell r \dots}(n_t, \varepsilon) \leq \eta$ for $(\ell, r \dots) \notin L(\varepsilon)$, because there are only *finitely* many sets of indices. From (1.3.10) it follows that, for $s = 1, \dots, k$,

$$\sum_{(\ell, r, \dots) \in L(\varepsilon)} \alpha_{\ell, r \dots}(n_t, \varepsilon) \geq \frac{1 - \overline{\lambda} - \varepsilon}{1 - \varepsilon} - \eta \cdot 2^k \quad s \in (\ell, r \dots). \tag{1.3.11}$$

Consider a code $(n_t, N', \delta)$ of maximal length for the compound channel with the sender's partial knowledge

$$K = \{(\ell, r, \dots) : (\ell, r, \dots) \in L(\varepsilon)\}.$$

For each $(\ell, r, \dots) \in L(\varepsilon)$ choose $N' \cdot \alpha_{\ell, r, \dots}(\varepsilon)$ indices from $1, \dots, N'$ (the choice is arbitrary, but different complexes which are in $L(\varepsilon)$ must correspond to disjoint sets of indices), and for these indices use as message sequences (i.e., $u_i$'s) *only* those message sequences which would have been used if the sender knew that the governing channel was in $(\ell, r, \dots)$. By (1.3.11) and Lemma 11 this leads to a code $(n_t, N, \overline{\lambda}')$ for $C^*_{n_t}$ of length

$$N(n_t, \overline{\lambda}') \geq \exp\left[n_t \cdot \inf\{C_{\ell r \dots} : (\ell, r, \dots) \in L(\varepsilon)\} - \text{const.}\sqrt{n_t}\right] \tag{1.3.12}$$

where $1-\overline{\lambda}' = \left(\frac{1-\overline{\lambda}-\varepsilon}{1-\varepsilon} - \eta \cdot 2^k\right)(1-\delta)$. Using the *same* $\alpha$'s for *all* $n$ sufficiently large, we get

$$N(n, \overline{\lambda}') \geq \exp\left[n \cdot \inf\{C_{\ell r \dots} : (\ell, r, \dots) \in L(\varepsilon)\} - \text{const.}\sqrt{n}\right]$$

and consequently

$$C^-(\overline{\lambda}') \geq \inf\{C_{\ell r \dots} : (\ell, r, \dots) \in L(\varepsilon)\}$$

Furthermore, $\overline{\lambda} = \lim_{\varepsilon,\eta,\delta\to 0} \overline{\lambda}'$, and therefore

$$C^-(\overline{\lambda}) \geq \lim_{\varepsilon\to 0} \inf\{C_{\ell r \dots} : (\ell, r, \dots) \in L(\varepsilon)\}$$

for every $\overline{\lambda}$ which is a continuity point of $C^-(\overline{\lambda})$. Using (1.3.9) we get

$$C^+(\overline{\lambda}) = C^-(\overline{\lambda}) = C(\overline{\lambda}) = \lim_{\varepsilon\to 0} \inf\{C_{\ell r \dots} : (\ell, r, \dots) \in L(\varepsilon)\} \qquad (1.3.13)$$

for all $\overline{\lambda}$ which are continuity points of $C^-(\overline{\lambda})$. However, $C^-(\overline{\lambda})$ is a monotonic function on [0, 1] and can therefore have only countably many discontinuities. It follows from (1.3.13) that $C^-(\overline{\lambda})$ takes only finitely many values on the set of its continuity points. Hence $C^-(\overline{\lambda})$, and therefore also $C(\overline{\lambda})$, have only finitely many discontinuities. This proves the theorem. □

From the definition of $C(\overline{\lambda})$, every point of continuity of $C(\overline{\lambda})$ is a point of continuity of $C^-(\overline{\lambda})$. From (1.3.13) and the fact that $C^-(\overline{\lambda})$ is a step function it follows that every point of continuity of $C^-(\overline{\lambda})$ is a point of continuity of $C(\overline{\lambda})$. Therefore, $C(\overline{\lambda})$ and $C^-(\overline{\lambda})$ have the same points of continuity.

Theorem 13 says that, except perhaps for at most finitely many points, $C(\overline{\lambda})$ is given by an expression

$$C_{\ell r \dots} = \max_{\pi} \inf_{s=\ell,r,\dots} R(\pi, s).$$

For different channels, $C(\overline{\lambda})$ may be given by different expressions. We now seek a formula for $C(\overline{\lambda})$ which does not depend on the channel. (The actual values taken by this formula will, of course, depend on the channel.)

We introduce the class of formulas

$$\tilde{f} = \left\{I : I \text{ is given by maxima and minima of } C_{\ell r \dots} = \max_{\pi} \inf_{s=\ell,r,\dots} R(\pi, s)\right\}. \qquad (1.3.14)$$

The value of a formula $I$ for $\mathcal{C}$ will be denoted by $I(\mathcal{C})$, A partial ordering is defined in $\tilde{f}$ by

$$I_1 \leq I_2 \quad \text{if and only if } I_1(\mathcal{C}) \leq I_2(\mathcal{C}) \quad \text{for all } \mathcal{C} \text{ with } |\mathcal{C}| = k. \qquad (1.3.15)$$

$\tilde{f}$ need not be totally ordered. It can happen that, for $I_1, I_2 \in \tilde{f}$ and two channels $\mathcal{C}_1, \mathcal{C}_2$, $I_1(\mathcal{C}_1) > I_2(\mathcal{C}_2)$ and $I_1(\mathcal{C}_1) < I_2(\mathcal{C}_2)$.

We start our considerations for a fixed $\mathcal{C}$ which has $k$ elements and develop an algorithm for the computation of $C(\overline{\lambda})$. For any real numbers $z_1$ and $z_2$ define $z_1 \cap z_2 = \min(z_1, z_2)$, $z_1 \cup z_2 = \max(z_1, z_2)$. Obviously,

$$\begin{aligned} C_{12\dots k} &\leq \underset{n\to\infty}{\to} \underline{\lim} \frac{1}{n} \log N(n, \overline{\lambda}) \\ &\leq \underset{n\to\infty}{\to} \overline{\lim} \frac{1}{n} \log N(n, \overline{\lambda}) \\ &\leq \bigwedge_{s=1,\dots,k} C_s \end{aligned} \tag{1.3.16}$$

Every term $C_{\ell r\dots}$ which is a possible value for $C(\overline{\lambda})$ for some value of $\overline{\lambda}$ therefore has to satisfy

$$C_{\ell r\dots} = C_{\ell r\dots} \bigwedge_{s \notin \{\ell, r, \dots\}} C_s \tag{1.3.17}$$

Every index $1, \dots, k$ appears in the right member of (1.3.17). We now write $C_{\ell r\dots}$ as

$$C_{\ell r\dots} = C_{\ell_1 r_1 \dots} \wedge C_{\ell_2 r_2 \dots} \wedge \dots \wedge C_{\ell_t r_t \dots}, \tag{1.3.18}$$

where

(i) no index can be added to any set $\{\ell_i r_i, \dots\}$ without violating (1.3.18),
(ii) no additional term can be added on the right without violating (1.3.18) or condition (i).

The representation (1.3.18) is therefore unique. Let the number of terms on the right of (1.3.18) be $t$. For $s = 1, \dots, k$ and $i = 1, \dots, t$ define

$$\delta(s, i) = 1 \quad \text{if } s \in (\ell_i, r_i, \dots)$$
$$\delta(s, i) = 0 \quad \text{if } s \notin (\ell_i, r_i, \dots)$$

Let $\alpha = (\alpha_1, \dots, \alpha_t)$ be a probability $t$-vector. We define

$$\overline{\lambda}(\ell, r, \dots) = 1 - \max_{\alpha} \min_{s} \sum_{i=1}^{t} \alpha_i \delta(s, i). \tag{1.3.19}$$

We will now prove that, for $\overline{\lambda} > \overline{\lambda}(\ell, r \dots)$,

$$\underset{n\to\infty}{\to} \underline{\lim} \frac{1}{n} \log N(n, \overline{\lambda}) \geq C_{\ell r\dots}. \tag{1.3.20}$$

Let $\alpha^*$ be the maximizing value of $\alpha$ in (1.3.19). Let $\varepsilon > 0$ be small enough. For suitable $m(\varepsilon) > 0$, we construct a code

$$(n, N = \exp_2\{nC_{\ell r \dots} - \sqrt{n}m(\varepsilon)\}, \varepsilon)$$

for the compound channel with the sender's partial knowledge

$$K = \{(\ell_1, r_1, \dots), \dots, (\ell_t, r_t, \dots)\}.$$

Let the code be written as

$$\left(u_i^{(1)}, \dots, u_i^{(t)}, D_i\right), \quad i = 1, \dots, N.$$

Consider the new code

$$\left(u_i^{(1)}, D_i\right), \quad i = 1, \dots, N \cdot \alpha_1^*$$
$$\left(u_i^{(2)}, D_i\right), \quad i = (N \cdot_1^* +1), \dots, N \cdot (\alpha_1^* + \alpha_2^*)$$
$$\dots \quad \dots \quad \dots \quad \dots \quad \dots \quad \dots$$
$$\left(u_i^{(t)}, D_i\right), \quad i = N \cdot (\alpha_1^* + \dots + \alpha_{t-1}^*) + 1, \dots, N.$$

For $s = 1, \dots, k$ the average error of this code is not greater than

$$1 - (1 - \varepsilon) \min_s \sum_{i=1}^t \alpha_i^* \delta(s, i).$$

When $\varepsilon$ is small enough, we obtain (1.3.20).

Now define

$$V_{\ell r \dots}(\overline{\lambda}) = \begin{cases} C_{\ell r \dots} & \text{for } \overline{\lambda} > \overline{\lambda}(\ell, r, \dots) \\ 0 & \text{otherwise} \end{cases} \tag{1.3.21}$$

and

$$V(\overline{\lambda}) = \max_{S'} \left\{ V_{\ell r \dots}(\overline{\lambda}) : S' = \{\ell, r, \dots\} \subset \mathcal{S} \right\}. \tag{1.3.22}$$

$V(\overline{\lambda})$ is a step function with at most finitely many jumps. It follows from (1.3.20) that

$$\underline{\lim}_{n \to \infty} \frac{1}{n} \log N(n, \overline{\lambda}) \geq V(\overline{\lambda}) \tag{1.3.23}$$

at every point of continuity of $V(\overline{\lambda})$.

Let $\overline{\lambda}$ be a point of continuity of $C(\overline{\lambda})$ and $V(\overline{\lambda})$. Let $\varepsilon_0 > 0$ be so small that $L(\varepsilon_0) = L(\varepsilon)$ for $0 < \varepsilon < \varepsilon_0$. From (1.3.9) we know that $C(\lambda^-)$ is the smallest, say $C_{\ell r \ldots}$, of a finite number of expressions of this type whose index sets belong to $L(\varepsilon_0)$. Passing to the limit in (1.3.11) we have, for $s = 1, \ldots, k$

$$\sum \alpha_\mu(\varepsilon_0) \geq \frac{1 - \overline{\lambda} - \varepsilon_0}{1 - \varepsilon_0} - \eta \cdot 2^k. \tag{1.3.24}$$

where the summation is over all index sets $\mu$ which contain $s$ and belong to $L(\varepsilon_0)$. Write $C_{\ell r \ldots}$ in the form (1.3.18) and suppose, without loss of generality, that (1.3.18) is the actual representation. Assign each element of $L(\varepsilon_0)$ to some *one* of the sets in the right member of (1.3.18) which contains this element, and define $\alpha^*(\varepsilon_0)$ of the latter set as the sum of the $\alpha(\varepsilon_0)$ of the sets assigned to it; $\alpha^*(\varepsilon_0)$ will be zero for a set to which no sets have been assigned. A fortiori, for $s = 1, \ldots, k$,

$$\sum_{i=1}^{t} \delta(s, i)\alpha^*_{\ell_i r_i \ldots}(\varepsilon_0) \geq \frac{1 - \overline{\lambda} - \varepsilon_0}{1 - \varepsilon_0} - \eta \cdot 2^k. \tag{1.3.25}$$

Letting $\eta$ and $\varepsilon_0$ approach zero we obtain from (1.3.19) and (1.3.25) that

$$\overline{\lim_{n\to\infty}} \frac{1}{n} \log N(n, \overline{\lambda}) \leq V(\overline{\lambda}). \tag{1.3.26}$$

From (1.3.23) and (1.3.26) we obtain that

$$C(\overline{\lambda}) = V(\overline{\lambda}) \tag{1.3.27}$$

at the points of continuity of both functions. $C(\overline{\lambda})$ is defined and continuous at all but a finite number of points, and monotonic. $V(\overline{\lambda})$ is defined everywhere and monotonic. Both are step functions. Hence the two functions are identical at every point of continuity of $C(\overline{\lambda})$.

We now have (1.3.18), (1.3.19), (1.3.21), and (1.3.22) determine an algorithm for the computation of $C(\overline{\lambda})$. (See 1.3.5 for applications.)

It follows from (1.3.21) and (1.3.22) that any point of discontinuity $\lambda_i$ of $C(\overline{\lambda})$ must be one of the set

$$\{\overline{\lambda}(\ell, r, \ldots) : (\ell, r \ldots) \subset \mathcal{S}\} \tag{1.3.28}$$

Now $\overline{\lambda}(\ell, r, \ldots)$ depends upon the representation (1.3.18). However, it does not depend on the actual values $C$ which enter into that representation, but only upon the indices which enter into the right member of (1.3.18). All possible sets of such indices are finite in number. Moreover, for any given $\mathcal{C}$ with $|\mathcal{S}| = d$, the set of indices in the right member of (1.3.18) depends *only on the ordering according to size of the various $C$'s* of $\mathcal{C}$, and not at all on the actual values taken by them. When $|\mathcal{S}| = d$ there are a fixed (finite) number of expressions of the form $C_{\ell r \ldots}$. A finite number of channels with $|\mathcal{S}| = d$ and alphabets of sufficient length will produce all the possible orderings of these expressions. Call one such set of channels

$$\mathcal{Q} = \{T_1, \dots, T_q\}. \tag{1.3.29}$$

We have therefore proved:

For any channel $\mathcal{C}$ with $|\mathcal{S}| = d$, the set of discontinuity of its function $C(\overline{\lambda})$ coincides with the set of points of discontinuity of the function $C(\overline{\lambda})$ of $T(\mathcal{C})$, where $T(\mathcal{C})$ is that member of $\mathcal{Q}$ whose $C$'s have the same ordering according to size as those of $\mathcal{C}$, and (1.3.30)

The set $D^*$ of all possible points of discontinuity of $C(\overline{\lambda})$ for all $\mathcal{C}$ with $|\mathcal{S}| = d$ consists of all points of the form (28), and can be evaluated by the algorithm implied by (19) and (18), and (1.3.31)

Two channels, $\mathcal{C}_1$ and $\mathcal{C}_2$, say, both with $|\mathcal{S}| = d$, have the same points of discontinuity for their respective functions $C(\overline{\lambda})$ if the set $\{C_{\ell r \dots} : (\ell, r, \dots) \subset \mathcal{S}\}$ has the same ordering according to size for both $\mathcal{C}_1$ and $\mathcal{C}_2$. (1.3.32)

The representation (1.3.18) is defined for a fixed $\mathcal{C}$. To indicate the dependence on $\mathcal{C}$ we write

$$C(\overline{\lambda}, \mathcal{C}), C_{\ell_1 r_1 \dots}(\mathcal{C}), \dots, C_{\ell_t r_t}(\mathcal{C}).$$

Suppose now that, for a fixed $\overline{\lambda}$ not in $D^*$,

$$C_{\ell r \dots}(\mathcal{C}) = C_{\ell_1 r_1 \dots}(\mathcal{C}) \wedge C_{\ell_2 r_2 \dots}(\mathcal{C}) \wedge \dots \wedge C_{\ell_t r_t \dots}(\mathcal{C}) = C(\overline{\lambda}, \mathcal{C}) \tag{1.3.33}$$

and for channel $T_1$

$$C_{\ell^* r^* \dots}(T_1) = C_{\ell_1^{(1)} r_1^{(1)} \dots}(T_1) \wedge \dots \wedge C_{\ell_t^{(1)} r_t^{(1)} \dots}(T_1) = C(\overline{\lambda}, T_1) \tag{1.3.34}$$

In (1.3.25) let $\alpha^*$ correspond to channel $\mathcal{C}$ and $\alpha^{**}$ correspond to channel $T_1$. Both $\{\alpha^*\}$ and $\{\alpha^{**}\}$ satisfy (1.3.25). Hence, by the argument which follows (1.3.11) we have

$$C(\overline{\lambda}, \mathcal{C}) \geq C_{\ell_1^{(1)} r_1^{(1)} \dots}(\mathcal{C}) \wedge \dots \wedge C_{\ell_t^{(1)} r_t^{(1)} \dots}(\mathcal{C}). \tag{1.3.35}$$

Hence, from (1.3.33) and (1.3.35),

$$C(\overline{\lambda}, \mathcal{C}) = \left[C_{\ell_1 r_1 \dots}(\mathcal{C}) \wedge \cdots\right] \vee \left[C_{\ell_1^{(1)} r_1^{(1)} \dots}(\mathcal{C}) \wedge \cdots\right]. \tag{1.3.36}$$

Repeating this argument we obtain

$$C(\overline{\lambda}, \mathcal{C}) = \left[C_{\ell_1 r_1 \dots}(\mathcal{C}) \wedge \cdots\right] \vee \bigvee_{i=1}^{q} \left[C_{\ell_1^{(i)} r_1^{(i)} \dots}(\mathcal{C}) \wedge \cdots\right] \tag{1.3.37}$$

where, for $i = 1, \ldots, q$,

$$C_{\ell_1^{(i)} r_1^{(i)} \ldots} \wedge \cdots$$

is the representation (1.3.18) of $C(\overline{\lambda}, T_i)$ in terms of the $C$'s of channel $T_i$.

Assume temporarily that we can show that

$$C(\overline{\lambda}, \mathcal{C}) = \bigvee_{i=1}^{q} \left[ C_{\ell_1^{(i)} r_1^{(i)} \ldots} (\mathcal{C}) \wedge \cdots \right]. \tag{1.3.38}$$

We would then regard (1.3.38) as an identity in the "free variable" (argument) $\mathcal{C}$ (with $|\mathcal{S}| = d$) if we could show that the system of subscripts of the $C$'s which occurs in the right member of (1.3.33) does not depend on $\mathcal{C}$. (It may, and actually does, depend on the fixed $\overline{\lambda}$.) To prove this it is sufficient to see that the system of subscripts is determined by

$$C(\overline{\lambda}, T_1), \ldots, C(\overline{\lambda}, T_q). \tag{1.3.39}$$

Write the points of $D^*$ as

$$a_1 < a_2 < \cdots < a_{Z(k)-1}. \tag{1.3.40}$$

Also write $a_0 = 1$, $a_{Z(k)} = 1$. Suppose $a_z < \overline{\lambda} < a_{z+1}$. Then clearly (1.3.38) is valid for all points in the interval $(a_z, a_{z+1})$, because both members are constant in the interval.

The formula (1.3.38) depends upon the interval $(a_z, a_{z+1})$; there may be a different formula for a different interval. However, since $C(\overline{\lambda}, \mathcal{C})$ is monotonic in $\overline{\lambda}$ for *any* $\mathcal{C}$, the different right members of (1.3.38) for different intervals are monotonic for *any* $\mathcal{C}$, and thus are totally ordered.

It remains to prove that we can omit the first bracket on the right of (1.3.37). The subscripts in it are determined by the representation (1.3.18) of

$$C_{\ell r \ldots}(\mathcal{C}) = C(\overline{\lambda}, \mathcal{C})$$

in terms of the $C$'s of $\mathcal{C}$. We have already seen, in (1.3.30), that this representation is the same as that in terms of the $C$'s of $T(\mathcal{C})$. Hence the first bracket on the right of (1.3.37) is already included among the square brackets in $\bigvee_{i=1}^{q}[\quad]$ in the right member of (1.3.37). This proves (1.3.38).

We sum up our results in:

**Theorem 14** *For any integer $k$ there is a finite set $D^*$, described in* (1.3.31). *The points of discontinuity of $C(\overline{\lambda})$ for any $\mathcal{C}$ with $|\mathcal{S}| = d$ belong to $D^*$. The right member of* (1.3.38) *is constant in any $\overline{\lambda}$—interval between two consecutive points of $D^*$, and is determined by this interval. (Different such intervals in general determine different right members of* (1.3.38).*) $C(\overline{\lambda})$ is given by* (1.3.38).

*Remark*

1. It is not possible to use only formulas of $\tilde{f}$ which are built up only by minima. In the second example of Sect. 1.3.5, for instance, we have

$$\begin{aligned} C(\overline{\lambda}) &= (C_{12} \vee C_{13} \vee C_{23}) \wedge C_1 \wedge C_2 \wedge C_3 \\ &= (C_{12} \wedge C_3) \vee (C_{13} \wedge C_2) \vee (C_{23} \wedge C_1) \quad \text{for } \overline{\lambda} \in \left(\frac{1}{2}, \frac{2}{3}\right) \end{aligned}$$

   Suppose $C_{12} \wedge C_3 > C_{13} \wedge C_2, C_{23} \wedge C_1$ then $C(\overline{\lambda}) = C_{12} \wedge C_3$. Permuting the indices we would get $C(\overline{\lambda}) \neq C_{12} \wedge C_3$.
2. It is not true that any two terms in square brackets on the right of (1.3.38) can be transformed into each other by permutation of indices, as can be seen from the third example in Sect. 1.3.5 for $\overline{\lambda} \in \left(\frac{3}{5}, \frac{2}{3}\right)$.

### *1.3.4 The Relationships of* $C(\lambda_R)$*,* $C(\overline{\lambda}_R)$*, and* $C(\overline{\lambda})$

**Theorem 15**

$$C(\lambda_R) = C(\overline{\lambda}) = C(\overline{\lambda}_R) \quad \textit{for } \lambda_R = \overline{\lambda} = \overline{\lambda}_R,$$

*at the points of continuity of* $C(\overline{\lambda})$*. [*$C(\overline{\lambda})$ *has only finitely many points of discontinuity.]*

*Proof* The proof will be given in several steps.

For any positive integer $n$ there exists a random code for $C_n^*$

$$\{(p_i, D_i) : i = 1, \ldots, N\} \tag{1.3.41}$$

which satisfies, for any $s \in \mathcal{S}$,

$$\frac{1}{N} \sum_{i=1}^{N} \sum_{x^n \in \mathcal{X}^n} p_i(x^n) P^n(D_i | x^n | s) \geq 1 - \overline{\lambda}_R, \tag{1.3.42}$$

and which is of maximal length $N(n, \overline{\lambda}_R)$. Define, for $i = 1, \ldots, N$

$$B^i_{\ell r \ldots}(\varepsilon) = \{x^n | P^n(D_i | x^n | s) > \varepsilon \quad \text{for} \quad s = \ell, r, \ldots, \text{ and no other index}\} \tag{1.3.43}$$

and also

$$B^i_0(\varepsilon) = \{x^n | P^n(D_i | x^n | s) \leq \varepsilon \quad \text{for every index} \quad s \in \mathcal{S}\} \tag{1.3.44}$$

There are $2^k$ possible index sets $\{\ell, r, \ldots\}$. Denote these sets in some order by $\rho_1, \ldots, \rho_{2^k}$. For every $i = 1, \ldots, N$ $\left\{B^i_{\rho_j} : j = 1, \ldots, 2^k\right\}$ is a disjoint partition of $\mathcal{X}^n$. Define the column vector

$$B_{\rho_j}(\varepsilon) = \begin{bmatrix} B^1_{\rho_j}(\varepsilon) \\ \vdots \\ B^N_{\rho_j}(\varepsilon) \end{bmatrix} \tag{1.3.45}$$

and the matrix

$$B(\varepsilon) = \left(B^i_{\rho_j}(\varepsilon)\right)_{\substack{i=1,\ldots,N \\ j=1,\ldots,2^k}} . \tag{1.3.46}$$

Henceforth, we operate only on the matrix $B(\varepsilon)$. Define

$$\begin{aligned} C^+(\lambda_R) &= \overline{\lim_{n\to\infty}} \frac{1}{n} \log N(n, \lambda_R) \\ C^-(\lambda_R) &= \underline{\lim_{n\to\infty}} \frac{1}{n} \log N(n, \lambda_R) \\ C^+(\overline{\lambda}_R) &= \overline{\lim_{n\to\infty}} \frac{1}{n} \log N(n, \overline{\lambda}_R) \\ C^-(\overline{\lambda}_R) &= \underline{\lim_{n\to\infty}} \frac{1}{n} \log N(n, \overline{\lambda}_R) \end{aligned} \tag{1.3.47}$$

Let $n_1, n_2$ be a sequence such that

$$\lim_{t\to\infty} \frac{1}{n_t} \log N(n_t, \overline{\lambda}_R) = C^+(\overline{\lambda}_R).$$

Assume now that for every $n = 1, 2, \ldots$ a random code $(n, N, \overline{\lambda}_R)$ with maximal length $N(n, \overline{\lambda}_R)$ is given. To indicate the dependence on $n$ we now write $B^i_{\rho_j}(\varepsilon, n)$. Denote by $\beta_{\rho_j}(\varepsilon, n)$ the number of components (rows) of $\beta_{\rho_j}(\varepsilon, n)$ which are non-empty sets. We say that the index set $\rho_j$ is $\varepsilon$-essential if

$$\overline{\lim_{t\to\infty}} \left\{ \left[ N\left(n_t, \overline{\lambda}_R\right) \right]^{-1} \beta_{\rho_j}(\varepsilon, n) \right\} = \beta_{\rho_j}(\varepsilon) > 0. \tag{1.3.48}$$

Let $M(\varepsilon)$ be the set of $\varepsilon$-essential index sets $\rho_j$. It follows from the definitions (1.3.47) and (1.3.48) and from the strong converse for compound channels (Theorem 4.4.1 of [45]) that

$$C^+(\overline{\lambda}_R) \leq C_{\rho_j}, \quad \rho_j \text{ in } M(\varepsilon).$$

Hence

$$C^+(\overline{\lambda}_R) \leq \inf \left\{ C'_{\rho_j} : \rho_j \text{ in } M(\varepsilon) \right\}.$$

This is true for every $\varepsilon > 0$. Hence, when $\lambda_R = \overline{\lambda}_R$,

$$C^+(\lambda_R) \leq C^+(\overline{\lambda}_R) \leq \inf_{\varepsilon>0} \inf \left\{ C_{\rho_j} : \rho_j \text{ in } M(\varepsilon) \right\}, \tag{1.3.49}$$

the first inequality being obvious.

We now prove the converse. Since there are only finitely many indices $\rho_j$ we can conclude the following for any $\eta > 0$: There exists an $n_0(\eta)$ such that, for $n_t \geq n_0(\eta)$,

$$\beta_{\rho_j}(n_t, \varepsilon) \leq \eta, \quad \rho_j \text{ not in } M(\varepsilon). \tag{1.3.50}$$

Then, for $n$ sufficiently large, in the matrix (1.3.46) for a code $(n_t, N, \overline{\lambda}_R)$, we delete column $B_0(\varepsilon)$ and all *columns* $B_{\rho_j}(\varepsilon)$ for which $\rho_j$ is not in $M(\varepsilon)$. As a result of this the average error of the resulting code is less than

$$\overline{\lambda}_R + 2^k \cdot \eta + \varepsilon \tag{1.3.51}$$

Now take an $(n_t, N', \delta)$ code

$$\left\{ \left( u_i(\rho_j), D_i^* \right) : i = 1, \ldots, N'; \rho_j \text{ in } M(\varepsilon) \right\}$$

of length

$$N' \geq \exp\left[ \inf \left\{ C_{\rho_j} : \rho_j \in M(\varepsilon) \right\} \cdot n_t - K'(\delta)\sqrt{n} \right] \tag{1.3.52}$$

for the compound channel with the sender's partial knowledge

$$K = \{\rho_j : \rho_j \text{ in } M(\varepsilon)\}.$$

For any $\ell \in \{1, 2, \ldots, N'\}$ define

$$p_i(u_\ell(\rho_j)) = p_i\left(B^i_{\rho_j}\right) \quad \text{for } i = 1, \ldots, N; \ \rho_j \in M(\varepsilon). \tag{1.3.53}$$

Also define $\delta_{\rho_j s} = 1$ when $s \in \rho_j$ and 0 when $s \notin \rho_j$. Then we can conclude that

$$\sum_{\rho_j \in M(\varepsilon)} p_i(u_\ell(\rho_j))\delta_{\rho_j s} P(D^*_\ell | u_\ell(\rho_j)|s) \geq [1-\delta] \sum_{\rho_j \in M(\varepsilon)} \sum_{x^n \in B^i_{\rho_j}} p_i(x^n) P(D_i | x^n | s) - \varepsilon \tag{1.3.54}$$

for $i = 1, \ldots, N; s \in \mathcal{S}; \ell = 1, \ldots, N'$.

It follows from (1.3.51) and (1.3.54) that

$$\begin{aligned}&\frac{1}{N}\sum_{i=1}^{N}\sum_{\rho_j\in M(\varepsilon)} p_i(u_\ell(\rho_j))\delta_{\rho_j s}P(A_\ell^*|u_\ell(\rho_j)|s)\\ &\geq [1-\delta]\frac{1}{N}\sum_{i=1}^{N}\sum_{\rho_j\in M(\varepsilon)}\sum_{x^n\in B^i_{\rho_j}} p_i(x^n)P(D_i|x^n|s)-\varepsilon\\ &\geq \Big[1-\overline{\lambda}_R-2^k\cdot\eta-\varepsilon\Big][1-\delta]-\varepsilon \quad \text{for} \quad s\in\mathcal{S} \quad \text{and } \ell=1,\dots,N'.\end{aligned} \tag{1.3.55}$$

Defining now

$$P(\rho_j)=\frac{1}{N}\sum_{i=1}^{N}p_i\left(B^i_{\rho_j}\right) \tag{1.3.56}$$

for $\rho_j\in M(\varepsilon)$, we conclude, using (1.3.55), that

$$\begin{aligned}&\sum_{\rho_j\in M(\varepsilon)} p(\rho_j)\delta_{\rho_j s}P(A_\ell^*|u_\ell(\rho_j)|s)\\ &\geq \Big[1-\overline{\lambda}_R-2^k\eta-\varepsilon\Big][1-\delta]-\varepsilon \quad \text{for } \ell=1,\dots,N';\, s\in\mathcal{S}.\end{aligned} \tag{1.3.57}$$

Thus we have a random code with *maximal* error $\lambda'$ defined by

$$1-\lambda'=\Big(1-\overline{\lambda}_R-2^k\eta-\varepsilon\Big)(1-\delta)-\varepsilon$$

and length given by (1.3.52).

Now define

$$\alpha_j=[p(\rho_j)\cdot N'] \quad \text{for } \rho_j\in M(\varepsilon).$$

If necessary we renumber the elements of $M(\varepsilon)$ so that

$$M(\varepsilon)=\{\rho_j : j=1,\dots,k^*(\varepsilon)\}.$$

Consider the nonrandom code

$$\begin{aligned}&(u_1(\rho_1),D_1^*),\dots,(u_{\alpha_1}(\rho_1),D_{\alpha_1}^*),\\ &(u_{\alpha_1+1}(\rho_2),D_{\alpha_1+1}^*),\dots,(u_{N'}(\rho_{k^*(\varepsilon)}),D_{N'}^*)\end{aligned} \tag{1.3.58}$$

It is a consequence of (1.3.57) that this code has an average error less than $\lambda'$. Hence, passing to the limit with $\varepsilon$, $\eta$, and $\delta$ we obtain, just as in the argument which led to (1.3.13), that

$$C^-(\overline{\lambda}) \geq \inf_{\varepsilon>0} \inf \left\{C_{\rho_j} : \rho_j \text{ in } M(\varepsilon)\right\} \tag{1.3.59}$$

at the continuity points of $C^-(\overline{\lambda})$, and

$$C^-(\lambda_R) \geq \inf_{\varepsilon>0} \inf \left\{C_{\rho_j} : \rho_j \text{ in } M(\varepsilon)\right\} \tag{1.3.60}$$

at the continuity points of $C^-(\lambda_R)$. From (1.3.49) and (1.3.60), we obtain that $C(\lambda_R)$ exists at the points of continuity of $C^-(\lambda_R)$ and that there

$$C(\lambda_R) = C^+(\overline{\lambda}_R), \quad \lambda_R = \overline{\lambda}_R. \tag{1.3.61}$$

From (1.3.49) and (1.3.59), we obtain that at the points of continuity of $C^-(\overline{\lambda})$,

$$C^-(\overline{\lambda}_R) \geq C^-(\overline{\lambda}) \geq C^+(\overline{\lambda}_R), \quad \overline{\lambda} = \overline{\lambda}_R, \tag{1.3.62}$$

the first inequality being obvious.

Finally, from (1.3.13), (1.3.61), and (1.3.62) we obtain that, at the points of continuity of $C^-(\overline{\lambda})$ and of $C^-(\lambda_R)$ we have

$$C(\overline{\lambda}) = C(\overline{\lambda}_R) = C(\lambda_R), \quad \overline{\lambda} = \lambda_R = \overline{\lambda}_R. \tag{1.3.63}$$

Since $C(\overline{\lambda})$ and $C^-(\overline{\lambda})$ have the same points of continuity, we have that

$$C(\overline{\lambda}_R) = C(\lambda_R) = C(\overline{\lambda}), \quad \overline{\lambda}_R = \lambda_R = \overline{\lambda} \tag{1.3.64}$$

at the points of continuity of $C(\overline{\lambda})$ and $C^-(\lambda_R)$.

Earlier we proved that $C(\overline{\lambda})$ has only finitely many points of discontinuity, takes on the set of continuity points only finitely many values, and is monotonic. The function $C^-(\lambda_R)$ is monotonic, and hence has at most enumerable many points of discontinuity. If it had a point of discontinuity which is not a point of discontinuity of $C(\overline{\lambda})$ this would result in a contradiction of (1.3.64). Hence every point of continuity of $C(\overline{\lambda})$ is a point of continuity of $C^-(\lambda_R)$.

Theorem 15 follows from this and (1.3.63). □

### *1.3.5 Evaluation of* $C(\overline{\lambda})$ *in Several Examples*

#### 1.3.5.1 Example $\mathcal{S} = \{1, 2\}$

We shall show that then

$$C(\overline{\lambda}) = \begin{cases} \max\limits_{\pi} \inf\limits_{s=1,2} R(\pi, s) & \text{for } 0 < \overline{\lambda} < \frac{1}{2} \\ \inf\limits_{s=1,2} \max\limits_{\pi} R(\pi, s) & \text{for } \frac{1}{2} < \overline{\lambda} < 1 \end{cases}$$

*Proof* That $C(\overline{\lambda}) \geq \max_\pi \inf_{s=1,2} R(\pi, s)$ for $0 < \overline{\lambda} < \frac{1}{2}$ follows from Theorem 4.3.1 of [45] (coding theorem for compound channels). On the other hand, given a $(n, N, \overline{\lambda})$ code for a $\overline{\lambda} < \frac{1}{2}$, we choose $\varepsilon > 0$ such that $2(\overline{\lambda} + \varepsilon) < 1$. Application of Lemma 8 with $d = 2$ guarantees the existence of a code with length $\left[\frac{\varepsilon}{1+\varepsilon} N\right]$ and maximal error $2(\overline{\lambda} + \varepsilon)$. Hence, from Theorem 4.4.1 of [45] (strong converse for compound channels) it follows that

$$C(\overline{\lambda}) \leq \max_\pi \inf_{s=1,2} R(\pi, s) \quad \text{for} \quad 0 < \overline{\lambda} < \frac{1}{2}.$$

*Case* $\frac{1}{2} < \overline{\lambda} < 1$. Choose $\varepsilon < \overline{\lambda} - \frac{1}{2}$. $\{(u_i(1), u_i(2), D_i) : i = 1, \dots, N\}$ be a code with maximal error $\varepsilon$ for the compound channel with complete knowledge by the sender. Then

$$\{(u_j(1), D_j) : j = 1, \dots, [N|2]\} \cup \{(u_j(2), D_j) : j = [N|2] + 1, \dots, N\}$$

is a code for $\mathcal{C}^*$ with average error less than $\overline{\lambda}$. It follows from Theorem 4.5.3 of [45] that $C(\overline{\lambda}) \geq \inf_{s=1,2} \max_\pi R(\pi, s)$ for $\frac{1}{2} < \overline{\lambda} < 1$, and from Lemma 9 that

$$C(\overline{\lambda}) \leq \inf_{s=1,2} \max_\pi R(\pi, s).$$

#### 1.3.5.2 Example $\mathcal{S} = \{1, 2, 3\}$

We shall show that

$$C(\overline{\lambda}) = \begin{cases} C_{123} & \text{for } 0 < \overline{\lambda} < \frac{1}{3} \\ C_{12} \wedge C_{13} \wedge C_{23} & \text{for } \frac{1}{3} < \overline{\lambda} < \frac{1}{2} \\ (C_{12} \vee C_{13} \vee C_{23} \wedge C_1 \wedge C_2 \wedge C_3 & \text{for } \frac{1}{2} < \overline{\lambda} < \frac{2}{3} \\ C_1 \wedge C_2 \wedge C_3 & \text{for } \frac{2}{3} < \overline{\lambda} < 1. \end{cases}$$

*Proof Case* $0 < \overline{\lambda} < \frac{1}{3}$. Use the coding theorem for compound channels with maximal error (Theorem 4.3.1 of [45]) for proving $C(\overline{\lambda}) \geq C_{123}$, and Lemma 8 and the strong converse for compound channels (Theorem 4.4.1 of [45]) for proving $C(\overline{\lambda}) \leq C_{123}$.

*Case* $\frac{1}{3} < \overline{\lambda} < \frac{1}{2}$. Choose $\varepsilon < \overline{\lambda} - \frac{1}{3}$. Let $\{u_i(12), u_i(13), u_i(23), D_i : i = 1, \dots, N\}$ be a $(n, N, \varepsilon)$ code for $\mathcal{C}^*$, where the sender has the partial knowledge $K = (\{12\}, \{13\}, \{23\})$. Then

$$u_1(12), \dots, u_{\left[\frac{N}{3}\right]}, u_{\left(\left[\frac{N}{3}\right]+1\right)}(13), \dots, u_{\left[2\cdot\frac{N}{3}\right]}(13), u_{\left(\left[2\cdot\frac{N}{3}\right]+1\right)}(23), \dots, u_N(23);$$

$D_1, \ldots, D_N$ is a $(n, N, \overline{\lambda})$ code for $\mathcal{C}^*$. Application of the coding theorem for compound channels (Theorem 4.3.1 of [45]) gives $C(\overline{\lambda}) \geq C_{12} \wedge C_{13} \wedge C_{23}$. Suppose now, w.l.o.g. $C_{12} = C_{12} \wedge C_{13} \wedge C_{23}$, then $C(\overline{\lambda}) \leq C_{12}$ by the first example.

*Case* $\frac{1}{2} < \overline{\lambda} < \frac{2}{3}$. Choose $\varepsilon < \overline{\lambda} - \frac{1}{2}$ and assume, w.l.o.g., that $(C_{12} \vee C_{13} \vee C_{23}) \cap C_1 \wedge C_2 \wedge C_3 = C_{12} \wedge C_3$. Then define $K = (\{12\}, \{3\})$. Apply Theorem 4.3.1 of [45] and select $u_1(12), \ldots, U_{\left[\frac{N}{2}\right]}(12), u_{\left[\frac{N}{2}\right]+1}(3), \ldots, u_N(3)$. By the usual procedure we finally get $C(\overline{\lambda}) \geq C_{12} \wedge C_3$. For proving the converse part we use the result for averaged channels. If $C_3 = C_{12} \wedge C_3$, obviously $C(\overline{\lambda}) \leq C_3$. Assume therefore that $C_{12} = C_{12} \wedge C_3$ [$\geq C_{23}, C_{13}$]. An $(n, N, \overline{\lambda})$ code for $\mathcal{C}^*$ is an $(n, N, \overline{\lambda})$ code for the averaged channel

$$(P^n(\cdot|\cdot), n = 1, 2, \ldots) = \left( \sum_{s=1}^{3} \frac{1}{3} P^n(\cdot \mid \cdot \mid s), n = 1, 2, \ldots \right).$$

Therefore $C_a(\overline{\lambda}) = C_a(\lambda) \geq C(\overline{\lambda})$, if $\lambda = \overline{\lambda}$ and not equal to 0, $\frac{1}{3}$, $\frac{2}{3}$, or 1. We get for $\frac{1}{2} < \overline{\lambda} < \frac{2}{3}$, that $C_a(\lambda) = C_{12}$, since $C_{12} \geq C_{23}, C_{13}$. Hence $C(\overline{\lambda}) \leq C_{12}$. This proves the desired result.

*Case* $\frac{2}{3} < \overline{\lambda} < 1$. Choose $\varepsilon < \overline{\lambda} - \frac{2}{3}$ and define $K = (\{1\}, \{2\}, \{3\})$. Apply Theorem 4.3.1 of [45] and select

$$u_1(1), \ldots, u_{\left[\frac{N}{3}\right]}(1), u_{\left[\frac{N}{3}\right]+1}(2), \ldots, u_{\left[\frac{2N}{3}\right]}(2), u_{\left[\frac{2N}{3}\right]+1}(3), \ldots, u_N(3).$$

Prove $C(\overline{\lambda}) \geq C_1 \wedge C_2 \wedge C_3$ as usual. $C(\overline{\lambda}) \leq C_1 \wedge C_2 \wedge C_3$ is obvious.

The converse parts could have been proved in all four cases by using suitable averaged channels. This will be illustrated in

#### 1.3.5.3 Example $\mathcal{S} = \{1, 2, 3, 4\}$

$$C(\overline{\lambda}) = \begin{cases} C_{1234} \quad \text{for } \overline{\lambda} \in \left(0, \frac{1}{4}\right) \\ C_{123} \wedge C_{124} \wedge C_{234} \wedge C_{234} \quad \text{for } \overline{\lambda} \in \left(\frac{1}{4}, \frac{1}{3}\right) \\ \bigvee\limits_{i \neq j \neq h \neq \ell} (C_{ijh} \wedge C_{ij\ell} \wedge C_{h\ell}) \quad \text{for } \overline{\lambda} \in \left(\frac{1}{3}, \frac{2}{5}\right) \\ (C_{123} \vee C_{124} \vee C_{134} \vee C_{234}) \wedge C_{12} \wedge C_{13} \\ \quad \wedge C_{14} \wedge C_{23} \wedge C_{24} \wedge C_{34} \quad \text{for } \overline{\lambda} \in \left(\frac{2}{5}, \frac{1}{2}\right) \end{cases}$$

$$\begin{cases} (C_{12} \wedge C_{34}) \vee (C_{13} \wedge C_{24}) \vee (C_{14} \wedge C_{23}) \\ \vee(C_{123} \wedge C_4) \vee (C_{124} \wedge C_3) \vee C_{134} \wedge C_2) \\ \vee(C_{234} \wedge C_1) \quad \text{for } \overline{\lambda} \in \left(\frac{1}{2}, \frac{3}{5}\right) \\ \\ (C_{12} \wedge C_{13} \wedge C_{23} \wedge C_4) \vee (C_{12} \wedge C_{14} \wedge C_{24} \wedge C_3) \dots \\ \vee(C_{12} \wedge C_{34}) \vee \dots \quad \text{for } \overline{\lambda} \in \left(\frac{3}{5}, \frac{2}{3}\right) \\ \\ (C_{12} \vee C_{13} \vee C_{14} \vee C_{23} \vee C_{24} \vee C_{34}) \\ \wedge C_1 \wedge C_2 \wedge C_3 \wedge C_4 \quad \text{for } \overline{\lambda} \in \left(\frac{2}{3}, \frac{3}{4}\right) \\ \\ C_1 \wedge C_2 \wedge C_3 \wedge C_4 \quad \text{for } \overline{\lambda} \in \left(\frac{3}{4}, 1\right) \end{cases}$$

*Proof* *Case* $\left(0, \frac{1}{4}\right)$. Obviously $C(\overline{\lambda}) \geq C_{1234}$. Use the averaged channel $P^n(\cdot|\cdot) = \sum_{s=1}^4 \frac{1}{4} P^n(\cdot|\cdot|s)$ for proving $C(\overline{\lambda}) \leq C_{1234}$.

*Case* $\left(\frac{1}{4}, \frac{1}{3}\right)$. Choose $\varepsilon < \overline{\lambda} - \frac{1}{4}$. Let $\{u_i(123), u_i(124), u_i(134), u_i(234), D_i : i = 1, \dots, N\}$ be an $(n, N, \varepsilon)$ code for $\mathcal{C}^*$, where the sender has partial knowledge

$$K = (\{123\}, \{124\}, \{134\}, \{234\}).$$

Then $\{u_1(123), \dots, u_{\left[\frac{N}{4}\right]}(123), u_{\left[\frac{N}{4}\right]+1}(124), \dots, u_N(234), D_1, \dots, D_N\}$ is an $(n, N, \overline{\lambda})$ code for $\mathcal{C}^*$. Application of Theorem 4.3.1 of [45] gives $C(\overline{\lambda}) \geq C_{123} \wedge C_{124} \wedge C_{134} \wedge C_{234}$. We want to prove the converse in $\left(\frac{1}{4}, x_0\right)$. Assume the infimum is taken for $C_{123}$.

We introduce an averaged channel

$$P^n(\cdot|\cdot) = \sum_{s=1}^{4} p_s P^n(\cdot|\cdot|s)$$

for which

(i) $p_1 + p_2 + p_3 \geq 1 - \frac{1}{4}$
(ii) $p_1 + p_2 + p_4, p_1 + p_3 + p_4, p_2 + p_3 + p_4 \leq 1 - x$

and $x_0$ is the maximal value of $x$ for which a solution of (i), (ii) exists.

We use the solution

$$p_1 = p_2 = p_3 = \frac{1}{3}, p_4 = 0$$

$$x_0 = \frac{1}{3}$$

It follows that $C_a(\overline{\lambda}) = C_{123}$ for $\overline{\lambda} \in \left(\frac{1}{4}, \frac{1}{3}\right)$, and therefore $C(\overline{\lambda}) \leq C_{123}$.

*Case* $\left(\frac{1}{3}, \frac{2}{5}\right)$. Assume that the maximum is taken for $C_{123} \wedge C_{124} \wedge C_{34}$. Then $C(\overline{\lambda}) \geq C_{123} \wedge C_{124} \wedge C_{34}$ follows as usual by taking

$$\begin{aligned} &\tfrac{1}{3} \quad \text{of the } \{u_i(123) : i = 1, \dots, N\}, \\ &\tfrac{1}{3} \quad \text{of the } \{u_i(124) : i = 1, \dots, N\}, \\ \text{and } &\tfrac{1}{3} \quad \text{of the } \{u_i(34) : i = 1, \dots, N\}. \end{aligned}$$

In the future we shall say shortly that we use a $\left(\frac{1}{3}, \frac{1}{3}, \frac{1}{3}\right)$-*fraction* (or in general a $(\alpha_1, \dots, \alpha_n)$-fraction).

If now $C_{34} = C_{123} \wedge C_{124} \wedge C_{34}$, then we use an average $p = (p_1 \dots, p_4) = \left(0, 0, \frac{1}{2}, \frac{1}{2}\right)$ and obtain the desired result.

We can therefore assume w.l.o.g. that

$$\begin{aligned} C_{123} &\leq C_{124} \wedge C_{34} \\ C_{123} &\leq C_{134} \wedge C_{24} \\ C_{123} &\leq C_{234} \wedge C_{14}. \end{aligned}$$

If $C_{134} \wedge C_{24} = C_{24}$ [or $C_{234} \wedge C_{14} = C_{14}$], we immediately get $C(\overline{\lambda}) \leq C_{24}$ by using an average $p = \left(0, \frac{1}{2}, 0, \frac{1}{2}\right)$ [or $p = \left(\frac{1}{2}, 0, 0, \frac{1}{2}\right)$.]

It remains to consider

$$\begin{aligned} C_{123} &\leq C_{124}, C_{34} \\ C_{123} &\geq C_{234}, C_{134}. \end{aligned}$$

In order to get an averaged channel with $C_a(\overline{\lambda}) = C_{123}$ in $\left(\frac{1}{3}, x_0\right)$, $p = (p_1, \dots, p_4)$ must satisfy $p_1 + p_2 + p_3 \geq 1 - \frac{1}{3}$

$$\begin{aligned} p_1 + p_2 + p_4 &\leq 1 - x \\ p_1 + p_3 &\leq 1 - x \\ p_2 + p_3 &\leq 1 - x \\ p_4 + p_3 &\leq 1 - x \end{aligned}$$

Let $x_0$ be the maximal $x$ for which a solution exists. We get $x_0 = \frac{2}{5}$, $p_1 = p_2 = p_4 = \frac{1}{5}$; $p_3 = \frac{2}{5}$ as a solution.

*Case* $\left(\frac{2}{5}, \frac{1}{2}\right)$. We can assume the infimum $= C_{123} \wedge C_{14} \wedge C_{24} \wedge C_{34}$. Use the fraction $\left(\frac{2}{5}, \frac{1}{5}, \frac{1}{5}, \frac{1}{5}\right)$ for $K = \{(123), (14), (24), (341)\}$ to prove

$$C(\overline{\lambda}) \geq C_{123} \wedge C_{14} \wedge C_{24} \wedge C_{34}$$

If the infimum is taken for $C_{123}$ use $p = \left(\frac{1}{4}, \frac{1}{4}, \frac{1}{4}, \frac{1}{4}\right)$ and if the infimum is taken for $C_{14}$, for instance, use $p = \left(\frac{1}{2}, 0, 0, \frac{1}{2}\right)$. In either case, we get $C(\overline{\lambda}) \leq C_{123} \wedge C_{14} \wedge C_{24} \wedge C_{34}$.

*Case* $\left(\frac{1}{2}, \frac{3}{5}\right)$. That the expression given above in the third Example is a lower bound, is trivial; take the fraction $\left(\frac{1}{2}, \frac{1}{2}\right)$. To prove that the expression given is an upper bound, we consider first the case

1. The maximum is taken by $C_{123} \wedge C_4$.

   *Subcase (i).* $C_{123} \wedge C_4 = C_{123}$

   Thus $C_{123} \geq C_{12} \wedge C_{34}, C_{13} \wedge C_{24}, C_{23} \wedge C_{14}, C_{12}, C_{13}, C_{23} \geq C_{123}$ implies $C_{123} \geq C_{34}, C_{24}, C_{14}$. We can assume that $C_{123} \geq C_{jkl}$, because if for instance $C_{134} > C_{123}$, then $C_2 = C_{123}$ and we can use the average $(0, 1, 0, 0)$.

   We have therefore finally $C_4 \geq C_{123} \geq C_{34}, C_{24}, C_{14}, C_{jkl}$. Now define $p = \left(\frac{1}{5}, \frac{1}{5}, \frac{1}{5}, \frac{2}{5}\right)$. Then $C(\overline{\lambda} \leq C_a(\overline{\lambda}) = C_{123}$.

   *Subcase (ii).* $C_4 \leq C_{123}$

   Use $p = (0, 0, 0, 1)$.
2. The maximum is taken by $C_{12} \wedge C_{34}$. W.l.o.g. $C_{12} = C_{12} \wedge C_{34}$. W.l.o.g. $C_{13} \leq C_{12}$.

   *Case (i).* $C_{23} \leq C_{12}$

   Use $p = \left(\frac{1}{3}, \frac{1}{3}, \frac{1}{3}, 0\right)$.

   *Case (ii).* $C_{14} \leq C_{12}$

   Use $\left(\frac{2}{5}, \frac{1}{5}, \frac{1}{5}, \frac{1}{5}\right)$. $C_{\ell h} > C_{12}$ implies $\ell, h \neq 1$, but then $p(\ell) + p(h) = \frac{2}{5} < 1 - \overline{\lambda}$. $C_{\ell hn} > C_{12}$ implies $\{\ell, h, n\} = \{2, 3, 4\}$. But $C_{234} \wedge C_1 \leq C_{12}$ implies that $C_1 = C_{12}$. Use $p = (1, 0, 0, 0)$.

*Case* $\left(\frac{3}{5}, \frac{2}{3}\right)$.

1. The maximum is attained by $(*) = C_{12} \wedge C_{13} \wedge C_{23} \wedge C_4$ and by no term $(C_{lh} \wedge C_{nu})$.

   Use fraction $\left(\frac{1}{5}, \frac{1}{5}, \frac{1}{5}, \frac{2}{5}\right)$ to prove $C(\overline{\lambda}) \geq (*)$. If $C_4 = (*)$, then the converse is obvious: $p = (0, 0, 0, 1)$. Assume therefore, w.l.o.g., that $C_{12} = (*)$. It follows that $C_{34} < C_{12}$ and also $C_{24}, C_{14} < C_{12}$. Use $p = \left(\frac{1}{3}, \frac{1}{3}, 0, \frac{1}{3}\right)$ to prove $C(\overline{\lambda}) \leq C_a(\overline{\lambda}) \leq (*)$.
2. W.l.o.g. assume $C_{12} \wedge C_{34} = (*)$. Use the fraction $\left(\frac{1}{2}, \frac{1}{2}\right)$ to prove $C(\overline{\lambda}) \geq (*)$.

Assume $C_{12} \leq C_{34}$. W.l.o.g. $C_{13} \leq C_{12}$.

*Case (i).* $C_{23} \leq C_{12}$

Use $p = \left(\frac{1}{3}, \frac{1}{3}, \frac{1}{3}, 0\right)$.

*Case (ii).* $C_{14} \leq C_{12}$

therefore $C_{14}, C_{13} \leq C_{12} \leq C_{34}$. Again, two cases:

a. $C_{34} > C_{12}$.
$(C_{34} \wedge C_{23} \wedge C_{24} \wedge C_1) \leq C_{12} \wedge C_{34}$ implies either $C_1 = C_{12}$, and we are finished, or $C_{23} \wedge C_{24} \leq C_{12}$, and therefore w.l.o.g. $C_{23} \leq C_{12}$. We have $C_{13}, C_{14}, C_{23} \leq C_{12} \leq C_{34}$. Use $p = \left(\frac{1}{3}, \frac{1}{3}, \frac{1}{3}, 0\right)$.

b. $C_{34} = C_{12}$.
Therefore $C_{34} = C_{12} \geq C_{13}, C_{14}$. Use $p = \left(\frac{1}{3}, 0, \frac{1}{3}, \frac{1}{3}\right)$.

$C_a(\overline{\lambda}) = C_{34} \geq C(\overline{\lambda})$.

*Case* $\left(\frac{2}{3}, \frac{3}{4}\right)$. W.l.o.g. let the value $I$ of the formula be $C_{12} \wedge C_3 \wedge C_4$. Use the fraction $\left(\frac{1}{3}, \frac{1}{3}, \frac{1}{3}\right)$ to get $C(\overline{\lambda}) \geq C_{12} \wedge C_3 \wedge C_4$. Suppose $I = C_{12}$, use $p = \left(\frac{1}{4}, \frac{1}{4}, \frac{1}{4}, \frac{1}{4}\right)$. Suppose $I = C_3$ for instance, then use $p = (0, 0, 1, 0)$.
*Case* $\left(\frac{3}{4}, 1\right)$. is obvious.

## 1.4 Lecture on Algebraic Compositions (Rings and Lattices) of Channels

We present in a form, which we consider mathematically elegant, known and some new models of channels.

Well-known are addition channels (or disjoint union channels) AC, product channels PC, thus combinations to rings of channels RC, min channels MinC (also known as simultaneous channels or compound channels CC), min product channels MinPC (known as arbitrarily varying channels AVC).

New are min addition channels MinAC (related to CC with side information at the sender) and compositions involving max-operations beginning with the max channel MaxC (the "dual" to CC, also called optimistic channel in [1]), the max addition channels MaxAC, and finally the max product channels MaxPC (related to infinite alphabet channels)—see also Shannon's lattice of sources.

We give the definitions for classes of discrete general time-structure channels GDC

$$\mathcal{W}^n(\mathcal{S}) = \{W^n(\cdot|\cdot|s) : s \in \mathcal{S}\} \tag{1.4.1}$$
$$W^n(\cdot|\cdot|s) : \mathcal{X}^n \to \mathcal{Y}^n$$

and for classes of discrete memoryless channels DMC with

$$\mathcal{W}^n(\mathcal{S}) = \{W^n(\cdot|\cdot|s) : s \in \mathcal{S}\} \tag{1.4.2}$$

$$W(\cdot|\cdot|s) : \mathcal{X} \to \mathcal{Y}, \quad W^n(y|x^n|s) = \prod_{t=1}^{n} W(y_t|x_t|s_t)$$

and

$$\mathcal{W}^n(\mathcal{S}^n) = \{\prod W(\cdot|\cdot|s^n) : s^n \in \mathcal{S}^n = \prod_1^n C\} \tag{1.4.3}$$

$$W^n(y^n|x^n|s^n) = \prod_{t=1}^{n} W(y_t|x_t|s_t).$$

For the analysis it is easier and more didactically to give results for DMC's for the cases of constant channels (1.4.2) and of (letter-wise) variable channels.

## *1.4.1 A Ring of Channels*

We are given stochastic matrices $\mathcal{W}(\mathcal{S})$ with $W(\cdot|\cdot|s) : \mathcal{X}(s) \to \mathcal{Y}(s)$ $(s \in \mathcal{S})$. We start with the well-known AC studied already by Shannon.

### 1.4.1.1 Additive Channels

**Definition 1** We assume that the $\mathcal{X}(s)$'s and also the $\mathcal{Y}(s)$'s are disjoint and define the AC channel $(W^n)_{n=1}^{\infty}$ by $W : \mathcal{X} \to \mathcal{Y}$, where $\mathcal{X} = \bigcup_s \mathcal{X}(s)$, $Y = \bigcup_s \mathcal{Y}(s)$ with

$$W(y|x) = W(y|x|s) \quad \text{for } x \in \mathcal{X}(s), y \in \mathcal{Y}(s)$$

and

$$W^n(y^n|x^n) = \prod_{t=1}^{n} W(y_t|x_t), \quad (n \in \mathbb{N}, x^n \in \mathcal{X}^n, y^n \in \mathcal{Y}^n)$$

**Theorem 16** (Shannon) *For $|\mathcal{S}| < \infty$ the capacity of $(W^n)_{n=1}^{\infty}$ equals*

$$C(W) = \log\left(\sum_{s \in \mathcal{S}} e^{C(s)}\right).$$

### 1.4.1.2 Product Channels

**Definition 2** We define for $\mathcal{W}(\mathcal{S})$ (as defined above) and $S = |\mathcal{S}| < \infty$ a product channel $(W^{\times n})_{n=1}^{\infty}$ by $W^{\times} : \prod_{s=1}^{S} \mathcal{X}(s) \to \prod_{s=1}^{S} \mathcal{Y}(s)$ with

$$W^{\times}(y(1), \dots, y(s)|x(1), \dots, x(s)) = \prod_{s=1}^{S} W(y(s)|x(s)|s)$$

for

$$(y(1), \dots, y(s)) \in \prod_{s=1}^{S} \mathcal{Y}(s), \quad (x(1), \dots, x(s)) \in \prod_{s=1}^{S} \mathcal{X}(s)$$

and

$$W^{\times n}(y_1(1), \dots, y_1(S); \dots; y_n(1), \dots, y_n(S)|x_1(1), \dots, x_1(S); \dots; x_n(1), \dots, x_n(S))$$
$$= \prod_{t=1}^{n} \prod_{s=1}^{S} W(y_t(s)|x_t(s)|s).$$

**Theorem 17** *For $S < \infty$ the capacity of $(W^{\times})_{n=1}^{\infty}$ equals*

$$C(W^{\times}) = \sum_{s=1}^{S} C(s)$$

### 1.4.1.3 Comparison of These Channels

Usually, $C(W^{\times})$ is much larger than $C(W)$ as can be seen for $|\mathcal{S}| = 2$ and $C_1 = C_2$, because $C_1 + C_2 = 2C_1 > \log(e^{C_1} + e^{C_2}) = C_1 + \log 2$, if $C_1 > \log 2$. However, the opposite inequality occurs here for $C_1 < \log 2$. Moreover, it occurs always if $C_1 > 0$ and $C_2 = 0$ because $C_1 + C_2 = C_1 < \log(e^{C_1} + e^0) = \log(e^{C_1} + 1)$. So the second completely noisy channel adds rate because $\mathcal{X}(2)$ gives rise to a letter added to $\mathcal{X}(1)$.

If we have two given channels, it is possible to form a single channel from them in two natural ways which we call the sum and product of the two channels. The *sums* of two channels is the channel found by using inputs from either of the two given channels with the same transition probabilities to the set of output letters consisting of the union of the two disjoint output alphabets. Thus the sum channel is defined by a transition matrix of one channel below and to the right, of that for the other channel and filling the remaining two rectangles with zeros. If $(W(j|i))_{\substack{i=1,\dots,a \\ j=1,\dots,b}}$ and $\left(W'(j|i)\right)_{\substack{i=1,\dots,a' \\ j=1,\dots,b'}}$ are the individual matrices, the sum has the following matrix:

$$\begin{matrix} W(1/1) & \cdots & W(b/1) & 0 & \cdots & 0 \\ \vdots & & \vdots & \vdots & & \vdots \\ W(1/a) & \cdots & W(b/a) & 0 & \cdots & 0 \\ 0 & \cdots & 0 & W'(1/1) & \cdots & W'(b'/1) \\ \vdots & & \vdots & \vdots & & \vdots \\ 0 & \cdots & 0 & W'(1/a') & \cdots & W'(b'/a') \end{matrix}$$

The *product* of two channels is the channel whose input alphabet consists of $\mathcal{X} \times \mathcal{X}'$, where output alphabet consists of $\mathcal{Y} \times \mathcal{Y}'$, and whose transmission probabilities are $W(j|i)W'(j'|i')$ $(i, j) \in \mathcal{X} \times \mathcal{X}'$, $(j, j') \in \mathcal{Y} \times \mathcal{Y}'$.

The sum of channels corresponds physically to a situation where either of two channels may be used (but not both), a new choice being made for each transmitted letter. The product channel corresponds to a situation where both channels are used each unit of time. It can be done for the same channel $K$ also $m$ times and thus one gets the channel $K^m$.

Combining the sum and product operations it is possible to build, for example, a channel of polynomial form $\sum_{j=1}^J a_j K^j$, where the $a_j$ are nonnegative integers and $K$ is a channel.

**Lemma 12** *If $W$ has capacity $C$ and $W'$ has capacity $C'$, then*

*(i)* $W \oplus W'$ *has capacity* $\log(e^C + e^{C'})$
*and*
*(ii)* $W \times W'$ *has capacity* $C + C'$.

*Proof* (ii) is a trivial consequence of the coding theorem for parallel channels for instance.

In order to show (i) we write $w = w(\cdot|\cdot|1)$, $w' = w(\cdot|\cdot|2)$, $S = \{1, 2\}$, $\mathcal{P}(\cdot|\cdot|s^n) = \prod_{t=1}^n w(\cdot| \cdot |s_t^*)$, $s^n = (s_1, \dots, s_n) \in S^n$.

We know from the Coding Theorem for NMDC that

$$\log N(n, \lambda, s^n) = \sum_{t=1}^n C_{s_t} + O(\sqrt{n}).$$

Now for the sum channel

$$N(n, \lambda) = \sum_{s^n \in S^n} N(n, \lambda, s^n) \tag{1.4.4}$$

and hence

$$N(n, \lambda) = \sum_{k=0}^{n} \binom{n}{k} e^{kC_1} \cdot e^{(n-k)C_2} e^{O(\sqrt{n})}$$
$$= (e^{C_1} + e^{C_2})^n.$$

Therefore, $\lim_{n\to\infty} \frac{\log N(n,\lambda)}{n} = \log(e^{C_1} + e^{C_2})$. □

## 1.4.2 *Min Channels Known as Compound Channels CC*

Given again the class $\mathcal{W}(\mathcal{S})$ with matrices $W(\cdot|\cdot|s) : \mathcal{X}(s) \to \mathcal{Y}(s)$, we define the class of DMC's $\mathcal{W}^n(\mathcal{S}) = \{W^n(\cdot|\cdot|s) : s \in \mathcal{Y}\}$, where

$$W(y^n|x^n|s) = \prod_{t=1}^{n} W(y_t|x_t|s).$$

**Definition 3** (Blackwell, Breiman, Thomasian [17], and Wolfowitz) Consider $(\mathcal{W}^n(\mathcal{S}))_{n=1}^{\infty}$ with $\mathcal{X}(s) = \mathcal{X}, \mathcal{Y}(s) = \mathcal{Y}$ $(s \in \mathcal{Y})$. The MinC associated with this class is $\left(\bigwedge_{s\in\mathcal{S}} W^n(\cdot|\cdot|s)\right)_{n=1}^{\infty}$ given by

$$\bigwedge_{s\in\mathcal{S}} W^n(D|x^n|s) \quad (D \subset \mathcal{Y}^n, x^n \in \mathcal{X}^n),$$

for which we use the abbreviation $\Lambda^n(D|x^n)$.

This definition is motivated by its application to the study of $(n, N, \lambda)$ codes which serve *simultaneously* for all DMC's in the class, that is, set of pairs $\{(u_i, D_i) : 1 \le i \le N\}$, $u_i \in \mathcal{X}^n$, $D_i \subset \mathcal{Y}^n$ $(1 \le i \le N)$, $D_i \cap D_j = \emptyset$ $(i \ne j)$ with

$$\Lambda^n(D_i|u_i) \ge 1 - \lambda \quad (1 \le i \le N).$$

**Theorem 18** *For finite $\mathcal{X}$ and $\mathcal{Y}$ the capacity of $(\Lambda^n)_{n=1}^{\infty}$ equals*

$$\max_{P\in\mathcal{P}(\mathcal{X})} \inf_{s\in\mathcal{S}} I(P, W(\cdot|\cdot|s)).$$

This is proved in the direct part in the next Sect. 1.4.3.

Wolfowitz introduced and treated the CC with side information about $s$ specifying a channel $W^n(\cdot|\cdot|s)$ selected for instance by a malevolent being, like a jammer, the original case of the MinC can be described as the case where neither sender nor receiver knows $s$, in short $(S^-, R^-)$. Indicating the knowledge of $s$ by a plus there are the three remaining cases $(S^-, R^+)$, $(S^+, R^-)$, and $(S^+, R^+)$.

It turns out that here the receiver's knowledge does not increase capacities. The remaining case $(S^+, R^-)$ is important.

**Theorem 19** *For the CC with side information* $(S^+, R^-)$ *the capacity equals*

$$\inf_{s \in \mathcal{S}} C(s).$$

An $(n, N)$-code meeting the constraints is a system of pairs $\{(u_i(s), D_i) : 1 \leq i \leq N, s \in \mathcal{S}\}$, $u_i(s) \in \mathcal{X}^n$, $D_i \subset \mathcal{Y}^n$ pairwise disjoint. It is an $(n, N, \lambda)$-code if $W^n(D_i|u_i(s)|s) > 1 - \lambda$ $(s \in \mathcal{S}, 1 \leq i \leq N)$. Using a block length $l$ with $\frac{l}{n} \to 0$ $(n \to \infty)$, the communicators ca use the channel so that the receiver can test the channel and learn $s$ with probability arbitrarily close to 1, if $|\mathcal{S}| < \infty$. In the remaining time $n - l$ a code can be set up for $W^n(\cdot | \cdot |s)$ of a rate arbitrarily close to $C(s)$. Clearly, this way the theorem is proved for $|\mathcal{S}| < \infty$.

### *1.4.3 Compound Channels Treated with Maximal Coding*

An (abstract) channel $W : J \to (\mathcal{O}, \mathcal{E})$ is a set of PDs $W = \{W(\cdot|i) : i \in T\}$ on $(\mathcal{O}, \mathcal{E})$. So for every input symbol $i$ and every (measurable) $E \in \mathcal{E}$ of output symbols $W(E|i)$ specifies the probability that a symbol in $E$ will be received, if symbol $i$ has been sent.

The set $J$ does not have to carry additional structure.

Of particular interest are discrete channels with "time structure," that means symbols are words over an alphabet, say $\mathcal{X}$ for the inputs and $\mathcal{Y}$ for the outputs. Here $\mathcal{X}^n = \prod_{t=1}^n \mathcal{X}_t$ with $\mathcal{X}_t = \mathcal{X}$ $(t \in \mathbb{N})$ are the input words and $\mathcal{Y}^n = \prod_{t=1}^n \mathcal{Y}_t$, $\mathcal{Y}_t = \mathcal{Y}$ $(t \in \mathbb{N})$ are the output words of length $n$.

So we can define a (constant block length) channel by the set of stochastic matrices $\mathcal{K} : \{W^n : \mathcal{X}^n \to \mathcal{Y}^n, n \in \mathbb{N}\}$.

The compound channel is defined as the class $(\mathcal{K}_s)_{s \in \mathcal{S}}$ of such channels and $\{(u_i, D_i) : 1 \leq i \leq M\}$ is an $(n, N, \lambda)$ code for the CC $(\mathcal{K}_s)_{s \in \mathcal{S}}$, if $u_i \in \mathcal{X}^n$ and the disjoint $D_i \subset \mathcal{Y}^n$ satisfy

$$W^n(D_i|u_i|s) \geq 1 - \lambda, \qquad 1 \leq i \leq M. \tag{1.4.5}$$

We consider first $|\mathcal{S}| < \infty$. Recall the maximal code inequality for standard abstract discrete channels whose formulation requires some definitions.

Let $Q'$ be a PD on $J$, $Q$ be a PD on $J \times \mathcal{O}$ defined by

$$Q(x, y) = Q'(x)W(y|x) \quad (x \in J, y \in \mathcal{O})$$

and

$$Q^n(y) = \sum_x Q(x, y).$$

Furthermore we define the information density

$$i(x, y) = \log \frac{Q(x, y)}{Q'(x)Q''(y)}.$$

**Lemma 13** (Feinstein 1954) *Let $\theta > 0$, $0 < \lambda < 1$ be arbitrary otherwise and let $Q'$ be any PD on $J$. Using it in the definition of $I(x, y)$ we define*

$$F = \{(x, y) : I(x, y) > \theta\}$$

*and $F^c$ as its complement in $J \times \mathcal{O}$. There exists an $(M, \lambda)$-code for the channels such that*

$$M > 2^{\theta}[\lambda - Q(F^c)].$$

*Proof* Let us see how the derivation of the coding theorem for the DMC is. We use i.i.d. pairs of RV's $(X^n, Y^n)$ with joint distributions $Q' = P_{X^n} = \prod_1^n P_X$, $Q = P_{X^n Y^n} = Q'(x^n)W(y^n|x^n)$, and $Q'' = P_{Y^n} = \prod_1^n P_Y$ and the RV $i(X^n, Y^n) = \sum_{t=1}^n i(X_t, Y_t)$. If now $P_X^*$ is a maximizing input distribution for the channel $W$, then $\mathbb{E}i(X, Y) = C$, and the variance $\mathbb{E}(i(X, Y) - C)^2$ is bounded by $\sigma^2 < 2(\log|\mathcal{X}|)^2$ (see p. 79 of Wolfowitz). By Chebychev's inequality

$$\Pr\{i(X^n, Y^n) \leq nC - d\sigma\sqrt{n}\} < \lambda' < \lambda$$

for every $n$, if $d = d(\lambda')$ is large enough. Now with the choice $\theta = nC - d\sigma\sqrt{2}$ $Q(F^c) \leq \lambda'$ and there exists an $(N, M, \lambda)$-code for

$$M > 2^{nC - d\sigma\sqrt{n} + \log(\lambda - \lambda')} \tag{1.4.6}$$

□

Turning now to the CC we follow the idea to apply Lemma 13 to a suitable associated averaged channel. We do this first in abstract for the CC

$$\{V(\cdot|\cdot|s) : s \in \mathcal{S}\}$$

without time structure.

Define the averaged channel

$$V^*(y|x) = \frac{1}{|\mathcal{S}|}\sum_{s \in \mathcal{S}} V(y|x). \tag{1.4.7}$$

$$\text{An } \left(M, \frac{\lambda}{|\mathcal{S}|}\right)\text{-code for } V^* \text{ is an } (M, \lambda)\text{-code for CC.} \tag{1.4.8}$$

Define now

$$G = \left\{ (x, y) : \sum_s V(y|x|s) \le 2^\theta \sum_s Q''_{(s)}(y) \right\},$$

where for $Q'$ on $\mathcal{X}$ and $Q'_{(s)}(x, y) = Q'(x)V(y|x|s)$, $Q''_{(s)}(y) = \sum_x Q_{(s)}(x, y)$.

**Lemma 14** *Let $\theta, \theta'$ and $\lambda < 1$ be arbitrary positive numbers, then there exists an $(M, \lambda)$-code for the CC such that*

$$M > \frac{2^Q}{|\mathcal{S}|} \left[ \lambda - \sum_s Q_{(s)}(G) \right] \tag{1.4.9}$$

*Proof* This follows immediately from Lemma 13 because $F^c$ for $V^*$ equals

$$G = \left\{ (x, y) : \log \frac{\frac{1}{|\mathcal{S}|} \sum_s V(y|x|s)}{\frac{1}{|\mathcal{S}|} \sum_s Q''_{(s)}(y)} \le \theta \right\} \tag{1.4.10}$$

and we can use (1.4.8). □

**Lemma 15**

$$\sum_s Q_{(s)}(G) \le |\mathcal{S}|^2 2^{-\theta'} + \sum_s Q_{(s)}\{(x, y) : i_{(s)}(x, y) < \theta + \theta'\} \tag{1.4.11}$$

*Proof* Consider the set

$$G_s = \{(x, y) : i_s(x, y) \ge \theta + \theta'\} \cap G \quad (s \in \mathcal{S}).$$

For any $(x, y) \in G_s$ we have

$$2^\theta \sum_s Q''_{(s)}(y) \ge \sum_s V(y|x|s) \ge V(y|x|s) \ge 2^{\theta+\theta'} Q''_{(s)}(y) \tag{1.4.12}$$

and hence

$$Q''_{(s)}(y) \le 2^{-\theta'} \sum_s Q''_{(s)}(y). \tag{1.4.13}$$

Summing (1.4.12) over all $y$ for which there exists an $x$ such that $(x, y) \in G_s$ we obtain

$$Q_{(s)}(G_{(s)}) \le t 2^{-\theta'}. \tag{1.4.14}$$

Now for $s \in \mathcal{S}$

$$Q_{(s)}(G) \le Q_{(s)}(G_s) + Q_{(s)}\{(x, y) : i_{(s)}(x, y) < \theta + \theta'\} \tag{1.4.15}$$

and (1.4.11) follows from (1.4.14). □

Combining now Lemmas 14 and 15 we get

**Theorem 20** *Let $\theta, \theta'$ and $\lambda < 1$ be arbitrary positive numbers, then there exists an $(M, \lambda)$-code for the CC such that*

$$M > \frac{2^{\theta}}{|\mathcal{S}|}\left[\lambda - |\mathcal{Y}|^2 2^{-\theta'} - \sum_s Q_{(s)}\{(x, y) : i_{(s)}(x, y) \le \theta + \theta'\}\right]. \tag{1.4.16}$$

It is again easy to apply Theorem 20 to CC with DMC's.

Let $C = \max_{P_X} \min_s \mathbb{E}i_{(s)}(X, Y) = \max_{P_X} \min_s I(X \wedge Y|s)$, choose $\theta = nC - 2d\sqrt{n}$, $\theta' = d\sqrt{n}$ and $d$ a suitable constant. If $\lambda' < \frac{\lambda}{2}$ and $n$ is sufficiently large, then by Chebychev's inequality and (1.4.16)

$$M > 2^{nC - 2d\sqrt{n} - \log|\mathcal{S}|}\left[\lambda - \frac{|\mathcal{S}|^2}{2^{d\sqrt{n}}} - \lambda'\right] \tag{1.4.17}$$

$$> 2^{nC - f(\lambda)\sqrt{n}} \tag{1.4.18}$$

**Theorem 21** *For a CC, discrete memoryless and arbitrary set of states $\mathcal{S}$ there exists an $(M, N, \lambda)$-code with*

$$M \ge 2^{nC - g(\lambda)\sqrt{n}}.$$

*Proof* Use inequality (1.4.18) and the Approximation Lemma to reduce the case $|\mathcal{S}| = \infty$ to the finite case. □

In Wolfowitz [45] there is a maximal code lemma also for the case where the sender knows $s$. However, there is a simpler proof of the coding theorem in this case. Reduce the case $|\mathcal{S}| = \infty$ again to the finite case $\mathcal{S}' \subset \mathcal{S}$ and let the receiver test the channels to learn $s$ with probability close to 1 and then sender and receiver use the DMC $W(\cdot|\cdot|s)$. This gives capacities

$$C_{S^+,R^-} = C_{S^+,R^-} = \min_{s \in \mathcal{S}} C(s). \tag{1.4.19}$$

### *1.4.4 For Comparison: The Method of Random Codes*

Consider an abstract $W : \mathcal{X} \to \mathcal{Y}$. Let $U_i$ $(1 \le i \le M)$ be i.i.d. RV's with values in $\mathcal{X}$ and distributions $P_U$ on $\mathcal{X}$.

Define decoding sets

$$D_i(U_1, \dots, U_M) = \{y \in \mathcal{Y} : W(y|U_i) > W(y|U_j) \text{ for } j \neq i\}. \tag{1.4.20}$$

Consequently

$$D_i(U_1, \dots, U_M) = \left\{ y \in \mathcal{Y} : \log \frac{W(y|U_i)}{P_Y(y)} > \log \frac{W(y|U_j)}{P_Y(y)} \text{ for } i \neq j \right\},$$

where

$$P_Y(y) = \sum_x P_U(x) W(y|x).$$

**Theorem 22** *For arbitrary* $\alpha > 1$ *we have*

$$\frac{1}{M}\mathbb{E}\left\{ \sum_{i=1}^{M} \sum_{y \in D_i^c} W(y|U_i) \right\} \leq \frac{1}{\alpha} + P_{UY}\{(x, y) : i(Y_i, y) \leq \log \alpha M\} \tag{1.4.21}$$

(The restriction $\alpha > 1$ only avoids the trivial case in (1.4.21).)

*Proof* By symmetry the LHS in (1.4.21) equals

$$\mathbb{E} \sum_{y \in D_i^c} W(y|U_i), \tag{1.4.22}$$

which it suffices to consider.

Suppose that $U_1 = x$. In order that $y \in D_1$ we must have

$$i(x, y) > i(U_j, y), \quad j \neq 1. \tag{1.4.23}$$

The probability of this relation (for fixed $x$ and $y$) is the $(M - 1)$th power of

$$\Pr(i(x, y) > i(U_2, y), \tag{1.4.24}$$

which we abbreviate as $1-\beta(x, y)$ with $\beta(x, y) = P_Y(\{y \in \mathcal{Y} : i(x, y) \leq i(U_2, y)\})$. Consequently

$$\mathbb{E} \sum_{y \in D_i^c} W(y|U_1) = \sum_{x,y} P_{UY}(x, y)(1 - \beta(x, y))^{M-1}.$$

Define now $A = \{(x, y) : i(x, y) > \log \alpha M\}$, then, since $(1-\gamma)^{M-1} > 1 - M\gamma$ for $0 < \gamma < 1$, we have

$$\mathbb{E} \sum_{y \in D_1^c} W(y|U_1) \leq P_{UY}(A^c) + M \sum_{(x,y) \in A} P_{UY}(x, y)\beta(x, y). \tag{1.4.25}$$

We now bound $\beta(x, y)$ for $(x, y) \in A$. If

$$i(x', y) \geq i(x, y) > \log \alpha M, \tag{1.4.26}$$

then

$$\frac{P_{UY}(x', y)}{P_Y(y)} > \alpha M P_U(x'). \tag{1.4.27}$$

Now we sum both sides in (1.4.27) over all $x' \in \mathcal{X}$ satisfying (1.4.27), then the LHS sum is at most 1 and the RHS sum is $\alpha M \beta(x, y)$. Hence

$$\beta(x, y) < \frac{1}{\alpha M} \quad \text{for } (x, y) \in A \tag{1.4.28}$$

and the desired result (1.4.21) follows from (1.4.25) and (1.4.28).

**Corollary 3** *If $\alpha$ and $M$ are such that the RHS in (1.4.21) is bounded by $\frac{\lambda}{2} < \frac{1}{2}$, then there exists an $\left(\lfloor \frac{M}{2} \rfloor, \lambda\right)$-code $\left\{(u_{ij}, D_{ij}) : 1 \leq j \leq \frac{M}{2}\right\}$ with $W(D_{ij}|u_{ij}) \geq 1 - \lambda$, $1 \leq j \leq \frac{M}{2}$.*

(Use the pigeon-hole principle.)
The application to, for instance a DMC, proceeds as the application of the maximal code theorem.

*Remark* Actually the proof only used that the $U_i$'s are pairwise independent or even only uncorrelated. This plays a role in choosing linear codes at random.

### 1.4.5 The New Channel ΛC

Our systematic algebraic modeling of channels leads to another seemingly natural channel, the general ΛC.

We can drop the assumptions $\mathcal{X}(s) = \mathcal{X}$, $\mathcal{Y}(s) = \mathcal{Y}$ $(s \in \mathcal{S})$. Keeping the sets distinct the receiver knows $s$ upon receiving a $y^n \in \mathcal{Y}(s)$! However, the sender not knowing $s$ can always only use letters in $\bigcap_{s \in \mathcal{S}} \mathcal{X}(s)$ (giving nothing new, if the intersection equals $\mathcal{X} = \mathcal{X}(s)$ $(s \in \mathcal{S})$). But he can do more (even if the intersection is empty)! He can, for instance, encode message $i$ by a string of blocks $u_1^{n_1}(1)u_i^{n_2}(2), \ldots, u_i^{n_2}(s)$, where $\sum_{s=1}^{S} n_s = n$. Moreover, he ca use the incidence structure of the $(\mathcal{X}(s))_{s \in \mathcal{S}}$ more efficiently by defining sets

$$\mathcal{X}(E) = \bigcap_{s \in E} \mathcal{X}(s), \quad E \subset \mathcal{S},$$

and now encode message $i$ more generally by a string of blocks

$$\left[u_i^{n(E)}(\mathcal{X}(E))\right]_{E\subset\mathcal{S}}, \quad \sum_{E\subset\mathcal{S}} n(E) = n.$$

*Assuming that* $\mathcal{S}$ *is finite*, we can, by using products of finitely many factors of codes known to exist, show that the following rates are achievable:

For $P \in \mathcal{P}(\mathcal{X}(E))$ set

$$R(P, (W(\cdot|\cdot|s))_{s\in E}) = \inf_{s\in E} I(P, W(\cdot|\cdot|s), \tag{1.4.29}$$

$$C(E) = \max_{P\in\mathcal{P}(\mathcal{X}(E))} R(P, (W(\cdot|\cdot|s)_{s\in E}) \tag{1.4.30}$$

and

$$R(s,\nu) = \sum_{E\subset\mathcal{S}} \nu_E C(E) 1_E(s), \quad \sum_{E\subset\mathcal{S}} \nu_E = 1.$$

Finally, define the optimal rate of transmission

$$\max_{\nu} \min_{s\in\mathcal{S}} R(s,\nu).$$

**Theorem 23** (Ahlswede, Oct. 26, 2008) *The (strong) capacity of the general* $\Lambda C$ *equals*

$$C(\Lambda) = \max_{\nu\in\mathcal{P}(2^{\mathcal{S}})} \min_{s\in\mathcal{S}} \sum_{E\subset\mathcal{S}} \nu_E 1_E(s) \max_{P\in\mathcal{P}(\mathcal{X}(E))} \min_{s\in E} I(P, W(\cdot|\cdot|s)).$$

**Corollary 4** *For* $\mathcal{X}(s) = \mathcal{X}$ $(s\in\mathcal{S})$, $\mathcal{X}(\mathcal{S}) = \mathcal{X}$, $\nu(\mathcal{S}) = 1$ *gives the capacity*

$$\begin{aligned} C(\Lambda) &= \min_{s\in\mathcal{S}} 1_{\mathcal{S}}(s) \max_{P\in\mathcal{P}(\mathcal{X})} \min_{s\in\mathcal{S}} I(P, W(\cdot|\cdot|s) \\ &= \max_{P\in\mathcal{P}(\mathcal{X})} \min_{s\in\mathcal{S}} I(P, W(\cdot|\cdot|s)) \end{aligned}$$

**Research Problem** Extend the capacity formula of the AVC, where the sender knows the state sequence $s^n$ to the case of different alphabets $\mathcal{X}(s)$!

### 1.4.5.1 Combination of AC, PC, and $\Lambda$C

Haemers result shows that for $W(\cdot|\cdot|1)W(\cdot|\cdot|2)$ the capacity $C$ can exceed $C_{01}+C_{02}$. If it holds for zero-error capacity, then also for AVC with maximal error probability. Now notice that already CC *are not additive* because $\inf a_s + \inf b_s$ can be smaller than $\inf(a_s + b_s)$.

$$\begin{aligned}&\max_P \inf_s I(P, W(\cdot|\cdot|s) + \max_Q \inf(Q, V(\cdot|\cdot|s))\\ &\leq \max_{P\times Q} \inf_s (I(P, W(\cdot|\cdot|s) + H(Q, V(\cdot|\cdot|s))\end{aligned}$$

**Research Problem** Extend the Körner zero-error compound capacity result to channel $\Lambda$C.

*Remark* The $\Lambda$ channel can also be practically important. Since we do not know the channel in use (indexed by $s$), we do not know its alphabet $\mathcal{X}(s)$. By an appropriate time sharing for all of them the desired messages get transmitted. Here if we send a letter $x$ and $x \notin \mathcal{X}(s)$ at this time instant the channel $W(\cdot|\cdot|s)$ is just blocked, but the sender does not notice it (it does not accept the letter). A realistic situation arises if we want to speak to a person speaking *only one of the languages* English, Chinese, or Spanish, and we do not know which one. So by time-sharing we can get sentences across to this person! The $\Lambda$ channel is conceivable as also relevant for universal translating machines.

## *1.4.6 The ΛC with Partial Knowledge of the Sender*

A basic tool for determining $C(\overline{\lambda})$ for CC is based on the model with partial side information

$$K = \{(S_1, \dots, S_k) : S_i \subset S, i = 1, \dots, k\},$$

which means that the sender knows that the governing channel has an index $i$ with a set $S_i$ specifying the actual possible status of the CC. We speak about $\mathcal{W}(K)$ to indicate that the CC $\mathcal{W}$ is supplied with partial knowledge $K$. Its capacity $C(\mathcal{W}K)$ equals

$$\inf_{i=1,\dots,k} \max_{\pi} \inf_{s\in S_i} R(\pi, s).$$

Now we can also supply the $\Lambda$C with $K$ and denote its capacity by $C(\Lambda K)$. It can be found as a combination of the formulas for $C(\Lambda)$ and $C(\mathcal{W}K)$.

Replacing $\mathcal{S}$ in Theorem 23 by $\mathcal{S}_i$ we denote the capacity by $C(\Lambda, \mathcal{S}_i)$ instead of $C(\Lambda)$. We then have

**Theorem 24** (Ahlswede, 2008)

$$C(\mathcal{W}K) = \inf_{i=1,\dots,k} C(\Lambda, \mathcal{S}_i)$$

# 1.5 Lecture on Discrete Finite State Channels

## *1.5.1 Definitions and Examples*

The discrete channels with memory treated in Chap.3 are defined by a function $P : \Omega \times \mathcal{B}'$ with the properties

(i) $P(w, E')$ is a PD on $\mathcal{B}'$ $(w \in \Omega)$
(ii) $P(w, E')$ is a $\mathcal{B}$-measurable function $(E' \in \mathcal{B}')$.

### 1.5.1.1 The Finite State Channel (FSC)

It is convenient to call the pair $(-w_{-1}, w_o, w_1, w_2, \dots, w_t; w'_{-1}, w'_0, w'_1, w'_2, \dots, w'_t)$ of input and output sequence up to a given time $t$ the state $s_t$ of the channel at time $t$. If the channel is *non-anticipatory*, then the transmission of the next letter is specified by probabilities conditional on state $s_t$ and the current input letter $w_{t+1}$

$$P(w'_{t+1}, s_{t+1}|w_{t+1}, s_t). \tag{1.5.1}$$

For physical channels with memory it is often more desirable to introduce meaningful parameters (such as the fading level of a fading transmission path) as state and specify the transmission probability

$$P(w'_{t+1}|w_{t+1}, s_t),$$

but the new state $s_{t+1}$ is determined possibly in other ways. It is important here that it is also natural to work with a finite set $\mathcal{S}$ of states, which fortunately also makes the mathematics more accessible, especially if the new conditional probability assignment is

$$P(y_t, s_t|x_t, s_{t-1}) \quad (x_t \in \mathcal{X}, y_t \in \mathcal{Y}, s_{t-1}, s_t \in \mathcal{S}). \tag{1.5.2}$$

We then speak about a finite state channel (DFSC). It is *stationary* and relates infinite sequences $(\dots, x_{-1}, x_0, x_1, \dots)$, $(\dots, y_{-1}, y_0, y_1, \dots)$, and $(\dots, s_{-1}, s_0, s_1, \dots)$.

### 1.5.1.2 Examples

$$P(y_t, s_t|x_t, s_{t-1}) = P(y_t|x_t, s_{t-1})Q(s_t|x_t, s_{t-1}) \tag{1.5.3}$$

Memory occurs in practice often in the form of *fading*, where increasing rate results in more noise and forces to slow down (see Gallager [26], pp. 431–441) and or as

intersymbol interference (Gallager [26], p. 408, Open problem) where output letters depend on more tha one letter sent caused by filtering.
Channel model with 7 boxes (cf. Gallager, p. 72)

#### 1.5.1.3 Fading (Bursty) Channel

$$(W(s'|s))_{s',s\in\mathcal{S}} = \begin{pmatrix} \alpha & 1-\alpha \\ 1-\beta & \beta \end{pmatrix}$$

$q$ is independent of $x$ and $\alpha$ and $\beta$ are close to but smaller than 1.

The channel tends to stay in his state and

$$W(\cdot|\cdot|0) = \begin{pmatrix} 1-\varepsilon & \varepsilon \\ \varepsilon & 1-\varepsilon \end{pmatrix}, \quad \varepsilon \text{ small (good)}$$

$$W(\cdot|\cdot|1) = \begin{pmatrix} 1-\delta & \delta \\ \delta & 1-\delta \end{pmatrix}, \quad \varepsilon < \delta < \frac{1}{2} \text{ (bad)}$$

#### 1.5.1.4 Intersymbol Interference

Here the changing of the state depends on input $x$ deterministically in the following way:

$$q(s'|x, s) = 1, \quad \text{if } s' = x \tag{1.5.4}$$

(independent of $s$).

The channel state changes with the input letter and

$$W(\cdot|\cdot|0) = \begin{pmatrix} \alpha & 1-\alpha \\ 1-\beta & \beta \end{pmatrix}, \quad W(\cdot|\cdot|1) = \begin{pmatrix} 1-\beta & \beta \\ \alpha & 1-\alpha \end{pmatrix}.$$

So the state entered at any instant $t$ is the same as the input at that instant. Therefore $P(y_t|x_t, s_{t-1}) = P(y_t|x_t, x_{t-1})$ with dependency on the current and previous input.

The probability for $y_t \neq x_t$ is larger when $x_t \neq x_{t-1}$ than when $x_t = x_{t-1}$. Moreover, when $x_{t-1}$ is chosen with PD $Q$, then we get $P(y_t|x_t) = P(0)P(y_t|x_t, 0) + Q(1)P(y_t|x_t, 1)$ and notice that $P(y_t|x_t)$ depends on the channel input PD and does not depend on the channel matrix alone. This is true for all DFSC, where the state sequence depends statistically on the input sequence.

The fading channel above, where $q(s_t|x_t, s_{t-1})$ is independent of $x_t$, is also called DFSC without *intersymbol interference*. Likewise the channel defined in (1.5.4), where $q(s_t|x_t, s_{t-1})$ takes on only the values 1 and 0 and depends only on $x_t$ is called DFSC with only intersymbol interference memory. In the first case the memory is due to noise alone, and in the second case, to previous inputs alone. The general DFSC involves both effects.

### 1.5.2 *Two Performance Criteria: Lower and Upper Maximal Information Rates for the FSC*

The concepts in this title were used by Gallager in his work on DFSC [26]. We give a concise presentation and advise the reader to consult [26], Chap. 5 and [G64] for some more complex proofs.

We draw attention to the statement in lines 6,7 on p. 100 of [26]: "The capacity of a DFSC is a quantity that can be reasonably defined in several different ways."

**Definition 4** The lower capacity $\underline{C}$ of a DFSC is defined as

$$\underline{C} = \lim_{n\to\infty} \underline{C}_n, \tag{1.5.5}$$

where

$$\underline{C}_n = \frac{1}{n} \max_{P^n \in \mathcal{P}(\mathcal{X}^n)} \min_{s_0 \in \mathcal{S}} I(X^n \wedge Y^n | s_0) \tag{1.5.6}$$

$$I(X^n \wedge Y^n | s_0) = \sum_{x^n} \sum_{y^n} P^n(x^n) P^n(y^n | x^n, s_0) \log \frac{P^n(y^n | x^n, s_0)}{\sum_{x'^n} P(y^n | x'^n, s_0)} \tag{1.5.7}$$

and (instructively)

$$P^n(y^n, s_n | x^n, s_0) = \sum_{s_{n-1}} P(y^n, s_n | x_n, s_{n-1}) P_{n-1}(y^{n-1}, s^{n-1} | x^{n-1}, s_0), \tag{1.5.8}$$

$$P^n(y^n | x^n, s_0) = \sum_{s_n} P^n(y^n, s_n | x^n, s_0) \tag{1.5.9}$$

Similarly, the upper capacity $\overline{C}$ of a DFSC is defined as

$$\overline{C} = \lim_{n\to\infty} \overline{C}_n, \tag{1.5.10}$$

where

$$\overline{C}_n = \frac{1}{n} \max_{P^n \in \mathcal{P}(\mathcal{X}^n)} \max_{s_0 \in \mathcal{S}} I(X^n \wedge Y^n | s_0). \tag{1.5.11}$$

**Theorem 25** (Gallager [26], Yudkin [47]) *For the* GFSC *with* $|\mathcal{S}| = S$

$$\lim_{n\to\infty} \underline{C}_n = \sup_n \left[ \underline{C}_n - \frac{\log S}{n} \right]$$

$$\lim_{n\to\infty} \overline{C}_n = \inf_n \left[ \overline{C}_n - \frac{\log S}{n} \right]$$

*Proof* Appendix 4 A of [25].

Obviously from the definitions $\underline{C}_n \leq \overline{C}_n$ $(n \in \mathbb{N})$ and the theorem implies the

**Corollary 5**

$$-\frac{\log S}{n} + \underline{C}_n \leq \underline{C} \leq \overline{C} \leq \overline{C}_n + \frac{\log S}{n} \quad (n \in \mathbb{N})$$

**Research Problems**

1. This relationship is of some use to calculate $\underline{C}$ and $\overline{C}$, especially when they are equal. One can look for an $n$ how far lower and upper bounds are apart. Unfortunately there is no estimate on the speed, so one does not know a priori how large an $n$ to choose. This should be investigated.
2. In Cover/Gopinath [20] we stated the problem to get bounds on the speed of approximation of capacity regions for multi-way channels via non-single letter characterizations. Can the present weaker type of solution be established?

## 1.5.3 *Two Further Instructive Examples of Channels*

*Example* A channel with "ambiguous capacity"
Let

$$W(\cdot|\cdot|0) = \begin{pmatrix} 1 & 0 & 0 & 0 \\ 0 & 1 & 0 & 0 \\ 0 & 0 & 1 & 0 \\ 0 & 0 & 1 & 0 \end{pmatrix}, \quad W(\cdot|\cdot|1) = \begin{pmatrix} 1 & 0 & 0 & 0 \\ 1 & 0 & 0 & 0 \\ 0 & 0 & 1 & 0 \\ 0 & 0 & 0 & 1 \end{pmatrix}.$$

The two states change periodically

$$q(1|0) = q(0|1) = 1.$$

$\overline{C}$ corresponds to the case where the beginning $s_0$ (or the phase) of the periodic channel is known. Using alternatively the PDs $\left(\frac{1}{3}, \frac{1}{3}, p_3, p_4\right)$ and $\left(p_1, p_2, \frac{1}{3}, \frac{1}{3}\right)$ (or looking also as the product of channels) we get $\overline{C} = \log 3$.

$\underline{C}$ concerns the case where the beginning state $s_0$ (the phase) is unknown. We get a CC with two periodic channels of capacity $\underline{C} = \frac{3}{2}$.
(We are just dealing with two different channels!)

*Example* $q(s|s) = 1$ for $s \in \mathcal{S} = \{0, 1\}$
No matter how $W(\cdot|\cdot|0)$ and $W(\cdot|\cdot|1)$ are defined, here $\overline{C} = \max_{s=1,2} C(s)$ and $\underline{C}$ equals the compound capacity.

*Example* Panic-button channel

$$q(1|x,1)=1 \qquad (x\in\{0,1,2\})$$
$$q(0|0,0)=q(0|1,0)=1$$
$$q(1|2,0)=1$$
$$W(\cdot|\cdot,0)=\begin{matrix}0\\1\\2\end{matrix}\begin{pmatrix}1&0&0\\0&1&0\\1&0&0\end{pmatrix},\quad W(\cdot|\cdot,1)=\begin{matrix}0\\1\\2\end{matrix}\begin{pmatrix}1&0&0\\1&0&0\\1&0&0\end{pmatrix}.$$

Input letter 2 is the "panic-button" and its use incapacitates the channel for all future times. Clearly now $\underline{C}=0$. On the other hand, avoiding letter 2, we get $\overline{C}=1$ bit. This and to some extent also the other examples are pathological in the sense that the effect of the starting is strong and remains strong as time goes on. We are looking for some kind of "ergodic behavior," where there is a more regular behavior in the sense of a disappearing initial effect (indecomposable channels). Before going into that we derive two kinds of converses for information transmission theorems for FSC's, one for $\overline{C}$ and one for $\underline{C}$.

## *1.5.4 Independently Selected States*

For some parts of the following three sections, especially 1.5.4, familiarity with nonstationary DMS is required.

We are given a finite set of stochastic $a\times b$-matrices $\mathcal{W}=\{w(\cdot|\cdot|s):s\in\mathcal{S}\}$ and we call the elements of $S$ "states." The understanding is that if the transmission system is in state $S$ the transmission is governed by $w(\cdot|\cdot|s)$. For the channels studied earlier the states may change in a rather arbitrarily manner. Now it is assumed—and only then we talk about a finite-state channel—that the states are selected according to some random mechanism (described by probabilistic laws), which may depend on letters sent and received and on previous states.

We study now channels for which the transmission matrices (or states), which govern the transmission of letters, are selected at any time instant according to some PD. Selections at different time instants are independent of each other. (They are also independent of previously transmitted and received letters.) Different communication situations arise and they depend on the knowledge which the communicators have about the states.

### 1.5.4.1 No Side Information ($\mathbf{S}^-,\mathbf{R}^-$)

If neither sender nor receiver knows the matrix which governs the transmission of any letter, then the situation is obviously as if we were given a DMC with matrix

$$\overline{w}(y|x) = \sum_{s\in\mathcal{S}} q(s)w(y|x|s),\ x \in \mathcal{X}, y \in \mathcal{Y}. \tag{1.5.12}$$

and no new problem arises.

#### 1.5.4.2 Side Information ($\mathbf{S^-}$, $\mathbf{R^+}$)

If the receiver only knows the sequence of states (before he decodes) one can view the situation like this:

If the letter $x$ is sent, the receiver receives a letter $y \in \mathcal{Y}$ and also a state $s \in \mathcal{S}$, thus he receives a "letter" $z = (y, s) \in \mathcal{Z} = \mathcal{Y} \times \mathcal{S}$.

The transmission matrix $w^*$ is given by

$$w^*(z|x) = q(s)w(y|x|s),\ x \in \mathcal{X}, z \in \mathcal{Z}, z = (y, s). \tag{1.5.13}$$

Thus again the problem reduces to the study of a DMC The capacity is readily computed and equals

$$\max_p R\big(p, w^*(\cdot|\cdot)\big).$$

#### 1.5.4.3 Side Information ($\mathbf{S^+}$, $\mathbf{R^+}$) and ($\mathbf{S_+}$, $\mathbf{R^+}$)

"$S^+$" stands for the case where the sender knows the sequence of states, before he sends a word and "$S_+$" indicates that the sender knows the state and all previous states before he sends a letter. We explain now why in both cases the capacity has to equal

$$\sum_s q(s) \max_p R\big(p, w(\cdot| \cdot |s)\big) = \sum_s q(s)C_s.$$

In case $(S^+, R^+)$ both know the sequence $s^n$ of states in advance. With high probability $s, s \in \mathcal{S}$, occurs $nq(s) + k(q)(\sqrt{n})$ times. For any $s^n$ such a structure $P(\cdot|\cdot|s^n)$ defines a nonstationary DMC for which the maximal $\lambda$-codes have a length

$$\begin{aligned}\log N(n, \lambda|s^n) &= \sum_s C_s\big(nq(s) + 0(\sqrt{n})\big)\\ &= \sum_s q(s)C_s \cdot n + 0(\sqrt{n}).\end{aligned}$$

In the case $(S_+, R^+)$ one can proceed as follows:

Choose $K(q)$ such that with high probability (tending to ! as $n \to \infty$) $s, s \in \mathcal{S}$, occurs at least $nq(s) - K(q)\sqrt{n}$ times as component in $s^n$.

Thus, find a code $n', N, \lambda)$ of maximal length, where

$$n' = \sum_s n\big(q(s)\big) - K(q)\sqrt{n} = n - K(q)\sqrt{n},$$

for the channel with first $nq(1) - K(q)\sqrt{n}$ $w(\cdot|\cdot|1)$'s, second $nq(2) - K(q)\sqrt{n}$ $w(\cdot|\cdot|2)$'s etc. Let the code be $\{(u_i, D_i)|i = 1, \dots, N(n', \lambda)\}$ and $s^{*n'}$ the sequence of states described. Clearly, $\log N(n', \lambda) = \sum_s q(s)C_S n + 0(\sqrt{n})$.

Finally, suppose message $i$ is to be encoded, we describe the encoding inductively. After $t$ letters have been sent, the sender sends at time $t+1$, knowing that $s$ is the state and that $s$ has now occurred into $K$ times, the letter $u_i^j$ if $s_j^* = s$ and $\mathcal{S}^{*j}$ contains exactly $K$ components with value $s$.

If $K > nq(s) - K(q)\sqrt{n}$ the sender may do anything and the receiver knowing the sequence of states will disregard the letters received in those components.

#### 1.5.4.4 Side Information ($S_+$, $R^-$) and ($S^+$, $R^-$)

In the previous subsection we learned that the senders "partial" knowledge of the states is essentially as good as a "full" knowledge provided that the receiver knows the sequence of states. The same phenomenon occurs if the receiver does not have any knowledge about the sequence of states. This later observation seems to be new, is not hard to prove, but does not seem to be quite as obvious as the analog result of Shannon considered all the cases treated here except for the case $(S^+, R^-)$. His solution in case $(S_+, R^-)$ is based on a pretty idea which we now repeat.

Since the sender knows $s$ when sending letter we can make the letter $x$ *depend on* $s$ so that choosing a letter really means to choose a vector $\big(x(s)\big)_{s\in\mathcal{S}}$. Then the input alphabet becomes $\mathcal{X}^* = \prod_1^{|\mathcal{S}|} \mathcal{X}$ and we can define a DMC with transmission matrix

$$w^*(y|x^*) = \sum_{s\in\mathcal{S}} q(s) w\big(y|x(s)|s\big).$$

Everything one can do with respect to coding with this channel one can certainly do with the original channel. Therefore, its capacity is greater than, or equal to,

$$C^* = \max_p R(p, w^*).$$

In defining the DMC we have ignored the option to choose letters at time $t$ depending on $s^t$ rather than just on $s_t$.

We show now that this does not decrease the capacity (even though $N(n, \lambda)$ is effected).

**Theorem 26** *The capacities of the finite-state channels with independently selected states and side information* $(S^+, R^-)$ *or* $(S_+, R^-)$ *equals*

$$C^* = \max_p R(p, w^*).$$

*Strong converse in the case* $(S^+, R^+)$.

Set $V_t = \log \frac{\sum_{s \in S} q(s) w\left(y_t | f_{tm}(Z^{(t)}, s)|s\right)}{q^*(y_t)}$, where $Z^{(t)} = Z_1, \dots, Z_{t-1}, Z_{t+1}, \dots, Z_n$, $q^*$ is defined as above and again

$$E(V_t|Y_1, \dots, Y_{t-1}, Z^{(t)}) \leq C^* = E(V_t|Y_1, \dots, Y_{t-1}, Z^n)$$

because $V_t$ depends only on $Y_t$ and $Z^{(t)}$. Critical $\|E(V_t|Y_1, \dots, Y_{t-1}, Z^n) \leq C^*$.

Since

$$E(V_t|Z^{(t)}) \leq C^*.$$

*Proof* After our previous remarks it suffices to prove the weak converse in case $(S^+, R^+)$. We describe first the encoding in this situation. Suppose there is given a finite set of messages $\mathbb{N} = \{1, \dots, N\}$ one of which will be presented to the sender for transmission. Message $m \in \mathbb{N}$ is encoded by an encoding (vector valued) function $Z^t$

$$f_n(m) = \left[f_{1m}(Z^n), f_{2m}(Z^n), \dots, f_{nm}(Z^n)\right]$$

where $\Pr(Z^n = s^n) = \prod_{t=1}^n (s_t)$, $s^n \in \mathcal{S}^n$. The encoding is as usual a partition of $\mathcal{Y}^n$ into $N$ sets $D_1, \dots, D_N$.

Suppose an $(n, N, \lambda)$-code is given to us. We shall again apply Fano's Lemma. Let $\mathcal{U}$ be a random variable with values in the set of messages $\mathbb{N} = \{1, \dots, N\}$ and $\Pr(\mathcal{U} = u) = \frac{1}{N}$, $m \in \mathbb{N}$. It is assumed that $\mathcal{U}$ is independent of $Z^n = (Z_1, \dots, Z_n)$.

Let $Y^n = (y_1, \dots, y_n)$ be the received sequence, when the message is selected by $U$ and, encoded as above and transmitted over the channel. The random variables $Y_t, t = 1, \dots, n$, for fixed $U$ and $Z^n$, distributed in dependently of each other with the following distributions:

$$\Pr\{Y_t = y_t | U = m, Z^n = s^n\} = w\left(y_t | f_{tm}(Z^n)|s_t\right).$$

We have to prove that

$$H(U) - H(U|Y^n) \leq nC^*. \tag{1.5.14}$$

From this and Fano's Lemma the result follows immediately.

We prove this by choosing that $H(U|Y^{t-1}) - H(U|V^t) \leq C^*$ for $t = 1, \dots, n$. (Here, $V^o$ is a constant and therefore $H(U|V^o) = H(U)$).

*The strong converse in case* $(S_+, R^+)$.

Set $V_t = \log\left(\sum_{s \in \mathcal{S}} \frac{(Y_t | f_{tm}(Z_2^{t-1})|=s)}{q^*(Y_t)}\right)$, where $q^*$ is the maximum output distribution for $w^*$ and therefore by Shannon's Lemma

$$E(V_t|Y_1, \dots, Y_{t-1}, Z_1, \dots, Z_{t-1}) \leq C^*$$

write

$$W_t = V_t - E(V_t|Y_1, \dots, Y_{t-1}Z_t, \dots, Z_{t-1}).$$

The $E(W_t|Y_1, \dots, Y_{t-1}, Z_1, \dots, Z_{t-1}) = 0$. Since $W_{t'_{(t'<t)}}$ is a function of $Y_1, \dots,$
$Y_{t'}, Z_1, \dots, Z_{t'}$ it is also a function of $Y_1, \dots, Y_{t-1}, Z_1, \dots, Z_{t-1}$ and hence $E(W_t|W_{t'}) = 0$ for $t' < t$. Hence $E(W_t W^{t'}) = 0$ for $t \neq t'$. The proof can now be completed.

**Problem** Does the strong converse hold for $(S^+, R^+)$?
What is the capacity?

### *1.5.5 The Finite State Channel with State Calculable by both Sender and Receiver*

$\mathcal{W} = \{w(\cdot|\cdot|s) : s \in \mathcal{S}\}$, $|\mathcal{S}| < \infty$. It is assumed that the state $s_t$ is uniquely determined as a function of $s_{t-1}$ and the letter $x_{t-1}$ send at time $t-1$

$$s_t = \mathcal{C}(s_{t-1}, x_{t-1}), t = 2, 3, \dots$$

The state $s_t$ is known by both sender and receiver. Furthermore, we assume that for $s, s' \in \mathcal{S}$ there exists a sequence $x(1), \dots, x(k)$ such that, if $s$ is the initial state of the channel and one sends over it, the letters $x(1), \dots, x(k)$, the channel is in the state $s'$ after $x(k)$ has been sent.

Clearly, $ak \leq |\mathcal{X}||\mathcal{S}|$ always suffices.

Until now we have not said anything about how the initial state of the channel is to be determined. Since we are interested in behavior for large $n$ and since $n \gg |\mathcal{X}|\mathcal{S}|$, it follows from (*) that it does not matter how the initial state is determined. For simplicity we assume that the initial state is always chosen at pleasure by the sender and is known to the receiver.

For any sequence of states $s^n = (s^1, \dots, s^n)$ there is a set $\mathcal{X}^n(s^n)$ of elements of $\mathcal{X}^n$ which are comparable with $s^n$, that is, for $x^n = (x_1, \dots, x_n) \in \mathcal{X}^n(s^n)$

$$\mathcal{C}(s_t, x_t) = s_{t+1}, \quad t = 1, \dots, n-1.$$

Denote by $\mathcal{X}_t(s_t, s_{t+1})$ the elements of $\mathcal{X}_t$ such that

$$\mathcal{C}(s_t, x_t) = s_{t+1}$$

$$\mathcal{X}^n(s^n) = \prod_{t=1}^{n-1} \mathcal{X}_t(s_t, s_{t+1})\mathcal{X}_n.$$

For $x^n \in \mathcal{X}^n(s^n)$ we have

$$\mathcal{P}(y^n|x^n|s^n) = \prod_{t=1}^{n} w(y_t|x_t|s_t),$$

that is, the transmission probabilities of a nonstationary DMC

Denote by $\mathbb{N}(s^n, \lambda)$ the maximal length of a code with blocklength $n$ and error $\lambda$. Since $\mathcal{X}^n(s^n)^{x\{s^n\}} \cap \mathcal{X}^n(s'^n)^{x\{s'^n\}} = \varnothing$ for $s^n \neq s'^n$ every code for our original channel is built up out of subcodes corresponding to the different $s^n$. Therefore,

$$N(n, \lambda) = \sum_{s^n \in \mathcal{S}^n} N(s^n, \lambda).$$

Denote by $C(s', s)$ the capacity of the matrix

$$\big(w(\cdot|x|s')\big)_{x \in \mathcal{X}(s',s)},$$

By the coding theorem for the DMS follows

$\log N(s^n, \lambda) = \sum_{t=1}^{n-1} C(s_t, s_{t+1}) + 0(\sqrt{n})$ if $\mathcal{X}^n(s^n) = \varnothing$, otherwise set $N(s^n, \lambda) = 0$.

The finite-state channel with state calculable only by the sender ([45], 3.6).

### 1.5.6 Stochastically Varying Channels

Suppose we are given a finite set of $|\mathcal{X}| \times |\mathcal{Y}|$-stochastic matrices $\mathcal{C} = \big\{w(\cdot|\cdot|s) : s \in \mathcal{S}\big\}$ and a PD $g$ on $\mathcal{S}$. The elements of $\mathcal{S}$ are called states.

Assume that at any time instant the state (and therefore also the corresponding matrix) is selected by channel by the PD $g$.

It is furthermore assumed that this selection is independent of all transmitted letters, of all previously received letters and of all previous states. $g$ is known to both communicators.

Sender and receiver may have some knowledge about the states selected. In [40] the following situation was considered:

The sender, but not the receiver, knows the state which governs the transmission of each letter. He knows the state which governs the transmission of any letter before that letter is sent, but only after all letters of the word have been sent. We describe this situation briefly by $(S_*^+, R^-)$. In [45], in 4.6, a more systematical discussion of the various combinations with respect to senders and receivers knowledge is presented. Let us indicate by $S^+$ the situation in which the sender knows the whole sequence of states before any letter is sent.

The DFSC with state calculable only by the sender ([45], 3.6).

With respect to the receivers knowledge it makes no difference whether he knows the states letter by letter or in advance for the whole word to be transmitted, since he will not decode a received letter until the whole word has been received.

Thus one has the following cases:

I. $(S^-, R^-)$
II. (i) $(S^+_*, R^+)$ (ii) $(S^+, R^+)$
III. $(S^-, R^+)$
IV. (i) $(S^+_*, R^-)$ (ii) $(S^+, R^-)$

The capacities in the cases I, II (i), (ii) and III are readily determined. Shannon [40] found the capacity in case IV (i). Case (ii) has not been considered before and we show that the capacity is the same as that in IV (i).

### 1.5.6.1 Partial Side Information

In the above cases states are either exactly known or unknown. One might concave of a situation where a communicator has some knowledge about a state without knowing it exactly. Let $\mathcal{A} = \{A_1, \dots, A_a\}$ and $\mathcal{L} = \{B_1, \dots, B_b\}$ be partitions of $\mathcal{S}$. The sender (resp. receiver) is assumed to decide in which set of the partition $\mathcal{A}$ (resp. $\mathcal{L}$) a a randomly chosen $s$ is contained. Depending on the two alternatives the sender has or has not knowledge of the future we distinguish between the two cases $(\mathcal{S}^{\mathcal{A}}, R^{\mathcal{L}})$ and $(\mathcal{S}^{\mathcal{A}}_*, R^{\mathcal{L}})$.

Perfect knowledge corresponds to the partition $\mathcal{A} = \big\{\{s\} : s \in \mathcal{S}\big\}$ and ignorance to the partition $\mathcal{A} = \{\mathcal{S}\}$.

More general: stochastic knowledge.

**Problem** $(S^+, R^-)$ Shannon

$W(y|x_1, \dots, x_s) = \sum_{s=1}^{c} g(s)w(y|x_s|s)$

$UZV$, $\Pr(U = i) = \frac{1}{N}$

$Y^n = (Y_1, \dots, Y_n)$ $Y^{t-1} = (Y_1, \dots, Y_{t-1})$

$H(U) = \log N$

Claim: $H(U) = H(U|Y^n) \leq nC$

(i) $H(U) - H(U|Y^n) = \sum_{t=1}^{n} \big(H(U|Y^{t-1}) - H(U|Y^t)\big)$

Sufficient:

(ii) $H(U|Y^{t-1}) - H(U|Y^t) \leq C$ $t = 1, 2, \dots, n$

(iii) $\Pr(Y + t = y|U = i, \mathcal{S}^{t-1} = s^{t-1}) = \Pr(Y_t = y|U = i), \mathcal{S}^{t-1} = s^{t-1}, Y^{t-1} = y^{t-1})$.

$\Rightarrow$

(iv) $H(Y_t|U, \mathcal{S}^{t-1}) = H(Y_t|U, \mathcal{S}^{t-1}, Y^{t-1}) \leq H(Y_t|U, Y^{t-1})$

(v) $H(U|Y^{t-1}) = H(U|Y^{t-1}, Y_t) = H(Y_t|Y^{t-1}) - H(Y_t|U, Y^{t-1}) \leq H(Y_t) - H(Y_t|U, Y^{t-1}) \leq H(Y_t) - H(Y_t|U, \mathcal{S}^{t-1})$

Therefore:
$H(U|Y^{t-1}) - H(U|Y^t) \leq H(Y_t) - H(Y_t|U, \mathcal{S}^{t-1}) \leq C$ (!).
$W_1, W_2$-**Lemma.** (new)

$$\begin{aligned} H(U) - H(U|Y^2) &= H(U) - H(U|Y_1) + H(U|Y_1) - H(U|Y^2) \\ &= H(Y_1) - H(Y_1|U) + H(Y_2|Y_1) - H(Y_2|U, Y_1) \\ &\leq H(Y_1) - H(Y_1|UmY_2) + H(H(Y_2) - H(Y_2|U, Y_1) \\ &\leq H(Y_2) - H(Y_2|U, \mathcal{S}_1) \leq C. \end{aligned}$$

### *1.5.7 Random Codes and Weakly Varying Channels*

In the sequel we introduce some new code concepts, which are meaningful also if there are no assumptions on the time structure of the channels. Let $\mathcal{X}$ be the input and $\mathcal{Y}$ the output alphabet and denote by $W(\mathcal{X}, \mathcal{Y})$ the set of all channels with those alphabets. A channel $w \in W(\mathcal{X}, \mathcal{Y})$ can be used for communication from one person, the sender, to another person, the receiver. There is given in advance a finite set of messages $\mathbb{N} = \{1, \dots, N\}$, one of which will be presented to the sender for transmission. The sender may use a "randomized encoding" and the receiver may use a "randomized decoding." More precisely, the sender encodes the messages by an encoding channel $E \in W(\mathbb{N}, \mathcal{X})$ with $E(v, x)$ being the probability that input $x$ is given to channel $w$ when message $v$ is presented to the sender for transmission. When the receiver observes the output $y$ of the transmission channel $w$, he decodes it by a decoding channel $D \in \mathcal{W}(\mathcal{Y}, \mathbb{N})$ with $D(y, \mu)$ being the probability that the receiver will decide that message $\mu$ is intended.

The matrix $e = e(E, D, w) = E \cdot w \cdot D \in \mathcal{W}(\mathbb{N}, \mathbb{N})$ is the error matrix of code $(E, D)$ for channel $w$. Its channel $e(v, \mu)$ gives the probability that, when $v$ is presented to the sender the receiver will decide that message $\mu$ is intrude, when code $(E, D)$ is used on channel $w$. The *average* error probability over all messages in the set $\mathbb{N}$ is therefore

$$\overline{\lambda}_1(E, D, w) = 1 - \frac{1}{N} \text{ trace } e(E, D, w).$$

One can also define the maximal error

$$\lambda_1(E, D, w) = \max_{v=1,\dots,N} \big(1 - e(v, v)\big).$$

Of course, the maximal error for a code $(E, D)$ is greater than, or equal to, the average error. If we restrict the receiver to using nonrandomized decoding only, then $D(y, \mu)$ has only 0 and 1 as elements. Similarly, if the sender uses only nonrandomized encoding, then $E(v, x)$ has only 0 and 1 as elements.

Further specialization leads to the definition: A code $(E, D)$ is pure if only 0 and 1 occurs as elements of $E, D$. Pure codes are exactly the codes familiar to us: $\{(u_i, D_i) | i = 1, \dots, N\}$ etc.

Let us denote the set of all pure codes of length $\mathbb{N}$ by $\mathcal{P}(\mathbb{N}, \mathcal{X}, \mathcal{Y})$. A probability distribution $r$ over $\mathcal{P}(\mathbb{N}, \mathcal{X}, \mathcal{Y})$ is a *correlated random code*, "correlated," because both sender and receiver have to know the outcome of the *same* random experiment in order to use the code.

We show now that every $\mathbb{N}, \mathcal{X}, \mathcal{Y})$ code $(E, D)$ is equivalent to some correlated random $(\mathbb{N}, \mathcal{X}, \mathcal{Y})$ code $r$ in the sense that

$$\overline{\lambda}_1(w, r) = \overline{\lambda}_1(w, E, D) \quad \text{for } all\ w \in \mathcal{W}(\mathcal{X}, \mathcal{Y}).$$

(The converse is not true.)

Namely, consider all matrices

$$\mathcal{E}^0 = \{E^0 : E^0(v, x) = \text{ stoch. } 0 - 1 \text{ matrix and } E^0(v, x) = 0 \text{ if } E(v, x) = 0\}.$$

Define $r(E^0) = \prod_{(v,x):E^0(v,x)=1} E(v, x)$.

Similarly, define

$$\mathcal{D}^0 = \{D^0 : D^0 \text{ stoch. } 0 - 1 \text{ matrix and } D^0(y, v) = 0 \text{ if } D(y, v) = 0\}$$

and $\nu''(D^0) = \prod_{(y,v):D^0(y,v)=1} D(y, v)$.

Finally, define the set of pure codes $\mathcal{E}^0 \times \mathcal{D}^0$ and set

$$\nu(E^0, D^0) = \nu'(E^0) \cdot \nu''(D^0) \quad \text{for all } (E^0, D^0) \in \mathcal{E}^0 \times \mathcal{D}^0.$$

We call a general $(E, D)$-code a code of type $K_1$, the special codes of type $K_1$ which use only nonrandomized decoding one of type $K_2$ and those which use only nonrandomized encoding one of type $K_3$.

A pure code will be called of type $K_4$ and a correlated random code one of type $K_5$. For each type $K_i (i = 1, \dots, 5)$ let $\lambda_i$ denote the maximal error and $\overline{\lambda}_i$ the average error.

A coding problem is completely described when we state which code type and which error the communicators are allowed to use. The combinations $(K_i, \overline{\lambda}_i)$, $(K_i, \lambda_i)$, $i = 1, \dots, 5$ are all possible, but not every possible combination corresponds to a problem of practical interest. The variables of basic interest to us are $\mathbb{N}_i(n, \lambda)\big(\text{resp. } \overline{\mathbb{N}}_i(n, \lambda)\big) =$ maximal cardinality of a set of messages $\mathbb{N}$ for which we can find a $K_i$-code with maximal error not greater than $\lambda$ (resp. with average error not greater than $\lambda$).

For a channel described by one matrix $w \in \mathcal{W}(\mathcal{X}, \mathcal{Y})$ there is no advantage in any kind of randomization:

**Lemma 16** *For* $w \in \mathcal{W}(\mathcal{X}, \mathcal{Y})$:

$$\overline{\mathbb{N}}_i(\lambda) = \overline{N}_4(\lambda) \quad \textit{for } i = 1, 2, 3, 4, 5.$$

*Proof* It suffices to show that $\overline{N}_5(\lambda) = \overline{N}_4(\lambda)$. This is true simply because in the ensemble of pure codes over which we average with $\nu$ there is at least one pure code with an average error not exceeding the average error of the random code.

For maximal errors our standard method yields for instance

$$N_5(\lambda) \le \overline{N}_5(\lambda) = \overline{N}_4(\lambda) \le 2N_4(2\lambda).$$

The notions introduced become relevant only for channels defined by classes of matrices such as VC and AVC However, also in this more general context *only for the weakly varying systems* the differences become significant. This can be seen as follows. A correlated random code of length $\mathbb{N}$ is given by PD $\nu$ over $\mathcal{P}(\mathbb{N}, \mathcal{X}, \mathcal{Y})$, that is, we are given a system of pure codes $\{(u_i^{(\alpha)}, D_i^{(\alpha)})|i = 1 \cdot N3, \alpha \in \mathcal{P}(\mathbb{N}, \mathcal{X}, \mathcal{Y})$, and a.PD $\nu$.

Let now $\mathcal{W}^* \subset W(\mathcal{X}, \mathcal{Y})$ be a set of matrices and let us suppose that those transmission matrices can very *strongly*, that is, the matrix chosen may depend on the code word (the $u_i^{(\alpha)}$!) send. Then the average error is given by

$$\sum_{\alpha} \nu(\alpha) \frac{1}{N} \sum_{i=1}^{N} \min_{w \in W^*} w\big(D_i^{(\alpha)} | u_i^{(\alpha)}\big)$$

$$\ge \frac{1}{N} \sum_{i=1}^{N} \min_{w \in W^*} w\big(D_i^{(\alpha_0)^c} | u_i^{(\alpha_0)}\big), \quad \text{for a suitable } \alpha_0, \text{ and}$$

$$\ge \frac{1}{2} \max_{j=1,\dots,\frac{N}{2}} \min_{w \in W^*} w\big(D_{ij}^{(\alpha_0)^c} | u_{ij}^{(\alpha_0)}\big) \quad \text{for } \frac{N}{2} \text{ suitable picked indices.} \qquad (1.5.15)$$

Thus we have come to the *important conclusion*:

*For the more robust channel model of strongly varying transmission probabilities* (as for instance SVC and SAVC) *there is no option to choose error or code concepts*, one has to use maximal error and pure codes (or at least does not gain anything by doing otherwise).

We now turn our attention to correlated coding. There is certainly some drawback in using this concept, because the outcome of the random experiment has to be made known to both communicators. This procedure is repeated at each messages. However, "physically one can imagine choices placed at the transmitting end and at the receiving end which involve random but not necessarily independent elements. For example, they may obtain their random choices from tapes which were prepared together with certain correlations. Physically this would be a perfectly feasible process" (Shannon).

#### 1.5.7.1 WAVC Under Random Coding

The set of stochastic matrices $\mathcal{W} = \{w(\cdot|\cdot|s) : s \in \mathcal{S}\}$, where $S$ is an arbitrary index set, defines an AVC channel $\mathcal{A} = (\mathcal{W}^n)_{n\geq 1}$, where $\mathcal{W}^n = \{W(\cdot| \cdot |s^n) : s^n \in \mathcal{S}^n\}$ and the $P(\cdot| \cdot |s^n)$ are defined as usual. Denote by $\overline{\mathcal{W}}$ the (ordinary) convex hull of $\mathcal{W}$.

Define $\gamma = \max_p \min_{w\in\overline{\mathcal{W}}} R(p, w)$.
By the min-max theorem

$$\gamma = \min_{w\in\overline{\mathcal{W}}} \max_p \mathcal{R}(p, w). \tag{1.5.16}$$

We give now a very short proof [11] of a theorem due to Blackwell, Breiman, and Thomasian [18].

**Theorem 27** *For the WAVC* $\mathcal{A}$

$$\lim_{n\to\infty} \frac{1}{n} \log \overline{N}_5(n, \lambda) = \gamma, 0 < \lambda < 1,$$

*that is, coding theorem and strong converse hold for average error and correlated coding.*

*Proof* The capacity cannot be greater than $\gamma$, because a $\lambda$-code for is $\lambda$-code for $\overline{\mathcal{A}} = (\overline{\mathcal{W}}^n)_{n\geq 1}$ and therefore also for the DMC with matrix $\overline{w} : \max_p \mathcal{R}(p, \overline{w}) = \gamma$. It therefore remains only to prove the coding theorem.

Let $q^n$ be a finite PD on $S^n$. We shall prove that for any given $\lambda$, $0 < \lambda < 1$, there is an $(n, \overline{N}, \lambda)$code for $P(\cdot| \cdot |q^n) = \sum_{s^n\in\mathcal{S}^n} P(\cdot| \cdot |s^n)q^n(s^n)$, whose length $\overline{N}$ satisfies, for any $\varepsilon > 0$ and $n$ large enough independent of $q^n$,

$$N > \exp\{n(\gamma - \varepsilon)\}.$$

This is the lemma on p. 564 of [18] and constitutes most of the proof of [18].
Let $p^*$ be a PD on $\mathcal{X}$ such that

$$\gamma = \min_{w\in\overline{W}} \mathcal{R}(p^*, w).$$

Let

$$X^n = (X_1, \dots, X_n)$$

be a sequence of independent, identically distributed random variables with the common distribution $p^*$. Let

$$Y^n = (Y_1, \dots, Y_n)$$

be a sequence of random variables, with values in $\mathcal{Y}$, defined on the same sample space as $X^n$, and such that $Y^n$ can be thought of as the random sequence received when $X^n$ is sent over the channel above. Of course the conditional distribution of $Y^n$, given $Y^n = y^n$, depends on $y^n$ and $q^n$.

Define the following (conditional information density) function for $t = 1, \dots, n$:

$$\begin{aligned} &I^{(t)}(x, y)|x_1, \dots, x_{t-1}, y_1, \dots, y_{t-1}) \\ &= \log \Pr(X_t = x, Y_t = t|X^{t-1} = x^{t-1}, Y^{t-1} = y^{t-1}) \\ &\quad - \log \Pr(X_t = x|X^{t-1} = x^{t-1}, Y^{t-1} = y^{t-1}) \\ &\quad - \log \Pr(Y_t = y|X^{t-1} = x^{t-1}, Y^{t-1} = y^{t-1}) \end{aligned}$$

provided that non of the three expressions is zero, and 0 otherwise.

Write, for $t = 1, \dots, n$,

$$V_t = I^{(t)}(X_t, Y_t|X^{t-1}, Y^{t-1}) - E\{I^{(t)}(X_t, Y_t|X^{t-1}, Y^{t-1})|X^{t-1}, Y^{t-1}\}.$$

Then $EV_t = 0$, $V_t V_{t'} = 0, t \neq t'$, and $EV_t^2 < c$, a constant independent of $t$ and $n$.

For any set of values $X^{t-1}$ and $Y^{t-1}$ we have

$$E\{I^{(t)}(X_t, Y_t|X^{t-1}, Y^{t-1})|X^{t-1}, Y^{t-1}\} \geq \gamma \tag{1.5.17}$$

and

$$\frac{\sum_{t=1}^n V_t}{n} \text{ converges stochastically to } 0 \quad \text{as } n \to \infty. \tag{1.5.18}$$

For any $\varepsilon > 0$ we have,

$$\Pr\left\{\sum_{t=1}^{n} I^{(t)}(X_t, Y_t|X^{t-1}, Y^{t-1}) > n(\gamma - \varepsilon)\right\} \to 1$$

as $n \to \infty$, uniformly in $q^n$.

The described result now follows immediately from Shannon's random code Lemma.

We now complete the proof (exactly as in [18]) by a game theoretic argument. Let $\mathcal{S}_1$ be the set of all random codes with length $\mathbb{N} = \left[\exp\{n(\gamma - \varepsilon)\}\right]$ and blocklength $n$, and let $\mathcal{S}_2 = \{q^n|q^n \text{ finite PD on } S^n\}$.

Choose $\mathcal{C}_1$ as set of strategies for player I in a two-person 0-game, and let $\mathcal{C}_2$ be the set of strategies of player II. Finally, let the average error of the code $\nu$ for channel $P(\cdot|\cdot|q^n)$:

$$\overline{\lambda}(\nu, \mathcal{P}(\cdot|\cdot|q^n))$$

be the playoff to player I.

$\overline{\lambda}$ is linear in both variables, $\mathcal{C}_2$ is convex and compact (simplex of PDs) and $\mathcal{C}_2$ is convex. The minmax theorem (see [30]) imply that there exists a random code, i.e., one of type $K_5$, whose average probability of error is at most $\lambda$ for every $q^n$ and therefore for every $s^n$. This completes the proof of the coding theorem and hence of Theorem 27. □

#### 1.5.7.2 WAVC with Side Information: The Receiver Knows the States

We now prove

**Theorem 28** *The weak capacity of the channel $\mathcal{A}$ under correlated coding equals*

$$\beta = \max_{p} \inf_{s \in S} \mathcal{R}\big(p, w(\cdot | \cdot | s)\big),$$

*in case the receiver knows the sequence of stated $s^n$ before we decode.*

*("Weak" capacity means that we prove the coding theorem and weak converse, i.e., the converse only for $\overline{\lambda}_5$ sufficiently small.)*

*Remark*

1. Comparing Theorems 27 and 28 we see that the receiver knowing the states does *increase* the capacity in the present situation. For WVC this was not the case.
2. It is also interesting to observe the effect of correlated coding with average error (or of one of the two) on the capacities of various channels. For an ordinary (one state) channel there is no effect.
   For WVC the capacity does not change the use of average error alone or of randomized encoding alone already makes the $\lambda$-capacity increase. And now we learned that for WAVC, with $|\mathcal{Y}| = 2$ for instance, the capacity changes from $\max_p \min_{w \in \overline{\overline{\mathcal{W}}}} \mathcal{R}(p, w)$ to $\max_p \min_{w \in \overline{\mathcal{W}}} \mathcal{R})p, w)$.

*Proof of Theorem 28* We begin with the direct part. It is important to notice and we therefore emphasize that in the random coding method one deals with an ensemble of codes one of which is selected random. Why not take this correlated code right away and avoid the game theoretic argument completely? Indeed in the present situation the approach gives the solution immediately:

Since the receiver knows $\delta^n$ he can decode max likelihood with respect to $\mathcal{P}(\cdot | \cdot |s^n)$, which by itself is a nonstationary DMC Lemma applies, the various are uniformly bounded and

$$\sum_{t=1}^{n} R\big(p, w(\cdot | \cdot |s^t)\big) \geq n \inf_{s \in S} \mathcal{R}\big(p, w(\cdot | \cdot |s)\big).$$

Therefore the average error probability $\overline{\lambda}_5(s^n)$ can be made small for all $s^n \in S^n$.

*Remark* We shall see later that this idea works also if there is no side information and one chooses a suitable decoding.

This approach can also be used to give a direct and simple proof of the coding theorem for VC with receiver's knowledge, because one can eliminate the randomization in this case. Actually, one can proceed and eliminate the effect of the receiver's knowledge and thus get a new proof of the coding theorem for VC.

*Proof of Converse.* For any $\mu > 0$ there exists a finite subset of $S(\mu)$ of $S$ such that

$$\left| \max_{p} \inf_{s \in S(\mu)} R\big(p, w(\cdot | \cdot |s)\big) - \max_{p} \inf_{s \in S} \mathcal{R}\big(p, w(\cdot | \cdot |s)\big) \right| \le \mu.$$

We shall prove that, for $\lambda$ sufficiently small, say $< \lambda_0$, and $n$ sufficiently large, say $> n_0$, any code of the type given in the statement of the theorem must have length $N$ such that

$$N < \exp\big\{n(\beta + 2\mu)\big\}. \tag{1.5.19}$$

Set now $d = |S(\mu)|$, and choose $\lambda_0$ and $\varepsilon_0$ positive and so small that

$$d\big[d(\lambda_0 + \varepsilon_0) + \varepsilon_0\big] < 1 \tag{1.5.20}$$

Suppose a random code of length $N$ with average error $\lambda_0$ is given, then

$$\frac{1}{N} \sum_{i=1}^{N} \sum_{\rho \in R} r(\rho) P\big(\rho D_i(s^n) | \rho u_i | s^n\big) \ge 1 - \lambda_0 \quad \text{for all } s^n \in S^n. \tag{1.5.21}$$

By considering only sequences $s^n(s, \dots, s)$ [case of VC], where $s \in S(\mu)$, we obtain that

$$\inf_{\substack{s^n = (s,\dots,s) \\ s \in S(\mu)}} \frac{1}{N} \sum_{i=1}^{N} \sum_{s \in \mathcal{R}} r(\rho) \mathcal{P}\big(\rho D_i(s^n) | \rho u_i | s^n\big) \ge 1 - \lambda_0 \tag{1.5.22}$$

and therefore

$$\inf_{\substack{s^n = (s,\dots,s) \\ s \in S(\mu)}} \sum_{s \in \mathcal{R}} r(\rho) \frac{1}{N} \sum_{i=1}^{N} \mathcal{P}\big(\rho D_i(s^n) | \rho u_i | s^n\big) \ge 1 - \lambda_0. \tag{1.5.23}$$

Now apply Lemma 16.
So the random variables

$$Z_s(\rho) = 1 - \frac{1}{N} \sum_{i=1}^{N} \mathcal{P}\big(\rho D_i(s^n) | \rho u_i | s^n = (s, \dots, s), s \in S(\mu),$$

defined on the probability space $R$ which has probability measure $r$.

Clearly,

$$EZ_s \le \lambda_0.$$

Hence, from Lemma 16 we obtain that there exists an element $\rho^* \in R$ such that

$$1-\frac{1}{N}\sum_{i=1}^{N} P\big(\rho^* D_i(s^n)|\rho^* u_i|s^n = (s,\dots,s)\big) \le \mathrm{d}(\lambda_0+\varepsilon_0)\text{for } s \in S(\mu). \quad (1.5.24)$$

We now apply Lemma 16 to the simple space $\{1,\dots,N\}$ with PD $\mathcal{P}^*(i) = \frac{1}{N}$ for $i = 1,\dots,N$, and random variables

$$Z_s^*(i) = 1 - P(\rho^* D_i(s^n)|\rho^* u_i|s^n = (s,\dots,s)), s \in S(\mu).$$

Then

$$EZ_s^* \le d(\lambda_0 + \varepsilon_0), s \in S(\mu).$$

Hence

$$P^*\{Z_s^* \le d[\mathrm{d}(\lambda_0 + \varepsilon_0) + \varepsilon_0], s \in S(\mu)\} \ge \frac{\varepsilon_0}{\mathrm{d}(\lambda_0 + \varepsilon_0) + \varepsilon_0} \quad (1.5.25)$$

and, from the definition of $P^*$, the number of elements in the set $B^*$ is not less than

$$\frac{N\varepsilon_0}{\mathrm{d}(\lambda_0 + \varepsilon_0) + \varepsilon_0} \ge N\varepsilon_0 = N_1 \text{ (say)}.$$

Denote the elements of $B^*$ by $i_v$, $v = 1,\dots,N_1$. It follows from the definition of $B^*$ that

$$P\big(\rho^* D_{i_v}(s^n)|\rho^* u_{i_v}|s^n = (s,\dots,s)\big) \ge 1 - d\big[\mathrm{d}(\lambda_0 + \varepsilon_0) + \varepsilon_0\big]$$

for $v = 1,\dots,N_1$.

#### 1.5.7.3 Remarks About WAVC and Uncorrelated Randomized Encoding and Decoding

We recall the definitions of $\mathcal{P}(x)$, $x \in \mathcal{X}$, and

$$\overline{\overline{\mathcal{W}}} = \{w | w(\cdot|x) \in \mathcal{T}(x), x \in \mathcal{X}\}.$$

The difference between the row convex closure $\overline{\overline{\mathcal{W}}}$ and the usual convex closure $\overline{\mathcal{W}}$ of a set of matrices $\mathcal{W}$ lies in the fact that for each word we take a possibly different linear combination of its elements to obtain $\overline{\overline{\mathcal{W}}}$.

We study first the effect of randomization in the encoding alone. We saw in Section that it increases the $\lambda$-capacity of AWVC. Here we have

**Example** (Randomization in the encoding increases the capacity of AWAVC).
Let $|\mathcal{X}| = |\mathcal{Y}| = 3$.

$$w_1 = \begin{pmatrix} 1\,0\,0 \\ 0\,1\,0 \\ 0\,0\,1 \end{pmatrix}, w_2 = \begin{pmatrix} 0\,1\,0 \\ 0\,0\,1 \\ 1\,0\,0 \end{pmatrix}$$

$\mathcal{W} = (w_1, w_2)$.

Obviously $\mathcal{T}(x) \cap \mathcal{T}(x') \neq \varnothing$ for $x \neq x'$. It follows therefore from Lemma 16, that the capacity is 0 in case $K_4$ (max. error, pure codes).

Randomization in the encoding can be interpreted as an enlargement of the set of possible input sequences for a channel.

Instead of the set of input $n$-sequences $\mathcal{X}^n$ we have the set $P(\mathcal{X}^n) =$ set of all PD on $\mathcal{X}^n$ available for the encoding. We shall make use only of the subset $\mathcal{P}^*(\mathcal{X}^n) =$ set of all product distributions on $\mathcal{X}^n$. Actually we shall use only all sequences $q^n = q_1 \times q_2 \times \cdots \times q_n$, which have as compounds $q_t$ either

$$\delta : \delta(1) = 1, \delta(2) = \delta(3) = 0$$

or

$$q : q(1) = 0, q(2) = q(3) = \frac{1}{2}.$$

This means that we restrict ourselves to special letter by letter randomizations. Randomization per letter means convex combination of rows in our matrices.

To find optimal codes using only $q^n$ means therefore to find optimal codes for

$$\mathcal{W}^* = \left\{ w_1 = \begin{pmatrix} 1\;0\;0 \\ 0\;\frac{1}{2}\;\frac{1}{2} \end{pmatrix}, w_2 = \begin{pmatrix} 0\;1\;0 \\ \frac{1}{2}\;0\;\frac{1}{2} \end{pmatrix} \right\}.$$

Now $\mathcal{T}(1) \cap \mathcal{T}(2) = \varnothing$ and we can transmit over our original channel with a positive rate if we use randomized encoding.

Dobrushin [23, 24] was the first to use code concept $\overline{K}_1$ for WAVC. The main reason is of course that one tries to avoid the correlated coding, which has some drawbacks after all. It was asserted in [23, 24] see in particular ([23], Theorem 1, Remark 3) ([24], end of paragraph following Eq. (4)) that for the WAVC $\mathcal{A} = (\mathcal{W}^n)_{n \geq 1,\ldots}$ in case $\overline{K}_1$ (average error, random code and decoding) equals

$$C_0 = \max_p \min_{w \in \overline{\overline{\mathcal{W}}}} R(p, w).$$

Taken literally this statement is incorrect, because an example (see [11]) shows that $C_D$ can be 0 and still the capacity is positive. However, this is actually only a minor technical point. By restricting the sets $\mathcal{W}$ to those which are already row-convex closed, $\mathcal{W} = \overline{\overline{\mathcal{W}}}$, the assertion might well be true and it is one of the unsolved problems in this area to prove (or disprove) this modified assertions.

In this context the results about feedback AVC and listcodes for AVC (see [5]) are certainly of interest.

There is also still the following question.

If the capacity for $\mathcal{A} = (\overline{\overline{\mathcal{W}^n}})_{n\geq 1}$ in case $\overline{K}_4$ (pure codes, average error) equal to $C_D$?

The following example shows only that in case $\overline{K}_4$ $\max\limits_{p} \min\limits_{w\in\overline{W}} \mathcal{R}(p, w)$ is too big for $\mathcal{A} = (\mathcal{W}^n)_{n\geq 1}$ or $\mathcal{A} = (\overline{\mathcal{W}}^n)_{n\geq 1}$.

## 1.5.8 Side Information

In the previous section it was assumed that neither sender nor receiver had any knowledge of $\mathcal{P}(\cdot| \cdot |s)$ governing the transmission. The definition of a code was therefore independent of $s$. In case the sender or the receiver (or both) know $s$, the code used may depend on this $s$. If only the sender knows $(S^+)$ the transmission probabilities, the $a_i$'s but not the $D_i$'s may depend on $s$.

An $(n, N, \lambda)$ code $\{(u_i(s), D_i)|i = 1, \dots, N\}$ must now satisfy

$$P\big(D_i|u_i(s)\big) \geq 1 - \lambda \quad \text{for } i = 1, \dots, N;\, s \in S. \tag{1.5.26}$$

If only the receiver knows $(R^+)$ $s$, the $D_i$'s but not the $u_i$'s may depend on $s$, and finally in case both knows $(R^+, S^+)$ the code words and the decoding sets may depend.

In case $(R^+, S^+)$ the capacity is simply $\min\limits_{s} \max\limits_{p} \mathcal{R}\big(p, w(\cdot| \cdot |s)\big)$, the smallest of the individual capacities. Also, in case $(R^+, S^-)$ the converse proof of the previous section $(R^-, S^-)$ literally applies and thus in this case the capacity equals $\max\limits_{p} \min\limits_{s\in S} \mathcal{R}(p, \dots)$ again.

Then the receiver's knowledge does *not* increase the capacity (explain intuitively why). However, the sender's knowledge does increase capacity:

**Theorem 29** *For the* SVC *we have in case* $(S^+, R^-)$*:*

$$N(n, \lambda) = exp\left\{\min_{s} C_s \cdot n + 0(\sqrt{n})\right\}.$$

*Proof* Since the converse direction is trivial we turn to the coding part. This follows again by a simple argument: We use an approximating channel $\nu(n)$ of $t(n) = 2^{\sqrt{n}}$, and all $C_s > \delta > 0$ $s = 1, \dots, t(n)$, therefore also

$$\underline{C} = \max_p \min_w \mathcal{R}(p, w) > 0$$

and we can transmit over this channel $t(n)$ messages.

We transmit the index of channel which is best approximation to the given one.

Then both, sender and receiver, know $S^+$, and we use code of length $\log N(n, \lambda) = \min C_s \cdot n - K(\lambda)\sqrt{n}$. □

For other proofs see [45], Chaps. 4 and 7.

### *1.5.9 Sources with Arbitrarily Varying Letter Probabilities*

Let $\mathcal{P}$ be any set of probability distributions on the finite set $\mathcal{X}$, $\overline{\mathcal{T}}$ the convex closed hull of $\mathcal{T}$,

$$\mathcal{T}^n = \{q^n : q^n = q_1 \times \cdots \times q_n \quad \text{with } q_k \in \mathcal{T}\}$$

and

$$M(n, \varepsilon) = \min_{E^n \in \mathcal{X}^n} \{p^n(E^n) : q^n(E^n) \geq 1 - \varepsilon \quad \text{for all } q^n \in \mathcal{T}^n\},$$

where $0 < \varepsilon < 1$ and $p^n(E^n) = |E^n|$.

Furthermore, let

$$H = \max_{q \in \mathcal{T}^*} H(q).$$

**Theorem 30** $\log M(n, \varepsilon) = nH + o(n)$.

*Proof* We can assume without loss of generality that for all $x \in \mathcal{X}$:

$$\sup_{q \in \mathcal{P}} q(x) > o,$$

because otherwise we could omit this letter.

Let now $q^* \in \mathcal{P}^*$:

$$H(q^*) = H.$$

Again, it is easy to show that $q^*(x) > 0 \quad$ for all $x \in \mathcal{X}$. We need

**Lemma 17** *With $g(x) = \log q^*(x)$, $x \in \mathcal{X}$,*

$$\sum_x g(x) q^*(x) \leq \sum_x g(x) q(x) \quad \textit{for all } q \in \mathcal{P}^*. \tag{1.5.27}$$

(Recall: $\sum_x q^*(x) \log q^*(x) \geq \sum_x q^*(x) \log q(x) \quad$ for all PD)
Equation (1.5.27) is a different inequality and not true for general $p, q$).

*Proof* Let $\mathcal{P}_1$ be any convex compact subset of $\mathcal{P}^*$ such that $q^* \in \mathcal{P}_1$ and $q(x) > o$ for all $q \in \mathcal{P}_1$ and $x \in \mathcal{X}$.

The sets of "strategies" $\mathcal{P}_1$, $\mathcal{P}_1$ and the payoff function

$$\sum_x q'(x) \log q(x).$$

(Convex-concave, and finite by definition of $\mathcal{P}_1$.)

Define a zero sum two-person game, where $q$ is a strategy of player I and $q'$ is a strategy of player II.

Since for $q \neq q'$:

$$\sum_x q'(x) \log q(x) < \sum_x q'(x) \log q'(x)$$

player I always does best by playing the same strategy as player II and player II maximizes his playoff (the negative of the playoff to player I), by choosing $q^*$.

Therefore

$$\sum_x q^*(x) \log q^*(x) \leq \sum_x q'(x) \log q^*(x) \quad \text{for all } q' \in \mathcal{P}_1.$$

By an approximation one gets the inequality for all $q' \in \mathcal{P}$. □

*Continuation of Proof of Theorem* 29.

Let $\delta > 0$ and define

$$E^n = \left\{ x^n | \sum_{k=1}^{n} g(x_k) > n \sum_x g(x) q^*(x) - n\delta \right\} \tag{1.5.28}$$

Since for any $q^n = q_1 \times \cdots \times q_n \in \mathcal{P}^{*n}$ $\frac{1}{n} \sum_{t=1}^{n} q_t \in \mathcal{P}^*$ and since by Lemma 17

$$\sum_x g(x) q^*(x) \leq \sum_x g(x) \left( \frac{1}{n} \sum_{t=1}^{n} q_t \right)$$

it follows from (1.5.28) that

$$\begin{aligned} q^n(E^n) &\geq q^n \left\{ \sum_{k=1}^{n} g(x_k) > n \sum_x g(x) \frac{1}{n} \sum_{t=1}^{n} q^t = n\delta \right\} \\ &= q^n \left\{ \frac{1}{n} \sum_{t=1}^{n} \left[ g(x_t) - \sum_x g_t(x) \right] > -\delta \right\} \geq 1 - \varepsilon \end{aligned}$$

for $n$ large, *uniformly* in $q^n$, because the variances are bounded.

On the other hand,

$$p^n(E^n) = p^n\left\{\sum_1^n g(x_k) > n\sum_x g(x)q^*(x) - n\delta\right\}$$
$$\leq \exp\left(-n\sum_x g(x)q^*(x) + n\delta\right)\left(\sum_x e^{g(x)}p_1(x)\right)^n.$$

(bound with $s = 1$ and $A = n\sum_x g(x)q^*(x) - n\delta$.)

Since $p_1(x) = 1$, $\sum_x e^{g(x)} = \sum_x q^*(x) = 1$, and therefore

$$\log M(n, \varepsilon) = \log p^n(E^n) \leq nH + n\delta. \tag{1.5.29}$$

From the source coding theorem follows

$$\log M(n, \varepsilon) \geq \log\min\{p^n(E^n)|q^{*n}(E^n) \geq 1 - \varepsilon\} = nH + o(n). \tag{1.5.30}$$

Equations (1.5.29) and (1.5.30) imply the theorem.
Starting with $\mathcal{P}^n$ one can pass to $\mathcal{P}^{*n}$ and still

$$\log M^*(n, \varepsilon) = nH + o(n).$$

Replacing $\mathcal{P}^{*n}$ by the convex hull of $\mathcal{P}^{*n}$ the answer is again the same as can seen immediately from the definition.

Naturally the question arises how many PDs can be added without changing the parameter $H$.

Consider $Q^n = \{q^n | q^n | \mathcal{X}_t \in \mathcal{P}^*\}$ is the answer still the same?

**Counterexample** $\mathcal{X} = \{0, 1\}, \mathcal{P}^* \left\{(\alpha, \beta)|\alpha \leq \frac{1}{4}\right\} q^n = \frac{1}{2}p^n + \frac{1}{2} + \delta^n$, where $p^n$ on $\mathcal{X}^n$ and $\delta^n(1, 1, \dots, 1) = 1$. Then $q^n \in Q^n$ but for $\varepsilon < \frac{1}{2} : M(n, \varepsilon) \geq e^{n\log 2 + o(n)}$.

**Theorem 31** *Let $\mathcal{P}$ be a set of PDs on $\mathcal{X}$, $\mathcal{P}^*$ its closed convex hull and let $\overline{\mathcal{P}^n}$ be the set of PDs on $\mathcal{X}^n$ such that the conditional probability distribution*

$$\big(\Pr(X_t = x_t | X_1 = x_1, \dots, X_{t-1} = x_{t-1})\big)_{x_t \in \mathcal{X}_t}$$

*are elements of $\mathcal{P}^*$ for all $t = 1, 2, \dots, n-1$ and all vectors $(x_1, \dots, x_{t-1})$.*

*Then*

$$\overline{M}(n, \varepsilon) = \min_{E^n \in \mathcal{X}^n}\left\{|E^n| : q^n(E^n) \geq 1 - \varepsilon \forall q^n \in \overline{\mathcal{P}^n}\right\} = e^{nH + o(n)},$$

*where*

$$H = \max_{q \in \mathcal{P}^*} H(q).$$

*Proof* For any $q^n \in \overline{\mathcal{P}^{*n}}$ : $H(q^n) \leq nH(q^*)$, with equality iff $q^n = q^{*n}$. Then, Lemma yields:

$$\sum_{x^n}\sum_{k=1}^{n} g(x_k)q^n(x^n) \geq \sum_{x} g(x)q^*(x)$$

and therefore we have for $E^n$,

$$q^n(E^n) \geq q^n \left\{ \sum_{k=1}^{n} g(x_k) > \sum_{x^n}\sum_{k=1}^{n} g(x_k)q^n(x^n) - n\delta \right\}.$$

Set $Y_j = \log q^*(X_j)$ $Z_j = Y_j - E(Y_j|X_1, \dots, X_{j-1})$, then $E(Z_j)X_1, \dots, X_{j-1}) = 0$ and since the $Z_t(1 \leq t \leq j-1)$ are functions of $X_s(1 \leq s \leq j-1)$ also $E(Z_j|Z_1, \dots, Z_{j-1}) = 0$ and $EZ_jZ_k = 0$, $j \neq k$.

Obviously, there exists a (upper) bound $d$ on $\{EZ_j^2 | j = 1, 2, \dots\}$ and therefore by Chebycheff's inequality

$$q^n \left\{ \sum_{t=1}^{n} Z_t > -n\delta \right\} \geq 1 - \varepsilon, \text{ for suitable } \delta, \text{ uniformly in } q^n \in \overline{\mathcal{P}^{*n}}. \quad (1.5.31)$$

Since by assumption

$$\left(q^n(X_j = x_j | X_1 = x_1, \dots, X_{j-1} = x_{j-1})\right)_{x_j \in \mathcal{X}_j} \in \mathcal{P}^*.$$

$$\begin{aligned} E(Y_j|X_1, \dots, X_{j-1}) &= X_{x_j} q^n(X_j = x_j | X_1, \dots, X_{j-1}) \log q^*(X_j = x_j) \\ &\geq \sum_{x} q^*(x) \log q^*(x). \end{aligned}$$

Now this implies

$$q^n \left\{ \sum_{t=1}^{n} Y_t > n \sum_{x} q^*(x) \log q^*(x) - n\delta \right\} \geq 1 - \varepsilon.$$

The remainder of the proof is exactly as in the proof of Theorem 29. □

### 1.5.10 Channels with Varying Transmission Probabilities

Let $S$ be any set and let $W = \{w(\cdot| \cdot |s)|s \in S\}$ be a set of stochastic $|\mathcal{X}| \times |\mathcal{Y}|$-matrices. For every $s \in S$ we define

$$P(y^n|x^n|s) = \prod_{t=1}^{n} w(y_t|x_t|s) \quad \text{for every } x^n \in X^n, y^n \in Y^n. \tag{1.5.32}$$

For every $n \in \mathbb{N}$ define $\mathcal{W}^n$ by

$$\mathcal{W}^n = \{P(\cdot|\cdot|s) : s \in S\}. \tag{1.5.33}$$

The channel with varying transmission probabilities (VC) $\mathcal{W}$ is defined by the sequence $(\mathcal{W}^n)_{n\in\mathbb{N}}$. Its study was initiated in [17, 46], and it has been called "a class of channels," "simultaneous channels," and "compound channel." We find the present more descriptive and it also indicates a relationship to the so called channels with arbitrarily varying transmission probabilities, which we discuss later. The VC is a model for a transmission system which has several states $S$ and the system may be in any state it wants. This stays either the same over a fixed blocklength, independently of the code word sent, or it stays the same for any particular code word only, and it may be different for another code word. In the first case we talk about AWVC and in the second case about an SVC, where in these abbreviations the "w" refers to "weakly" and the "s" refers to "strongly". An $(n, N)$-code $\{(u_i, D_i)|i = 1, \dots, N\}$ is an $(n, N, \lambda)$-code for the s.w.ch., if

$$P(D_i|u_i|s) \geq 1 - \lambda \text{ for } i = 1, \dots, N \quad \text{and all} \quad s \in S. \tag{1.5.34}$$

The only meaningful definition of an average error probability would be in this case

$$\lambda^* = \frac{1}{N} \sum_{i=1}^{n} \sum_{s\in S} P(D_i^c|u_i|s). \tag{1.5.35}$$

From here we always can pass to an $\left(n, \frac{N}{2}, 2\overline{\lambda}\right)$-code with maximal error probability $2\overline{\lambda}$ and for the asymptotic theory there is no point in using this for average error probability. For the channel WVC the situation is quite different, because here the average error probability can be defined as

$$\overline{\lambda} = \sup_{s\in S} \frac{1}{N} \sum_{i=1}^{n} \mathcal{P}(D_i|u_i|s). \tag{1.5.36}$$

Let us denote the maximal length of a code with parameters $n$ and $\lambda$ (resp. $\overline{\lambda}$) by $N_{\mathcal{W}}(n, \lambda)$ (resp. $\overline{N_{\mathcal{W}}}(n, \overline{\lambda})$).

Whereas the capacities are the same in both cases, we shall see below that for $N_{\mathcal{W}}(n, \lambda)$ there is a coding theorem and a strong converse, for $\overline{N_{\mathcal{W}}}(n, \overline{\lambda})$ only a weak converse holds. For more complex channels the differences between the two notions of error probability becomes even more important. We prove now first a coding theorem (for both channels) in case $|S| < \infty$. The proof is based on the following simple idea: An $\left(n, N, \frac{\lambda}{t}\right)$-code for the *average channel* (AC) $\overline{P}(\cdot|\cdot) = \frac{1}{t}\sum_{s=1}^{t} \mathcal{P}(\cdot| \cdot |s)$ is an $(n, N, \lambda)$-code for the SVC.

We prove now first a coding theorem (and also a weak converse) for this auxiliary channel, proceed than to the channels SVC and WVC in case $|S| < \infty$, and finally we remove the supposition $|S| < \infty$.

#### 1.5.10.1 The Average Channel

The following definitions are needed for the theorem below: for a probability distribution $p^n = p \times \cdots \times p$ on $\mathcal{X}^n$ set

$$
\begin{aligned}
P_s = (x^n, y^n) &= p^n(x^n), P(y^n|x^n|s), s \in S, q^n(y^n|s) \\
&= \sum_{x^n} p^n(x^n) P(y^n|x^n|s), s \in S,
\end{aligned}
$$

$$\overline{P}(x^n, y^n) = p^n(x^n)\overline{P}(y^n|x^n)\overline{q}^n(y^n) = \frac{1}{t}\sum_{s=1}^{t} q^n(y^n|s).$$

$$I_s(x^n, y^n) = \begin{cases} \log \dfrac{P(y^n|x^n|s)}{q^n(y^n|s)}, & \text{if } q^n(y^n|s) > 0 \\ 0 & \text{otherwise} \end{cases}$$

and

$$\overline{I}(x^n, y^n) = \begin{cases} \log \dfrac{\overline{P}(y^n|x^n)}{\overline{q}^n(y^n)}, & \text{if } \overline{q}^n(y^n) > 0 \\ 0 & \text{otherwise.} \end{cases}$$

The later quantities are frequently called "information densities."

We obtain that there exists a code of length $N^+$ and average error probability $\overline{\lambda}$ for this channel.

## 1.6 Lecture on Gallager's Converse for General Channels Including DFMC and FSC

### *1.6.1 Lower Bounding Error Probabilities of Digits*

Let us be given a discrete source with time structure and finite alphabet $\mathcal{U}$, $(U^l)_{l=1}^{\infty}$, where RV $U^l$ takes values in $\mathcal{U}^l$. We know that for a stationary source $\frac{1}{l}H(U^l)$ is nonincreasing with $l$ and approaches a limit $H_\infty$ as $l \to \infty$, also called the mean entropy. Clearly for a DMS $\frac{1}{l}H(U^l) = H(U)$ $(l \in \mathbb{N})$ encoded into $X^n = f(U^l)$

(the channel input) and producing a channel output $Y^n$, which gets decoded into $V^l = g(Y^n)$ with values in $\mathcal{X}^l$, the aim is to reproduce in some sense $U^l$.

If $U_t \neq V_t$ then an error has occurred on the $t$th digit of the transmission. The probability of such an error $\lambda_t$ is determined by the joint ensemble $U^l V^l$. The average error probability $\langle\lambda\rangle$ over the sequence of $l$ digits is defined as

$$\langle\lambda\rangle = \frac{1}{l}\sum_{t=1}^{l}\lambda_t \tag{1.6.1}$$

and the expected number of errors in the sequence is $l\langle\lambda\rangle$.

We are interested in showing that reliable communication is impossible if the source entropy $H_\infty$ is larger than "capacity."

Thus, even if we showed that the probability of error $\Pr(U^l \neq V^l)$ is 1 (or close to 1) for large $l$, this would only guarantee one digit error out of $l$, or $\langle\lambda\rangle \geq \frac{1}{l}$. Thus, to show that $\langle\lambda\rangle$ is bounded away from 0 in the limit as $l \to \infty$, we must consider $\langle\lambda\rangle$ directly rather than the sequence error probability.

We use Fano's Lemma in the form: for two RV's $U$, $V$ with a joint PD and the same spaces of $M$ values let

$$\lambda = \Pr(U \neq V) = \sum_u \sum_{v \neq u} P_{UV}(u, v).$$

Then

$$\lambda \log(M-1) + h(\lambda) \geq H(U|V). \tag{1.6.2}$$

Instead of $\lambda$ we use now $\langle\lambda\rangle$.

**Theorem 32** *Let $(U^l, V^l) = (U_1, \dots, U_l; V_1, \dots, V_l)$ where each $U_t$ and $V_t$ take values in the same space $\mathcal{U}$ of $M$ values. Then*

$$\langle\lambda\rangle \log(M-1) + h(\langle\lambda\rangle) \geq \frac{1}{l} H(U^l|V^l). \tag{1.6.3}$$

*Proof* Using the chain rule for conditional entropy we get

$$\begin{aligned} H(U^l|V^l) &= H(U_1|V^l) + H(U_2|U_1V^l) + \dots + H(U_l|V^lU_1\cdots U_{l-1}) \\ &\leq \sum_{t=1}^{l} H(U_t|V_t), \end{aligned} \tag{1.6.4}$$

because adding variables in the condition decreases entropy.

Applying (1.6.2) to each term in (1.6.4) we get

$$H(U^l|V^l) \leq \sum_{t=1}^{l}[\lambda_t \log(M-1) + h(\lambda_t)]$$

and

$$\frac{1}{l}H(U^l|V^l) \leq \langle\lambda\rangle \log(M-1) + \frac{1}{l}\sum_{t=1}^{l} h(\lambda_t). \tag{1.6.5}$$

By the convexity of entropy $h$

$$\frac{1}{l}\sum_{t=1}^{l} h(\lambda_t) \leq h(\langle\lambda\rangle)$$

and (1.6.3) follows. □

So far we have lower bounded $\langle\lambda\rangle$ in terms of the equivocation $H(U^l|V^l)$. Now the channel has to enter via the joint ensemble $U^l X^n Y^n V^l$. It has the properties that $Y^n$ is conditionally independent of $U^l$ given $X^n$, and $V^l$ is conditionally independent of $U^l$ and $X^n$ given $Y^l$. These conditions can be written as

$$P(y^n|x^n u^l) = P(y^n|x^n)$$

and

$$P(v^l|y^l x^n, u^l) = P(v^l|y^l).$$

They say that the channel output depends statistically on the source sequence only through the channel input, and the decoded output depends on the source output and the channel input only through the channel output. In other words there is no subsidiary "hidden" channel passing information about $U^l$ to the decoder.

If the source has memory, this definition is less innocuous than it appears. If successive blocks of $l$ source digits are transmitted to the destination, a *decoder could be constructed which made use of blocks of received letters in decoding others*. The above definitions rule out such decoders, *but finally this problem disappears when* $l \to \infty$. Next, since $X^n = f(U^l)$ and $V^l = g(Y^n)$ the data-processing inequality gives

$$I(U^l \wedge V^l) \leq I(X^n \wedge Y^n). \tag{1.6.6}$$

Combining this with the previous theorem we get

$$\begin{aligned}\langle\lambda\rangle \log(M-1) + h(\langle\lambda\rangle) \geq \frac{1}{l}H(U^l|V^l) &= \frac{1}{l}H(U^l) - \frac{1}{l}I(U^l \wedge V^l)\\ &\geq \frac{1}{l}H(U^l) - \frac{1}{l}I(X^n \wedge Y^n).\end{aligned} \tag{1.6.7}$$

If the channel is a DMC, we have $I(X^n \wedge Y^n) \leq nC$ and therefore

$$\langle\lambda\rangle \log(M-1) + h(\langle\lambda\rangle) \geq \frac{1}{l}H(U^l) - \frac{n}{l}C. \tag{1.6.8}$$

Since the transmission shall run synchronously in time, choosing a time interval $T_s$ between source letters and a time interval $T_c$ between channel letters we get for the number of channel uses $n = \lfloor lT_s/T_c \rfloor$.

**Theorem 33** (Converse to the Coding Theorem) *Let a discrete stationary source with alphabet size $M$ have entropy $H_\infty = \lim_{l\to\infty} \frac{1}{l} H(U^l)$ and produce letters at a rate of one letter each $T_s$ seconds. Let a DMC have capacity $C$ and be used at a rate of one letter each $T_c$ seconds. Let finally a source sequence of length $l$ be connected to a destination through a sequence of $n$ channel uses where $n = \left\lfloor \frac{lT_s}{T_c} \right\rfloor$, then for any $l$, the error probability per source digit $\langle\lambda\rangle$, satisfies*

$$\langle\lambda\rangle \log(M-1) + h(\langle\lambda\rangle) \geq H_\infty - \frac{T_s}{T_c} C. \tag{1.6.9}$$

*Proof* Since $H_\infty \leq \frac{1}{l} H(U^l)$ this follows from (1.6.8). □

#### 1.6.1.1 Discussion

Using total time $LT_s$, no matter what coding is used for the source, the average error probability per source digit must satisfy (1.6.9) and is thus bounded away from zero if $H_\infty$ (the source rate) is greater than $\frac{T_s}{T_c} C$ (the channel capacity per source digit). However, the theorem says *nothing about the individual error probabilities $\lambda_t$*. By appropriate design of the coder and decoder, we can make $\lambda_t$ small for some values of $t$ and large for others.

Next note that we used the property of a DMC only while going from (1.6.7) to (1.6.8). For more general channels we must find a way to define a joint $X^n$, $Y^n$ ensemble and "relate to a capacity $C$" (see below).

A side effect we get if the channel is noiseless. Then (1.6.9) gives a converse to the source coding theorem. It differs from our previous result in that it bounds the error probability per digit rather than just the block error probability. Actually the bound in (1.6.9) is quite weak when the source alphabet size $M$ is large. This weakness is unavoidable as can be seen from the following construction. Let $C = 0$ and let the source be memoryless with

$$\Pr(U = a_1) = 1 - \varepsilon, \quad \Pr(U = a_j) = \frac{\varepsilon}{M-1} \quad \text{for } 2 \leq j \leq M.$$

Then, if the decoder decodes each digit as $a_1$, it is readily seen that $\langle\lambda\rangle = \varepsilon$. On the other hand

$$H_\infty = (1-\varepsilon)\log\frac{1}{1-\varepsilon} + (M-1)\frac{\varepsilon}{M-1}\log\frac{M-1}{\varepsilon}.$$

For any $\varepsilon > 0$ we can make $H_\infty$ as large as desired by making $M$ large. This communicative system satisfies (1.6.9) with equality.

### 1.6.2 Application to FSC

As in the previous section we consider a source $(U^l)_{l=1}^{\infty}$, which is stationary and produces letters at a rate of one letter each $T_s$ seconds. The source sequence $U^l$ is after appropriate processing to be transmitted now over a FSC (rather than a DMC) and reproduced as a sequence $V^l$ at the destination. Having again a channel to be used once each $T_c$ seconds so that

$$n = \left\lfloor \frac{lT_s}{T_c} \right\rfloor . \tag{1.6.10}$$

The source probabilities, the channel transmission probabilities, the initial state and the processors now determine a joint ensemble $U^l, X^n, Y^n, V^l$, which depends on the initial state $s_0$, which for the time being we treat as deterministic.

As earlier we get now corresponding properties

$$P^n(y^n|x^n, s_0) = P^n(y^n|x^n, s_0, u^l) \tag{1.6.11}$$

$$P^n(v^l|y^n, s_0) = P^n(v^l|y^n, s_0, x^n, u^l). \tag{1.6.12}$$

By data processing

$$I(U^l \wedge V^l|s_0) \leq I(X^n \times Y^n|s_0). \tag{1.6.13}$$

Again we get an average error probability per source letter $\langle\lambda(s_0)\rangle$ and by Fano's inequality

$$\begin{aligned}\langle\lambda(s_0)\rangle \log(M-1) + h(\langle\lambda(s_0)\rangle) &\geq \frac{1}{l} H(U^l|V^l, s_0)\\ &\geq \frac{1}{l}(H(U^l|s_0) - I(U^l \wedge V^l|s_0))\\ &\geq \frac{1}{l}(H(U^l|s_0) - I(X^n \times Y^n|s_0)).\end{aligned} \tag{1.6.14}$$

We now make the assumption that the source probabilities are independent of $s_0$ and obtain

$$\frac{1}{l} H(U^l|s_0) \geq H_\infty = \lim_{l' \to \infty} \frac{1}{l'} H(U^{l'}). \tag{1.6.15}$$

Substituting this and $n$ by (1.6.10) into (1.6.14) we get

$$\langle\lambda(s_0)\rangle \log(M-1) + h(\langle\lambda(s_0)\rangle) \geq H_\infty - \frac{T_s}{nT_c} I(X^n \times Y^n|s_0). \tag{1.6.16}$$

This can be related to $\underline{C}$ and $\overline{C}$. By definition of $\overline{C}_n$

$$\frac{1}{n} I(X^n \wedge Y^n | s_0) \leq \overline{C}_n \quad (s_0 \in \mathcal{S}) \tag{1.6.17}$$

and then letting $l \to \infty, n \to \infty$

$$\langle \lambda(s_0) \rangle \log(M-1) + h(\langle \lambda(s_0) \rangle) \geq H_\infty - \frac{T_s}{T_c} \overline{C}. \tag{1.6.18}$$

Notice that no assumption concerning independence of $X^n$ from $s_0$ was made in the derivation.

This means validity of (1.6.18) is guaranteed even if the encoder knows and uses $s_0$. Similarly (1.6.18) is also valid if the decoder knows and uses $s_0$.

To relate (1.6.16) to $\underline{C}$ we must *assume* that $X^n$ is independent of $x_0$. Then from the definition of $\underline{C}_n$ we have for some initial state $s_0$

$$\frac{1}{n} I(X^n \wedge Y^n | s_0) \leq \underline{C}_n.$$

Since $\underline{C}_n \leq \underline{C}_n + \frac{\log S}{n}$ we have for any $n$ and some $s_0$

$$\langle \lambda(s_0) \rangle \log(M-1) + h(\langle \lambda(s_0) \rangle) \geq H_\infty - \frac{T_s}{T_c} \left[ \underline{C} + \frac{\log S}{n} \right]. \tag{1.6.19}$$

Now, if there is a PD $q_0$ on $\mathcal{S}$, then $\langle \lambda(s_0) \rangle$ can be averaged to get an overall error probability per source letter

$$\langle \lambda \rangle = \sum_{s_0} q_0(s_0) \langle \lambda(s_0) \rangle.$$

If $q_{\min}$ gives the PD with smallest error probability then $\langle \lambda \rangle \geq q_{\min} \langle \lambda(s_0) \rangle$ for each initial state and (1.6.19) can be further bounded by

$$\frac{\langle \lambda \rangle}{q_{\min}} \log(M-1) + h\left(\frac{\langle \lambda \rangle}{q_{\min}}\right) \geq H_\infty - \frac{T_s}{T_c} \left[ \underline{C} + \frac{\log S}{n} \right]. \tag{1.6.20}$$

**Theorem 34** *With the same assumptions on the source and time performances of source and channel, that is,* $n = \left\lfloor \frac{lT_s}{T_c} \right\rfloor$, *then, independent of the initial state,* (1.6.18) *holds.*

*If in addition the channel input* $X^n$ *is independent of the initial state, then for each* $n$ *there is some initial state for which* (1.6.19) *is satisfied.*

*If there is also a PD on* $\mathcal{S}$ *with a smallest probability* $q_{min}$, *then* (1.6.20) *is satisfied.*

### 1.6.3 Indecomposable Channels

Roughly speaking an FCS is indecomposable if the effect of the initial state dies away with time. Formally, let

$$q_n(s_n|x^n, s_0) = \sum_{y^n} P^n(y^n, s_n|x^n, s_0).$$

A FSC is indecomposable if, for every $\varepsilon > 0$, no matter how small, there exists an $n_0$ such that for all $n \geq n_0$

$$|q_n(s_n|x^n, s_0) - q_n(s_n|x^n, s_{o'})| \leq \varepsilon. \tag{1.6.21}$$

for all $s_n$, $x^n$, $s_0$, and $s_0'$.

The channel with "ambiguous capacity" and the panic-button channel are not indecomposable.

On the other hand in the channel with intersymbol interference the LHS in (1.6.21) is 0 and therefore the channel indecomposable.

Now we draw conclusions.

For a fixed input sequence $(X_1, X_2, \dots)$ we can regard the state sequence $(S_0, S_1, S_2, \dots)$ as a nonhomogeneous Markov chain.

**Convention** Omitting $x^n$ in $q_n(s_n|x^n, s_0)$ we write $q(s_n|s_0)$. We study the dependence on $s_0$ for large $n$.

As measure of dependence we define the distance $d_n(s_0', s_0'')$ as

$$d_n(s_0', s_o'') = \sum_{s_n} |q(s_n|s_0') - q(s_n|s_0'')|. \tag{1.6.22}$$

For $n = 0$ and $s_0' \neq s_0''$ we take $d_0(s_0', s_0'') = 2$.

**Lemma 18** *For any given input $x_1, x_2, \dots$ and any given $s_0', s_0'', \dots$ the distance $d_n(s_0', s_0'')$ as defined above is nonincreasing in $n$.*

*Proof* For $n \geq 1$ we have

$$\begin{aligned} d_n(s_0', s_0'') &= \sum_{s_n} |q(s_n|s_0') - q(s_n|s_0'')| \\ &= \sum_{s_n} \left| \sum_{s_{n-1}} q(s_n|s_{n-1})[q(s_{n-1}|s_0') - q(s_{n-1}|s_0'')] \right| \qquad (1.6.23) \\ &\leq \sum_{s_n} \sum_{s_{n-1}} q(s_n|s_{n-1})|q(s_{n-1}|s_0') - q(s_{n-1}|s_0'')| \\ &= \sum_{s_{n-1}} |q(s_{n-1}|s_0') - q(s_{n-1}|s_0'')| = d_{n-1}(s_0', s_0''). \end{aligned}$$

□

**Lemma 19** *Suppose that for some* $m > 0$, *some* $\delta > 0$, *and each* $n \geq 0$, *there is some choice of* $s_{n+m}$ *with*

$$q(s_{n+m}|s_n) \geq \delta \quad (s_n \in \mathcal{S}). \tag{1.6.24}$$

*Then* $d_n(s_0'|s_0'') \leq 2(1-\delta)n/m^{-1}$ *and hence approaches 0 exponentially fast with* $n$.

*Proof*

$$\begin{aligned} d_{n+m}(s_0'|s_0'') &= \sum_{s_{n+m}} |q(s_{n+m}|s_0') - q(s_{n+m}|s_0'')| \\ &= \sum_{s_{n+m}} \left| \sum_{s_n} q(s_{n+m}|s_n)[q(s_n|s_0') - q(s_n|s_0'')] \right|. \end{aligned} \tag{1.6.25}$$

Define

$$a(s_{n+m}) = \min_{s_n} q(s_{n+m}|s_n). \tag{1.6.26}$$

Observing that

$$\sum_{s_n} a(s_{n+m})[q(s_n|s_0') - q(s_n|s_0'')] = 0$$

we can rewrite (1.6.25) as

$$d_{n+m}(s_0', s_0'') = \sum_{s_{n+m}} \left| \sum_{s_n} [q(s_{n+m}|s_n) - a(s_{n+m})] \cdot [q(s_n|s_0) - q(s_n|s_0'')] \right|. \tag{1.6.27}$$

Again by the triangle inequality and the fact that $q(s_{n+m}|s_n) - a(s_{n+m}) \geq 0$, we obtain

$$d_{n+m}(s_0', s_0'') \leq \sum_{s_{n+m}} [q(s_{n+m}|s_n) - a(s_{n+m})] |q(s_n|s_0') - q(s_n|s_0'')|. \tag{1.6.28}$$

Summing over $s_{n+m}$ this becomes

$$\begin{aligned} d_{n+m}(s_0', s_0'') &\leq \left[ 1 - \sum_{s_{n+m}} a(s_{n+m}) \right] \sum_{s_n} |q(s_n|s_0') - q(s_n|s_0'')| \\ &= \left[ 1 - \sum_{s_{n+m}} a(s_{n+m}) \right] d_n(s_0', s_0''). \end{aligned}$$

By hypothesis, $a(s_{n+m}) \geq \delta$ for at least one value of $s_{n+m}$, and thus

$$d_{n+m}(s_0', s_0'') \leq (1-\delta)d_n(s_0', s_0'').$$

Applying this result for $n = 0$, then for $n = m$, then $n = 2m$, and so forth, and recalling that $d_0(s_0', s_0'') = 2$, we obtain

$$d_{rm}(s_0', s_0'') \leq 2(1-\delta)^r.$$

Since $d_n(s_0', s_0'')$ is nonincreasing in $n$, this completes the proof. □

The following theorem provides a test for whether or not an FSC is indecomposable.

**Theorem 35** (Thomasian [41]) *Necessary and sufficient for an* FSC *to be indecomposable is that for some fixed m and each $x$ there exists a choice for the mth state, say $s_m$, such that*

$$q(s_m|x^n, s_0) > 0 \quad \textit{for all } s_0 \tag{1.6.29}$$

*($s_m$ can depend on $x^n$). Furthermore, if the channel is indecomposable, m above can always be taken as less than $2^{S^2}$, where S is the number of channel states.*

*Proof Sufficiency.* If (1.6.29) holds for some $m$, then, since $s_0'$ and $x^m$ can only take on finitely many values, there is some $\delta > 0$ such that

$$q(s_m|x^m, s_0) \geq \delta \tag{1.6.30}$$

for all $s_0$, all $x^m$, and some $s_m$ depending on $x^m$. Also, since the channel probabilities are independent of time, we also have, for all $n$ and some $s_{n+m}$ depending on $x_{n+1}, \ldots, x_{n+m}$

$$q(s_{n+m}|x_{n+1}, \ldots, x_{n+m}, s_n) > \delta \quad \text{for all } s_n. \tag{1.6.31}$$

Thus the conditions of the previous lemma are satisfied, and

$$\sum_{s_n} |q(s_n|x^m, s_0') - q(s_n|x^m, s_0')|$$

approaches zero exponentially as $n \to \infty$ uniformly in $x^m$, $s_0'$, and $s_0''$. Thus (1.6.21) is satisfied for sufficiently large $n$ and the channel is indecomposable.

*Necessity.* Pick $\varepsilon < \frac{1}{S}$ and $n$ large enough so that (1.6.21) is satisfied, and for a given $s_0$ and $x^n$, pick $s_n$ such that $q(s_n|x^n, s_0) \geq \frac{1}{S}$. Then from (1.6.21) $q(s_n|x^n, s_0') > 0$ for all $s_0'$, and the condition of the theorem is satisfied for $m$ equal to $n$.

*Proof That* $m < 2^{S^2}$. For a given $m$, $x^m$ define the connectivity matrix

$$T_{m,x^m}(s_0, s_m) = \begin{cases} 1 & \text{if } q(s_m|x^m s_0) > 0 \\ 0 & \text{if } q(s_m|x^m s_0) = 0. \end{cases} \tag{1.6.32}$$

This is an $S \times S$ matrix of 1's and 0's with rows labeled by values of $s_0$, columns by values of $s_m$. A given matrix is 1 if that $s_m$ can be reached from that $s_0$ with the given $x^m$. Since

$$q(s_m|x^m, s_0) = \sum q(s_m|x_m, s_{m-1})q(s_{m-1}|x^{m-1}, s_0)$$

we can express $T_{m,x^m}(s_0, s_m)$ in terms of $T_{m-1,x^m}$ by

$$T_{m,x^m}(s_0, s_m) = \begin{cases} 1 & \text{if } T_{m-1,x^m}(s_0, s_{m-1})q(s_m|x_m s_{m-1}) \quad \text{for some } s_{m-1} \\ 0 & \text{otherwise.} \end{cases} \tag{1.6.33}$$

Since there are only $2^{S^2} - 1$ nonzero $S$ by $S$ matrices with binary entries the sequence of matrices $(T^{(n)}_{m,x^m}(s_0, s_m))_{n=1}^{2^{S^2}}$ must contain two matrices that are the same, say for $i < j \leq 2^{S^2}$. If for this $(x_1, \ldots, x_j)$ we choose $x_{j+n} = x_{i+n}$ for all $n \geq 1$, then from (1.6.33) we see that $T_{j+n,x^n} = T_{i+n,x^n}$ for all $n \geq 1$. For this choice of $x^n$, if $T_{m,x}$ has no column of 1's for $m \leq j$, it will have no column of 1's for any larger $m$. But this means that, for this $x^n$, there is no $m$, $s^m$ for which $q(s_m|x^n, s_0) > 0$ for all $s_0$, and thus the channel is not indecomposable. Thus, for the channel to be indecomposable, for each $x^n$, $T_{m,x^n}$ must have a column of 1's for some $m < 2^{S^2}$.

Finally, if $T_{m,x^n}$ has a column of 1's for some $m$, it follows from (1.6.32) that it has a column of 1's for all larger $m$, and thus, for an indecomposable channel there is some smallest $m \leq 2^{S^2}$ for which $T_{m,x^n}$ has columns of 1's for all $x^n$. □

For the fading channel discussed in the beginning we shall see below that it is indecomposable.

**Theorem 36** *For an indecomposable* FSC

$$\underline{C} = \overline{C}. \tag{1.6.34}$$

*Proof* For arbitrary $n$, let $Q_n(x^n)$ and $s\prime_0$ be the input distribution and initial state that maximize $I(X^n \wedge Y^n|s_0)$ and let $s''_0$ denote the initial state minimizing $I(X^n \wedge Y^n|s''_0)$ for the same input distribution. Then

$$\overline{C}_n = \frac{1}{n} I(X^n \wedge Y^n|s\prime_0) \tag{1.6.35}$$

$$\underline{C}_n = \frac{1}{n} I(X^n \wedge Y^n|s''_0). \tag{1.6.36}$$

Now let $m + l = n$, where $m, l \in \mathbb{Z}^+$. Let $X^m = (X_1, \dots, X_m)$ and $X^{m+1,n} = (X^{m+1}, \dots, X^n)$ with PD $Q_n(x^n)$. Similarly, let $Y^m$ and $Y^{m+1,n}$ be defined, conditional on $s\prime_0$. We then have

$$\overline{C}_n = \frac{1}{n}\Big[(X^m \wedge Y^m Y^{m+1,n}) + I(X^{m+1,n} \wedge Y^M | X^m, s\prime_0) \\ + I(X^{m+1,n} \wedge Y^{m+1,n} | X^m, Y^m, s\prime_0)\Big] \quad (1.6.37)$$

*Remark*

1. While indecomposability of a FSC is sufficient for $\underline{C} = \overline{C}$, it turns out that also many decomposable FSCs have $\underline{C} = \overline{C}$. Channels with intersymbol interference only are particularly easy to analyze in this respect.
2. The indecomposable FSC introduced by Blackwell, Breiman, and Thomasian [16] constitute the same class.
3. FSC with unknown initial state can be treated as compound channels. Gallager derived error bounds in Sect. 5.9 of his book [25] (Theorem 5.9.1).
4. The coding theorem for finite state channels was first proved under somewhat more restrictive conditions by Gallager [25] and, in stronger form, by Blackwell, Breiman, and Thomasian [16]. The random coding exponent for finite state channels is due to Yudkin [47]

   The first sphere-packing bound established for FSC is for a class of binary channels treated by Kennedey [31].

   We conclude our treatment of FSC with the results of Kieffer [34].

### 1.6.4 A Lower Bound on the Probability of Decoding Error for the FSC

We continue with the model of a FSC discussed in Sect. 1.6.1. We also use Gallager's [26, p. 100], definition of informational lower capacity $\underline{C}$ and upper capacity $\overline{C}$ given in Sect. 1.6.2.

As a dual to the random coding bound of Yudkin, Kieffer presented a lower bound for error probabilities of this channel using the method of Arimoto. For rates $R > \overline{C}$ it gives a lower bound $1 - e^{-\alpha n}$ for the error probabilities. If $\underline{C} = \overline{C}$ it gives a strong converse and thus especially for indecomposable channels it includes Wolfowitz's result. Moreover, the strong converse for indecomposable FSC holds not only for maximal, but also for average probability of error, whereas that is not the case for CC (see [1, 12]). Let $\lambda_n(R, s)$ be the smallest error probability of $(n, e^{Rn})$ codes, *when the initial state is* $s$. Analogously we define $\overline{\lambda}_n(R, s)$ as the smallest average error probability.

For a PD $P^n$ on $\mathcal{X}^n$, $s \in \mathcal{S}$, and $\beta > 0$ define

$$E_n(\beta, P^n, s) = -\frac{1}{n} \ln \left[ \sum_{y^n} \left( \sum_{x^n} P^n(x^n) P^n(y^n|x^n, s)^{\frac{1}{\beta}} \right)^{\beta} \right] \tag{1.6.38}$$

The following coding theorem is due to Yudkin [47].

**Theorem 37** *For*

$$F_1(\beta) = \lim_{n \to \infty} \max_{P^n} \min_{s} E_n(\beta, P^n, s), \quad \beta \geq 1,$$

*define*

$$E_1(R) = \sup_{1 \leq \beta \leq 2} [F_1(\beta) - (\beta - 1)R].$$

*If $R < \underline{C}$, then $E_1(R) >)$; furthermore, if $0 < \alpha < E_1(R)$, then*

$$\lambda_n(R, s) \leq e^{-n\alpha} \quad (s \in \mathcal{S}).$$

For a proof we refer to the original paper. However, we shall prove

**Theorem 38** *For*

$$F(\beta) = \lim_{n \to \infty} \min_{P^n, s} E_n(\beta, P^n, s), \quad 0 < \beta \leq 1,$$

*define*

$$E(R) = \sup_{0 < \beta \leq 1} [F(\beta) - (\beta - 1)R].$$

*If $R > \overline{C}$, then $E(R) > 0$; furthermore, if $0 < \alpha < E(R)$, then*

$$\lambda_n(R, s) \geq 1 - e^{-\alpha n} \quad (s \in \mathcal{S})$$

*and n large.*

This says that reliable transmission at rates above $\overline{C}$ is not possible even if the initial state can be set arbitrarily before transmission starts, and is known to both, sender and receiver.

#### 1.6.4.1 Auxiliary Results from Analysis

We shall make use of

- Hölder's inequality

- Minkowski's inequality
- Fekete's lemma

which can all be found in Gallager's book on pp. 523, 524, and 112, and in our discussion of exponents arising in error bounds.

**Lemma 20** *For a sequence* $(f_n)_{n=1}^{\infty}$ *of nonnegative, nonincreasing continuous functions defined on a compact subset* $K$ *of* $\mathbb{E}^1$

$$\lim_{n\to\infty} \max_{r\in K} f_n(r) = \sup_{x\in K} \left(\lim_{n\to\infty} f_n\right).$$

Finally, we need a basic result from Kuhn-Tucker theory [[Hadley], p. 202, Eqs. (49), (52)]

**Lemma 21** *Let* $\mathcal{X}_1, \mathcal{X}_2, \mathcal{Y}_1, \mathcal{Y}_2$ *be finite sets and let*

$$f_i : \mathcal{X}_i \times \mathcal{Y}_i \to \mathbb{R}_+ \quad \text{for } i = 1, 2.$$

*If* $Q$ *is a PD on* $\mathcal{X}_1 \times \mathcal{X}_2$ *and* $Q_1$ *is a PD on* $\mathcal{X}_1$ *defined as marginal PD*

$$Q_1(x_1) = \sum_{x_2} Q(x_1, x_2) \quad (x_1 \in \mathcal{X}_1),$$

*then there exists a PD* $Q_2$ *on* $\mathcal{X}_2$ *such that*

$$\sum_{y_1,y_2} \left[ \sum_{x_1,x_2} Q(x_1, x_2) f_1(x_1, y_1) f_2(x_2, y_2) \right]^{\beta} \tag{1.6.39}$$

$$\leq \sum_{y_1} \left[ \sum_{x_1} Q_1(x_1) f_1(x_1, y_1) \right]^{\beta} \sum_{y_2} \left[ \sum_{x_2} Q_2(x_2) f_2(x_2, y_2) \right]^{\beta} \quad \text{for } 0 < \beta \leq 1.$$

*Proof* Let

$$\mathcal{M} = \left\{ P \in \mathcal{P}(\mathcal{X}_1 \times \mathcal{X}_2) : \sum_{x_2} P(x_1, x_2) = Q_1(x_1) \quad \text{for } x_1 \in \mathcal{X}_1 \right\}.$$

The function $f : \mathcal{M} \to \mathbb{R}_+$ defined by

$$f(P) = \sum_{y_1,y_2} \left[ \sum_{x_1,x_2} P(x_1, x_2) f_1(x_1, y_1) f_2(x_2, y_2) \right]^{\beta}$$

is concave. The cited result implies that $P$ maximizes $f$ over $\mathcal{M}$ if for each $x_1 \in \mathcal{X}_1$ such that $Q_1(x_1) > 0$, it is true that $\frac{\partial f}{\partial P(x_1,x_2)} \leq \gamma(x_1)$ for each $x_2 \in \mathcal{X}_2$, where

$$\gamma(x_1) = Q_1(x_1)^{-1} \sum_{x_2} P(x_1, x_2) \left[ \frac{\partial f}{\partial P(x_1, x_2)} \right].$$

Using this criterion, it is easily seen that *product measure* $P = Q_1 \times Q_2$ maximizes $f$ over $\mathcal{M}$ if

$$\sum_{y_2} \left[ \sum_{x_2} Q_2(x_2) f_2(x_2, y_2) \right]^{\beta-1} f_2(x_2', y_2) \leq \sum_{y_2} \left[ \sum_{x_2} Q_2(x_2) f_2(x_2, y_2) \right]^{\beta} \quad \text{for every } x_2' \in \mathcal{X}_2 \tag{1.6.40}$$

Consider the function $g : \mathcal{P}(\mathcal{X}_2) \to \mathbb{R}_+$ defined by

$$g(P) = \sum_{y_2} \left[ \sum_{x_2} P(x_2) f_2(x_2, y_2) \right]^{\beta}.$$

Choose $Q_2 \in \mathcal{P}(\mathcal{X}_2)$ that maximizes the concave function $g$. It follows from the Kuhn-Tucker theory, Theorem 4.4.1, that the condition $\frac{\partial g}{\partial (Q_2(x_2))} \leq \gamma$ holds for each $x_2 \in \mathcal{X}_2$, where

$$\gamma = \sum_{x_2} Q_2(x_2) \frac{\partial g}{\partial Q_2(x_2)}.$$

This condition simplifies to inequality (1.6.40). Therefore $Q_2 \times Q_2$ maximizes $f$ over $\mathcal{M}$. in particular $f(Q) \leq f(Q_1 \times Q_2)$, which is precisely the inequality (1.6.39). □

#### 1.6.4.2 Auxiliary Results for Error Bounds

**Lemma 22** *Define* $E_n(\beta) = \min_{P^n, s} E_n(\beta, P^n, s)$. *Then*

$$\lambda_n(R, s) \geq 1 - exp\left[-n(E_n(\beta) - (\beta - 1)R)\right], \quad 0 < \beta \leq 1 \tag{1.6.41}$$

*Proof* Arimoto's bound applies to an (abstract) channel (without time structure), see inequality (9) of [15]. □

**Lemma 23** *For fixed* $P^n$ *and* $s$, $E_n(\beta, P^n, s)$ *is a concave nonincreasing function of* $\beta$, $\beta > 0$. *Furthermore*

$$-\ln |\mathcal{Y}| \leq E_n(\beta, P^n, s) \leq 0 \quad for \ 0 < \beta \leq 1; \tag{1.6.42}$$

*and*

$$\frac{\partial}{\partial \beta} E_N(\beta, P^n, s)|_{\beta=1} = n^{-1} I(P^n, s).$$

**Lemma 24** *Define* $F_n(\beta) = E_n(\beta) + (\beta - 1)n^{-1} \ln |\mathcal{S}|$. *If* $l + m = n$, *then*

$$nF_n(\beta \geq lF_l(\beta) + mF_m(\beta), \quad 0 < \beta \leq 1. \tag{1.6.43}$$

*Proof* It suffices to show that

$$\begin{aligned} &\sup_{s, P^n} \exp[-nE_n(\beta, P^n, s)] \\ &\leq |\mathcal{S}|^{\beta-1} \sup_{s, P^m} \exp[-mE_m(\beta, P^m, s)] \cdot \sup_{s, \lambda} \exp[-lE_l(\beta, P^l, s)] \end{aligned} \tag{1.6.44}$$

Fix $s' \in \mathcal{S}$ and $Q \in \mathcal{P}(\mathcal{X}^n)$. Then

$$\begin{aligned} \exp[-nE_n(\beta, Q, s')] &= \sum_y \left[ \sum_x Q(x) P^n(y|x, s')^{1/\beta} \right]^\beta \\ &= \sum_{y_1, y_2} \left[ \sum_{x_1, x_2} Q(x_1, x_2) \left[ \sum_s P^m(y_1, s|x_1, s') P^l(y_2|x_2, s) \right]^{1/\beta} \right]^\beta \end{aligned}$$

where in these sums $x \in \mathcal{X}^n$, $y \in \mathcal{Y}^n$, $x_1 \in \mathcal{X}^m$, $y_1 \in \mathcal{Y}^m$, $x_2 \in \mathcal{X}^l$, and $y_2 \in \mathcal{Y}^l$, a convention that will be followed throughout the proof.

The preceding sum is not greater than

$$\sum_{y_1, y_2} \sum_s \left[ \sum_{x_1, x_2} Q(x_1, x_2) P^m(y_1, s|x_1, s')^{1/\beta} P^l(y_2|x_2, s)^{1/\beta} \right]^\beta$$

by a variant of Minkowski's inequality involving PDs ([26] p. 524) let $Q'$ be the PD on $\mathcal{X}^m$ such that

$$Q'(x_1) = \sum_{x_2} Q(x_1, x_2), \quad x_1 \in \mathcal{X}^m.$$

By Lemma 21 for each $s \in \mathcal{S}$ there exists a PD $Q_s$ on $\mathcal{X}^l$ such that

$$\begin{aligned} &\sum_{y_1, y_2} \left[ \sum_{x_1, x_2} Q(x_1, x_2) P^m(y_1 s|x_1 s')^{1/\beta} P^l(y_2|x_2, s)^{1/\beta} \right]^\beta \\ &\leq \sum_{y_1} \left[ \sum_{x_1} Q'(x_1) P^m(y_1, s|x_1, s')^{1/\beta} \right]^\beta \cdot \sum_{y_2} \left[ \sum_{x_2} Q_s(x_2) P^l(y_2|x_2, s)^{1/\beta} \right]^\beta \end{aligned}$$

Therefore

$$\exp[-nE_n(\beta, Q, s')] \\ \leq \sum_s \sum_{y_1} \left[ \sum_{x_1} Q'(x_1) P^m(y_1, s|x_1, s')^{1/\beta} \right]^\beta \cdot \sup_{P^l, s} \exp[-lE_l(\beta, P^l, s)].$$

Now by Hölder's inequality for nonnegative functions $\varphi$ and $\psi$ defined on $\mathcal{S}$

$$\sum_s \varphi(s)^\beta = |\mathcal{S}| \sum_s |\mathcal{S}|^{-1} \varphi(s)^\beta \leq |\mathcal{S}| \left[ \sum_s |\mathcal{S}|^{-1} \varphi(s) \right]^\beta = |\mathcal{S}|^{1-\beta} \left[ \sum_s \varphi(s) \right]^\beta$$

and

$$\sum_s \psi(s)^{1/\beta} \leq \left[ \sum_s \psi(s) \right]^{1/\beta}$$

for $0 < \beta \leq 1$. These inequalities imply that

$$\sum_s \sum_{y_1} \left[ \sum_{x_1} Q'(x_1) P^m(y_1, s|x_2, s')^{1/\beta} \right]^\beta \\ \leq |\mathcal{S}|^{1-\beta} \sum_{y_1} \left[ \sum_s \sum_{x_1} Q'(x_1) P^m(y_1, s|x_1, s')^{1/\beta} \right]^\beta \\ \leq |\mathcal{S}|^{1-\beta} \sum_{y_1} \left[ \sum_{x_1} Q'(x_1) \left[ \sum_{\hat{s}} P^m(y_1, s|x_1, s') \right]^{1/\beta} \right]^\beta \\ = |\mathcal{S}|^{1-\beta} \exp[-mE_m(\beta, Q', s')].$$

The result easily follows. □

**Lemma 25** *Define* $F(\beta) = \sup_n F_n(\beta)$. *Then for each* $\beta$, $0 < \beta \leq 1$,

$$\lim_{n \to \infty} E_n(\beta) = F(\beta).$$

*The convergence is uniform over compact subintervals of the interval* $0 < \beta \leq 1$.

*Proof* For $0 < \beta \leq 1$ $(F_n(\beta))_{n=1}^\infty$ is a bounded sequence, since by Lemma 23 $-\ln |\mathcal{Y}| \leq E_n(\beta) \leq 0$. Therefore, since $nF_n(\beta) \geq lF_l(\beta) + mF_m(\beta)$ for $l + m = n$ by Fekete's Lemma (see also [26, p. 112],) $\lim_{n\to\infty} E_n(\beta)$ exists and equals $F(\beta)$. Moreover, Lemma 23 states that $E_n(\beta, P^n, s)$ is concave in $\beta$ and therefore

$$\frac{\partial}{\partial \beta} E_n(\beta, P^n, s) \leq \frac{E_n(\beta, P^n, s) - E_n(\beta', P^n, s)}{\beta - \beta'}$$

where $0 < \beta' < \beta \le 1$. Since

$$-\ln|\mathcal{Y}| \le E_n(\beta, P^n, s) \le 0, \quad \text{for } 0 < \beta \le 1,$$

we have

$$\frac{\partial}{\partial\beta} E_n(\beta, P^n, s) \le \frac{-E_n(\beta', P^n, s)}{\beta - \beta'} \le \frac{\ln|\mathcal{Y}|}{\beta - \beta'}.$$

Letting $\beta' \to 0$, we see that

$$\frac{\partial}{\partial\beta} E_n(\beta, P^n, s) \le \beta^{-1} \ln|\mathcal{Y}|.$$

Let $\beta_0 > 0$. Applying the mean value theorem, we see that

$$0 \le E_n(\beta) - E_n(\beta') \le \beta_0^{-1}(\beta - \beta') \ln|\mathcal{Y}|$$

for $\beta_0 \le \beta' \le \beta \le 1$ and all $n$. This implies the uniform convergence of $E_n(\beta)$ on the compact interval $\beta_0 \le \beta \le 1$. □

**Lemma 26**

$$\limsup_{\beta \nearrow 1} \frac{F(\beta)}{\beta - 1} \le \overline{C}$$

*Proof* For fixed $P^n, s$, the concavity of $E_n(\beta, P^n, s)$ implies that

$$\frac{E_n(\beta, P^n, s)}{\beta - 1} \le \frac{\partial}{\partial\beta} E_n(\beta, P^n, s), \quad 0 < \beta < 1.$$

Thus

$$\frac{E_n(\beta)}{\beta - 1} \le \sup_{P^n, s} \frac{\partial}{\partial\beta} E_N(\beta, P^n, s).$$

Since $\beta - 1 \le 0$ we then have

$$\frac{F(\beta)}{\beta - 1} \le \frac{F_n(\beta)}{\beta - 1} = n^{-1} \ln|\mathcal{S}| + \frac{E_n(\beta)}{\beta - 1}$$
$$\le n^{-1} \ln|\mathcal{S}| + \sup_{P^n, s} \frac{\partial}{\partial\beta} E_n(\beta, P^n, s).$$

Let $\beta \nearrow 1$, with Lemma 20 we conclude that

$$\begin{aligned}
&\lim_{\beta\nearrow 1}\sup_{P^n,s}\frac{\partial}{\partial\beta}E_n(\beta,P^n,s)\\
&=\sup_{P^n,s}\lim_{\beta\nearrow 1}\frac{\partial}{\partial\beta}E_n(\beta,P^n,s)\\
&=\sup_{P^n,s} n^{-1}I(P^n,s);
\end{aligned}$$

we obtain

$$\lim_{\beta\nearrow 1}\sup\frac{F(\beta)}{\beta-1}\leq n^{-1}\ln|S|+sup_{P^n,s}n^{-1}I(P^n,s).$$

Taking the infimum over $n$, the lemma follows. □

*Remark* It can be shown that $\lim_{\beta\nearrow 1}F(\beta)/\beta-1=\overline{C}$, but this stronger result is not needed.

We are now ready for the next step

*Proof of Theorem 38.* It follows easily from Lemma 26 that $E(R)>0$. Let $\alpha<E(R)$ and then choose $0<\beta_0<1$ so that

$$\sup_{\beta_0\leq\beta\leq 1}\left[F(\beta)-(\beta-1)R\right]>\alpha.$$

Since $E_n(\beta)\to F(\beta)$ uniformly for $\beta_0\leq\beta\leq 1$, we have that $\sup_{\beta_0\leq\beta\leq 1}[F(\beta)-(\beta-1)R]>\alpha$ for $N$ sufficiently large. We now apply Lemma 22.

#### 1.6.4.3 The Finite-State Channel with State Calculable only by the Sender

Consider the channel of the previous section but now with the side information only at the sender.

The treatment is such that blocks of length $l$ are used and capacity $C(l)$ of the corresponding DMC yielding as capacity $C=\lim_{l\to\infty}\frac{C(l)}{l}$ and

$$\begin{aligned}
&\frac{C(l)}{l}-C\leq\frac{d\log a}{d+l},\quad C=\inf_l\frac{C(l)}{l},\ C\leq\frac{C(l)}{l}\\
&\left|\frac{C(l)}{l}-C\right|\leq\frac{d\log a}{d+l}
\end{aligned}$$

and can be computed, at least in principle to within any specified degree of accuracy.

## 1.7 Lecture on Information and Control: Matching Channels

The transmission problem for noisy channels is usually studied under the condition that the decoding error probability $\lambda$ is small and is sometimes studied under the condition that $\lambda = 0$. Here we just require that $\lambda < 1$ and obtain a problem which is equivalent to a coding problem with small $\lambda$ for the "Deterministic Matching channel." In this new model, a cooperative person knows the code word to be sent and can choose (match) the state sequence of the channel. There are interesting connections to combinatorial matching theory and extensions to the theory of identification as well as to multi-way channels. In particular there is a surprising connection to Pinsker's coding theorem for the deterministic broadcast channel.

### *1.7.1 New Concepts and Results*

#### 1.7.1.1 The Matching Channel

Let $\mathcal{X}$ serve as input alphabet and let $\mathcal{Y}$ serve as output alphabet. By adding dummy letters we can always assume that $\mathcal{X} \subset \mathcal{Y}$. The transmission of letters is ruled by a class $\mathcal{W}$ of stochastic matrices with $|\mathcal{X}|$ rows and $|\mathcal{Y}|$ columns as follows. In addition to sender and receiver, there is a third person (or device) called controller who decides which matrix $W \in \mathcal{W} = \{w(\cdot|\cdot|s) : s \in \mathcal{S}\}$ shall govern the transmission of a letter by the sender. The controller knows which code word the sender wants to transmit. The receiver has no knowledge about the actions of the controller. As code concept appropriate for this situation we introduce a matching code (MC). We call $\{(u_i, \mathcal{D}_i) : 1 \leq i \leq M\}$ an $(n, M, \lambda)$-MC-code for $\mathcal{W}$, if

$$u_i \in \mathcal{X}^n, \mathcal{D}_i \subset \mathcal{Y}^n \quad \text{for } i = 1, 2, \ldots, M, \tag{1.7.1}$$

$$u_i \neq u_j \text{ and } \mathcal{D}_i \cap \mathcal{D}_j = \varnothing \quad \text{for } i \neq j, \tag{1.7.2}$$

and if for every $i$ there is a sequence $s^n(i) = \big(s_1(i), \ldots, s_n(i)\big) \in \mathcal{S}^n = \prod_1^n \mathcal{S}$ with

$$W^n\big(\mathcal{D}_i|u_i|s^n(i)\big) \geq 1 - \lambda, \tag{1.7.3}$$

if $W^n\big(y^n|x^n|s^n(i)\big) = \prod_{t=1}^n W\big(y_t|x_t|s_t(i)\big)$ for $x^n = (x_1, \ldots, x_n) \in \mathcal{X}^n$ and $y^n = (y_1, \ldots, y_n) \in \mathcal{Y}^n$. Let $C(\mathcal{W})$ be the capacity of the matching channel $\mathcal{W}$.

As usual we denote by $X, S, Y$ RVs with values in $\mathcal{X}$, $\mathcal{S}$, and $\mathcal{Y}$, resp. Let $P_{XS}$ be the joint distribution of $(X, S)$ and $P_{XSY}(x, s, y) = P_{XS}(x, s)W(y|x, s)$ for $x \in \mathcal{X}$, $s \in \mathcal{S}$, and $y \in \mathcal{Y}$.

**Theorem 39** *The capacity of the matching channel is given by*

$$C(\mathcal{W}) = \max_{P_{XS}} \min\big(H(X), I(XS \wedge Y)\big).$$

Notice that the quantity $C = \max_{P_{XS}} I(XS \wedge Y)$ is the capacity of the corresponding discrete memoryless channel $(\mathcal{X} \times \mathcal{S}, \mathcal{Y}, W')$ (or for the model, where the controller knows not only the code word but even the message to be send) and that therefore $C(\mathcal{W}) \leq C$.

The minimization with $H(X)$ reflects the fact that only pairs $(x^n, s^n)$, which are all different in the first component are permitted in the encoding. Obviously, choosing $M \sim \exp\{n \min(H(X), I(XS \wedge Y)\}$ of such pairs, say $\{(x_i^n, s_i^n) : 1 \leq i \leq M\}$, independently with distribution $P_{XS}^n$ results with high probability in a code, for which most $x_i^n$'s are different—and those we keep! This gives the direct part of Theorem 39 and the converse part is also obvious.

*Remark*

1. In case the controller is restricted to choose only state sequences $(s, s, \dots, s)$, $s \in S$, we are led to the "optimistic" channel of [1] (see Sect. 1.2).
2. Jim Massey conveyed the following interpretation to us:
   One can speak of "Coding with a Barrister" for the following reason. In the British system of law there are two kinds of lawyers, solicitors and barristers. The solicitor is the lawyer who prepares the case, but only the barrister is permitted to argue the case before the court. In the American system of law, the same lawyer usually performs both functions. Previously in coding theory, the "encoder" performed like an American lawyer, both mapping the message into a code word then transmitting this code word over the channel. The new feature of the present model is that the "encoder" acts like a "solicitor," only mapping the message into a code word. It is then the "barrister" who transmits the code word over the channel. Of course, if the barrister knew the message, there would be nothing new.

#### 1.7.1.2 The Deterministic Matching Channel

It is instructive to consider the case where $\mathcal{W} = \mathcal{W}_0$ contains only 0-1-matrices.

Then Theorem 39 has the following specialization.

**Theorem 40**

$$C(\mathcal{W}_0) = \max_{P_{XS}} \min\big(H(X), H(Y)\big).$$

Clearly, by the definition of an $(n, M, \lambda)$-MC-code we can assume now that $|\mathcal{D}_i| = 1$ and thus $\mathcal{D}_i = \{v_i\}$. Also if $\lambda < 1$, then it can actually be chosen to equal 0. So the distinct $u_i$'s are matched with distinct $v_i$'s. In determining the capacity $C(\mathcal{W}_0)$ we are thus led to a *new probabilistic coding theory*, whose mathematical structure is interesting and natural: *a novel combinatorial matching theory for products of bipartite graphs*.

It is convenient to work with an equivalent formulation of coding problems for $\mathcal{W}_0$ in terms of an *associated* DMC $W$:

$$W(\cdot|x) = \sum_{s\in\mathcal{S}} Q(s)W(\cdot|x|s), \qquad (1.7.4)$$

where $Q$ is any probability distribution on $\mathcal{S}$ with $Q(s) > 0$ for $s \in \mathcal{S}$.

Since for the DMC $W^n(y^n|x^n) = \prod_{t=1}^n W(y_t|x_t)$, one notices that for any $(n, M, 0)$
-MC-code $\{(u_i, \mathcal{D}_i) : 1 \leq i \leq M\}$ for $\mathcal{W}_0$ the condition (1.7.3) can equivalently be described in the "dummy" formulation by

$$W^n(\mathcal{D}_i|u_i) > 0 \quad \text{for} \quad i = 1, 2, \ldots, M. \qquad (1.7.5)$$

It is mathematically and aesthetically quite appealing that by weakening the requirements on the error performance we are led from zero-error codes [4], where

$$W^n(\mathcal{D}_i|u_i) = 1 \quad \text{for} \quad i = 1, \ldots, M, \qquad (1.7.6)$$

to the most familiar $\lambda$-error codes ($\lambda > 0$) with

$$W^n(\mathcal{D}_i|u_i) \geq 1 - \lambda \quad \text{for} \quad i = 1, \ldots, M, \qquad (1.7.7)$$

to MC-codes for $\mathcal{W}_0$ in the "dummy" formulation.

The corresponding capacities $C_0(W)$, $C(W)$, and $C(\mathcal{W}_0)$ satisfy of course

$$C_0(W) \leq C(W) \leq C(\mathcal{W}_0). \qquad (1.7.8)$$

There is a code concept between the first two, namely that of an erasure code, for which in addition to (1.7.7) we also have

$$W^n(\mathcal{D}_j|u_i) = 0 \quad \text{for } i \neq j. \qquad (1.7.9)$$

The zero-error erasure capacity $C_{er}(W)$ has been studied in several papers and recently quite intensively by several authors (c.f. [8]). Until now no "single-letter" formula exists.

Quite analogously we can also require (1.7.9) in conjunction with (1.7.5). This gives exactly the MDC-code defined in part D in terms of $\mathcal{W}_0$.

It is interesting that coding for $\mathcal{W}_0$ is equivalent with matching in products of bipartite graphs (see 1.7.3). This connection leads to a combinatorial version of Theorem 40, which is stated as Theorem 44 in Sect. 1.7.4. It has a nice direct proof with König's Minimax Theorem. Beyond this result on the asymptotic behavior of matching numbers under products, we give an exact result for two factors in Theorem 45 (1.7.5). This enables us to get also exact results for powers of certain bipartite graphs (Theorem 46 in Sect. 1.7.6). Finally, in Sect. 1.7.7 we underline with two examples the significance of Theorems 45, 46.

### 1.7.1.3 Multi-way Matching Channels

The concept of a matching DMC has straightforward extensions to several sender and receiver models. A highlight in Sect. 1.7.8 is the solution of the *general* broadcast problem in this matching theory. As a special case of Theorem 49 we obtain Pinsker's [38] capacity region for the deterministic broadcast channel.

The corresponding Theorems 47 and 48 for compound and multiple-access channels are stated without their (routine) proofs.

### 1.7.1.4 The Controller Falls Asleep

It seems to us that the interplay between information transfer and controlling certain channels deserves more and deeper investigations. Channels with control aspects are the permuting relay channels of [22, 24] and the outputwise varying channels of [10], which arose in the study of rewritable storage media. We also draw attention to [13] for still another philosophy: controlling by creating order.

Now we are more specific. In the model described in A. above the controller is not only assumed to be cooperative, but he also acts perfectly. Next the communicators safeguard against mistakes of the controller and even against malicious operations (jamming) by using matching zero-error detection codes (MDC) $\{(u_i, \mathcal{D}_i) : 1 \leq i \leq M\}$ which in addition to (1.7.3) (automatically with $\lambda = 0$) satisfy

$$W^n(\mathcal{D}_j|u_i|s^n) = 0 \quad \text{for } i \neq j \quad \text{and all } s^n \in \mathcal{S}^n. \tag{1.7.10}$$

To determine its capacity $C_{\text{mde}}(\mathcal{W})$ is a formidable task. For the deterministic matching channel $\mathcal{W}_0$ results and relations to other zero-error capacities are contained in Sect. 1.7.9. Section 1.7.14 contains instructive examples and Theorem 55 as the main contribution on the relation. For one genuine channel, the $\binom{\alpha}{\beta}$-uniform hypergraph channel $W_{\alpha,\beta}$, we succeeded in determining the capacity in Sect. 1.7.10.

As in the classical AWAC system we assume here that there is a noiseless feedback channel or just an active feedback channel on which the receiver can ask for retransmission. The frequency of such retransmissions depends on the error frequency (the sleeping habits) of the controller.

We emphasize again that in A it makes a big difference whether the controller knows the message (and thus the same word can be used in conjunction with different state sequences to represent the different messages) or only the code word (and thus no word can represent different messages).

In the present situation it makes no sense to assume that the controller knows in addition to the code word the messages, because the code words have to be different, anyhow, to cope with a sleeping controller.

### 1.7.1.5 Matching Zero-Error Detection Codes with Feedback for $\mathcal{W}_0$ (MDCF)

We mention first that feedback or also randomization in the encoding increases the capacity of the matching channel $\mathcal{W}$ to the effect that the term $H(X)$ has to be dropped in the formula of Theorem 39. This is stated in (1.7.101) and proved in Sect. 1.7.11.

We turn now to the MDCF.

The feedback is now really used in the design of the code. There is given a finite set of messages $\mathcal{M} = \{1, 2, \dots, M\}$. One of these messages is to be sent over the channel. Message $i \in \mathcal{M}$ is encoded by a (vector-valued) function

$$f_i^n = [f_{i1}, f_{i2}, \dots, f_{in}]$$

where, for $t \in \{2, \dots, n\}$, $f_{it}$ is defined on $\mathcal{Y}^{t-1}$ and takes values in $\mathcal{X}$. $f_{i1}$ is an element of $\mathcal{X}$. It is understood that after the received elements $Y_1, \dots, Y_{t-1}$ have been made known to the sender by the feedback channel, the sender transmits $f_{it}(Y_1, \dots, Y_{t-1})$. At $t = 1$ the sender transmits $f_{i1}$. Again, we assume that the controller knows only the encoding functions, but not the messages, and therefore $f_i^n \neq f_j^n$ if $i \neq j$. The distribution of the RVs $Y_t$ $(t = 1, 2, \dots, n)$ is determined by $f_i^n$ and $W(\cdot | \cdot | s^n)$. We denote the probability of receiving $y^n = (y_1, \dots, y_n) \in \mathcal{Y}^n$, if $i$ has been encoded and the controller uses $s^n$, by

$$W^n(y^n | f_i^n | s^n) = W(y_1 | f_{i1} | s_1) W\big(y_2 | f_{i2}(y_1) | s_2\big) \cdots W\big(y_n | f_{in}(y_1, \dots, y_{n-1}) | s_n\big).$$

In an $(n, M, \lambda)$ matching zero-error detection feedback code $\big\{(f_i, \mathcal{D}_i, s_i^n) : 1 \leq i \leq M\big\}$ the $\mathcal{D}_i$ are disjoint subsets of $\mathcal{Y}^n$ and

$$W^n(\mathcal{D}_i | f_i | s_i^n) \geq 1 - \lambda \quad \text{for } i = 1, \dots, N; \; W^n(\mathcal{D}_j | f_i | s^n) = 0 \quad \text{for all } s^n \in \mathcal{S}^n, j \neq i. \tag{1.7.11}$$

We are interested in the capacity $C_{m\,\mathrm{def}}(\mathcal{W}_0)$. Here we can assume the $\mathcal{D}_i$'s to have one element, say the $v_i$'s. Our optimism for finding a nice formula was originally just speculative: in case of feedback Shannon found also a nice formula for his zero-error capacity! Indeed we have a surprising result, which is proved in Sect. 1.7.12.

**Theorem 41** *(i)* $C_{m\,\mathrm{def}}(\mathcal{W}_0) = \begin{cases} \max\limits_{P_{XS}} I(XS \wedge Y) \\ or \\ 0 \end{cases}$

*(ii)* $C_{m\,\mathrm{def}}(\mathcal{W}_0) = 0$ *exactly if all columns have positive or zero entries only.*

The astute reader may notice that Shannon's formula or the alternate formula of [4] (asked for by Shannon in [39]) has also a dichotomy relative to positivity. The formula in [4] describes the capacity of a jamming problem, namely that of an arbitrarily varying channel with feedback. Our formula for $C_{m\,\mathrm{def}}(\mathcal{W}_0)$ also settles a feedback problem involving jamming.

**Problem** Is there a common generalization of both jamming problems?

#### 1.7.1.6 Identification

We emphasize that a systematic analysis of code concepts is still rewarding. By giving up the disjointness of the decoding sets and by requiring in addition to (1.7.7)

$$W^n(\mathcal{D}_j|u_i) \leq \lambda \quad \text{for } i \neq j \tag{1.7.12}$$

we get the concept of a (nonrandomized) identification code [9]. Randomization means here that instead of $u_i \in \mathcal{X}^n$ we allow $Q_i \in \mathcal{P}(\mathcal{X}^n)$, the set of PDs on $\mathcal{X}^n$.

In [8] we assumed (1.7.12) in conjunction with (1.7.6), that is, zero-error probability for misrejection and found that here the second order identification capacity equals the (first order) erasure capacity $C_{\text{er}}$.

Now we combine for instance (1.7.5) and (1.7.9), that is, identification with zero probability of misacceptance. Actually, we analyze all possible capacity concepts in Theorems 52, 53, 54 in Sect. 1.7.13.

#### 1.7.1.7 Further Code Concepts Leading to New Combinatorial Extremal Problems

Finally, we present and analyze in Definitions 5–7 in Sect. 1.7.15 pairwise zero-error detection codes, component-pairwise zero-error detection codes and pseudo-matching zero-error detection codes (Theorem 56). We comment also on other concepts. *With the only exception of* Sect. 1.7.11 *we consider from now on the deterministic* $\mathcal{W}_0$ *or an associated DMC.*

### *1.7.2 Definitions, Known Facts, and Abbreviations*

We use essentially the terminology of [9].

1. **Sets, Channels, Types, Generated Sequences** Script capitals $\mathcal{X}, \mathcal{Y}, \ldots,$ denote finite sets. The cardinality of a set $\mathcal{A}$ is denoted by $|\mathcal{A}|$. $\binom{\mathcal{A}}{k}$ is the family of all $k$ element subsets of the set $\mathcal{A}$. The letters $P, Q$ always stand for probability distributions on finite sets. $X, Y, \ldots$ denote random variables (RVs). The functions "log" and "exp" are understood to be to the base 2. For a stochastic $|\mathcal{X}| \times |\mathcal{Y}|$-matrix $W$ we have already defined the transmission probabilities $W^n$ of a DMC, and we have also introduced $\mathcal{P}(\mathcal{X}^n)$ as the set of PDs on $\mathcal{X}^n$. We abbreviate $\mathcal{P}(\mathcal{X})$ as $\mathcal{P}$. $\mathcal{V}$ denotes the set of all channels $V$ with input alphabet $\mathcal{X}$ and output alphabet $\mathcal{Y}$. For positive integers $n$ we set

$$\mathcal{P}_n = \{P \in \mathcal{P} : P(x) \in \{0, 1/n, 2/n, \dots, 1\} \quad \text{for all } x \in \mathcal{X}\}.$$

For any $P \in \mathcal{P}_n$, called ED or $n$-ED, we define the set

$$\mathcal{V}_n(P) = \left\{V \in \mathcal{V} : V(y \mid x) \in \left\{0, \frac{1}{nP(x)}, \frac{2}{nP(x)}, \dots 1\right\}, \quad x \in \mathcal{X}, y \in \mathcal{Y}\right\}.$$

For $x^n \in \mathcal{X}^n$ we define for every $x \in \mathcal{X}$

$$P_{x^n}(x) = \frac{1}{n} \cdot \text{ (number of occurrences of } x \text{ in } x^n).$$

$P_{x^n}$ is a member of $\mathcal{P}_n$ by definition. $P_{x^n}$ is called ED of $x^n$. Similarly, we define the ED $P_{x^n y^n}$ for pairs $(x^n, y^n) \in \mathcal{X}^n \times \mathcal{Y}^n$. For $P \in \mathcal{P}$ the set $\mathcal{T}_P^n$ of all P-typical sequences in $\mathcal{X}^n$ is given by

$$\mathcal{T}_P^n = \{x^n : P_{x^n} = P\}.$$

For a RV $Z$ with distribution $P_Z$ we abbreviate $\mathcal{T}_{P_Z}^n$ as $\mathcal{T}_Z^n$, and when we emphasize that $P$ is a distribution on $\mathcal{Z}$ and that $\mathcal{T}_Z^n \subset \mathcal{Z}^n$, then we write $\mathcal{Z}^n(P)$ instead of $\mathcal{T}_Z^n$.

For $V \in \mathcal{V}$, a sequence $y^n \in \mathcal{Y}^n$ is said to be V-generated by $x^n$ if, for all $x \in \mathcal{X}$, $y \in \mathcal{Y}$,

$$P_{x^n y^n}(x, y) = P_{x^n}(x) \cdot V(y \mid x).$$

The set of those sequences is denoted by $\mathcal{T}_V^n(x^n)$. Notice that $\mathcal{T}_P^n \neq \varnothing$ if and only if $P \in \mathcal{P}_n$ and $\mathcal{T}_V^n(x^n) \neq \varnothing$ if and only if $V \in \mathcal{V}_n(P_{x^n})$. For the pair of RV's $(X, Y)$ with $\Pr(Y = y|X = x) = V(y|x)$ we write also $\mathcal{T}_{Y|X}^n(x^n)$ instead of $\mathcal{T}_V^n(x^n)$. For $P \in \mathcal{P}$, $V \in \mathcal{V}$ we write PV for the PD on $\mathcal{Y}$ given by

$$PV(y) = \sum_x P(x)V(y \mid x), \ y \in \mathcal{Y}.$$

$\mathcal{T}_{PV}^n$ is the set of PV-typical sequences in $\mathcal{Y}^n$.

**2. Entropy and Information Quantities** Let $X$ be an RV with values in $\mathcal{X}$ and distribution $P \in \mathcal{P}$, and let $Y$ be an RV with values in $\mathcal{Y}$ such that the joint distribution of $(X, Y)$ on $\mathcal{X} \times \mathcal{Y}$ is given by

$$\Pr(X = x, Y = y) = P(x) \cdot V(y \mid x), \ V \in \mathcal{V}.$$

We write $H(P)$, $H(V \mid P)$, and $I(P, V)$ for the entropy $H(X)$, the conditional entropy $H(Y \mid X)$, and the mutual information $I(X \wedge Y)$, respectively. For $P$, $\widetilde{P} \in \mathcal{P}$

$$D(\widetilde{P} \| P) = \sum_x \widetilde{P}(x) \log \frac{\widetilde{P}(x)}{P(x)}$$

denotes the relative entropy and for $V, \widetilde{V} \in \mathcal{V}$ the quantity

$$D(\widetilde{V} \| V \mid P) = \sum_x P(x) D\big(\widetilde{V}(\cdot|x) \| V(\cdot|x)\big)$$

for the conditional relative entropy.

**3. Elementary Properties of Typical Sequences and Generated Sequences**

$$|\mathcal{P}_n| \le (n+1)^{|\mathcal{X}|}$$
$$|\mathcal{V}_n(P)| \le (n+1)^{|\mathcal{X}|\cdot|\mathcal{Y}|}$$
$$(n+1)^{-|\mathcal{X}|} \cdot \exp\{nH(P)\} \le |\mathcal{T}_P^n| \le \exp\{nH(P)\} \tag{1.7.13}$$

for $P \in \mathcal{P}_n$

$$|\mathcal{T}_V^n(x^n)| \ge (n+1)^{-|\mathcal{X}|\cdot|\mathcal{Y}|} \cdot \exp\{nH(V|P)\}$$
$$|\mathcal{T}_V^n(x^n)| \le \exp\{nH(V|P)\}$$

for $P \in \mathcal{P}_n$, $V \in \mathcal{V}_n(P)$, $x^n \in \mathcal{T}_P^n$, and

$$W^n\big(\mathcal{T}_V^n(x^n)|x^n\big) \le \exp\{-nD(V \| W|P)\},$$
$$W^n\big(\mathcal{T}_V^n(x^n)|x^n\big) \ge (n+1)^{-|\mathcal{X}|\cdot|\mathcal{Y}|} \cdot \exp\{-nD(V \| W|P)\}$$

for $P \in \mathcal{P}_n$, $V \in \mathcal{W}_n(P)$, $x^n \in \mathcal{T}_P^n$, $y^n \in \mathcal{T}_V^n(x^n)$, and $W \in \mathcal{V}$.

Let $T_{V,\delta}^n(x^n) = \big\{y^n \in T_W^n(x^n) : \sum_{x,y} |W(y|x) - V(y|x)| < n\delta\big\}$.

For $V \in \mathcal{V}_n(P_{x^n})$ one can always find a sequence $(\delta_n)_{n=1}^{\infty}$ with $\lim_{n\to\infty} \delta_n = 0$ and $\lim_{n\to\infty} \sqrt{n}\delta_n = \infty$ such that

$$V^n\big(T_{V,\delta_n}^n(x^n)|x^n\big) \to 1 \quad \text{as } n \to \infty.$$

Moreover for any pair of RVs $(X, Y)$ with $T_{XY}^n \ne \varnothing$, we always have $T_{XY}^n = \dot{\bigcup}_{x^n \in T_X^n} \{x^n\} \times T_{Y|X}^n(x^n)$ and therefore

$$|T_{XY}^n| = |T_X^n||T_{Y|X}^n(x^n)| \quad \text{for all } x^n \in T_X^n. \tag{1.7.14}$$

It is used in Sect. 1.7.8.

### *1.7.3 The Deterministic Matching Channel and Matching in Products of Bipartite Graphs*

We have shown in the Introduction that one can associate with the deterministic matching channel $\mathcal{W}_0$ a DMC $W$ given by (1.7.4) so that any $(n, M, 0)$-MC-code

$\{(u_i, D_i) : 1 \leq i \leq 1\}$ for $\mathcal{W}_0$ satisfies (1.7.5) for $W$. This condition means that the (correct) *decoding probability* is *positive*.

In codes with this property there are $v_i \in \mathcal{D}_i$ for $i = 1, 2, \ldots, M$ with

$$W^n(v_i|u_i) > 0. \tag{1.7.15}$$

So it suffices to study $\mathcal{D}_i'$ with one element or sets of code words $\mathcal{U}$ with an injective map $f : \mathcal{U} \to \mathcal{Y}^n$ such that for $u \in \mathcal{U}$

$$W^n\big(f(u)|u\big) > 0. \tag{1.7.16}$$

Such an $f$ is called a matching and $(\mathcal{U}, f)$ is a matching code. Their study obviously concerns only the support of $W^n$ (i.e., the set of positive entries of $W^n$). This set can be viewed as the edge set $\mathcal{E}_n = \mathcal{E}_n(W)$ in the bipartite graph $\mathcal{G}(W^n) = \big(\mathcal{X}^n, \mathcal{Y}^n, \mathcal{E}_n(W)\big)$, where $(x^n, y^n) \in \mathcal{E}_n$ iff $W(y^n|x^n) > 0$.

Clearly, for the $f$ above $\{(u, f(u)) : u \in \mathcal{U}\}$ is exactly a set of nonintersecting edges in $\mathcal{G}(W^n)$, that is, a matching in the terminology of graph theory. Conversely, such a matching is a matching in the corresponding matching code $(\mathcal{U}, f)$, where $\mathcal{U}$ is the set of vertices in $\mathcal{X}^n$, which are matched to vertices in $\mathcal{Y}^n$. So we have reduced the study of the matching channel $\mathcal{W}_0$ via $W$ to the study of maximal matchings in the bipartite graph $\mathcal{G}(W^n)$.

At first we notice that this graph is an $n$th power $\mathcal{G}^{\otimes n} = \mathcal{G}\otimes\mathcal{G}\otimes\cdots\otimes\mathcal{G}$ of the graph $\mathcal{G}(W) = (\mathcal{X}, \mathcal{Y}, \mathcal{E})$, if the product $\mathcal{G}_1 \otimes \mathcal{G}_2$ of two bipartite graphs $\mathcal{G}_i = (\mathcal{X}_i, \mathcal{Y}_i, \mathcal{E}_i)$; $i = 1, 2$; is defined as $\left(\prod_{i=1}^2 \mathcal{X}_i, \prod_{i=1}^2 \mathcal{Y}_i, \mathcal{E}\right)$ with

$$\mathcal{E} = \{(x^2, y^2) : (x_i, y_i) \in \mathcal{E}_i \quad \text{for} \quad i = 1, 2\}. \tag{1.7.17}$$

So we can write

$$\mathcal{G}(W^n) = \mathcal{G}^{\otimes n}(W). \tag{1.7.18}$$

Shannon looked at these graphs in his study of the zero-error capacity problem (= problem of determining the vertex-independence number).

For any graph $\mathcal{G} = (\mathcal{V}, \mathcal{E})$ the size of a largest matching is called the matching number of $\mathcal{G}$ and is denoted by $\nu(\mathcal{G})$. A matching of the bipartite graphs $(\mathcal{V}_1, \mathcal{V}_2, \mathcal{E})$ is called a matching of $\mathcal{V}_1$ into $\mathcal{V}_2$, if every $v \in \mathcal{V}_1$ is endpoint of an edge in the matching. A matching is perfect, if it is both, a matching of $\mathcal{V}_1$ into $\mathcal{V}_2$ and of $\mathcal{V}_2$ into $\mathcal{V}_1$.

A vertex cover of $\mathcal{G}$ is a subset $S \subset \mathcal{V}$ such that each edge from $\mathcal{E}$ has an endpoint in $S$. The cardinality of a smallest vertex cover of $\mathcal{G}$ is the vertex cover number $\tau(\mathcal{G})$.

We introduce for every $v \in \mathcal{V}$

$$\Gamma_{\mathcal{G}}(v) = \{v' : (v, v') \in \mathcal{E}\} \tag{1.7.19}$$

and for every $S \subset \mathcal{V}$

$$\Gamma_{\mathcal{G}}(S) = \bigcup_{v \in S} \Gamma_{\mathcal{G}}(v). \tag{1.7.20}$$

The degree of $v$ is

$$d_{\mathcal{G}}(v) = |\Gamma_{\mathcal{G}}(v)|. \tag{1.7.21}$$

Two of the first and most basic results in matching theory (see [36]) are

**Theorem 42** (Hall's marriage theorem, [28, 36]) *A bipartite graph $\mathcal{G} = (\mathcal{V}_1, \mathcal{V}_2, \mathcal{E})$ has a matching of $\mathcal{V}_1$ into $\mathcal{V}_2$ iff*

$$|\Gamma(S)| \geq S \quad \textit{for all } S \subset \mathcal{V}_1 \tag{1.7.22}$$

and

**Theorem 43** (König's minimax theorem, [35, 36]) *For every bipartite graph $\mathcal{G}$*

$$\tau(\mathcal{G}) = \nu(\mathcal{G}).$$

These theorems can easily be derived from each other. We need here a consequence of Theorem 43.

**Corollary 6** *If $\mathcal{G} = (\mathcal{V}_1, \mathcal{V}_2, \mathcal{E})$ satisfies for two numbers $d_{\mathcal{V}_1}$, $d_{\mathcal{V}_2}$ and for $i = 1, 2$*

$$d_{\mathcal{G}}(v) = d_{\mathcal{V}_i} \quad \textit{for all } v \in \mathcal{V}_i,$$

*then*

$$\tau(\mathcal{G}) = \nu(\mathcal{G}) = \min_{i=1,2} |\mathcal{V}_i|.$$

*Proof* W.l.o.g. we can assume $|\mathcal{V}_1| \leq |\mathcal{V}_2|$. By the hypothesis we have also $|\mathcal{E}| = d_{\mathcal{V}_i}|\mathcal{V}_i|$ for $i = 1, 2$ and thus $d_{\mathcal{V}_1} \geq d_{\mathcal{V}_2}$.

Hence, each vertex of $\mathcal{G}$ covers at most $d_{\mathcal{V}_1}$ edges and therefore

$$\tau(\mathcal{G})d_{\mathcal{V}_1} \geq |\mathcal{E}| = d_{\mathcal{V}_1}|\mathcal{V}_1|,$$

which gives the result.

*Remark* We draw attention to the fact that matching and covering in $\mathcal{G}(W^n)$ are different from packing and covering by edges for Cartesian products of hypergraphs (c.f. [6]).

### *1.7.4 Main Results on Matching in Products of Bipartite Graphs*

For any bipartite graph $\mathcal{G} = (\mathcal{X}, \mathcal{Y}, \mathcal{E})$ we study the asymptotic behavior of the matching number $\nu(\mathcal{G}^{\otimes n})$. A key idea is to extend the König-Hall condition (1.7.22), which is in terms of cardinalities as measure of sets, to pairs of PDs associated with $\mathcal{G}$. The matching capacity of $\mathcal{G}$ is

$$\gamma(\mathcal{G}) = \lim_{n\to\infty} \frac{1}{n} \log \nu(\mathcal{G}^{\otimes n}). \tag{1.7.23}$$

We define the set

$$\mathcal{K}(\mathcal{G}) = \{(P, Q) : P \in \mathcal{P}(\mathcal{X}), Q \in \mathcal{P}(\mathcal{Y}), P(S) \leq Q(\Gamma_{\mathcal{G}}(S)) \ \forall\, S \subset \mathcal{X}\} \tag{1.7.24}$$

and call its members König-Hall pairs of distributions.

Moreover, in the sequel we assume that all graphs have no *isolated vertices.*

**Theorem 44** *For every bipartite graph $\mathcal{G} = (\mathcal{X}, \mathcal{Y}, \mathcal{E})$*

$$\gamma(\mathcal{G}) = \max_{(P,Q)\in\mathcal{K}(\mathcal{G})} \min\big(H(P), H(Q)\big).$$

*Proof* Recall the definitions of typical sequences and ED's in Sect. 1.7.2. We shall decompose $\mathcal{G}^{\otimes n}$ into subgraphs

$$\mathcal{G}_n(P, Q) = \big(\mathcal{X}^n(P), \mathcal{Y}^n(Q), \mathcal{E}_n(P, Q)\big),$$

where $P \in \mathcal{P}_n(\mathcal{X})$, $Q \in \mathcal{P}_n(\mathcal{Y})$, and

$$\mathcal{E}_n(P, Q) = \mathcal{E}_n \cap \big(\mathcal{X}^n(P) \times \mathcal{Y}^n(Q)\big).$$

Clearly, since $\mathcal{G}_n(P, Q)$ is a subgraph of $\mathcal{G}^{\otimes n}$,

$$\tau\big(\mathcal{G}_n(P, Q)\big) \leq \tau(\mathcal{G}^{\otimes n}). \tag{1.7.25}$$

On the other hand, if $C_n(P, Q)$ is a cover of $\mathcal{G}_n(P, Q)$ of smallest size, then

$$\bigcup_{(P,Q)\in\mathcal{P}_n(\mathcal{X})\times\mathcal{P}_n(\mathcal{Y})} C_n(P, Q)$$

is a cover of $\mathcal{G}^{\otimes n}$ and thus

$$\tau(G^{\otimes n}) \leq \sum_{(P,Q)\in\mathcal{P}_n(\mathcal{X})\times\mathcal{P}_n(\mathcal{Y})} \tau\big(G_n(P, Q)\big). \tag{1.7.26}$$

Now, since $|\mathcal{P}_n(\mathcal{X})|$ and $|\mathcal{P}_n(\mathcal{Y})|$ grow only polynomially in $n$, (1.7.25) and (1.7.26) imply

$$\lim_{n\to\infty}\frac{1}{n}\log\tau(\mathcal{G}^{\otimes n}) = \lim_{n\to\infty}\frac{1}{n}\log\max_{(P,Q)\in\mathcal{P}_n(\mathcal{X})\times\mathcal{P}_n(\mathcal{Y})}\tau\big(\mathcal{G}_n(P,Q)\big). \tag{1.7.27}$$

Next observe that for $(P,Q)\in\mathcal{P}_n(\mathcal{X})\times\mathcal{P}_n(\mathcal{Y})$ with $\mathcal{E}_n(P,Q)\neq\varnothing$ $\mathcal{G}_n(P,Q)$ satisfies the hypothesis of Corollary 6, because for any $x^n, x'^n\in\mathcal{X}^n(P)$ there is a permutation $\pi$ on $\{1,2,\dots,n\}$ with $\pi x^n = \big(x_{\pi(1)},\dots,x_{\pi(n)}\big) = x'^n$ and by the invariance of $\mathcal{Y}^n(Q)$ under $\pi$

$$|\Gamma_{\mathcal{G}_n(P,Q)}(x^n)| = |\prod_{i=1}^n\Gamma_{\mathcal{G}}(x_t)\cap\mathcal{Y}^n(Q)| = |\prod_{t=1}^n\Gamma_{\mathcal{G}}(x_{\pi(t)})\cap\mathcal{Y}^n(Q)| = |\Gamma_{\mathcal{G}_n(P,Q)}(x'^n)|$$

(and, symmetrically this holds for $y^n, y'^n\in\mathcal{Y}^n(Q)$). We conclude with Corollary 6 that for these $P, Q$

$$\nu\big(\mathcal{G}_n(P,Q)\big) = \tau\big(\mathcal{G}_n(P,Q)\big) = \min\big(|\mathcal{X}^n(P)|, |\mathcal{Y}^n(Q)|\big). \tag{1.7.28}$$

By Theorem 43 also

$$\nu(\mathcal{G}^{\otimes n}) = \tau(\mathcal{G}^{\otimes n}). \tag{1.7.29}$$

Now, from (1.7.27) we conclude with (1.7.28), (1.7.29) and (1.7.13)

$$\lim_{n\to\infty}\frac{1}{n}\log\nu(\mathcal{G}^{\otimes n}) = \lim_{n\to\infty}\max_{P,Q:\mathcal{E}_n(P,Q)\neq\varnothing}\min\big(H(P), H(Q)\big). \tag{1.7.30}$$

The final step is based on a result of interest on its own.

**Lemma 27** *(i) For all n*

$$(P,Q)\in\mathcal{K}(\mathcal{G})\cap\big(\mathcal{P}_n(\mathcal{X})\times\mathcal{P}_n(\mathcal{Y})\big)\Leftrightarrow\mathcal{E}_n(P,Q)\neq\varnothing. \tag{1.7.31}$$

*(ii) If $P(x)>0$ for all $x\in\mathcal{X}$, then for all $\varepsilon>0$, $(P,Q)\in\mathcal{K}(\mathcal{G})$ and sufficiently large n, there exists $(P',Q')\in\mathcal{K}(\mathcal{G})\cap\big(\mathcal{P}_n(\mathcal{X})\times\mathcal{P}_n(\mathcal{Y})\big)$ such that*

$$\sum_{x\in\mathcal{X}}|P(x)-P'(x)|, \sum_{x\in\mathcal{Y}}|Q(y)-Q'(y)| < \varepsilon.$$

*Proof* (i) Fix an $x^n\in\mathcal{X}^n(P)$. By symmetry it will not matter which one. Clearly,

$$\mathcal{E}_n(P,Q)\neq\varnothing\Leftrightarrow d_{\mathcal{G}_n(P,Q)}(x^n)>0. \tag{1.7.32}$$

We give first another characterization for $\mathcal{E}_n(P,Q)\neq\varnothing$ in terms of a matching property of another bipartite graph $\mathcal{G}_{(n)}(P,Q)$.

This graph has the vertex sets $\mathcal{X}_{(n)} = \{x_1, x_2, \dots, x_n\}$ and $\mathcal{Y}_{(n)} = \{y_1, \dots, y_n\}$, where $\mathcal{X}_{(n)}$ contains $n\ P(x)$ "copies" of each $x \in \mathcal{X}$ and $\mathcal{Y}_{(n)}$ contains $n\ Q(y)$ "copies" of each $y \in \mathcal{Y}$.

It has the edge set

$$\mathcal{E}_{(n)} = \big\{(x^*, y^*) : x^* \text{ is copy of } x \in \mathcal{X}, y^* \text{ is copy of } y \in \mathcal{Y}, \text{ and } (x, y) \in \mathcal{E}\big\}.$$

By the definition 5 of $\mathcal{G}^{\otimes n}$, $\mathcal{G}_n(P, Q)$, and $\mathcal{G}_{(n)}(P, Q)$, $x^n = (x_1, \dots, x_n)$ is adjacent with at least one vertex in $\mathcal{G}_n(P, Q)$ iff $\mathcal{G}_{(n)}(P, Q)$ has a perfect matching or by (1.7.32)

$$\mathcal{E}_n(P, Q) \neq \varnothing \Leftrightarrow \mathcal{G}_{(n)}(P, Q) \text{ has a perfect matching.} \tag{1.7.33}$$

A fortiori (1.7.31) is equivalent to

$$(P, Q) \in \mathcal{K}(\mathcal{G}) \cap \mathcal{P}_n(\mathcal{X}) \times \mathcal{P}_n(\mathcal{Y}) \Leftrightarrow \mathcal{G}_{(n)}(P, Q) \text{ has a perfect matching.} \tag{1.7.34}$$

To show this, let us start with a $(P, Q) \in \mathcal{K}(\mathcal{G}) \cap \mathcal{P}_n(\mathcal{X}) \times \mathcal{P}_n(\mathcal{Y})$.

Now, every $S^* \subset \mathcal{X}_{(n)}$ is associated with a subset $S$ of $\mathcal{X}$, where

$$x \in S \Leftrightarrow x \text{ has a copy } x_i \in S^*. \tag{1.7.35}$$

By the definitions of $\mathcal{G}_{(n)}$, $\mathcal{X}^n(P)$, and $\mathcal{K}(\mathcal{G})$ we have now

$$|S^*| \leq \sum_{x \in S} n\ P(x) = n\ P(S) \leq n\ Q\big(\Gamma_{\mathcal{G}}(S)\big) = \sum_{y \in \Gamma_{\mathcal{G}}(S)} n\ Q(y) = |\Gamma_{\mathcal{G}_{(n)}}(S^*)|. \tag{1.7.36}$$

This Hall condition and Theorem 42 imply that $\mathcal{G}_{(n)}(P, Q)$ has a perfect matching. Conversely, let us assume now that $\mathcal{G}_{(n)}(P, Q)$ has a perfect matching. For any $S \subset \mathcal{X}$ define

$$S^{**} = \{x_i : x_i \text{ is copy of some } x \in S\}, \tag{1.7.37}$$

a subset of $\mathcal{X}_{(n)}$. Then

$$n\ P(S) = \sum_{x \in S} n\ P(x) = |S^{**}| \tag{1.7.38}$$

and since $\mathcal{G}_{(n)}$ has a perfect matching, by Theorem 42

$$|S^{**}| \leq |\Gamma_{\mathcal{G}_{(n)}}(S^{**})|. \tag{1.7.39}$$

Also, by (1.7.37)

$$|\Gamma_{\mathcal{G}_{(n)}}(S^{**})| = \sum_{y\in\Gamma_{\mathcal{G}}(S)} n\, Q(y) = n\, Q\big(\Gamma_{\mathcal{G}}(S)\big),$$

and finally, this and (1.7.38), (1.7.39) imply $P(S) \leq Q\big(\Gamma_{\mathcal{G}}(S)\big)$ and so $(P, Q) \in \mathcal{K}(\mathcal{G})$.

(ii) We proceed by induction on $|\mathcal{X}|$. For $|\mathcal{X}| = 1$ the statement is trivial.

$\underline{|\mathcal{X}| > 1}$:

We say that a distribution $P^*$ on $\mathcal{Z}$ is $\delta$-approximated by $\hat{P}$ if $\sum_{z\in\mathcal{Z}} |P^*(z) - \hat{P}(z)| < \sigma$.

*Case 1.* For all $\phi \neq S \subsetneq \mathcal{X}$,

$$P(S) < Q(\Gamma_{\mathcal{G}} S). \tag{1.7.40}$$

Let $\delta' = \frac{1}{4}\min_{\phi\neq S\subsetneq\mathcal{X}} \big(Q(\Gamma_{\mathcal{G}}(S)) - P(S)\big)$ and let $\delta = \min(\delta', \varepsilon)$. Then by (1.7.40), $\delta' > 0$. When $n$ is sufficiently large, we always can choose $P' \in \mathcal{P}_n(\mathcal{X})$ and $Q' \in \mathcal{P}_n(\mathcal{Z})$ $\delta$-approximating $P$ and $Q$, respectively. Moreover, $(P', Q') \in \mathcal{K}(\mathcal{G})$, because for all $S \subset \mathcal{X}$ $|P(S) - P'(S)| < \delta$ and $|Q\big(\Gamma_{\mathcal{G}}(S)\big) - Q'\big(\Gamma_{\mathcal{G}}(S)\big)| < \delta$. This completes the proof in this case.

*Case 2.* There exists an $\mathcal{X}_0 \subset \mathcal{X}$ with $0 < |\mathcal{X}_0| < |\mathcal{X}|$ (and so by assumption $0 < P(\mathcal{X}_0) < 1$), such that

$$P(\mathcal{X}_0) = Q(\Gamma_{\mathcal{G}}\mathcal{X}_0). \tag{1.7.41}$$

Let $\mathcal{Y}_0 = \Gamma_{\mathcal{G}}(\mathcal{X}_0)$, $\mathcal{X}_1 = \mathcal{X}\setminus\mathcal{X}_0$ and $\mathcal{Y}_1 = \mathcal{Y}\setminus\mathcal{Y}_0$, and introduce two sub-bipartite graphs $\mathcal{G}_0 = (\mathcal{X}_0, \mathcal{Y}_0, \mathcal{E}_0)$ and $\mathcal{G}_1 = (\mathcal{X}_1, \mathcal{Y}_1, \mathcal{E}_1)$ of $\mathcal{G}$, where $\mathcal{E}_i = \big\{(x, y) : x \in \mathcal{X}_i, y \in \mathcal{Y}_i, (x, y) \in \mathcal{E}\big\}$ for $i = 1, 2$.

Then by (1.7.41)

$$P(\mathcal{X}_i) = Q(\mathcal{Y}_i) \quad \text{for } i = 0, 1 \tag{1.7.42}$$

and therefore, since $(P, Q) \in \mathcal{K}(\mathcal{G})$, $\big(P(\cdot|\mathcal{X}_0), Q(\cdot|\mathcal{Y}_0)\big) \in \mathcal{K}(\mathcal{G}_0)$. Since for any $\delta > 0$ $\overline{P} = \big(P(\mathcal{X}_0), P(\mathcal{X}_1)\big)$, which equals $\overline{Q} = \big(Q(\mathcal{Y}_o), Q(\mathcal{Y}_1)\big)$, can be $\delta$-approximated by $m$-ED's for sufficiently large $m$, (ii) follows from the induction hypothesis, if we can show that $\big(P(\cdot|\mathcal{X}_1), Q(\cdot|\mathcal{Y}_1)\big) \in \mathcal{K}(\mathcal{G}_1)$.

Indeed, this must be true, since otherwise one could find $S \subset \mathcal{X}_1$ such that $P(S|\mathcal{X}_1) < Q\big(\Gamma_{\mathcal{G}_1}(S)|\mathcal{Y}_1\big)$, and therefore by (1.7.42), such that

$$P(S) < Q\big(\Gamma_{\mathcal{G}_1}(S)\big) \leq Q\big(\Gamma_{\mathcal{G}}(S)\big),$$

which contradicts $(P, Q) \in \mathcal{K}(\mathcal{G})$.

*Remark* Lemma 27 shows that the definition of $\mathcal{K}(\mathcal{G})$ is symmetrical in the vertex sets, that is, we have also

$$(P, Q) \in \mathcal{K}(\mathcal{G}) \Leftrightarrow \forall\, T \subset \mathcal{Y} \quad Q(T) \leq P\big(\Gamma_{\mathcal{G}}(T)\big).$$

Lemma 27 has an immediate consequence:

**Corollary 7** *For all $P \in \mathcal{P}_n(\mathcal{X})$, $Q \in \mathcal{Q}_n(\mathcal{Y})$, $W \in \mathcal{V}$, and $x^n \in T_P^n$:*
*$W^n\big(T_Q^n(x^n)|x^n\big) > 0$ iff $P(S) \leq Q\big(\{y : w(y|x) > 0 \text{ for some } x \in S\}\big)$ for all $S \subset \mathcal{X}$.*

## 1.7.5 Matching in Products of Non-identical Bipartite Graphs

The result of this answers a natural combinatorial question, but its main motivation was to extend our coding theorems for the deterministic matching channel to the nonstationary situation. In terms of the associated discrete memoryless channel this means that we are given a sequence $(W_t)_{t=1}^{\infty}$ of $|\mathcal{X}| \times |\mathcal{Y}|$-stochastic matrices and the transmission for words of length $n$ is governed by $W^n = \prod_{t=1}^{n} W_t$. In this situation an approach with typical sequences is very clumsy, but our approach via König-Hall pairs of distributions goes rather smoothly.

The heart of the matter is the case of two factors: $\mathcal{G}_i = (\mathcal{X}_i, \mathcal{Y}_i, \mathcal{E}_i)$ $(i = 1, 2)$. For $P_i \in \mathcal{P}(\mathcal{Z}_i)$, where $\mathcal{Z}_i$ is finite and $i = 1, 2$, we define the product distribution $P_1 \times P_2$ by

$$P_1 \times P_2(z_1, z_2) = P_1(z_1)P_2(z_2) \quad \text{for } z_i \in \mathcal{Z}_i \quad \text{and } i = 1, 2. \tag{1.7.43}$$

We introduce a product of König-Hall sets, namely

$$\mathcal{K}(\mathcal{G}_1) \times \mathcal{K}(\mathcal{G}_2) = \big\{(P_1 \times P_2, Q_1 \times Q_2) : (P_i, Q_i) \in \mathcal{K}(\mathcal{G}_i) \quad \text{for } i = 1, 2\big\}. \tag{1.7.44}$$

**Theorem 45** *For bipartite graphs $\mathcal{G}_i = (\mathcal{X}_i, \mathcal{Y}_i, \mathcal{E}_i)$ $(i = 1, 2)$*

$$\begin{aligned}\gamma(\mathcal{G}_1 \otimes \mathcal{G}_2) &= \max_{(P_1\times P_2, Q_1\times Q_2)\in\mathcal{K}(\mathcal{G}_1)\times\mathcal{K}(\mathcal{G}_2)} \min\big(H(P_1) + H(P_2), H(Q_1) + H(Q_2)\big) \\ &= \max_{\substack{(P_i, Q_i)\in\mathcal{K}(\mathcal{G}_i) \\ i=1,2}} \min\big(H(P_1) + H(P_2), H(Q_1) + H(Q_2)\big).\end{aligned} \tag{1.7.45}$$

*Proof* Obviously the second equation follows immediately from (1.7.44). We show now the first equation.

By Theorem 44 and Lemma 27 for this it suffices to prove that for

$$\mathcal{K}_n(\mathcal{G}_i) = \mathcal{K}(\mathcal{G}_i) \cap \big(\mathcal{P}_n(\mathcal{X}_i) \times \mathcal{P}_n(\mathcal{Y}_i)\big), i = 1, 2; \tag{1.7.46}$$

and

$$\mathcal{K}_n(\mathcal{G}_1 \otimes \mathcal{G}_2) = \mathcal{K}(\mathcal{G}_1 \otimes \mathcal{G}_2) \cap \big(\mathcal{P}_n(\mathcal{X}_1 \times \mathcal{X}_2) \times \mathcal{P}_n(\mathcal{Y}_1 \times \mathcal{Y}_2)\big) \tag{1.7.47}$$

we have for all $n$

$$\max_{(P,Q)\in\mathcal{K}_n(\mathcal{G}_1\otimes\mathcal{G}_2)} \min\big(H(P), H(Q)\big) \tag{1.7.48}$$

$$\leq \max_{(P_1\times P_2, Q_1\times Q_2)\in\mathcal{K}_n(\mathcal{G}_1)\times\mathcal{K}_n(\mathcal{G}_2)} \min\big(H(P_1) + H(P_2), H(Q_1) + H(Q_2)\big) \tag{1.7.49}$$

$$\leq \max_{(P,Q)\in\mathcal{K}_{n^2}(\mathcal{G}_1\otimes\mathcal{G}_2)} \min\big(H(P), H(Q)\big). \tag{1.7.50}$$

We need

$$\mathcal{K}_n(\mathcal{G}_1) \times \mathcal{K}_n(\mathcal{G}_2) \subset \mathcal{K}_{n^2}(\mathcal{G}_1 \otimes \mathcal{G}_2). \tag{1.7.51}$$

To verify this, by (1.7.31), we have to show that for all $(P_1 \times P_2, Q_1 \times Q_2) \in \mathcal{K}_n(\mathcal{G}_1) \times K_n(\mathcal{G}_2)$ $\mathcal{E}_{n^2}(P_1 \times P_2, Q_1 \times Q_2) \neq \varnothing$. Actually, for all $(v_i, v_i') \in \mathcal{E}_n(P_i, Q_i) \neq \varnothing; i = 1, 2$, $(v_1 v_2, v_1' v_2') \in \mathcal{E}_{n^2}(P_1 \times P_2, Q_1 \times Q_2)$.

Therefore (1.7.51) holds and the second inequality in (1.7.50) follows.

Finally, we have to prove the first inequality in (1.7.50). Suppose that $(\hat{P}, \hat{Q})$ achieve the maximum in the L.H.S. of (1.7.50) and that $\hat{P}_1, \hat{P}_2$ (resp. $\hat{Q}_1, \hat{Q}_2$) are the marginal distributions of $\hat{P}$ (resp. $\hat{Q}$).

Since clearly

$$H(\hat{P}) \leq H(\hat{P}_1) + H(\hat{P}_2) \quad \text{and} \quad H(\hat{Q}) \leq H(\hat{Q}_1) + H(\hat{Q}_2), \tag{1.7.52}$$

it suffices to prove that $(\hat{P}_i, \hat{Q}_i) \in \mathcal{K}(\mathcal{G}_i)$ for $i = 1, 2$.

Actually, one readily verifies that for all $S \subset \mathcal{X}_1$

$$\Gamma_{\mathcal{G}_1\otimes\mathcal{G}_2}(S \times \mathcal{X}_2) = \Gamma_{\mathcal{G}_1}(S) \times \mathcal{Y}_2, \tag{1.7.53}$$

and therefore $(\hat{P}, \hat{Q}) \in \mathcal{K}(\mathcal{G}_1 \otimes \mathcal{G}_2)$ implies $\hat{P}_1(S) = \hat{P}(S \times \mathcal{X}_2) \leq \hat{Q}\big(\Gamma_{\mathcal{G}_1}(S) \times \mathcal{Y}_2\big) = \hat{Q}_1\big(\Gamma_{\mathcal{G}_1}(S)\big)$ and hence $(\hat{P}_1, \hat{Q}_1) \in \mathcal{K}(\mathcal{G}_1)$. By the same reasons also $(\hat{P}_2, \hat{Q}_2) \in \mathcal{K}(\mathcal{G}_2)$.

## 1.7.6 *An Exact Formula for the Matching Number of Powers of "Stared" Bipartite Graphs*

We consider here bipartite graphs $\mathcal{G} = (\mathcal{X}, \mathcal{Y}, \mathcal{E})$, which can be presented in the following form:

There are sets of vertices $\mathcal{J} \subset \mathcal{X}$ and $\mathcal{K} \subset \mathcal{Y}$ such that

(i) every vertex in $\mathcal{J}$ (resp. $\mathcal{K}$) is adjacent with at least one vertex in $\mathcal{Y} \setminus \mathcal{K}$ (resp. $\mathcal{X} \setminus \mathcal{J}$).
(ii) every vertex in $\mathcal{X} \setminus \mathcal{J}$ (resp. $\mathcal{Y} \setminus \mathcal{K}$) is adjacent with exactly one vertex in $\mathcal{K}$ (resp. $\mathcal{J}$), and there is no edge between $\mathcal{X} \setminus \mathcal{J}$ and $\mathcal{U} \setminus \mathcal{K}$.

We speak of a *stared* bipartite graph. We also introduce the abbreviation

$$\mathcal{Z} = \mathcal{J} \cup \mathcal{K} \tag{1.7.54}$$

and for every $z \in \mathcal{Z}$ we define a *star with center* **z** as $S_z = \big\{\{z\}, \mathcal{V}_z, \mathcal{F}_z\big\}$, where

$$\mathcal{V}_z = \big\{v \in (\mathcal{X} \setminus \mathcal{J}) \cup (\mathcal{Y} \setminus \mathcal{K}) : (z, v) \in \mathcal{E}\big\}, \tag{1.7.55}$$

$$\mathcal{F}_z = \big\{(z, v) : v \in \mathcal{V}_z\big\}. \tag{1.7.56}$$

Of course, since $\mathcal{G}$ is bipartite, for $z = j \in \mathcal{J}$ (resp. $z = k \in \mathcal{K}$) necessarily $\mathcal{V}_j \subset \mathcal{Y} \setminus \mathcal{K}$ (resp. $\mathcal{V}_k \subset \mathcal{X} \setminus \mathcal{J}$). By the conditions (i) and (ii), obviously $\big\{\{z\} \cup \mathcal{V}_z : z \in \mathcal{Z}\big\}$ is a partition of $\mathcal{X} \cup \mathcal{Y}$.

Now we associate with every $z^n \in \mathcal{Z}^n = \prod_1^n \mathcal{Z}$ the complete bipartite graph

$$S_{z^n} = S_{z_1} \otimes S_{z_2} \otimes \cdots \otimes S_{z_n} \tag{1.7.57}$$

where $S_{z_t}$'s are stars defined by (1.7.55) and (1.7.56) for $z_t = z$.

Denote its vertex set by $\mathcal{V}^*_{z^n}$ and its edge set by $\mathcal{F}_{z^n} = \prod_{t=1}^n \mathcal{F}_{z_t}$. Notice that

$$|\mathcal{V}^*_{z^n} \cap \mathcal{X}^n| = \prod_{z_t \notin \mathcal{J}} |\mathcal{V}_{z_t}|. \tag{1.7.58}$$

This is the number of vertices of $S_{z^n}$ falling into $\mathcal{X}^n$ and will be denoted by $\omega_{\mathcal{X}}(z^n)$. Similarly we define

$$\omega_{\mathcal{Y}}(z^n) = \prod_{z_t \notin \mathcal{K}} |\mathcal{V}_{z_t}|. \tag{1.7.59}$$

We speak of the $\mathcal{X}$-weight and of the $\mathcal{Y}$-weight of $S_{z^n}$.

**Theorem 46** *For every stared bipartite graph $\mathcal{G}$ the matching number of its nth power is given by*

$$\nu(\mathcal{G}^{\otimes n}) = \sum_{z^n \in \mathcal{Z}^n} \min\big(\omega_{\mathcal{X}}(z^n), \omega_{\mathcal{Y}}(z^n)\big). \tag{1.7.60}$$

*Proof* Since $S_{z^n}$ is a complete bipartite graph with vertex sets of sizes $\omega_{\mathcal{X}}(z^n)$ and $\omega_{\mathcal{Y}}(z^n)$ it has obviously a matching of size $\min\big(\omega_{\mathcal{X}}(z^n), \omega_{\mathcal{Y}}(z^n)\big)$. Furthermore, by our definitions, for $z^n \neq z'^n$ an edge in $\mathcal{F}_{z^n}$ and an edge in $\mathcal{F}_{z'^n}$ have never a common vertex.

Therefore the matching corresponding to the different $z^n$'s can be taken together to form one matching. Thus

$$\nu(\mathcal{G}^{\otimes n}) \geq \sum_{z^n \in \mathcal{Z}^n} \min\big(\omega_{\mathcal{X}}(z^n), \omega_{\mathcal{Y}}(z^n)\big). \tag{1.7.61}$$

To show the opposite inequality, by Theorem 43 it suffices to find a vertex cover of $\mathcal{G}^{\otimes n}$ of size $\sum_{z^n \in \mathcal{Z}^n} \min\big(\omega_{\mathcal{X}}(z^n), \omega_{\mathcal{Y}}(z^n)\big)$. Our candidate is the set of vertices

$$\left(\bigcup_{z^n \in \mathcal{Z}^n : \omega_{\mathcal{X}}(z^n) \leq \omega_{\mathcal{Y}}(z^n)} (\mathcal{V}^*_{z^n} \cap \mathcal{X}^n)\right) \cup \left(\bigcup_{z^n \in \mathcal{Z}^n : \omega_{\mathcal{X}}(z^n) > \omega_{\mathcal{Y}}(z^n)} (\mathcal{V}^*_{z^n} \cap \mathcal{Y}^n)\right). \tag{1.7.62}$$

Clearly, it has the desired cardinality. It remains to be seen that it is a vertex cover for $\mathcal{G}^{\otimes n}$.

Suppose that $(x^n, y^n) \in \mathcal{E}_n$ is not covered. Then necessarily $T = \{t : \text{both}, x_t \text{ and } y_t, \text{ are centers of a star}\} \neq \varnothing$, because all edges in $\mathcal{F}_{z^n}$, $z^n \in \mathcal{Z}^n$, are covered for our candidate. Next we observe that for every $t \in \{1, 2, \ldots, n\} \setminus T$ $x_t$ and $y_t$ are in the same star. Therefore, if $x^n \in \mathcal{V}_{z^n}$, $y^n \in \mathcal{V}_{z'^n}$, then $z_t = z'_t$ for $t \in \{1, 2, \ldots, n\} \setminus T$ and thus

$$\omega_{\mathcal{X}}(z^n) = \omega_{\mathcal{X}}(z'^n)\left(\prod_{t \in T} |\mathcal{V}_{z'_t}|\right)^{-1}, \tag{1.7.63}$$

$$\omega_{\mathcal{Y}}(z^n) = \omega_{\mathcal{Y}}(z'^n)\left(\prod_{t \in T} |\mathcal{V}_{z_t}|\right). \tag{1.7.64}$$

(Since for all $t \in T$ $x_t$ and $y_t$ are centers of $S_{z_t}$ and $S_{z'_t}$, resp.).

Since by assumption $(x^n, y^n)$ is not covered, neither $x^n$ nor $y^n$ is in our candidate subset. By the construction of this subset

$$\omega_{\mathcal{X}}(z^n) > \omega_{\mathcal{Y}}(z^n) \quad \text{and} \quad \omega_{\mathcal{X}}(z'^n) \leq \omega_{\mathcal{Y}}(z'^n) \tag{1.7.65}$$

and (1.7.63)–(1.7.65) imply

$$\omega_{\mathcal{Y}}(z^n) \geq \omega_{\mathcal{X}}(z'^n)\left(\prod_{t\in T}|\mathcal{V}_{z_t}|\right) = \omega_{\mathcal{X}}(z^n)\left(\prod_{t\in T}|\mathcal{V}_{z'_t}|\right)\left(\prod_{t\in T}|\mathcal{V}_{z_t}|\right)$$
$$> \omega_{\mathcal{Y}}(z^n)\left(\prod_{t\in T}|\mathcal{V}_{z'_t}|\right)\left(\prod_{t\in T}|\mathcal{V}_{z_t}|\right).$$

This contradicts the definition of the graph, in which

$$|\mathcal{V}_z| \geq 1 \quad \text{for all } v \in \mathcal{Z} = \mathcal{J} \cup \mathcal{K}.$$

### *1.7.7 Two Examples Illustrating the Significance of Theorems 45 and 46*

**Example** $\gamma(\mathcal{G}_1 \otimes \mathcal{G}_2) > \gamma(\mathcal{G}_1) + \gamma(\mathcal{G}_2)$

Consider two complete bipartite graphs $\mathcal{G}_i = (\mathcal{X}_i, \mathcal{Y}_i, \mathcal{E}_i)$; $i = 1, 2$; with parameters $|\mathcal{X}_1| = |\mathcal{Y}_2| = \alpha < \beta = |\mathcal{X}_2| = |\mathcal{Y}_1|$. Obviously

$$\nu(\mathcal{G}_i^{\otimes n}) = \alpha^n \quad \text{for } i = 1, 2 \tag{1.7.66}$$

and therefore

$$\gamma(\mathcal{G}_i) = \log\alpha \quad \text{for } i = 1, 2. \tag{1.7.67}$$

However, by Theorem 45 or by direct reasoning

$$\nu\big((\mathcal{G}_1 \otimes \mathcal{G}_2)^{\otimes n}\big) = (\alpha\beta)^n, \tag{1.7.68}$$

because $\mathcal{G}_1 \otimes \mathcal{G}_2$ is a complete bipartite graph.

Thus we have

$$\gamma(\mathcal{G}_1 \otimes \mathcal{G}_2) = \log\alpha + \log\beta > 2\log\alpha = \gamma(\mathcal{G}_1) + \gamma(\mathcal{G}_2).$$

**Example** $\nu(\mathcal{G}^{\otimes(m_1+m_2)}) > \prod_{i=1}^{2} \nu(\mathcal{G}^{\otimes m_i})$ and $2^{n\gamma(\mathcal{G})} > \nu(\mathcal{G}^{\otimes n})$ can occur infinitely often and for arbitrarily large $m_1, m_2$, and $n$.

Consider the stared bipartite graph $\mathcal{G} = (\mathcal{X}, \mathcal{Y}, \mathcal{E})$ with $\mathcal{X} = \{x_i : i = 0, 1, \ldots, \alpha\}$, $\mathcal{Y} = \{y_j : j = 0, 1, \ldots, \alpha\}$ and $\mathcal{E} = \big\{(x_i, y_j) : i = 0 \text{ or } j = 0\big\}$.

By Theorem 46

$$\nu(\mathcal{G}^{\otimes n}) = \begin{cases} 2\left[\sum_{i=0}^{\frac{n}{2}} \binom{n}{i}\alpha^i + \frac{1}{2}\binom{n}{\frac{n}{2}}\alpha^{\frac{n}{2}}\right], & n \text{ even} \\ 2\sum_{i=0}^{\frac{n-1}{2}} \binom{n}{i}\alpha^i, & n \text{ odd.} \end{cases} \tag{1.7.69}$$

By Theorem 44 $\big(P(x_0) = Q(y_0) = \frac{1}{2}$, and $Q(y_j) = P(x_i) = \frac{1}{2\alpha}$ for $i = 1, \dots, \alpha\big)$ or directly by (1.7.69)

$$\gamma(\mathcal{G}) = 1 + \frac{1}{2}\log\alpha. \tag{1.7.70}$$

Therefore for *all n*

$$2^{\gamma(\mathcal{G})n} = 2^n\alpha^{\frac{n}{2}} > \nu(\mathcal{G}^{\otimes n}).$$

Moreover, (1.7.69) also shows that $\nu\big(\mathcal{G}^{\otimes(m_1+m_2)}\big) > \prod_{i=1}^{2}\nu(\mathcal{G}^{\otimes m_i})$.

## *1.7.8 Multi-way Deterministic Matching Channels*

We take first another look at the (one-way) deterministic matching channel in order to get a certain understanding of its structure, which helps us when dealing with more complex channels such as compound, multiple-access and broadcast deterministic matching channels.

### 1.7.8.1 Another Look at Theorem 40

A straightforward proof of its direct part by random coding is sketched in the Introduction. Actually this approach gives even the more general Theorem 39. Here we deal only with *deterministic* channels. Our first and detailed proof of Theorem 40 via an *extension* of *combinatorial matching theory* is contained in Sects. 1.7.3 and 1.7.4. It arose in an analysis in the "control" model of the "dummy" model (see (1.7.4), (1.7.5)).

Actually, there is a very simple direct path to Theorem 40 using König's Theorem 40 in Sect. 1.7.3. Indeed just consider the bipartite graph

$$\mathcal{G}_n = \mathcal{G}_n(P_{XS}, W) = \left(T_X^n, T_Y^n, \mathcal{E}_{P_{XS,W}}^n\right) \tag{1.7.71}$$

where $P_Y = P_{XS} \cdot W$ and $(x^n, y^n) \in \mathcal{E}^n_{P_{XS,W}}$ iff there exists an $(x^n, s^n) \in T^n_{XS}$ with $W(y^n|x^n|s^n) = 1$. We know (see 1.7.2) that

$$|T^n_X| = \exp\{H(X)n + o(n)\}, |T^n_Y| = \exp\{(H(Y)n + o(n)\},$$
$$d_{\mathcal{G}_n}(x^n) = \exp\{H(Y|X)n + o(n)\}, d_{\mathcal{G}_n}(y^n) = \exp\{H(X|Y)n + o(n)\}$$

and $d_{\mathcal{G}_n}(x^n)$ (resp. $d_{\mathcal{G}_n}(y^n)$) has the same value for all $x^n \in T^n_X$ (resp. $y^n \in T^n_Y$).

The vertex covering number

$$\tau\big(\mathcal{G}_n(P_{XS}, W)\big)$$

obviously equals $\min\big(H(X), H(Y)\big)$ and the proof of Theorem 40 follows with Theorem 43.

**Problem** Ca one give explicit constructions of matchings achieving the capacity in Theorem 40? Can it be done as orbit of a group of permutations?

### 1.7.8.2 Compound Channels

We are given now $c$ deterministic matching channels

$$W_j : \mathcal{X} \times \mathcal{S} \to \mathcal{Y}_j \ (j = 1, 2, \dots, c)$$

and ask for a simultaneous code, that is, one set of code words $\mathcal{U} \subset \mathcal{X}^n$ and decoding sets $\{\mathcal{D}_{ji} : 1 \leq i \leq |\mathcal{U}|\}$ for $j = 1, 2, \dots, c$.

Here the random choice described above works again. Details are left to the reader.

**Theorem 47** *For the compound deterministic matching channel the capacity equals*

$$\max_{XS} \min_{j=1,\dots,c} \min\big(H(X), H(Y_j)\big).$$

Here it was implicitly assumed that encoder and controller do not know the individual (implicitly) channel, but that the receiver does. However, it is easy to show that (as for classical compound channels) the capacity is not affected by the receivers knowledge.

More generally one can also describe the capacity region of the compound multiple-access deterministic matching channel

$$W : \mathcal{X}_1 \times \mathcal{X}_2 \times \dots \times \mathcal{X}_b \times \mathcal{S} \to \mathcal{Y}_1 \times \dots \times \mathcal{Y}_c,$$

where we consider codes $\mathcal{U}_j \subset \mathcal{X}^n_j$ for $j = 1, 2, \dots, b$ and decoding sets $\{\mathcal{D}_{jk}(i) : 1 \leq i \leq |\mathcal{U}_j|\}$ for $j = 1, 2, \dots, b$ and $k = 1, 2, \dots, c$.

### 1.7.8.3 Multiple-access Channels (MAC)

A matching MAC is given by a stochastic

$$W : (\mathcal{X} \times \mathcal{S}) \times (\mathcal{Y} \times \mathcal{T}) \to \mathcal{Z}.$$

It is understood that there are two controllers, $K_{\mathcal{S}}$ and $K_{\mathcal{T}}$. Observing input word $x^n$ controller $K_{\mathcal{S}}$ can choose $s^n = s^n(x^n)$. Similarly controller $K_{\mathcal{T}}$ responds to input $y^n$. $K_{\mathcal{S}}$ doesn't observe $y^n$ and $K_{\mathcal{T}}$ doesn't observe $x^n$.

Combining the sketch of proof of Theorem 39 with standard proofs for the MAC coding theorem gives the following result.

**Theorem 48** *The capacity region of the matching MAC contains*
$conv\{(R_{\mathcal{X}}, R_{\mathcal{Y}}) : 0 \le R_{\mathcal{X}}, R_{\mathcal{Y}}, \exists P_{XS}, P_{YT} \text{ with } R_{\mathcal{X}} \le \min(I(XS \wedge Z|YT), H(X)), R_{\mathcal{Y}} \le \min(I(YT \wedge Z|XS), H(Y)), R_{\mathcal{X}} + R_{\mathcal{Y}} \le I(XSYT \wedge Z)\}$.

Adding a time-sharing parameter (if necessary) gives the exact region.

**Problem** In another model there is only one controller $K_{\mathcal{X}\mathcal{Y}}$ who acts upon the independent inputs $x^n, y^n$. The pair $(R_{\mathcal{X}}, R_{\mathcal{Y}})$ is achievable, if for some
$P_{XYS} = P_{S|XY} \cdot P_X \cdot P_Y$ $R_{\mathcal{X}} \le \min(I(XS \wedge Z|Y), H(X))$, $R_{\mathcal{Y}} \le \min(I(YS \wedge Z|X), H(Y))$, $R_{\mathcal{X}} + R_{\mathcal{Y}} \le I(XYS \wedge Z)$.
Establish the capacity region!

**Problem** What are the capacity regions for matching zero-error detection codes for deterministic channels in both models?

### 1.7.8.4 Broadcast Channels

It is surprising that we can determine also here the capacity region! This means in the noisy channel terminology that under the (not realistic) condition that the error probability is strictly smaller than 1 we moved the rock!

Here is the result.

For the broadcast deterministic matching channel $W : \mathcal{X} \times \mathcal{S} \to \mathcal{Y} \times \mathcal{Z}$ an $(n, M, N)$ matching code is a family $\{(u_{ij}, \mathcal{D}_i(\mathcal{Y}), \mathcal{D}_j(\mathcal{Z})) : 1 \le i \le M, 1 \le j \le N\}$, where $u_{ij} \in \mathcal{X}^n$, $\mathcal{D}_i(\mathcal{Y}) \subset \mathcal{Y}^n$, $\mathcal{D}_j(\mathcal{Z}) \subset \mathcal{Z}^n$, $\mathcal{D}_i(\mathcal{Y}) \cap \mathcal{D}_{i'}(\mathcal{Y}) = \varnothing$ $(i \ne i')$, $\mathcal{D}_j(\mathcal{Z}) \cap \mathcal{D}_{j'}(\mathcal{Z}) = \varnothing$ $(j \ne j')$, and for every pair of messages $(i, j)$ there is a sequence $s^n(i, j) = (s_1(i, j), \dots, s_n(i, j)) \in \mathcal{S}^n$ with

$$W^n(\mathcal{D}_i(\mathcal{X})|u_{ij}|s^n(i, j)) = 1 \quad \text{and} \quad W^n(\mathcal{D}_j(\mathcal{Y})|u_{ij}|s^n(i, j)) = 1$$

for $1 \le i \le M$, $1 \le j \le N$, if $W^n(y^n, z^n|x^n|s^n) = \prod_{t=1}^n W(y_t, z_t|x_t|s_t)$.

**Theorem 49** *The broadcast deterministic matching channel has the set of all achievable pairs of rates $(R_{\mathcal{Y}}, R_{\mathcal{Z}})$ defined by the convex hull of the sets*

$$0 \le R_{\mathcal{Y}} \le H(Y),\ 0 \le R_{\mathcal{Z}} \le H(Z),\ 0 \le R_{\mathcal{Y}} + R_{\mathcal{Z}} \le \min\big(H(X), H(YZ)\big),$$

*where all RV's are induced by distributions $P_{XS}$ and the channel.*

*Remark*

1. Actually, more generally we also have a solution for the case with a common message set [37].
2. In case $|\mathcal{S}| = 1$ this yields Pinsker's characterization [38] of the capacity region for deterministic broadcast channels.

We use in our proof a certain binning idea, namely the

**Lemma 28** (Color carrier lemma, [14]) *For each hypergraph $\mathcal{H} = (\mathcal{V}, \mathcal{E})$ there is a coloring $\varphi : \mathcal{V} \to \mathcal{L} = \{1, \dots, L\}$ such that every color in $\mathcal{L}$ appears in every edge from $\mathcal{E}$ whenever*

$$L \le \left(\ln |\mathcal{E}| \min_{E \in \mathcal{E}} |E|\right)^{-1} \min_{E \in \mathcal{E}} |E|.$$

*Proof of Theorem 49.* The converse follows by a standard decomposition into subcodes corresponding to sets of typical sequences. The issue is the direct part. We give its flavor first in the case $|\mathcal{S}| = 1$, which is Pinsker's celebrated result.

#### 1.7.8.5 Embedding of $\mathbf{T^n_{YZ}}$ into $\mathbf{T^n_{XS}}$

Every pair of typical sequences $(y^n, z^n) \in T^n_{YZ}$ is the image of at least one pair $(x^n, s^n) \in T^n_{XS}$ under $W$. We select any such pair $\widetilde{(y^n, z^n)}$.

The matrix $\Omega = \widetilde{(y^n, z^n)}_{y^n \in T^n_Y, z^n \in T^n_Z}$ shall have entries "0" in all positions $(y^n, z^n) \notin T^n_{YZ}$.

#### 1.7.8.6 A Code Based on the Color Carrier Lemma

We can replace $XS$ by $X$. Now we have $\widetilde{(y^n, z^n)} \in T^n_{XZ}$ and $\Omega$ has $\sim \exp\{H(Y, Z)n\} \le \exp\{H(X)n\}$ nontrivial entries, $r \sim \exp\{H(Y)n\}$ rows and $c \sim \exp\{H(Z)n\}$ columns. Consider the hypergraph $\big(\{1, \dots, c\}, \{E_\rho : 1 \le \rho \le r\}\big)$, where $E_\rho$ is the set of nontrivial positions in row $\rho$, so $E_\rho \subset \{1, 2, \dots, c\}$.

Now message $\rho \in \{1, 2, \dots, r\}$ can be associated with the row index $\rho$ and message $\ell \in \{1, 2, \dots, L\}$ can be associated with any $\varphi^{-1}(\ell) \in E_\rho$, where $\varphi : \{1, 2, \dots, c\} \to \{1, 2, \dots, L\}$ has by the Color Carrier Lemma the property that $\{\ell : \varphi^{-1}(\ell) \in E_\rho\} = \{1, 2, \dots, L\}$. Moreover

$$L \sim \big(H(Y)n\big)^{-1} \exp\{H(Z|Y)n\} \sim \exp\{H(Z|Y)n\}. \tag{1.7.72}$$

Thus with $(r, L)$ we have achieved the rates $\big(H(Y), H(Z|Y)\big)$ and the direct part is completed by time-sharing.

*General case* $|\mathcal{S}| > 1$. Now we have to cope with the fact that now $T^n_{YZ}$ has to be an embedding in $T^n_{XS}$ such that every $x^n \in T^n_X$ is used only at most ones in conjunction with an $s^n$ (the controller has to choose always the same $s^n$ for $x^n$. Otherwise he transmits illegally information).

So let us consider any quadruple $(X, S, Y, Z)$ of RVs with joint distribution

$$P_{XSYZ}(x, s, y, z) = P_{XS}(x, s) W(y, z|x|s) \tag{1.7.73}$$

for $(x, s, y, z) \in \mathcal{X} \times \mathcal{S} \times \mathcal{Y} \times \mathcal{Z}$.

Clearly, for $(x^n, y^n, z^n) \in \mathcal{X}^n \times \mathcal{Y}^n \times \mathcal{Z}^n$

$$(y^n, z^n) \in T^n_{YZ|X}(x^n) \text{ implies } W^n(y^n, z^n|x^n|s^n) > 0 \text{ for some} s^n \in \mathcal{S}^n. \tag{1.7.74}$$

Moreover, by symmetry and time sharing it is sufficient to show that

$$(R_{\mathcal{Y}}, R_{\mathcal{Z}}) = \Big(\big(\min\big(H(X), H(Y)\big), \min\big(H(X), H(YZ)\big) - \min\big(H(X), H(Y)\big)\Big)$$

is achievable.

*Case* $H(X) \leq H(Y)$. Here $R_{\mathcal{Y}} = H(X)$, $R_{\mathcal{Z}} = 0$. For instance by Corollary 6 in Sect. 1.7.3 one can match $T^n_X$ into $T^n_Y$ in the bipartite graph $(T^n_X, T^n_Y, T^n_{XY})$ (where $T^n_{XY}$ serves as the edge set). Thus we get $\big\{(u_{i,1}, v_i) \in T^n_{XY} : 1 \leq i \leq |T^n_X|\big\}$, where for any $i \neq j$ $u_{i,1} \neq u_{j,1}$ and $v_i \neq v_j$. With the decoding sets $\mathcal{D}_i(\mathcal{Y}) = \{v_i\}$ for $i = 1, 2, \dots, |T^n_X|$ and $\mathcal{D}_1(\mathcal{Z}) = T^n_Z$ we see from (1.7.74) that

$$\big\{\big(u_{i,1}, \mathcal{D}_i(\mathcal{Y}), \mathcal{D}_1(\mathcal{Z})\big) : 1 \leq i \leq |T^n_X|\big\}$$

is a matching code. It has the pair of rates $\big(H(X), 0\big)$.

*Case* $H(Y) < H(X)$. We proceed in two steps.

First we find a subset $\hat{\mathcal{X}}_n \subset T^n_X$ and an injective mapping $g : \hat{\mathcal{X}}_n \to T^n_{YZ}$ such that for all $x^n \in \hat{\mathcal{X}}_n$

$$g(x^n) \in T^n_{YZ|X}(x^n) \tag{1.7.75}$$

and for all $y^n \in T^n_Y$ and any fixed $\theta \in (0, 1)$ the set

$$E_{y^n} \triangleq \big\{z^n : g(x^n) = (y^n, z^n) \quad \text{for some } x^n \in \hat{\mathcal{X}}_n\big\} \tag{1.7.76}$$

satisfies

$$|E_{y^n}| \geq \exp\big\{n(R_{\mathcal{Z}} - \theta)\big\}. \tag{1.7.77}$$

This describes an embedding of a subset of $T^n_{YZ}$ into $T^n_X$.

*Subcase* $H(Y) \leq H(YZ) \leq H(X)$. Here $R_{\mathcal{Y}} = H(Y)$ and $R_{\mathcal{Z}} = H(Z|Y)$. Obviously, the matching of $T^n_{YZ}$ into $T^n_X$ for the bipartite graph $(T^n_X, T^n_{YZ}, T^n_{XYZ})$ given by Corollary 6 provides a $g$ with all desired properties.

We are left with the crucial

*Subcase* $H(Y) < H(X) < H(YZ)$. Our concern are the rates

$$R_{\mathcal{Y}} = H(Y) \quad \text{and } R_{\mathcal{Z}} = H(X) - H(Y) > 0. \tag{1.7.78}$$

Originally we achieved them by a fairly lengthy (due to complications caused by the exponential sizes of $T_X^n$ and $T_{YZ}^n$) counting argument.

Now we use a large deviational argument based on Bernstein's version of Chebyshev's inequality. Its power lies in double exponential bounds, which proved to be very useful in Information Theory already in [7].

Let $U(x^n)$, $x^n \in T_X^n$, be a family of independent RV's and for each $x^n$ $U(x^n)$ has uniform distribution over $T_{YZ|X}^n(x^n)$. For fixed $y^n \in T_Y^n$ and $(y^n, z^n) \in T_{YZ}^n$, we introduce the events

$$E_1(y^n) \triangleq \left\{ \sum_{x^n \in T_X^n} \delta^*\big(y^n, U(x^n)\big) \leq \frac{1-e^{-1}}{2} \frac{|T_X^n|}{|T_Y^n|} \right\} \tag{1.7.79}$$

and

$$E_2\big((y^n, z^n)\big) \triangleq \left\{ \sum_{x^n \in T_X^n} \delta\big((y^n, z^n), U(x^n)\big) \geq \exp\left\{ n\frac{\theta}{2} \right\} \right\}, \tag{1.7.80}$$

where

$$\delta^*\big(y^n, (y'^n, z'^n)\big) = \begin{cases} 0 & \text{if } y^n \neq y'^n \\ 1 & \text{if } y^n = y'^n \end{cases}$$

and $\delta$ is Kronecker's delta, i.e.,

$$\delta\big((y^n, z^n), (y'^n, z'^n)\big) = \begin{cases} 0 & \text{if } (y^n, z^n) \neq (y'^n, z'^n) \\ 1 & \text{if } (y^n, z^n) = (y'^n, z'^n). \end{cases}$$

The definitions of $U(x^n)$, $E_1(y^n)$, $E_2\big((y^n, z^n)\big)$ and the Bernstein version of Chebyshev's inequality imply for all $y^n \in T_Y^n$

$$\begin{aligned}
\Pr\big(E_1(y^n)\big) &\leq e^{\frac{1-e^{-1}}{2}\frac{|T_X^n|}{|T_Y^n|}} \mathbb{E}\, e^{-\sum_{x^n \in T_X^n} \delta^*\left(y^n, U(x^n)\right)} \\
&\overset{1.}{=} e^{\frac{1}{2}(1-e^{-1})\frac{|T_X^n|}{|T_Y^n|}} \mathbb{E}\, e^{-\sum^n_{x^n \in T_{X|Y}^n(y^n)} \delta^*\left(y^n, U(x^n)\right)} \\
&\overset{2.}{=} e^{\frac{1}{2}(1-e^{-1})\frac{|T_X^n|}{|T_Y^n|}} \prod_{x^n \in T_{X|Y}^n(y^n)} \mathbb{E}\, e^{-\delta^*\left(y^n, U(x^n)\right)}
\end{aligned}$$

$$
\begin{aligned}
&= e^{\frac{1}{2}(1-e^{-1})\frac{|T_X^n|}{|T_Y^n|}} \prod_{x^n \in T_{X|Y}^n(y^n)} \left( \frac{|T_{Z|XY}^n(x^n, y^n)|}{|T_{YZ|X}^n(x^n)|} e^{-1} + 1 - \frac{|T_{Z|XY}^n(x^n, y^n)|}{|T_{YZ|X}^n(x^n)|} \right) \\
&\overset{3.}{=} e^{\frac{1}{2}(1-e^{-1})\frac{|T_X^n|}{|T_Y^n|}} \left( 1 - (1-e^{-1}) \frac{|T_X^n|}{|T_{XY}^n|} \right)^{|T_{X|Y}^n(y^n)|} \\
&\leq e^{(1-e^{-1})\left[ \frac{1}{2}\frac{|T_X^n|}{|T_Y^n|} - \frac{|T_X^n|}{|T_{XY}^n|} |T_{X|Y}^n(y^n)| \right]} \\
&\overset{4.}{=} \exp\left\{ -\frac{1}{2}(1-e^{-1}) \frac{|T_X^n|}{|T_Y^n|} \log e \right\} \\
&\overset{5.}{=} \exp\{-\exp(n(R_z + o(1))\}, \qquad (1.7.81)
\end{aligned}
$$

if $n$ is big enough, and for all $(y^n, z^n) \in T_{YZ}^n$

$$
\begin{aligned}
\Pr(E_2(y^n, z^n)) &\leq e^{-\exp\left\{\frac{1}{2}n\theta\right\}} E\, e^{\sum_{x^n \in T_X^n} \delta\left((y^n, z^n), U(x^n)\right)} \\
&\overset{6.}{=} e^{-\exp\left\{\frac{1}{2}n\theta\right\}} \mathbb{E}\, e^{\sum_{x^n \in T_{X|YZ}^n(y^n, z^n)} \delta\left((y^n, z^n), U(x^n)\right)} \\
&\overset{2.}{=} e^{-\exp\left\{\frac{1}{2}n\theta\right\}} \prod_{x^n \in T_{X|YZ}^n(y^n, z^n)} \mathbb{E}\, e^{\delta\left((y^n, z^n), U(x^n)\right)} \\
&= e^{-\exp\left\{\frac{1}{2}n\theta\right\}} \prod_{x^n \in T_{X|YZ}^n(y^n, z^n)} \left( \frac{e}{|T_{YZ|X}^n(x^n)|} + 1 - \frac{1}{|T_{YZ|X}(x^n)|} \right) \\
&\overset{7.}{=} e^{-\exp\left\{\frac{1}{2}n\theta\right\}} \left( 1 + (e-1) \frac{|T_X^n|}{|T_{XYZ}^n|} \right)^{|T_{X|YZ}^n(y^n, z^n)|} \\
&\leq e^{-\exp\left\{\frac{1}{2}n\theta\right\} + (e-1)\frac{|T_X^n||T_{X|YZ}^n(y^n, z^n)|}{|T_{XYZ}^n|}} \\
&\overset{8.}{\leq} \left\{ -\exp\left\{ \frac{1}{2}n\theta \right\} \log e \right\}, \qquad (1.7.82)
\end{aligned}
$$

for $n$ large enough, where the steps are justified as follows:

1. for all $x^n \notin T_{X|Y}^n(y^n)$, $(y'^n, z'^n) \in T_{YZ|X}^n(x^n)$, $y'^n \neq y^n$,
2. $U(x^n)$, $x^n \in T_X^n$, are independent
3. for all $(x^n, y^n) \in T_{XZ}^n$, $|T_{Z|XY}^n(x^n, y^n)||T_{XY}^n| = |T_{XYZ}^n|$ and for all $x^n \in T_X^n$, $|T_{YZ|X}^n(x^n)||T_X^n| = |T_{XYZ}^n|$ (c.f. (1.7.14))
4. for all $y^n \in T_Y^n$, $|T_Y^n||T_{X|Y}^n(y^n)| = |T_{XY}^n|$ (c.f. (1.7.14))
5. by (1.7.78)
6. for all $x^n \notin T_{X|YZ}^n(y^n, z^n)$, $(y'^n, z'^n) \in T_{YZ|X}^n(x^n)$, $(y'^n, z'^n) \neq (y^n, z^n)$
7. for all $x^n \in T_X^n$ $|T_{YZ|X}^n(x^n)||T_X^n| = |T_{XYZ}^n|$ (c.f. (1.7.14))
8. by 1.7.2 and the inequalities characterizing this subcase

$$(e-1)\frac{|T_X^n||T_{X|YZ}^n(y^n,z^n)|}{|T_{XYZ}^n|} = \exp\{n(H(X)+H(X|YZ)-H(XYZ)+o(1))\}$$
$$= \exp\{n(H(X)-H(YZ)+o(1))\} < 1.$$

Thus (1.7.81) and (1.7.82) imply for $n$ large enough,

$$\Pr\left(\left[\bigcup_{y^n\in T_Y^n} E_1(y^n)\right]\cup\left[\bigcup_{(y^n,z^n)\in T_{YZ}^n} E_2(y^n,z^n)\right]\right) < \frac{1}{2}. \tag{1.7.83}$$

So one can find a mapping $U : T_X^n \to T_{YZ}^n$, such that

$$U(x^n) \in T_{YZ|X}^n(x^n) \tag{1.7.84}$$

and for all $y^n \in T_Y^n$

$$|\{x^n : U(x^n) = (y^n, z^n) \quad \text{for some } z^n\}| > \frac{1-e^{-1}}{2}\frac{|T_X^n|}{|T_Y^n|} \tag{1.7.85}$$

and for all $(y^n, z^n) \in T_{YZ}^n$

$$|U^{-1}((y^n,z^n))| = |\{x^n : U(x^n) = (y^n,z^n)\}| < \exp\left\{\frac{1}{2}n\theta\right\}. \tag{1.7.86}$$

Now for each $(y^n, z^n)$ with $U^{-1}((y^n, z^n)) \neq \varnothing$, we take one $x^n \in U^{-1}((y^n, z^n))$ into our $\hat{\mathcal{X}}_n$ and define $U(x^n) = g(x^n)$ for this $x^n$. Then $g$ is injective and by (1.7.78), (1.7.84), (1.7.85) and (1.7.86), the relations (1.7.75), and (1.7.77) are satisfied.

Next we move to the second step and apply the Color Carrier Lemma to the hypergraph $\{T_Z^n, \{E_{y^n}\}_{y^n\in T_Y^n}\}$, to obtain a coloring $\varphi : T_Z^n \to \mathcal{L} = \{1, \dots, L\}$ for $L = \lceil \exp\{n(R_Z - 2\theta)\}\rceil$, and $n$ big enough.

Label the elements of $T_Y^n$ as $v_i$, $1 \le i \le M = |T_Y^n|$, (in any order), and for each $1 \le i \le M$ and $1 \le \ell \le L$ we choose a $z^n \in E_{v_i}$ with $\varphi(z^n) = \ell$ as $v'_{i,\ell}$.

Now the definition of $E_{y^n}$ and injectivity of $g$ allow us to choose $g^{-1}((v_i, v'_{i,\ell}))$, the inverse image of $(v_i, v'_{v,\ell})$ under $g$ as our $u_{i,\ell}$. Finally, set $D_i(\mathcal{Y}) = \{v_i\}$ for $1 \le i \le M$ and $D_\ell(Z) = \{v_{i,\ell} : 1 \le i \le M\}$. We get our matching code $\{(u_{i,\ell}\ D_i(\mathcal{Y}), D_\ell(Z)) \colon 1 \le i \le M, 1 \le \ell \le L\}$ with rate $(R_{\mathcal{Y}} + o(1), R_{\mathcal{Z}} - 2\theta)$, (by (1.7.78)).

### 1.7.9 *The Controller Falls Asleep—on Matching Zero-Error Detection Codes*

We consider now again the one-way deterministic matching channel $\mathcal{W}_0$. Now the communicators, sender and receiver, safeguard against mistakes of the controller and even against malicious operations (jamming) by using *matching zero-error detection codes* (MDC) $\{(u_i, v_i) : 1 \le i \le M\}$, which satisfy for some $s^n(i)(1 \le i \le M)$

$$W^n\big(v_i|u_i|s^n(i)\big) = 1 \tag{1.7.87}$$

and

$$W^n(v_j|u_i|s^n) = 0 \quad \text{for } i \neq j \quad \text{and all } s^n \in \mathcal{S}^n \tag{1.7.88}$$

$C_{\text{mde}}(\mathcal{W}_0)$ denotes the capacity of this channel.

Let $W(\cdot|\cdot)$ be as in (1.7.4) of Introduction B, associated with $\mathcal{W}_0$. Then (1.7.87) takes the form

$$W(v_i|u_i) > 0 \quad \text{for } i = 1, 2, \dots, M \tag{1.7.89}$$

and (1.7.88) takes the form

$$W(v_j|u_i) = 0 \quad \text{for } i \neq j. \tag{1.7.90}$$

These conditions are quite similar to those defining a zero-error detection code $\{u_i : 1 \le i \le M\}$

$$W(u_i|u_i) > 0 \quad \text{for } 1 \le i \le M, \; W(u_j|u_i) = 0 \quad \text{for } i \neq j. \tag{1.7.91}$$

In [8] its capacity was denoted by $C_{de}$. In another terminology this is called Sperner capacity [27].

*Remark* Just formally, one can skip the condition $W^n(u_i \mid u_i) > 0$ and arrive at another mathematically meaningful notion: zero-error pseudo-detection codes. Related concepts can be found in Sect. 1.7.15.

The connection is the following. Relevant for (1.7.91) are rows and columns of $W$ with indices $x, y$ in $\mathcal{X} \cap \mathcal{Y}$ and $W(y|x) > 0$, if $x = y$. This gives a square matrix $\widetilde{W}$ as restriction of $W$ to alphabets $\widetilde{\mathcal{X}} = \widetilde{\mathcal{Y}} \subset \mathcal{X} \cap \mathcal{Y}$, for which (1.7.91) can equivalently be stated. This matrix $\widetilde{W}$ corresponds to a directed graph $\mathcal{G} = (\mathcal{V}, \vec{\mathcal{E}})$ with

$$\mathcal{V} = \widetilde{\mathcal{X}} = \widetilde{\mathcal{Y}} \quad \text{and } (v, v') \in \mathcal{E} \Leftrightarrow W(v'|v) > 0. \tag{1.7.92}$$

In the directed product graph $\mathcal{G}^n(\mathcal{V}^n, \vec{\mathcal{E}^n})$ we have $(v^n, v'^n) \in \vec{\mathcal{E}^n} \Leftrightarrow W(v'^n|v^n) > 0$ and therefore (1.7.91) is equivalent to an independent set in $\mathcal{Y}^n$. The rate of the independence numbers equals $C_{de}$. Notice also that all loops are included in $\mathcal{G}^n$.

*Remark* In [27] zero-error detection codes are described in terms of the dual graph $(\mathcal{G}^n)^* = \big(\mathcal{V}^n, (\overrightarrow{\to} \mathcal{E}^n)^*\big)$, which contains exactly the directed edges which are not in $\overrightarrow{\to} \mathcal{E}^n$. Then for any $u_i, u_j$ in the code, there exists $t$ with $(u_{i_t}, u_{j_t}) \in \overrightarrow{\to} \mathcal{E}^*$.

Now we discuss MDC (see (1.7.89), (1.7.90)). For matrix $W$ consider support sets

$$\mathcal{Y}_W(x) = \{y \in \mathcal{Y} : W(y|x) > 0\} \tag{1.7.93}$$

and define an associated directed graph $\mathcal{G}(W) = \big(\mathcal{X}, \mathcal{E}(W)\big)$ by

$$(x, x') \in \mathcal{E}(W) \Leftrightarrow \mathcal{Y}_W(x) \supset \mathcal{Y}_W(x'). \tag{1.7.94}$$

So all loops are included and the MDC $\{(u_i^n, v_i^n) : 1 \le i \le M\}$ have the property: $\{u_i^n : 1 \le i \le M\}$ is an independent set in the directed product graph $\mathcal{G}^n(W) = \mathcal{G}(W^n)$. The converse is not true.

By (1.7.94) the associated graph has no directed cycles (if $\mathcal{Y}_W(x) \neq \mathcal{Y}_W(x')$ for $x \neq x'$) and therefore the class of these graphs is, again by (1.7.94), smaller than the class of graphs associated via (1.7.92).

Denote by $M^n_{\text{mde}}(W)$, $M^n_{\text{de}}(W)$ and $M^n_0(W)$ the largest sizes of $n$-length MDC, zero-error detection codes and zero-error codes for $W^n$, respectively. When $n = 1$ we write them as $M_{\text{mde}}(W)$, $M_{\text{de}}(W)$ and $M_0(W)$.

**Example** For

$$W_1 = \begin{array}{c} \\ 0 \\ 1 \\ 2 \end{array} \begin{array}{c} \begin{array}{ccc} 0 & 1 & 2 \end{array} \\ \begin{pmatrix} + & + & 0 \\ 0 & + & + \\ + & 0 & + \end{pmatrix} \end{array}$$

we have the independence number $I\big(\mathcal{G}(W_1)\big) = 3$, but

$M_{\text{mde}}(W_1) = 2$. (Here $M_{\text{de}}(W_1) = 1$ and for zero-error codes $M_0(W_1) = 1$). One can identify $x$'s with equal supports without loss in code length.

**Example** For

$$W_2 = \begin{pmatrix} + & + & 0 & 0 & 0 \\ 0 & + & + & 0 & 0 \\ 0 & 0 & + & + & 0 \\ 0 & 0 & 0 & + & + \\ + & 0 & 0 & 0 & + \end{pmatrix}$$

$M_0(W_2) = M_{\text{de}} = 2$ and $M_{\text{mde}}(W_2) = |\{(1, 1), (2, 3), (4, 4)\}| = 3$.

However, by considering all matrices, we can show that the matching zero-error detection coding problem is a proper special case of the zero-error detection coding problem.

We discuss it in Sect. 1.7.14.

### 1.7.10 *The Matching Zero-Error Detection Capacity* $C_{\text{mde}}$ *in a Genuine Example*

We introduce the $\binom{\alpha}{\beta}$-uniform complete hypergraph channel $W_{\alpha,\beta}$ as a channel with input alphabet $\mathcal{X} = \{1, 2, \dots, \alpha\}$, output alphabet $\mathcal{Y} = \binom{\mathcal{X}}{\beta}$, that is, the letters of $\mathcal{Y}$ are the $\beta$-element subsets of $\mathcal{X}$, and for $x \in \mathcal{X}$, $E \in \mathcal{Y}$

$$W_{\alpha,\beta}(E|x) > 0 \Leftrightarrow x \in E. \tag{1.7.95}$$

Therefore an MDC is a system $\{(u_i^n, E_i^n) : 1 \leq i \leq M\}$ with $u_i^n \in \mathcal{X}^n$, $E_i^n \in \binom{\mathcal{X}}{\beta}^n$, and $u_i^n \in E_j^n$ exactly if $i = j$. Its maximal size is $M_{\alpha,\beta}(n)$ and its capacity is

$$C_{\alpha,\beta} = \lim_{n\to\infty} \frac{1}{n} \log M_{\alpha,\beta}(n), \tag{1.7.96}$$

because MDC's can be concatenated. We are going to determine $M_{\alpha,\beta}(n)$ with an elegant method used by Blokhuis in [19]. Its main idea is that all polynomials in indeterminates $\xi_1, \dots, \xi_n$ over any field, with degree $(\xi_i) \leq d_i$ for $i = 1, \dots, n$, form a linear space $\mathcal{L}(d_1, \dots, d_n)$ of a dimension

$$\dim\big(\mathcal{L}(d_1, \dots, d_n)\big) = \prod_{i=1}^{n} (d_i + 1). \tag{1.7.97}$$

**Theorem 50** *For* $\binom{\alpha}{\beta}$*-channels*

$$M_{\alpha,\beta}(n) = (\alpha - \beta + 1)^n$$

*and*

$$C_{\alpha,\beta} = \log(\alpha - \beta + 1).$$

*Proof* Since we can take products of codes, it suffices to show that $M_{\alpha,\beta}(1) \geq \alpha - \beta + 1$ in order to establish the lower bound. Define for $i = 1, 2, \dots, \alpha - \beta + 1$

$$u_i = i \quad \text{and} \quad E_i = \{i, \alpha - \beta + 2, \dots, \alpha\}. \tag{1.7.98}$$

Then $\{(u_i, E_i) : 1 \leq i \leq \alpha - \beta + 1\}$ is an MDC for the $\binom{\alpha}{\beta}$-channel. Conversely, let now $\{(u_i^n, E_i^n) : 1 \leq i \leq M\}$ be an MDC for the $\binom{\alpha}{\beta}$-channel. Consider the polynomials in $X^n = (X_1, \dots, X_n)$

$$f_i(\xi^n) = \prod_{t=1}^{n} \prod_{x \notin E_{it}} (\xi_t - x), \ 1 \leq i \leq M, \tag{1.7.99}$$

where $(E_{i1}, \dots, E_{in}) = E_i^n$.

Clearly, $f_i \in \mathcal{L}\,(\alpha-\beta,\dots,\alpha-\beta)$ and for all $i, i'$

$$f_i(u^n_{i'}) \neq 0 \Leftrightarrow u^n_{i'} \in E^n_i \Leftrightarrow i = i'. \tag{1.7.100}$$

Therefore $f_1,\dots,f_n$ are linearly independent and hence

$$M \leq \dim\big(\mathcal{L}(\alpha-\beta,\dots,\alpha-\beta)\big) = (\alpha-\beta+1)^n.$$

*Remark* Comparison with [19] shows that our result is slightly stronger, because the algebraic method is better exploited.

### *1.7.11 Feedback and also Randomization Increase the Capacity of the Matching Channel*

We have already seen in Sect. 1.7.1 that in the formula for $C(\mathcal{W})$ in Theorem 39 $H(X)$ enters only, because words from $\mathcal{X}^n$ can be used at most once as a code word (in conjunction with an $s^n$) and if we drop this restriction, then we achieve the capacity of the DMC $W' : \mathcal{X} \times \mathcal{S} \to \mathcal{Y}$

$$C = \max_{P_{XS}} I(XS \wedge Y). \tag{1.7.101}$$

This leaves room to enlarge $C(\mathcal{W})$ in the following two ways.

a Suppose now that the controller knows in case of feedback the *encoding function for a message like earlier the code word for a message*, then the $C$ in (1.7.101) can still be achieved and is the capacity!

Clearly $C$ cannot be superceded, because feedback does not increase the capacity of a DMC. For the direct part, let us start with a set of code words $\mathcal{U} = \{u_1,\dots,u_M\}$ and decoding sets $\{\mathcal{D}_1,\dots,\mathcal{D}_M\}$ for $W : \mathcal{X} \times \mathcal{S} \to \mathcal{Y}$, where

$$u_i = (x^n_i, s^n_i), \quad 1 \leq i \leq M. \tag{1.7.102}$$

Let $\mathcal{S}(x^n)$ be the $s^n$'s appearing with $x^n$. This set grows at most exponentially in $n$. Its elements serve as "first names" for $x^n$.

Recalling that encoding with feedback is described by a family of functions $\{f^n_i\}_{i=1}^M$ with $f^n_i = [f_{i1},\dots,f_{in}]$ and $f_{it} : \mathcal{Y}^{t-1} \to \mathcal{X}$, we use now a simple trick.

We add $m$ positions before the word of length $n$. For these positions there are

$$|\mathcal{X}||\mathcal{X}|^{|\mathcal{Y}|}|\mathcal{X}|^{|\mathcal{Y}|^2}\cdots|\mathcal{X}|^{|\mathcal{Y}|^m}$$

many encoding functions $g^m$, which can be used as the beginnings of the encoding functions of length $n+m$. Since there are double exponentially in $m$ many $g^m$'s, they can be used as "nick names" for the first names.

The controller knows by assumption the nick names and therefore the first name $s_i^n$ in $(x_i^n, s_i^n)$, which he then uses in the set of positions $\{m+1, \dots, m+n\}$. Obviously $m$'s contribution to the loss in rate is negligible.

We summarize our findings for the matching capacity in case of feedback.

**Proposition 1** $C_f(\mathcal{W}) = \max_{P_{XS}} I(XS \wedge Y)$.

$\beta$. Now we have no feedback, but the encoder can use randomization, that is, he encodes message $i$ as $Q_i \in \mathcal{P}(\mathcal{X}^n)$. $Q_i$ is known to the controller before he chooses $s_i^n$. Then again the capacity of the matching channel equals $C$. Again we use the previous idea, which can now be realized even simpler. We just use one additional position ($m = 1$). This gives infinitely many nick names, because $|\mathcal{P}(\mathcal{X})| = \infty$.

Formally, we choose an $n$-length code $\{(u_i, \mathcal{D}_i) : 1 \leq i \leq M\}$ as in $\alpha$. and $M$ different distributions $P_i$ on $\mathcal{P}(\mathcal{X})$. For message $i$ the sender chooses and sends an $n+1$-length code word $xx_i^n$, if this is the outcome of random experiment $(Q_i, \mathcal{X}^{n+1})$, where $Q_i(xx_i^n) = P_i(x)$. The controller, who knows $i$, chooses $s_i^n$. Finally the sets $\{\mathcal{X} \times \mathcal{D}_1, \dots, \mathcal{X} \times \mathcal{D}_M\}$ are used as decoding sets.

## 1.7.12 *The Capacity for Matching Zero-Error Detection Codes with Feedback (MDCF) for $\mathcal{W}_0$*

A matching code with feedback $\{(f_i^n, \mathcal{D}_i) : 1 \leq i \leq M\}$ for $\mathcal{W}_0$ or its associated $W$ (in the sense of (1.7.4)) is specified by

$$f_i^n = [f_{i1}, \dots, f_{in}],\ f_{it} : \mathcal{Y}^{t-1} \to \mathcal{X} \tag{1.7.103}$$

and for all $i \in \mathcal{M} = \{1, 2, \dots, M\}$ exists a $v_i \in \mathcal{D}_i$ with

$$W(v_{i1} \mid f_{i1}) \prod_{t=2}^{n} W\big(v_{it} \mid f_{it}(v_{i1}, \dots, v_{i(t-1)})\big) > 0 \tag{1.7.104}$$

$$v_i \neq v_{i'}\ (i \neq i'). \tag{1.7.105}$$

This code is zero-error detecting, if for $i \neq i'$ exists a $t \geq 1$ with $W\big(v_{it} \mid f_{i't}(v_{i1}, \dots, v_{i(t-1)})\big) = 0$. Obviously, we can assume that $\mathcal{D}_i$ contains only one element.

We can assume again that $W$ is nontrivial: not all row supports are identical and for all $y \in \mathcal{Y}$ $W(y|x) > 0$ for some $x \in \mathcal{X}$. Further, a necessary condition for $C_{\text{mdef}}$ to be positive is the existence of two input letters $x_1, x_2$ and an output letter $y_0$ with

$$W(y_0 \mid x_1) > 0 \quad \text{and } W(y_0 \mid x_2) = 0. \tag{1.7.106}$$

**Proposition 2** $C_{\text{mdef}} \begin{cases} \geq C(\mathcal{W}_0) & \textit{if (1.7.106) holds} \\ \textit{or} & \\ = 0 & \textit{otherwise.} \end{cases}$

*Proof* We can assume that (1.7.106) holds. We start with a matching code $\{(u_i^n, v_i^n) : 1 \leq i \leq M\}$ and define a feedback encoding $f_i^{n+1}$ $(1 \leq i \leq M)$ as follows:
For $u_i^n = (u_{i1}, \dots, u_{in})$

$$f_{it}(y^{t-1}) = u_{it} \quad \text{for } t = 1, 2, \dots, n \quad \text{and all } y^n, \tag{1.7.107}$$

$$f_{i\,n+1}(y^n) = \begin{cases} x_1 & \text{if } y^n = v_i^n \\ x_2 & \text{otherwise.} \end{cases} \tag{1.7.108}$$

Define now for all $i$

$$v_i^{n+1} = v_i^n y_0 \tag{1.7.109}$$

and verify that $\{(f_i^{n+1}, v_i^{n+1}) : 1 \leq i \leq M\}$, where $f_i^{n+1} = (f_{i1}, \dots, f_{in}, f_{i\,n+1})$, is by (1.7.106) a matching zero-error detection code with feedback.

**Theorem 51** $C_{\text{mdef}} = \begin{cases} C_f(\mathcal{W}_0) = \max\limits_{P_{XS}} I(XS \wedge Y), & \textit{if (1.7.106) holds} \\ \textit{or} & \\ 0 & \textit{otherwise.} \end{cases}$

*Proof* Clearly, if (1.7.106) does not hold, then the nontrivial matrix has only positive entries and therefore $C_{\text{mdef}} = 0$.

Now, if (1.7.106) holds, then we start with a matching feedback code $\{(f_i^n, v_i^n) : 1 \leq i \leq M\}$ and extend these encoding functions as in the proof of Proposition 2:

$$f_{i\,n+1}(y^n) = \begin{cases} x_1 & \text{if } y^n = v_i^n \\ x_0 & \text{otherwise.} \end{cases}$$

Correspondingly we set $v_i^{n+1} = (v_i^n y_0)$. Finally we apply Proposition 1.

### 1.7.13 Identification for Matching Channels

For our DMC $W : \mathcal{X} \times \mathcal{S} \to \mathcal{Y}$ there are several models of identification with randomized encoding. Fix any code $\{(Q_i, \mathcal{D}_i) : 1 \le i \le N\}$. The $Q_i$'s and $\mathcal{D}_i$'s are assumed to be different. One can consider the following performance criteria:

I. $\sum_{x^n} Q_i(x^n) W^n(\mathcal{D}_i \mid x^n) > 1 - \lambda \quad \text{for all } i$

II. $\sum_{x^n} Q_i(x^n) W^n(\mathcal{D}_i \mid x^n) = 1 \quad \text{for all } i$

III. $\sum_{x^n} Q_i(x^n) W^n(\mathcal{D}_i \mid x^n) > 0 \quad \text{for all } i$

a. $\sum_{x^n} Q_j(x^n) W^n(\mathcal{D}_i \mid x^n) < \lambda \quad \text{for } j \neq i$

b. $\sum_{x^n} Q_j(x^n) W^n(\mathcal{D}_i \mid x^n) = 0 \quad \text{for } j \neq i$

c. $\sum_{x^n} Q_j(x^n) W^n(\mathcal{D}_i \mid x^n) < 1 \quad \text{for } j \neq i$

The classical work [9] concerns the combination (I,a) and in [8] the combination (II,a) was analyzed.

We settle (III,a), *the matching identification problem*, and also (III,c), in Theorem 52. Furthermore, the capacities for the cases (II,c) and (I,c) are characterized in Theorems 53 and 54, resp. Finally, condition b. implies that $\mathcal{D}_i$ can be replaced by $\mathcal{D}_i \smallsetminus \bigcup_{j \neq i} \mathcal{D}_j$ and so we have here disjoint decoding sets: Actually, identification is reduced to transmission. (II,b) gives Shannon's classical zero-error capacity problem. (I,b) gives the erasure problem (see [8]) for transmission, and (III, b), reduces to the matching zero-error detection problem discussed in Sects. 1.7.9 and 1.7.10. The discussion of all cases is complete.

*Remark* Some combinations are meaningful also in case of feedback.

We consider first the cases (III,a) and (III,c). Their capacities are denoted by $C(>0, <\lambda)$ and $C(>0, <1)$.

**Theorem 52** *We have for the second order identification capacities*

$$C(>0, <\lambda) = C(>0, <1) = \begin{cases} \log |\mathcal{Y}|, & \text{if } W \text{ is nontrivial} \\ 0 & \text{otherwise.} \end{cases}$$

*W is nontrivial, if not all rows are identical and no column has only 0's as entries.*

*Proof* Let $\overline{Y}$ take values in $\mathcal{Y}$ according to the uniform distribution and let $X$ take values in $\mathcal{X}$ such that for some small $\delta \in (0, 1)$ and $P_Y = P_X W$

$$\|P_Y - P_{\overline{Y}}\| \geq \delta. \tag{1.7.110}$$

This is possible, because $W$ is nontrivial.

Let

$$Q(x^n) = \begin{cases} \frac{n-1}{n} \cdot \frac{1}{|T_X^n|} & \text{if } x^n \in T_X^n \\ \frac{1}{n} \frac{1}{|\mathcal{X}^n| - |T_X^n|} & \text{otherwise.} \end{cases} \tag{1.7.111}$$

Then,

$$\sum_{x^n \in T_X^n} Q(x^n) W^n(y^n \mid x^n) > 0 \quad \text{for all } y^n \in \mathcal{Y}^n$$

and for every $\mathcal{D} \subset T_{\overline{Y}}^n$, by (1.7.110) (and the properties of typical and generated sequences stated in Sect. 1.7.2)

$$\sum_{x^n} Q(x^n) W^n(\mathcal{D} \mid x^n) < \varepsilon_n(\delta)$$

and $\lim_{n \to \infty} \varepsilon_n(\delta) = 0$.

Now just choose the $2^{|T_{\overline{Y}}^n|} - 1$ nonempty $\mathcal{D}$'s as decoding sets and choose the same number of arbitrary $Q$'s on $T_X^n$ as corresponding encoding distributions. Since

$$\log \log 2^{|T_{\overline{Y}}^n|} \sim n \log |\mathcal{Y}|,$$

the proof of the direct parts is complete. The converses follow from the fact that there are at most $2^{|\mathcal{Y}^n|}$ decoding sets.

We are left with the case (II,c) and its capacity $C(=1, <1)$.

**Theorem 53** *The second order capacity $C(=1, <1)$ equals the first order $C_{\text{mde}}$.*

*Proof of direct part.* Start with an MDC $\{(u_i, v_i) : 1 \leq i \leq M\}$:

$$W^n(v_i \mid u_i) > 0 \quad \text{and} \quad W^n(v_j \mid u_i) = 0 \quad \text{for all } i \neq j.$$

Define the supports

$$\mathcal{Y}^n(u) = \{v \in \mathcal{Y}^n : W^n(v|u) > 0\}, u \in \mathcal{X}^n, \tag{1.7.112}$$

and notice that for any $\mathcal{U} = \{u_i : 1 \leq i \leq M\} \subset \mathcal{X}^n$ $\{(u_i, v_i) : 1 \leq i \leq M\}$ is a set of code words for some $v_i$ if and only if

$$\mathcal{Y}^n(u) \not\subset \bigcup_{u' \in \mathcal{U} \smallsetminus \{u\}} \mathcal{Y}^n(u'). \tag{1.7.113}$$

Let $M$ be even and consider the $\frac{M}{2}$-element subsets of $\mathcal{U}$

$$\binom{\mathcal{U}}{\frac{M}{2}} = \left\{ A_\ell : 1 \le \ell \le \binom{M}{\frac{M}{2}} \right\}. \tag{1.7.114}$$

Let $Q_\ell$ be any distribution with support $A_\ell$ and define

$$\mathcal{D}_\ell = \bigcup_{u \in A_\ell} \mathcal{Y}^n(u). \tag{1.7.115}$$

Then obviously

$$\sum_{x^n} Q_\ell(x^n) W(\mathcal{D}_\ell \mid x^n) = 1 \quad \text{for } 1 \le \ell \le \binom{M}{\frac{M}{2}} \tag{1.7.116}$$

and, moreover, for all $\ell \neq \ell'$

$$A_\ell \smallsetminus A_{\ell'} \neq \varnothing$$

and so by (1.7.113) and (1.7.115) for $u_j \in A_\ell \smallsetminus A_{\ell'}$

$$\mathcal{Y}^n(u_j) \not\subset \bigcup_{i \neq j} \mathcal{Y}^n(u_i) \supset \bigcup_{u \in A'_\ell} \mathcal{Y}^n(u) = \mathcal{D}'_\ell. \tag{1.7.117}$$

Since $Q_\ell(u_j) > 0$, we have finally that

$$\sum_{u \in A_\ell} Q_\ell(u) W^n(\mathcal{D}'_\ell \mid u) < 1.$$

*Converse part.* Let $\{(Q_i, \mathcal{D}_i) : 1 \le i \le M\}$ be a (II,c) identification code. Define the support set of $Q_i$

$$S_i = \{u \in \mathcal{X}^n : Q_i(u) > 0\}.$$

W.l.o.g.

$$\mathcal{D}_i = \bigcup_{u \in S_i} \mathcal{Y}^n(u) \tag{1.7.118}$$

and by (II,c)

$$\mathcal{D}_i \not\subset \mathcal{D}_j \quad \text{for } i \neq j. \tag{1.7.119}$$

Let now $S'_i \subset S_i$ be minimal with the properties

$$\bigcup_{u \in S'_i} \mathcal{Y}^n(u) = \bigcup_{u \in S_i} \mathcal{Y}^n(u) = \mathcal{D}_i. \tag{1.7.120}$$

The minimality property implies that for all $i$ $S'_i$ satisfies (1.7.113) and therefore corresponds to an MDC. Consequently

$$|S'_i| \leq \exp\{n\, C_{\text{mde}}\}. \tag{1.7.121}$$

Furthermore, the sets $S'_i$ are different, because the $\mathcal{D}_i$'s are different and (1.7.120) holds. Therefore

$$M \leq |\mathcal{X}|^{n \cdot \exp\{n\, C_{\text{mde}}\}}$$

and finally

$$\frac{1}{n} \log\log M \leq C_{\text{mde}} + \frac{1}{n}\log n + \frac{1}{n}\log\log|\mathcal{X}|.$$

Finally, we consider the combination (I,c) and determine its second order identification capacity $C(> 1-\lambda, < 1)$. We call $W$ degenerate iff for some $y_0 \in \mathcal{Y}$

$$W(y_0|x) = 1 \quad \text{for all } x \in \mathcal{X}. \tag{1.7.122}$$

**Theorem 54**

$$C(> 1-\lambda, < 1) = \begin{cases} \log|\mathcal{Y}| & \textit{if } W \textit{ is non-degenerate and } |\mathcal{X}| > 1 \\ 0 & \textit{if } W \textit{ is degenerate or } |\mathcal{X}| = 1. \end{cases}$$

*Proof* For a degenerate $W$ we have for all $\mathcal{D} \subset \mathcal{Y}^n$ and all $x^n \in \mathcal{X}^n$

$$W^n(\mathcal{D}|x^n) = \begin{cases} 1 & \text{if } (y_0, \dots, y_0) \in \mathcal{D} \\ 0 & \text{if } (y_0, \dots, y_0) \notin \mathcal{D} \end{cases}$$

and so cannot have two decoding sets with the required conditions. For $|\mathcal{X}| = 1$ there is only one input distribution.

For non-degenerate $W$ and $|\mathcal{X}| > 1$ it is easy to see that there are distinct $x_1, x_2 \in \mathcal{X}$ and $y_0 \in \mathcal{Y}$ with

$$W(y_0|x_1) > 0 \quad \text{and} \quad W(y_0|x_2) < 1. \tag{1.7.123}$$

Indeed, if there is no $y_0$ satisfying (1.7.123), for all $y$ $W(y|x_1) > 0$ implies that $W(y|x_2) = 1$. However, because there is at most one $y \in \mathcal{Y}$ with $W(y|x_2) = 1$, this means that for some $y$ $W(y|x_1) = W(y|x_2) = 1$.

Now define $u_1 = (x_1, \dots, x_1)$; $u_2 = (x_2, \dots, x_2) \in \mathcal{X}^n$, $v_0 = (y_0, \dots, y_0) \in \mathcal{Y}^n$ and let $n$ be so large that

$$W^n(v_0 \mid u_2) < \frac{\lambda}{4} \qquad (1.7.124)$$

and

$$W^n(T^n_{W,\delta}(u_2) \mid u_2) > 1 - \frac{\lambda}{4} \qquad (1.7.125)$$

(in the terminology of 1.7.2). Set

$$\mathcal{E} = T^n_{W,\delta}(u_2) \setminus \{v_0\}, \mathcal{F} = \mathcal{Y}^n \setminus \big(\{v_0\} \cup T^n_{W,\delta}(u_2)\big).$$

Now, $|T^n_{W,\delta}(u_2)|$ is much smaller than $|\mathcal{Y}|^n$ and $\mathcal{F}$ has the same rate as $|\mathcal{Y}|^n$, that is, for $\mu \in (0, 1)$

$$\log |\mathcal{F}| > n\big(\log |\mathcal{Y}| - \mu\big) \qquad (1.7.126)$$

for $n$ large enough.

Now we choose as decoding sets the family

$$\big\{\mathcal{D}_i : \mathcal{D}_i = \mathcal{E} \cup \mathcal{F}'_i, \mathcal{F}'_i \subset \mathcal{F}, 1 \le i \le 2^{|\mathcal{F}|}\big\}.$$

It has cardinality $2^{|\mathcal{F}|}$ and $\log \log 2^{|\mathcal{F}|} > n(\log |\mathcal{Y}| - \mu)$. Finally define $Q_i$ with support $\{u_1, u_2\}$ and

$$0 < Q_i(u_1) < \frac{\lambda}{2}. \qquad (1.7.127)$$

Since $W^n(v_0 \mid u_1) > 0$ we get for all $i, j$

$$\sum_{x^n} Q_j(x^n) W^n(\mathcal{D}_i \mid x^n) < 1$$

and (1.7.125) and (1.7.127) imply

$$\sum_{x^n} Q_i(x^n) W^n(\mathcal{D}_i \mid x^n) > \left(1 - \frac{\lambda}{2}\right)^2 > 1 - \lambda$$

for all $i$.

The converse obviously holds.

### 1.7.14 Zero-Error Detection

**Lemma 29** *An MDC* $\{(u_i, v_i) : 1 \le i \le M\}$ *for* $W^n : \mathcal{X}^n \to \mathcal{Y}^n (n \ge 1)$ *can be viewed as zero-error detection code for* $V^n$*, where* $V$ *has input and output alphabet*

$$\mathcal{Z} = \{(x, y) : x \in \mathcal{X}, y \in \mathcal{Y}, W(y|x) > 0\} \tag{1.7.128}$$

*and*

$$V\big((x', y')|(x, y)\big) > 0 \Leftrightarrow W(y'|x) > 0. \tag{1.7.129}$$

*Conversely, for every zero-error detection code* $\{((u_{i1}, v_{i1}), \dots, (u_{in}, v_{in})) : 1 \le i \le M\}$ *for* $V^n$ $\{(u_i^n, v_i^n) : 1 \le i \le M\}$ *is an MDC for* $W^n$.

*Proof* These are immediate consequences of the definitions.

Next we identify MDC problems as a certain class of zero-error detection problems. For this we proceed as follows:

(i) We observe that, while considering MDC for $W^n$, letters $x, x' \in \mathcal{X}$ with equal row support sets, that is,

$$\mathcal{Y}_W(x) = \mathcal{Y}_W(x') \text{ (defined in (1.7.93))}, \tag{1.7.130}$$

can be contracted to one letter.

(ii) We also observe that letters $y, y' \in \mathcal{Y}$ with equal columns support sets

$$\mathcal{X}_W(y) \triangleq \{x \in \mathcal{X} : W(y|x) > 0\} = \mathcal{X}_W(y') \triangleq \{x \in \mathcal{X} : W(y'|x) > 0\}$$

can be contracted to one.

We call a matrix $W$ *irreducible*, if no contractions as described in (i) and (ii) are possible.

It is clear from the definitions of $\mathcal{Z}$, $V$ in (1.7.128) and (1.7.129) that

$$V(z|z) > 0 \quad \text{for all } z \in \mathcal{Z}. \tag{1.7.131}$$

It is also clear that row supports

$$\mathcal{Z}_{V,r}(z) = \{z' : V(z'|z) > 0\} \tag{1.7.132}$$

and columns supports

$$\mathcal{Z}_{V,c}(z) = \{z' : V(z|z') > 0\} \tag{1.7.133}$$

being equal simultaneously means that the corresponding letters can be contracted for zero-error detection codes for $V^n$. This leads to the notion of an irreducible $V$. Moreover, one readily verifies the next fact

**Lemma 30** *For an irreducible W the corresponding V is also irreducible and conversely.*

Next we characterize those matrices $V$ which can be obtained via (1.7.128), (1.7.129) from some irreducible $W$ and so for the maximal code sizes

$$M^n_{\text{mde}}(W) = M^n_{de}(V), \quad n \geq 1. \tag{1.7.134}$$

Using this characterization (Theorem 55) we then show by the first example in Sect. 1.7.14, that matching zero-error detection coding is indeed more special than zero-error detection coding.

A few more definitions are needed. Set

$$A_x = \{x\} \times \mathcal{Y}_W(x) = \{(x, y) : y \in \mathcal{Y}, (x, y) \in \mathcal{Z}\}, \tag{1.7.135}$$

$$B_y = \mathcal{X}_W(y) \times \{y\} = \{(x, y) : x \in \mathcal{X}, (x, y) \in \mathcal{Z}\}. \tag{1.7.136}$$

Clearly, $(A_x)_{x\in\mathcal{X}}$ and $(B_y)_{y\in\mathcal{Y}}$ are both partitions of $\mathcal{Z}$ and

$$a, a' \in A_x \Rightarrow \mathcal{Z}_{V,r}(a) = \mathcal{Z}_{V,r}(a') \tag{1.7.137}$$

$$b, b' \in B_y \Rightarrow \mathcal{Z}_{V,c}(b) = \mathcal{Z}_{V,c}(b'). \tag{1.7.138}$$

On the other hand, every $V : \mathcal{Z} \to \mathcal{Z}$ (not necessary generated by (1.7.128) and (1.7.129), whose row partition $(A_x)_{x\in\mathcal{X}}$ and column partition $(B_y)_{y\in\mathcal{Y}}$ satisfies (1.7.137), (1.7.138) (where $\mathcal{X}$ and $\mathcal{Y}$ serve only as index sets) has the following properties:

It partitions for all $x \in \mathcal{X}$ the set $\mathcal{Y}$ into $\mathcal{Y}^*(x)$ and $\mathcal{Y}^{*c}(x) = \mathcal{Y} \smallsetminus \mathcal{Y}^*(x)$ such that for all $a \in A_x$

$$\begin{aligned} \mathcal{Z}_{V,r}(a) &= \bigcup_{y\in\mathcal{Y}^*(x)} B_y, \\ \mathcal{Z}_{V,r}(a) \cap B_y &= \varnothing, \text{ if } y \in \mathcal{Y}^{*c}(x). \end{aligned} \tag{1.7.139}$$

Similarly, it partitions, for all $y \in \mathcal{Y}$, $\mathcal{X}$ into $\mathcal{X}^*(y)$ and $\mathcal{X}^{*c}(y) = \mathcal{X} \smallsetminus \mathcal{X}^*(y)$ such that for all $b \in B_y$

$$\begin{aligned} \mathcal{Z}_{V,c}(b) &= \bigcup_{x\in\mathcal{X}^*(y)} A_x, \\ \mathcal{Z}_{V,c}(b) \cap A_x &= \varnothing, \text{ if } x \in \mathcal{X}^{*c}(y). \end{aligned} \tag{1.7.140}$$

**Theorem 55** *For* $W : \mathcal{X} \to \mathcal{Y}$ *and corresponding* $\mathcal{Z}, V$ *(defined by* (1.7.128) *and* (1.7.129)*)*

*(i) the identity* (1.7.134) *holds*
*(ii)* (1.7.131) *and the identities*

$$|A_x| + |\mathcal{Y}^{*c}(x)| = |\mathcal{Y}| \quad \textit{for all } x \in \mathcal{X} \tag{1.7.141}$$

$$|B_y| + |\mathcal{X}^{*c}(y)| = |\mathcal{X}| \quad \textit{for all } y \in \mathcal{Y} \tag{1.7.142}$$

*hold, if* $A_x$, $B_y$, $\mathcal{Y}^*(x)$, $\mathcal{Y}^{*c}(x)$, $\mathcal{X}^*(y)$, *and* $\mathcal{X}^{*c}(y)$ *are defined as in* (1.7.135), (1.7.136), (1.7.139), *and* (1.7.140).

*(iii) Conversely, any irreducible matrix* $V : \mathcal{Z} \to \mathcal{Z}$ *satisfying* (1.7.131) *corresponds to an irreducible matrix* $W$ *so that* (1.7.134) *holds, if there are partitions* $(A_x)_{x\in\mathcal{X}}$ *and* $(B_y)_{y\in\mathcal{Y}}$*, for some index sets* $\mathcal{X}$ *and* $\mathcal{Y}$*, such that* (1.7.137), (1.7.138), *and* (1.7.141) *hold for* $\mathcal{Y}^{*c}(x)$ *defined by* (1.7.139). *Symmetrically, here* (1.7.141) *can be replaced by* (1.7.142).

*Proof* (i) Here Lemma 29 is just restated.
(ii) By (1.7.139) and (1.7.140)

$$\mathcal{Y}^{*c}(x) = \{y : W(y|x) = 0\} \quad \text{for all} \quad x \in \mathcal{X}$$

and

$$\mathcal{X}^{*c}(y) = \{x : W(y|x) = 0\} \quad \text{for all} \quad y \in \mathcal{Y}.$$

Thus (1.7.141) and (1.7.142) follow from (1.7.135) and (1.7.136), respectively.
(iii) Define $W : \mathcal{X} \to \mathcal{Y}$ as follows:

$$\text{For all } x \in \mathcal{X}, y \in \mathcal{Y} : W(y|x) > 0 \quad \text{iff} \quad y \in \mathcal{Y}^*(x), \tag{1.7.143}$$

In order to see that $\mathcal{Z}$ is generated by $W$ via (1.7.128), we have to show that there is a bijection from $\{(x, y) : w(y|x) > 0\} = \bigcup_{x\in\mathcal{X}}\{x\} \times \mathcal{Y}^*(x)$ to $\mathcal{Z}$. To achieve this goal, we first characterize $\mathcal{Y}^*(x)$ and $\mathcal{Y}^{*c}(x)$. Indeed we shall show that for all $x \in \mathcal{X}$ and $y \in \mathcal{Y}$

$$y \in \mathcal{Y}^{*c}(x) \quad \text{iff} \quad A_x \cap B_y = \varnothing \tag{1.7.144}$$

$$y \in \mathcal{Y}^{*}(x) \quad \text{iff} \quad |A_x \cap B_y| = 1. \tag{1.7.145}$$

We begin with the assumption $y \in \mathcal{Y}^{*c}(x)$. If now $A_x \cap B_y \neq \varnothing$ and $z \in A_x \cap B_y$, then by (1.7.131), (1.7.137), (1.7.138) and the definition of $\mathcal{Z}_{V,r}$, we have for all $a \in A_x$ $B_y \subset \mathcal{Z}_{V,r}(a)$ and therefore by (1.7.139) $y \in \mathcal{Y}^*(x)$, a contradiction.

Moreover, since $V$ is irreducible, we also know that

$$|A_x \cap B_y| \leq 1 \quad \text{for } x \in \mathcal{X}, y \in \mathcal{Y}. \tag{1.7.146}$$

So we know that

$$|A_x| \le |\mathcal{Y}^*(x)| = |\mathcal{Y}| - |\mathcal{Y}^{*c}(x)|.$$

The reverse implications in (1.7.144) and (1.7.145) follow immediately from our assumption (1.7.141), which gives also the direct implication in (1.7.144).

Now, (1.7.145) gives rise to a mapping defined on $\bigcup_{x\in\mathcal{X}}\{x\}\times\mathcal{Y}^*(x)$ and sending $(x, y)$ to the unique element in $A_x \cap B_y$. Further, by (1.7.144), $\{A_x \cap B_y : y \in \mathcal{Y}^*(x)\}$ is a partition of $\mathcal{Z}$ and this guarantees the mapping to be bijective. Thus we can rename the elements of $\mathcal{Z}$ by $\{(x, y) : W(y|x) > 0\} = \bigcup_{x\in\mathcal{X}}\{x\}\times\mathcal{Y}^*(x)$ and rewrite $z \in \mathcal{Z}$ as $(x, y)$, if $z \in A_x \cap B_y$.

So only (1.7.129) remains to be verified. For the new names $(x, y), (x', y') \in \mathcal{Z}$

$$V\big((x', y') \mid (x, y)\big) > 0 \overset{1.}{\Longleftrightarrow} A_{x'} \cap B_{y'} \subset \mathcal{Z}_{V,r}\big((x, y)\big) \overset{2.}{\Longleftrightarrow} y' \in \mathcal{Y}^*(x)$$
$$\overset{3.}{\Longleftrightarrow} W(y'|x) > 0,$$

that is, (1.7.129).

Here the equivalences are justified as follows:

1. Use definitions of $(x', y')$ and $\mathcal{Z}_{V,r}\big((x, y)\big)$
2. Use (1.7.139) and that by definition of $(x, y)$ necessarily $(x, y) \in A_x$
3. Use (1.7.143).

**Example** Consider $\mathcal{Z} = \{0, 1, 2\}$, $V = W_1$, defined in the first example of 1.7.9.

The only partitions satisfying (1.7.137) and (1.7.138) are $\{A_0, A_1, A_2\} = \big\{\{0\}, \{1\}, \{2\}\big\}$ and $\{B_0, B_1, B_2\} = \big\{\{0\}, \{1\}, \{2\}\big\}$. However, since $0 \in A_0$ and $\mathcal{Z}_{V,r}(0) = \{0, 1\} = B_0 \cup B_1$, $\mathcal{Y}^{*c}(0) = \{2\}$, we have $|A_0| + |\mathcal{Y}_0^{*c}| = 2 \ne 3 = |\mathcal{Y}|$. Thus by Theorem 50 no $W$ can generate $V$.

**Example** $W = W_1$, defined in the first example of 1.7.9, generates by (1.7.128) and (1.7.129),

$$\mathcal{Z} = \big\{(0, 0), (0, 1), (1, 1), (1, 2), (2, 0), (2, 2)\big\},$$

$$V = \begin{array}{c} \\ (0,0) \\ (0,1) \\ (1,1) \\ (1,2) \\ (2,0) \\ (2,2) \end{array} \begin{array}{c} \begin{array}{cccccc} (0,0) & (0,1) & (1,1) & (1,2) & (2,0) & (2,2) \end{array} \\ \begin{pmatrix} + & + & + & 0 & + & 0 \\ + & + & + & 0 & + & 0 \\ 0 & + & + & + & 0 & + \\ 0 & + & + & + & 0 & + \\ + & 0 & 0 & + & + & + \\ + & 0 & 0 & + & + & + \end{pmatrix} \end{array},$$

and $A_0 = \big\{(0, 0), (0, 1)\big\}$, $A_1 = \big\{(1, 1), (1, 2)\big\}$, $A_2 = \big\{(2, 0), (2, 2)\big\}$, $B_0 = \big\{(0, 0), (2, 0)\big\}$, $B_1 = \big\{(1, 1), (1, 2)\big\}$, and $B_2 = \big\{(1, 2), (2, 2)\big\}$.

### 1.7.15 A Digression: Three Further Code Concepts

At the moment our only motivation for the code concepts below is mathematical interest in searching for new combinatorial structures and their connections to the zero-error detection and matching zero-error detection codes. We use the abbreviations

$$\mathcal{Y}(x) = \mathcal{Y}_W(x), \mathcal{X}(y) = \mathcal{X}_W(y),$$
$$\mathcal{Y}^n(x^n) = \prod_{t=1}^{n} \mathcal{Y}(x_t), \mathcal{X}^n(y^n) = \prod_{t=1}^{n} \mathcal{X}(y_t). \tag{1.7.147}$$

**Definition 5** A *pairwise* zero-error detection code for $W^n$ is a set $\{u_i : 1 \leq i \leq M\} \subset \mathcal{X}^n$ such that for all pairs $(i, j)$ there is a $v_{ij} \in \mathcal{Y}^n$ with

$$W^n(v_{ij} \mid u_j) = 0 \quad \text{and} \quad W^n(v_{ij} \mid u_i) > 0. \tag{1.7.148}$$

For two sets $A, B$ we write

$$A \supset\!\subset B \quad \text{iff } A \subset B \quad \text{or } B \subset A$$

and we write

$$A \not\supset\!\subset B \quad \text{iff } A \not\subset B \quad \text{and } B \not\subset A. \tag{1.7.149}$$

In this terminology (1.7.148) is equivalent to

$$\mathcal{Y}^n(u_i) \not\supset\!\subset \mathcal{Y}^n(u_j) \quad \text{for all } i \neq j. \tag{1.7.150}$$

Hence, if we define $U : \mathcal{X} \to \mathcal{X}$ by

$$U(x|x') > 0 \quad \text{iff } \mathcal{Y}(x) \subset \mathcal{Y}(x') \tag{1.7.151}$$

we get the following characterization:

**Lemma 31** $\{u_i : 1 \leq i \leq M\} \subset \mathcal{X}^n$ *is a pairwise zero-error detection code for* $W^n$ *iff it is a zero-error detection code for* $U^n$.

*Of course, every zero-error detection code is also a pairwise zero-error detection code for the same channel.*

Since by (1.7.151) every $U$ generated by a $W$ must have a positive diagonal, we ask whether arbitrary $U$ with positive diagonal can be generated this way. This is not the case.

**Example** Choose again $U = W_1$ in the first example of Eq. 1.7.9. Every $W$ generating $U$ must satisfy

$$\mathcal{Y}(0) \supset \mathcal{Y}(1), \mathcal{Y}(1) \supset \mathcal{Y}(2) \quad \text{and } \mathcal{Y}(2) \supset \mathcal{Y}(0)$$

and hence $\mathcal{Y}(0) = \mathcal{Y}(1) = \mathcal{Y}(2)$. However, such a $W$ generates $\begin{pmatrix} + & + & + \\ + & + & + \\ + & + & + \end{pmatrix}$ and not $U$.

**Definition 6** A component-pairwise zero-error detection code for $W^n$ is a set $\{u_i : 1 \leq i \leq M\} \subset \mathcal{X}^n$ such that for all pairs $(u_i, u_j) = (u_{i1} \dots u_{in}, u_{j1} \dots u_{jn})$ $(i \neq j)$ there is a component $t = t(i, j)$ and there are letters $y, y'$ with

$$W(y \mid u_{it}) > 0, W(y' \mid u_{jt}) > 0 \quad \text{and } W(y' \mid u_{it}) = W(y \mid u_{jt}) = 0$$

(or equivalently $\mathcal{Y}(u_{it}) \not\supset\!\!\!\subset \mathcal{Y}(u_{jt})$).

For a suitable set $\mathcal{Y}'$ we define now $T : \mathcal{X} \to \mathcal{Y}'$ by requiring that for all $x, x' \in \mathcal{X}$

$$\mathcal{Y}(x) \not\supset\!\!\!\subset \mathcal{Y}(x') \quad \text{iff for all } y \in \mathcal{Y}' \; T(y \mid x) \cdot T(y \mid x') = 0. \tag{1.7.152}$$

One notices that $\{u_i : 1 \leq i \leq M\}$ is a component-pairwise zero-error detection code for $W^n$ iff it is a zero-error code (in Shannon's sense) for $T^n$.

**Example** That

$$T = \begin{array}{c} a_0 \\ a_1 \\ a_2 \\ b_0 \\ b_1 \\ b_2 \end{array} \begin{pmatrix} + & 0 & + & 0 & 0 \\ + & 0 & 0 & + & 0 \\ + & 0 & 0 & 0 & + \\ 0 & + & + & 0 & 0 \\ 0 & + & 0 & + & 0 \\ 0 & + & 0 & 0 & + \end{pmatrix}$$

cannot be generated by any $W$ via (1.7.152) can be seen as follows. Such a $W$ would have to satisfy

$\mathcal{Y}(a_0) \supset\!\!\!\subset \mathcal{Y}(b_0)$, $\mathcal{Y}(a_0) \supset\!\!\!\subset \mathcal{Y}(a_1)$ and $\mathcal{Y}(a_0) \supset\!\!\!\subset \mathcal{Y}(a_2)$. Since $\mathcal{Y}(b_0) \not\supset\!\!\!\subset \mathcal{Y}(a_1)$ and $\mathcal{Y}(b_0) \not\supset\!\!\!\subset \mathcal{Y}(a_2)$, we get $\mathcal{Y}(a_0) \supset \mathcal{Y}(b_0)$, $\mathcal{Y}(a_0) \supset \mathcal{Y}(a_1)$ and $\mathcal{Y}(a_0) \supset \mathcal{Y}(a_2)$ or

$$\mathcal{Y}(a_0) \subset \mathcal{Y}(b_0), \mathcal{Y}(a_0) \subset \mathcal{Y}(a_1) \quad \text{and } \mathcal{Y}(a_0) \subset \mathcal{Y}(a_2). \tag{1.7.153}$$

The corresponding relations for $a_1$ and $a_2$ are

$$\mathcal{Y}(a_1) \supset \mathcal{Y}(b_1), \mathcal{Y}(a_1) \supset \mathcal{Y}(a_2) \quad \text{and } \mathcal{Y}(a_1) \supset \mathcal{Y}(a_2)$$

or

$$\mathcal{Y}(a_1) \subset \mathcal{Y}(b_1), \mathcal{Y}(a_1) \subset \mathcal{Y}(a_2) \quad \text{and } \mathcal{Y}(a_1) \subset \mathcal{Y}(a_0), \tag{1.7.154}$$

and

$$\mathcal{Y}(a_2) \supset \mathcal{Y}(b_2), \mathcal{Y}(a_2) \supset \mathcal{Y}(a_0) \quad \text{and } \mathcal{Y}(a_2) \supset \mathcal{Y}(a_1)$$

or

$$\mathcal{Y}(a_2) \subset \mathcal{Y}(b_2), \mathcal{Y}(a_2) \subset \mathcal{Y}(b_2) \quad \text{and } \mathcal{Y}(a_2) \subset \mathcal{Y}(a_1), \tag{1.7.155}$$

respectively.

However, for $i \neq j$ $\mathcal{Y}(a_i) \neq \mathcal{Y}(a_j)$ because, for example, for $i = 0, j = 1$ $\mathcal{Y}(a_0) \not\supset\not\subset \mathcal{Y}(b_1)$ and $\mathcal{Y}(a_1) \supset\subset \mathcal{Y}(b_1)$.

So there is no way to satisfy (1.7.153)–(1.7.155) simultaneously (with $\mathcal{Y}(a_i) \neq \mathcal{Y}(a_j)$ for $i \neq j$).

**Definition 7** $\{(u_i, v_i) : 1 \le i \le M\} \subset \mathcal{X}^n \times \mathcal{Y}^n$ is a pseudo-matching zero-error detection code, if

$$W^n(v_j \mid u_i) = 0 \quad \text{for } i \neq j.$$

We have given up (1.7.89) in the definition of an MDC.

At first notice that for MDC necessarily $u_i \neq u_j$ and $v_i \neq v_j$ for $i \neq j$. For the new codes (without (1.7.89)) this no longer is true. We can study the following four conditions (of increasing strength)

$$(u_i, v_i) \neq (u_j, v_j) \quad \text{for} \quad i \neq j \tag{1.7.156}$$

$$u_i \neq u_j \quad \text{for} \quad i \neq j \tag{1.7.157}$$

$$v_i \neq v_j \quad \text{for} \quad i \neq j \tag{1.7.158}$$

$$u_i \neq u_j \text{ and } v_i \neq v_j \quad \text{for} \quad i \neq j. \tag{1.7.159}$$

We denote the corresponding maximal code sizes for $W^n$ by $M^n_{--}(W), M^n_{+-}(W), M^n_{-+}(W), M^n_{++}(W)$.

Our main, but simple, observations are:

I. All these quantities are generally different from the corresponding quantities for zero-error, zero-error detection, and matching zero-error detection codes for $W^n$.

II. All these quantities can be estimated at least asymptotically rather accurately, so that the capacities are known.

III. The bounds on the code sizes are almost trivial. However, an exact determination of the code sizes is perhaps challenging for a combinatorialist.

For the formulation of our results we need a few definitions. $A \times B$ with $A \subset \mathcal{X}$, $B \subset \mathcal{Y}$ is a zero-rectangle for $W$ if

$$W(y \mid x) = 0 \quad \text{for all} \quad x \in A, \quad y \in B. \tag{1.7.160}$$

Let $\mathcal{R}$ be the set of such rectangles, let $A^* \times B^*$ be a maximal rectangle in the sense

$$|A^*||B^*| = \max_{A\times B\in\mathcal{R}} |A||B|, \tag{1.7.161}$$

and let $A^{(n)} \times B^{(n)}$ be a maximal rectangle in the sense

$$\max_{A\times B\in\mathcal{R}} \left(\min\left[|\mathcal{X}|^{n-1}|A|, |\mathcal{Y}|^{n-1}|B|\right]\right) = \min[|\mathcal{X}|^{n-1}A^{(n)}, |\mathcal{Y}|^{n-1}B^{(n)}]. \tag{1.7.162}$$

**Theorem 56** *For every memoryless channel $W^n$ with $A^* \times B^* \neq \emptyset$*

*(i)* $|\mathcal{X}|^{n-1}|\mathcal{Y}|^{n-1}|A^*||B^*| \leq M^n_{--}(W) \leq |\mathcal{X}|^n|\mathcal{Y}|^n$

*(ii)* $|\mathcal{X}|^n - \left(\min_{y\in\mathcal{Y}} |\mathcal{X}(y)|\right)^n \leq M^n_{+-}(W) \leq |\mathcal{X}|^n - \left(\min_{y\in\mathcal{Y}} |\mathcal{X}(y)|\right)^n + 1$

*(iii)* $|\mathcal{Y}|^n - \left(\min_{x\in\mathcal{X}} |\mathcal{Y}(x)|\right)^n \leq M^n_{-+}(W) \leq |\mathcal{Y}|^n - \left(\min_{x\in\mathcal{X}} |\mathcal{Y}(x)|\right)^n + 1$

*(iv)* $\min\left[|\mathcal{X}|^{n-1}A^{(n)}, |\mathcal{Y}|^{n-1}B^{(n)}\right] \leq M^n_{++}(W) \leq \min\left(|\mathcal{X}|^n, |\mathcal{Y}|^n\right)$

Obviously, all quantities equal 1, if $A^* \times B^* = \emptyset$.

*Proof* The upper bounds in (i) and (iv) are trivial. The lower bounds in (i) is achieved by the code

$$\left\{(x^{n-1}x_n, y^{n-1}y_n) : x_n \in A^*, y_n \in B^*\right\}.$$

The lower bound in (iv) is achieved by a maximal matching of the complete bipartite graph with vertex sets

$$\mathcal{X}^{n-1} \times A^{(n)}, \mathcal{Y}^{n-1} \times B^{(n)}.$$

Finally, we have to prove only (ii), because (iii) is symmetrically the same. Suppose then that $\left\{(u_i, v_i) : 1 \leq i \leq M\right\}$ is pseudo-matching and satisfies (1.7.157). By Definition B3 $u_i \notin \mathcal{X}^n(v_1)$ for $i > 1$ and consequently we get the upper bound.

In conclusion we define $y_0 \in \mathcal{Y}$ by

$$|\mathcal{X}(y_0)| = \min_{y\in\mathcal{Y}} |\mathcal{X}(y)|$$

and observe that the lower bound in (ii) is achieved by the code

$$\left\{(x^n, y_0^n) : x^n \notin \mathcal{X}^n(y_0^n)\right\},$$

where $y_0^n = y_0 \dots y_0$.

*Remark*

1. The notion of pseudo-matching detection codes in the sense of (1.7.156) is paralleled by the notion of a pseudo-zero-error detection code, where in the definition of a zero-error detection code the condition

$$W^n(u_i \mid u_i) > 0,\ 1 \le i \le M, \tag{1.7.163}$$

is dropped.
Also, Theorem 55 is then paralleled by pseudo-codes: every pseudo-matching detection code for $W^n$ generates a pseudo-zero-error detection code for $V^n$. Further, a $\mathcal{Z}, V$ is generated this way by some $W$ iff $\mathcal{Z}$ has partitions $\{A_x\}_{x\in\mathcal{X}}$, $\{B_y\}_{y\in\mathcal{Y}}$ in the sense of (1.7.137), (1.7.138), and

$$|A_x \cap B_y| = 1 \quad \text{for all } x \in \mathcal{X}, y \in \mathcal{Y}. \tag{1.7.164}$$

2. In Theorem 56 (ii) (similarly (iii)) can be improved to

$$M^n_{+-}(W) = \max\big(L_n, M^n_{\text{mde}}(W)\big), \tag{1.7.165}$$

where $L_n$ is the lower bound in (ii). Indeed, let $\{(u_i, v_i) : 1 \le i \le M\}$ be a code achieving $M^n_{+-}(W)$, then in case $W^n(v_i \mid u_i) > 0$ for all $i$ we really have a MDC and thus $M \le M^n_{\text{mde}}(W)$. Otherwise, we can find an $i_0$ with $W^n(v_{i_0} \mid u_i) = 0$ for all $i$ and therefore $M \le L_n$.
Consequently, the lower bound $L_n$ is tight for "most" channels, for example for the $\binom{\alpha}{\beta}$-uniform complete hypergraph channels of 1.7.10, but for "some" channels like

$$W = \begin{pmatrix} + & 0 & 0 & + \\ 0 & + & 0 & + \\ 0 & 0 & + & + \end{pmatrix}$$

the upper bound and *not* the lower bound is tight.

3. Pseudo-matching zero-error detection codes for $W^n$ can be formulated also in graphtheoretic terminology. We begin with the bipartite graph $\mathcal{G}^{\otimes n} = (\mathcal{X}^n, \mathcal{Y}^n, \mathcal{E}_n)$ associated with $W^n$ as in Sect. 1.7.3 and finally, define a graph $\widetilde{\mathcal{G}}_n = (\mathcal{X}^n \times \mathcal{Y}^n, \widetilde{\mathcal{E}}_n)$, where

$$\begin{aligned} \widetilde{\mathcal{E}}_n = &\{\{(x^n, y^n), (x'^n, y^n)\} : x^n, x'^n \in \mathcal{X}^n, y^n \in \mathcal{Y}^n\} \\ &\cup \{\{(x^n, y^n), (x^n, y'^n)\} : x^n \in \mathcal{X}^n, y^n, y'^n \in \mathcal{Y}^n\} \\ &\cup \{\{(x^n, y^n), (x'^n, y'^n)\} : (x^n, y^n) \in \mathcal{E}_n\} \end{aligned}$$

Then a pseudo-matching zero-error detection code for $W^n$ corresponds to an independent set of $\widetilde{\mathcal{G}}_n$.

## References


1. R. Ahlswede, *Certain Results in Coding Theory for Compound Channels*, in Proceedings of the Colloquium Information Theory, (Debrecen (Hungary), 1967), pp. 35–60
2. R. Ahlswede, Beiträge zur Shannonschen Informationstheorie im Falle nichtstationärer Kanäle. Z. Wahrscheinlichkeitstheorie und verw. Geb. **10**, 1–42 (1968)
3. R. Ahlswede, The weak capacity of averaged channels. Z. Wahrscheinlichkeitstheorie und verw. Geb. **11**, 61–73 (1968)
4. R. Ahlswede, Channels with arbitrary varying channel probability functions in the presence of feedback. Z. Wahrscheinlichkeitstheorie und verw. Geb. **25**, 239–252 (1973)
5. R. Ahlswede, Channel capacities for list codes. J. Appl. Probab. **10**, 824–836 (1973)
6. R. Ahlswede, *On Set Coverings in Cartesian Product Spaces, Ergänzungsreihe 92-005, SFB 343 Diskrete Strukturen in der Mathematik* (Universität Bielefeld, 1992)
7. R. Ahlswede, A method of coding and an application to arbitrarily varying channels. J. Comb. Inf. Syst. Sci. **5**(1), 10–35 (1980)
8. R. Ahlswede, N. Cai, Z. Zhang, *Erasure, List and Detection Zero-Error Capacities for Low Noise and a Relation to Identification*, Preprint 93-068, SFB 343 Diskrete Strukturen in der Mathematik. Universität Bielefeld. IEEE Trans. Inf. Theory **42**(1), 55–62 (1996)
9. R. Ahlswede, G. Dueck, Identification via channels. IEEE Trans. Inf. Theory **35**(1), 15–29 (1989)
10. R. Ahlswede, G. Simonyi, Reusable memories in the light of the old varying and a new outputwise varying channel theory. IEEE Trans. Inf. Theory **37**(4), 1143–1150 (1991)
11. R. Ahlswede, J. Wolfowitz, Correlated decoding for channels with arbitrarily varying channel probability functions. Inf. Control **14**, 457–473 (1969)
12. R. Ahlswede, J. Wolfowitz, *The Structure of Capacity Functions for Compound Channels*, in Proceedings of the International Symposium on Probability and Information Theory, McMaster University, Canada, April 1968, pp. 12–54 (1969)
13. R. Ahlswede, J-p. Ye, Z. Zhang, Creating order in sequence spaces with simple machines. Inf. Comput. **89**(1), 47–94 (1990)
14. R. Ahlswede, Z. Zhang, Coding for write efficient memories. Inf. Comput. **83**(1), 80–97 (1989)
15. S. Arimoto, On the converse to the coding theorem for discrete memoryless channels. IEEE Trans. Inf. Theory **IT-19**, 357–359 (1973)
16. D. Blackwell, L. Breiman, A.J. Thomasian, Proof of Shannon's transmission theorem for finite-state indecomposable channels. Ann. Math. Stat. **29**, 1209–1220 (1958)
17. D. Blackwell, L. Breiman, A.J. Thomasian, The capacity of a class of channels. Ann. Math. Stat. **30**(4), 1229–1241 (1959)
18. D. Blackwell, L. Breiman, A.J. Thomasian, The capacities of certain channel classes under random coding. Ann. Math. Stat. **31**, 558–567 (1960)
19. A. Blokhuis, On the Sperner capacity of the cyclic triangle. J. Algebraic Comb. **2**, 123–124 (1993)
20. T.M. Cover, B. Gopinath, *Open Problems in Communication and Computation* (Springer, New York, 1987)
21. H.G. Ding, On the information stability of a sequence of channels. Theor. Probab. Appl. **7**, 258–269 (1962)
22. R.L. Dobrushin, *Arbeiten zur Informationstheorie IV* (VEB Deutscher Verlag der Wissenschaften, Berlin, 1963)
23. R.L. Dobrushin, Individual methods for transmission of information for discrete channels without memory and messages with independent components. Doklady Akad. Nauk SSSR **148**, 1245–48 (1963)
24. R.L. Dobrushin, Unified methods for the transmission of information: the general case. Doklady Akad. Nauk SSSR **149**, 16–19 (1963)
25. R.G. Gallager, in *Information Theory, Chapter 4 of the Mathematics of Physics and Chemistry*, ed. by H. Margenan, G.M. Murphy (Van Nostrand, Princeton, 1976)

26. R.G. Gallager, *Information Theory and Reliable Communication* (Wiley, New York, 1968)
27. L. Gargano, J. Körner, U. Vaccaro, Sperner capacities. Graphs Combin. **9**, 31–46 (1993)
28. P. Hall, On representatives of subsets. J. London Math. Soc. **10**, 26–30 (1935)
29. K. Jacobs, *Almost Periodic Channels* (Aarhus, Colloquium on Combinatorial Methods in Probability Theory, 1962), pp. 118–126
30. S. Kakutani, A generalization of Brouwer's fixed point theorem. Duke Math. J. **8**, 457–458 (1941)
31. R.S. Kennedy, *Finite State Binary Symmetric Channels*, Sc.D. Thesis, Department of Electrical Engineering, MIT Cambridge, Mass (1963)
32. H. Kesten, Some remarks on the capacity of compound channels in the semicontinuous case. Inf. Control **4**, 169–184 (1961)
33. J. Kiefer, J. Wolfowitz, Channels with arbitrarily varying channel probability functions. Inf. Control **5**, 169–184 (1962)
34. J.C. Kieffer, A lower bound on the probability of decoding error for the finite-state channel. IEEE Trans. Inf. Theory **20**, 549–551 (1974)
35. D. König, Über Graphen und ihre Anwendung auf Determinantentheorie und Mengenlehre. Math. Annalen **77**, 453–465 (1916)
36. L. Lovász, M.D. Plummer, *Matching Theory*, (North-Holland, 1986)
37. E.C. van der Meulen, Random coding theorems for the general discrete memoryless broadcast channel. IEEE Trans. Inf. Theory **IT 21**, 180–190 (1975)
38. M.S. Pinsker, Capacity region of noiseless broadcast channels, (in Russian). Problemi Peredachi Informatsii **14**(2), 28–32 (1978)
39. C.E. Shannon, *On the Zero-Error Capacity of the Noisy Channel*, Transactions on PGIT, IRE, pp. 8–19, Sept 1956
40. C.E. Shannon, Channels with side information at the transmitter. IBMJ Res. Dev. **2**(4), 289–293 (1958)
41. A.J. Thomasian, A finite criterion for indecomposable channels. Ann. Math. Stat. **34**, 337–338 (1963)
42. J. Wolfowitz, The coding of messages subject to chance errors. Illinois J. Math. **1**, 591–606 (1957)
43. J. Wolfowitz, Simultaneous channels. Arch. Rat. Mech. Anal. **4**(4), 371–386 (1960)
44. J. Wolfowitz, Channels without capacity. Inf. Control **6**(1), 49–54 (1963)
45. J. Wolfowitz, *Coding Theorems of Information Theory*, 1st edn, 1961; 2nd edn, 1964; 3rd edn, 1978 (Springer, Berlin)
46. J. Wolfowitz, *Memory Increases Capacity* in Colloquium on Information Theory (Kossuth Lajos University, Debrecen, Sept 1967), pp. 19–24
47. H. Yudkin, *On the Exponential Error Bound and Capacity for Finite State Channels*, in International Symposium on Information Theory, (San Remo, Italy, 1967)

## Further Reading

48. R. Ahlswede, *Coloring Hypergraphs: A New Approach to Multi-user Source Coding* Part I, J. Comb, Inf. Syst. Sc. **4**, 76–115, (1979); Part II, **5**, 220–268 (1980)
49. R. Ahlswede, N. Cai, On extremal set partitions in Cartesian product spaces. Comb. Probab. Comput. **2**, 211–220 (1993)
50. R. Ahlswede, N. Cai, *Cross Disjoint Pairs of Clouds in the Interval Lattice*, Preprint 93-038, SFB 343 Diskrete Sturkturen in der Mathematik, ed. by R.L. Graham, J. Nesetril, The Mathematics of Paul Erd"os; Algorithms and Combinatorics B, (Universität Bielefeld, Springer, Berlin, 1996), pp. 155–164
51. R. Ahlswede, N. Cai, Information and control: matching channels. IEEE Trans. Inf. Theory **44**(2), 542–563 (1998)

52. R. Ahlswede, G. Dueck, Good codes can be produced by a few permutations. IEEE Trans. Inf. Theory **IT–28**(3), 430–443 (1982)
53. R. Ahlswede, G. Dueck, Identification in the presence of feedback—a discovery of new capacity formulas. IEEE Trans. Inf. Theory **35**(1), 30–39 (1989)
54. R. Ahlswede, A. Kaspi, Optimal coding strategies for certain permuting channels. IEEE Trans. Inf. Theory **IT–33**(3), 310–314 (1987)
55. I. Anderson, *Combinatorics of Finite Sets* (Claredon Press, Oxford, 1987)
56. U. Augustin, Gedächtnisfreie Kanäle für diskrete Zeit. Zeitschrift f. Wahrscheinlichkeitstheorie u. verw. Geb. **6**, 10–61 (1966)
57. T.M. Cover, Broadcast channels. IEEE Trans. Inf. Theory **IT 18**, 2–14 (1972)
58. I. Csiszár, J. K"orner, *Information Theory: Coding Theorems for Discrete Memoryless Systems* (Academic Press, New York, 1981)
59. K. Kobayashi, Combinatorial structure and capacity of the permuting relay channel. IEEE Trans. Inf. Theory **33**(6), 813–826 (1987)
60. C.E. Shannon, *Two-way Communication Channels*, in Proceedings of 4th Berkeley Symposium on Mathematical Statistics and Probability, (University of California Press, Berkeley and Los Angeles), pp. 611–644 (1961)

# Chapter 2
# Algorithms for Computing Channel Capacities and Rate-Distortion Functions

## 2.1 Lecture on Arimoto's Algorithm for Computing the Capacity of a DMC

Computation of the capacity $C = C(W)$ of a DMC $W : \mathcal{X} \to \mathcal{Y}$ involves the solution of a convex programming problem. Generally, analytic solutions are unknown. However, in [1] Arimoto presented an iterative method of computing the capacity. It is simple and involves only elementary arithmetical operations, exponentials, and logarithms. The convergence to capacity is monotone and the approximation error does not exceed the reciprocal of the number of iteration times a known constant, sometimes it even decreases exponentially fast.

The first result concerning computation of channel capacity is due to Muroga [6], who developed a straightforward method, but it is restricted to the case $|\mathcal{X}| = |\mathcal{Y}|$ and a channel with nonvanishing determinant, optimizing input distributions positive for all $x \in \mathcal{X}$, and the following symmetry properties.

Then Meister and Oettli [5] proposed an iterative procedure based on concave programming, which converges to the capacity.

Shannon [7], and Shannon and Weaver [8] started to calculate his capacity

$$C = \max_{P \in \mathcal{P}(\mathcal{X})} -\sum P(x)W(y|x) \log \sum_{x} P(x)W(y|x) + \sum_{x,y} P(x)W(y|x) \log W(y|x) \tag{2.1.1}$$

by the method of Lagrange.

*Arimoto's procedure has similarities with that of Meister/Oettli, but originated from a concept of generalized equivocation.*



A. Ahlswede et al. (eds.), *Transmitting and Gaining Data*,
Foundations in Signal Processing, Communications and Networking 11,
DOI 10.1007/978-3-319-12523-7_2

### 2.1.1 Mutual Information and Equivocation

W.l.o.g. we can make the assumption that for every $y \in \mathcal{Y}$ there exists an $x \in \mathcal{X}$ with $W(y|x) > 0$, because otherwise we just omit a column of the matrix $W$.

In addition to the set $\mathcal{P}(\mathcal{X})$ of PD's on $\mathcal{X}$ we also use its subset $\mathcal{P}^+(\mathcal{X})$ of PDs with $P(x) > 0$ for all $x \in \mathcal{X}$.

Recall the notation of mutual information

$$I(W|P) = H(P) - H(W|P), \tag{2.1.2}$$

where

$$H(W|P) = -\sum_{y \in \mathcal{Y}} \sum_{x \in \mathcal{X}} W(y|x)P(x) \log \frac{W(y|x)P(x)}{\sum_{x' \in \mathcal{X}} W(y|x')P(x')}$$

is called also the equivocation, and of Shannon's capacity in this terminology

$$C(W) = \max_{P \in \mathcal{P}(\mathcal{X})} I(W|P).$$

A generalized equivocation can be defined by

$$J(W|P, V) = -\sum_{y \in \mathcal{Y}} \sum_{x \in \mathcal{X}} W(y|x)P(x) \log V(x|y),$$

where $V : \mathcal{Y} \rightsquigarrow \mathcal{X}$ is a stochastic matrix.

This specializes to the equivocation, if $V$ is defined by the Bayes formula

$$V(x|y) = \frac{W(y|x)P(x)}{\sum_x W(y|x)P(x)} = w^*(x|y). \tag{2.1.3}$$

It can easily be seen that

$$J(W|P, V) \geq J(W|P, W^*). \tag{2.1.4}$$

Hence, one obtains another characterization of the capacity

$$C(W) = \max_{P \in \mathcal{P}(\mathcal{X})} \max_{V : J \rightsquigarrow \mathcal{X}} (H(P) - J(W|P, V)), \tag{2.1.5}$$

where $V : \mathcal{Y} \rightsquigarrow \mathcal{X}$ is a stochastic matrix.

We also recall

**Lemma 33** *$P_0 \in \mathcal{P}(\mathcal{X})$ maximizes $I(W|P)$ iff*

$$\sum_y W(y|x) \log \frac{W(y|x)}{\sum_{x'} W(y|x')P_0(x')} \begin{cases} = I(W|P_0) = C(W), & P_0(x) > 0 \\ \leq I(W|P_0) = C(W), & P_0(x) = 0 \end{cases}$$

*Moreover, if $P_0, P_1, , \dots, P_{k-1} \in \mathcal{P}(\mathcal{X})$ all maximize $I(W|P)$, then any linear combination*

$$P_a = \alpha_0 P_0 + \cdots + \alpha_{a-1} P_{a-1},$$

*where $\alpha = (\alpha_0, \dots, \alpha_{a-1}) \in \mathcal{P}(\mathcal{X})$, $|\mathcal{X}| = a$, also maximizes $I(W|P)$ and in addition*

$$\sum_{x \in \mathcal{X}} W(y|x)P_0(x) = \cdots = \sum_{x \in \mathcal{X}} W(y|x)P_a(x). \tag{2.1.6}$$

## 2.1.2 The Algorithm and Its Convergence

Using the characterization (2.1.5) of the capacity, we define the following procedure for its evaluation based on the following steps:

Initially, choose an arbitrary PD $P_1 \in \mathcal{P}^+(\mathcal{X})$ (for instance the uniform distribution on $\mathcal{X}$). Then maximize $H(P_1) - J(W|P_1, V)$ with respect to $V$.

According to (2.1.4) maximizing is

$$V_1(x|y) = \frac{W(y|x)P_1(x)}{\sum_{x'} W(y|x')P_1(x')}, \tag{2.1.7}$$

that is,

$$I(W|P_1) = \max_V [H(P_1) - J(W|P_1, V)] = H(P_1) - J(W|P_1, V_1). \tag{2.1.8}$$

Now maximize $H(P) - J(W|P, V_1)$ with respect to $P \in \mathcal{P}(\mathcal{X})$ while fixing $V_1$. This gives $P_2$ and for $t = 2, 3, \dots$ having found $P_t$ and $V_t$ find $P_{t+1}$ by maximizing $H(P) - J(W|P, V_t)$ with respect to $P \in \mathcal{P}(\mathcal{X})$ and then $V_{t+1}$ via (2.1.7). This maximizing PD $P_{t+1}$ is given by

$$P_{t+1}(x) = \frac{r_t(x)}{\sum_{x' \in \mathcal{X}} r_t(x')}, \qquad r_t(x) = \exp\left[\sum_{y \in \mathcal{Y}} W(y|x) \log V_t(x|y)\right]. \tag{2.1.9}$$

Indeed, this statement is established by the following:

**Lemma 34** *For any fixed* $V : \mathcal{Y} \rightsquigarrow \mathcal{X}$

$$\max_{P \in \mathcal{P}(\mathcal{X})} H(P) - J(W|P, V) = H(\tilde{P}) - J(W|\tilde{P}, V)$$
$$= \log\left[\sum_{x \in \mathcal{X}} \exp\left\{\sum_{y \in \mathcal{Y}} W(y|x) \log V(x|y)\right\}\right] \leq C(W),$$

*where* $\tilde{P} \in \mathcal{P}(\mathcal{X})$ *is given by*

$$\tilde{P}(x) = \frac{r(x)}{\sum_{x' \in \mathcal{X}} r(x')}, \quad r(x) = \exp\left[\sum_{y \in \mathcal{Y}} W(y|x) \log V(x|y)\right], \quad \textit{for } x \in \mathcal{X}. \tag{2.1.10}$$

*Moreover,*

$$\max_{P \in \mathcal{P}(\mathcal{X})} H(P) - J(W|P, V) = \log\left(\sum_{x \in \mathcal{X}} r(x)\right). \tag{2.1.11}$$

*Proof* Using a Lagrange multiplier, let

$$F(P) = H(P) = J(W|P, V) + \lambda\left(1 - \sum_{x \in \mathcal{X}} P(x)\right). \tag{2.1.12}$$

It is known in nonlinear programming that a maximizing PD $\tilde{P} \in \mathcal{P}(\mathcal{X})$ satisfies the relation

$$\left.\frac{\partial F}{\partial P(x)}\right|_{P(x)=\tilde{P}(x)} = -1 - \log \tilde{P}(x) + \sum_{y \in \mathcal{Y}} W(y|x) \log V(x|y) - \lambda = 0, \quad \tilde{P}(x) > 0. \tag{2.1.13}$$

The RHS equality can be rewritten as

$$\tilde{P}(x) = \exp\left[-1 - \lambda + \sum_{y \in \mathcal{Y}} W(y|x) \log V(x|y)\right], \quad \tilde{P}(x) > 0. \tag{2.1.14}$$

This and $\sum_x \tilde{P}(x) = 1$ implies (2.1.10), for $\tilde{P}(x) > 0$. With respect to $\tilde{P}(x) = 0$ note that by the continuity of $F$ on $\mathcal{P}(\mathcal{X})$ there exists at least one $y$ with $V(x|y) = 0$ and $V(y|x) > 0$. With the conventions $\log 0 = -\infty$ and $0 \cdot \log 0 = 0$ (2.1.10) is valid even for $\tilde{P}(x) = 0$. □

### 2.1.3 The Convergence of the Algorithm

It is convenient to use the following notation:

$$C(t,t) = H(P_t) - J(W|P_t, V_t)$$
$$C(t+1,t) = H(P_{t+1}) - J(W|P_{t+1}, V_t)$$

**Theorem 57** *Let $P_1 \in \mathcal{P}^+(\mathcal{X})$, then the sequence $(C(t,t))_{t=1}^{\infty}$ converges monotonically from below to the capacity $C(W)$.*

*Proof* Since $C(t+1,t) = \max_{P \in \mathcal{P}(\mathcal{X})}(H(P) - J(W|P, V_t)$ it follows from (2.1.11) in Lemma 33 that

$$C(t+1,t) = \log \sum_{x \in \mathcal{X}} r_t(x) \tag{2.1.15}$$

and from the definitions of $C(t,t)$ and $C(t+1,t)$ that

$$C(1,1) \le C(2,1) \le C(2,2) \le, \cdots, \le C(t,t) \le C(t+1,t) \le \cdots \le C(W). \tag{2.1.16}$$

Let now $P_0$ be one of the input PD maximizing the mutual information, that is, $I(W|P_0) = C(W)$. Then, using the notations

$$Q_0(y) = \sum_{x \in \mathcal{X}} W(y|x) P_0(x) \tag{2.1.17}$$

$$Q_t(y) = \sum_{x \in \mathcal{X}} W(y|x) P_t(x) \tag{2.1.18}$$

(2.1.15) and Lemma 33, we get

$$\begin{aligned}
\sum_x P_0(x) \log \frac{P_{t+1}(x)}{P_t(x)} &= \sum_x P_0(x) \log \frac{1}{P_t(x)} \cdot \frac{r_t(x)}{\sum_{x'} r_t(x')} \\
&= -C(t+1,t) + \sum_x P_0(x) \left( \sum_{y \in \mathcal{Y}} W(y|x) \log W(y|x) Q_t(y) \right) \\
&= -C(t+1,t) + \sum_x \sum_y W(y|x) P_0(x) \cdot \left[ \log \frac{W(y|x)}{Q_0(y)} + \log \frac{Q_0(y)}{Q_t(y)} \right] \\
&= C(W) - C(t+1,t) + \sum_y Q_0(y) \log \frac{Q_0(y)}{Q_t(y)}.
\end{aligned} \tag{2.1.19}$$

Since here the last term is nonnegative, it follows that

$$C(W) - C(t+1,t) \le \sum_x P_0(x) \log \frac{P_{t+1}(x)}{P_t(x)}. \tag{2.1.20}$$

Consequently, for any integer $T \geq 1$

$$\sum_{t=1}^{T}[C(W) - C(t+1,t)] \leq \sum_{x} P_0(x) \log \frac{P_{T+1}(x)}{P_1(x)}$$
$$\leq \sum_{x} P_0(x) \log \frac{P_0(x)}{P_1(x)}. \qquad (2.1.21)$$

Also, since $P_1 \in \mathcal{P}^+(\mathcal{X})$ the RHS in (2.1.21) is a finite constant and therefore, since the summands are nonnegative, it follows that

$$\lim_{t\to\infty} C(t+1,t) = C(W) \qquad (2.1.22)$$

and by (2.1.16) also

$$\lim_{t\to\infty} C(t,t) = C(W). \qquad (2.1.23)$$

□

**Corollary 8** *The approximation error $e(t) = C(W) - C(t+1,t)$ is inversely proportional to the number of iterations. In particular, if $P_1$ is chosen as the uniform distribution, then*

$$C(W) - C(t+1,t) \leq [\log|\mathcal{X}| - H(P_0)]\frac{1}{t}. \qquad (2.1.24)$$

*Proof* This follows immediately from (2.1.16), the bound (2.1.21), and the choice of $P_1$. □

**Corollary 9**

$$\lim_{T\to\infty} Q_T = Q_0 \qquad (2.1.25)$$

*Proof* First, note that the convergence of

$$w(T) = \sum_{y\in\mathcal{Y}} Q_0(y) \log \frac{Q_0(y)}{Q_T(y)}$$

to zero implies (2.1.25). Therefore, we have to show that $\lim_{T\to\infty} w(T) = 0$. For this rewrite (2.1.19) as follows:

$$\sum_{t=1}^{T} w(t) = \sum_{t=1}^{T}\sum_{y} Q_0(y) \log \frac{Q_0(y)}{Q_t(y)}$$
$$\leq \sum_{t=1}^{T}\sum_{x} P_0(x) \log \frac{P_{t+1}(x)}{P_t(x)}$$

$$= \sum_x P_0(x) \log \frac{P_{T+1}(x)}{P_1(x)}$$

$$\leq \sum_x P_0(x) \log \frac{P_0(x)}{P_1(x)}, \tag{2.1.26}$$

where we have used that $C(W) - C(t+1, t)$ is nonnegative. Since $P_1 \in \mathcal{P}^*(\mathcal{X})$ the RHS of (2.1.26) is a finite constant and the summands must converge to zero. □

## 2.1.4 Speed of Convergence

In case the input distribution achieving capacity is unique and belongs to $\mathcal{P}^+(\mathcal{X})$ more results can be established. In particular, the speed of convergence goes considerably beyond that of Corollary 8.

**Theorem 58** (Arimoto 1972, [1]) *If the input distribution achieving capacity is unique, then the sequence of distributions $(P_t)_{t=1}^{\infty}$ converges to $P_0$ monotonically in the sense that*

$$D(P_0||P_t) = \sum_x P_0(x) \log \frac{P_0(x)}{P_t(x)} \geq \sum_x P_0(x) \log \frac{P_0(x)}{P_{t+1}(x)} \geq 0 \text{ and } \lim_{t\to\infty} D(P_0||P_t) = 0. \tag{2.1.27}$$

*Proof* The monotonicity follows directly from (2.1.19). To prove convergence note that $\mathcal{P}(\mathcal{X})$ is a compact subset of the $|\mathcal{X}|$-dimensional Euclidean space hence the sequence $(P_t)_{t=1}^{\infty}$ has at least one point of accumulation in $\mathcal{P}(\mathcal{X})$, say $\overline{P}$: Theorem 57 implies that $\overline{P}$ maximizes the mutual information and must equal $P_0$ by the uniqueness assumption. □

**Theorem 59** *If the input distribution $P_0$ achieving capacity is unique and belongs to $\mathcal{P}^+(\mathcal{X})$, then there exists an integer $T$ and a constant $0 < \sigma \leq 1$ such that for all $t \geq T$*

$$\sum_x P_0(x) \log \frac{P_0(x)}{P_t(x)} \leq (1-\sigma)^{t-T} \sum_x P_0(x) \log \frac{P_0(x)}{P_T(x)} \tag{2.1.28}$$

*and $T$ is independent of $\sigma$.*

*Proof* It follows from Theorem 58 that for any $\delta > 0$ there exists a positive integer $T = T(\delta)$ such that

$$\sum_x P_0(x) \log \frac{P_0(x)}{P_T(x)} < \delta. \tag{2.1.29}$$

Of course $\lim_{\delta\to\infty} P_{N(\delta)} = P_0$.

Consider the difference vector (or set function) $d = P_0 - P_T$ and note that

$$\sum_x P_0(x) \log \frac{P_0(x)}{P_T(x)} = -\sum_x P_0(x) \log \left(1 - \frac{d(x)}{P_0(x)}\right) = \frac{1}{2}\left(d, \mathbb{P}_0^{-1} d\right) + \sum_{x \in \mathcal{X}} O(d(x)^3), \tag{2.1.30}$$

$$\sum_{y \in \mathcal{Y}} Q_0(y) \log \frac{Q_0(y)}{Q_T(y)} = \sum_y -Q_0(y) \log \left(1 + \sum_x W(y|x) \frac{d(x)}{Q_0(y)}\right) = \frac{1}{2}\left(d, W' \mathbb{Q}_0^{-1} W d\right) + \sum_x O(d(x)^3), \tag{2.1.31}$$

where we used the inner product $(\cdot, \cdot)$ and the transpose $W'$ of matrix $W$, and $\mathbb{P}_0, \mathbb{Q}_0$ are diagonal matrices of the form

$$\mathbb{P}_0 = \begin{pmatrix} P_0(1) & & 0 \\ & \ddots & \\ 0 & & P_0(a) \end{pmatrix}, \quad \mathbb{Q}_0 = \begin{pmatrix} Q_0(1) & & 0 \\ & \ddots & \\ 0 & & Q_0(b) \end{pmatrix}.$$

If $\delta > 0$ is chosen sufficiently small, then from (2.1.30) and (2.1.31) one can choose an integer $T$ such that simultaneously the following inequalities hold:

$$\sum_x P_0(x) \log \frac{P_0(x)}{P_T(x)} \leq \frac{2}{3}(d, \mathbb{P}_0^{-1} d),$$

$$\sum_y Q_0(y) \log \frac{Q_0(y)}{Q_T(y)} \geq \frac{1}{3}(d, W' \mathbb{Q}_0^{-1} W d). \tag{2.1.32}$$

By the positive definiteness of $W' \mathbb{Q}_0^{-1} W$, shown below, there exists a positive number $\sigma > 0$ such that for every $d \in \mathbb{R}^a$

$$\frac{1}{3}(d, W' \mathbb{Q}_0^{-1} W d) \geq \sigma \frac{2}{3}(d, \mathbb{P}_0^{-1} d). \tag{2.1.33}$$

Substituting this inequality into (2.1.32) and noting that by Theorem 58 the same properties hold for all $t \geq T$, we get for all these $t$

$$\sum_y Q_0(y) \log \frac{Q_0(y)}{Q_t(y)} \geq \sigma \sum_x P_0(x) \log \frac{P_0(x)}{P_t(x)}. \tag{2.1.34}$$

On the other hand, it follows from (2.1.19) that

$$\sum_x P_0(x) \log \frac{P_0(x)}{P_{t+1}(x)} \leq \sum_x P_0(x) \log \frac{P_0(x)}{P_t(x)} - \sum_y Q_0(y) \log \frac{Q_0(y)}{Q_t(y)}. \tag{2.1.35}$$

Substituting (2.1.34) into (2.1.35) gives (2.1.28).
Finally, we establish now the claimed positive definiteness.

Assume that there exists a nonzero vector $\xi = (\xi(1), \ldots, \xi(a)) \in \mathbb{R}^a$ with $W\xi = 0$. Then $\xi$ clearly has a representation of the form

$$\xi = \alpha e + \beta d^1, \tag{2.1.36}$$

where $\alpha$ and $\beta$ are scalars, $e$ is the vector with components all equal to 1, and $d^1$ is orthogonal to $e$. Now $W\xi = 0$ implies

$$\begin{aligned} 0 &= \sum_y \sum_x W(y|x)\xi(x) = \sum_y \sum_x W(y|x)(\alpha(x) + \beta d^1(x)) \\ &= \sum_x \sum_y W(y|x)(\alpha + \beta d^1(x)) = \sum_x [\alpha + (e, d^1)] = a\alpha, \end{aligned}$$

saying that $\alpha = 0$ and hence that $W\xi = \beta W d^1 = 0$. This in turn implies $\beta = 0$ or $Wd^1 = 0$. If $Wd^1 = 0$ then for sufficiently small $\varepsilon > 0$ the vector $P = P_0 + \varepsilon d^1$ becomes also the capacity achieving PD in view of Lemma 33, since $Wd^1 = 0$ and $P \in \mathcal{P}^+(\mathcal{X})$. Thus by the uniqueness of the capacity achieving PD it follows that $\beta = 0$ or $d^1 = 0$, but this implies $\xi = 0$ in contradiction to the assumption. □

### 2.1.5 *Upper and Lower Bounds on the Capacity*

Consider

$$C(W, V) = \max_{P \in \mathcal{P}(\mathcal{X})} (H(P) - J(W|P, V)).$$

Its exact form is given in Lemma 34 and $C(W, V) \le C(W)$, $\max_V(W, W) = C(W)$.

**Theorem 60** *For two transmission matrices $W_1$ and $W_2 : \mathcal{X} \to \mathcal{Y}$ and a $V : \mathcal{Y} \to \mathcal{X}$, $0 \le \alpha \le 1$, the following inequality holds:*

$$C(\alpha W_1 + (1 - \alpha) W_2, V) \le \alpha C(W_1, V) + (1 - \alpha) C(W_2, V).$$

*Proof* Let

$$\beta_1(x) = \exp\left\{\sum_y w_1(y|x) \log V(x|y)\right\}$$

$$\beta_2(x) = \exp\left\{\sum_y w_2(y|x) \log V(x|y)\right\},$$

then

$$C(\alpha W_1 + (1-\alpha)W_2, V) = \log \sum_x \beta_1(x)^\alpha \beta_2(x)^{1-\alpha}.$$

By Hölder's inequality

$$\begin{aligned}\log \sum_x \beta_1(x)^\alpha \beta_2(x)^{1-\alpha} &\le \alpha \log \sum_x \beta_1(x) + (1-\alpha) \log \sum_x \beta_2(x) \\ &= \alpha C(W_1, V) + (1-\alpha) C(W_2, V),\end{aligned}$$

and the theorem follows. □

As an easy consequence one gets

**Corollary 10** (Shannon 1957, [7])

$$C(\alpha W_1 + (1-\alpha)W_2) \le \alpha C(W_1) + (1-\alpha) C(W_2).$$

**Theorem 61** *Define* $\Delta = \sum_y W(y|x_y)$, *where the* $x_y$ *are chosen arbitrarily from* $\mathcal{X}$, $|\mathcal{X}| = a$.

$$(i) \qquad C(W) \ge \log a - \left(1 - \frac{\Delta}{a}\right) \log(a-1) - h\left(\frac{\Delta}{a}\right)$$

$$(ii) \qquad C(W) \ge \log \sum_x \exp \sum_y W(y|x) \log \frac{W(y|x)}{\sum_{x'} W(y|x')}$$

$$(iii)\ C(W) \ge \log a - \frac{1}{a} \sum_{x \in \mathcal{X}} H(W(\cdot|x)) - \frac{1}{a} \sum_{y \in \mathcal{Y}} W(y|x) \log \sum_x W(y|x)$$

$$(iv) \qquad C(W) \le \log a + \max_x \left[ \sum_y W(y|x) \log \frac{W(y|x)}{\sum_{x'} W(y|x')} \right]$$

## 2.2 Lecture on Blahut's Algorithm for Computing Rate-Distortion Functions

### 2.2.1 *Basic Definitions and the Algorithm*

We remind the reader about the definition of the rate-distortion function $R : \mathcal{P}(\mathcal{X}) \times \mathbb{R}_+ \to \mathbb{R}_+$ for a DMS $(X_t)_{t=1}^\infty$ and distribution measure $d : \mathcal{X} \times \hat{\mathcal{X}} \to \mathbb{R}_+$ and its characterization by Shannon.

$$R(P, \Delta) = \min_{\substack{(X,\hat{\mathcal{X}}):P_X=P,\\ \mathbb{E}d(X,\hat{\mathcal{X}})\leq\Delta}} I(X, \hat{\mathcal{X}}) \tag{2.2.1}$$

for every $\Delta \in \mathbb{R}_+$ and $P \in \mathcal{P}(\mathcal{X})$. Since $P = P_X$ is now always kept fixed we abbreviate $R(P, \Delta)$ as $R(\Delta)$, for $W : \mathcal{X} \rightsquigarrow \hat{\mathcal{X}}$ $I(W|P)$ as $I(W)$, and $D(W|P) = \sum_x \sum_y P(x)W(\hat{\mathcal{X}}|X)d(X, \hat{\mathcal{X}})$ as $D(W)$. We can also write (2.2.1) in the form

$$R(\Delta) = \min_{\substack{W:\mathcal{X}\rightsquigarrow\hat{\mathcal{X}}\\ D(W)\leq\Delta}} I(W). \tag{2.2.2}$$

Setting $F_s = \inf_{W:\mathcal{X}\rightsquigarrow\hat{\mathcal{X}}}(I(W) - sD(W))$, the $R(\Delta)$ curve is the envelope of the lines with slope $s \leq 0$ and $R$-axis intercept $F_s$. It suffices therefore to compute $F_s$. Blahut's algorithm [2] is based on the observation that

$$F_s = \inf_{W,Q} F_s(W, Q), \tag{2.2.3}$$

where

$$F_s(W, Q) = \sum_{x\in\mathcal{X}} \sum_{\hat{x}\in\hat{\mathcal{X}}} P(x)W(\hat{x}, x) \log \frac{W(\hat{x}, x)}{Q(\hat{x})} - sD(W), \tag{2.2.4}$$

and it consists in iterative minimizations with respect to $W$ and $Q$.
For fixed $W$ or $Q$, $F_s(W, Q)$ is minimized by $Q(W)$ or $W(Q)$, respectively, defined by

$$Q(W)(\hat{x}) = \sum_x P(x)W(\hat{x}, x),$$

$$W(\hat{x}, x)(Q) = Q(\hat{x}) \exp\left\{ sd(x, \hat{x}) \left[ \sum_{\hat{x}} Q(\hat{x}) \exp_x sd(x, \hat{x}') \right]^{-1} \right\}. \tag{2.2.5}$$

In [1] this was proved by Lagrange multipliers, but can be established directly as follows:

$$\begin{aligned} F_s(W, Q) &= F_s(W, Q(W)) + I(Q(W)||Q) \\ &= F_s(W(Q), Q) + \sum_x P(x)I(W(\cdot|x)|W(\cdot|x)(Q)) \end{aligned} \tag{2.2.6}$$

and $I(Q||Q')$ denotes relative entropy, a nonnegative quantity.
Starting with an arbitrary $Q_1 \in \mathcal{P}^+(\hat{\mathcal{X}})$ set recursively $W_n = W(Q_{n-1})$, $Q_n = Q(W_n)$ for $n = 2, 3, \ldots$. Then by (2.2.6)

$$F_s(W_2, Q_1) \geq F_s(W_2, Q_2) \geq F_s(W_3, Q_2) \geq \cdots. \tag{2.2.7}$$

### 2.2.2 *Convergence of the Algorithm*

As in Sect. 2.1 use for $W$ and $Q = Q(W)$ the "backward channel"

$$W^*(x|\hat{x}) = \frac{P(x)W(\hat{x}|x)}{Q(\hat{x})} \tag{2.2.8}$$

and the associated PD's $W^*(\cdot|\hat{x})$ for $\hat{x}$ with $Q(\hat{x}) > 0$. We start with the easily checked identity

$$\begin{aligned} &F_s(W_n, Q_{n-1}) + \sum_{\hat{x}} Q(\hat{x}) I(W^*(\cdot|\hat{x})||W_n^*(\cdot|\hat{x})) - F_s(W, Q) \\ &= \sum_{\hat{x}} Q(\hat{x}) \log \frac{Q_n(\hat{x})}{Q_{n-1}(\hat{x})} \end{aligned} \tag{2.2.9}$$

where $W_n^*(\cdot|\hat{x})$ is defined as $W^*(\cdot|\hat{x})$ with $W_n$ playing the role of $W$, that is,

$$W_n^*(x|\hat{x}) = P(x)Q_{n-1}(\hat{x}) \exp(sd(x, \hat{x})) \left[ Q_n(\hat{x}) \sum_{\hat{x}'} Q_{n-1}(\hat{x}') \exp(sd(x, \hat{x}')) \right]^{-1}. \tag{2.2.10}$$

Now, if

$$F_s(W, Q) \leq \lim_{n\to\infty} F_s(W_n, Q_n) = \lim_{n\to\infty} F_s(W_n, Q_{n-1})$$

then (2.2.9) implies that for any $N > M \geq 1$

$$\begin{aligned} 0 &\leq \sum_{n=M+1}^{N} \left[ F_s(W_n, Q_{n-1}) - F_s(W, Q) \right] \\ &\leq \sum_{n=M+1}^{N} \sum_{\hat{x}} Q(\hat{x}) \log \frac{Q_n(\hat{x})}{Q_{n-1}(\hat{x})} \\ &= \sum_{\hat{x}} Q(\hat{x}) \log \frac{Q_N(\hat{x})}{Q_M(\hat{x})} \\ &= I(Q||Q_M) - I(Q||Q_N). \end{aligned} \tag{2.2.11}$$

This shows that the series

$$\sum_n \left[ F_s(W_n, Q_{n-1}) - F_s(W, Q) \right]$$

converges. Therefore,

$$\lim_{n\to\infty} \left[F_s(W_n, Q_{n-1}) - F_s(W, Q)\right] = 0 \tag{2.2.12}$$

if $I(Q||Q_1) < \infty$, proving

$$\lim_{n\to\infty} F_s(W_n||Q_n) = \lim_{n\to\infty} F_s(W_n, Q_{n-1}) = F_s = \inf F_s(W, Q) \tag{2.2.13}$$

provided that the infimum in (2.2.13) can be approached by a $Q$ with $I(Q||Q_1) < \infty$. The latter condition is easily checked to be true for finite reproducing alphabet $\hat{\mathcal{X}}$.

**Theorem 62** *If* $|\hat{\mathcal{X}}| < \infty$, *then there exists a* $Q_0$ *(possibly depending on* $Q_1$*) such that* $\lim_{n\to\infty} Q_n = Q_0$, $\lim_{n\to\infty} W_n = W_0 = W_0(Q_0)$, *and* $F_s(W_0, Q_0) = F_s$.

*Proof* Choose a convergent subsequence $Q_{n_i} \to Q_0$ of $(Q_n)_{n=1}^{\infty}$. Then $W_{n_i+1} = W(Q_{n_i}) \to W(Q_0) = W_0$ and $F_s(W_{n_i+1}, Q_{n_i}) \to F_s(W_0, Q_0)$. In view of this we have $F_s(W_0, Q_0) = F_s$. Thus $Q_0 = Q(W_0)$ and hence (2.2.11) applies for $W_0$ and $Q_0$. In particular $I(Q_0||Q_n)$ is a nonincreasing sequence. Since $Q_{n_i} \to Q_0$ implies $I(Q_0||Q_{n_i}) \to 0$, this means $I(Q_0||Q_n) \to 0$. Hence $Q_n \to Q_0$ and the proof is complete. □

*Remark* An extension to $|\hat{\mathcal{X}}| = \infty$ can be found in [3]. There is no estimate on the speed of convergence!

Randomization in the encoding does not improve code parameters $(n, N, \lambda)$ for

(i) the DMC,
(ii) the general time-structure channel.

The result to prove that for a DMC $W$ with $|\mathcal{X}| \geq |\mathcal{Y}|\ |\mathcal{X}| - |\mathcal{Y}|$ input letters can be eliminated without reducing the achievable triples $(n, N, \lambda)$. In particular, the capacity remains unchanged.

### Research Problem

- Prove or disprove the conjecture of Eisenberg.
- Establish an analog for the speed of convergence of Blahut's algorithm for computation of $R(\Delta)$.

### Further Reading

- Csiszár and Shields, Iterative Algorithms [4]
- S. Arimoto, Computation of random coding exponent functions.

## 2.3 Lecture on a Unified Treatment of the Computation of the $\Delta$-distortion of a DMS and the Capacity of a DMS under Input Cost Constraint

### 2.3.1 Preliminaries

We call $C(F) = C(W, \Gamma)$ the capacity constraint function. It was defined in the previous section. The $\Delta$-distortion rate was called the rate-distortion function $R(\Delta) = R(P, \Delta)$. It is positive if and only if

$$\Delta > \underset{\sim}{\Delta} = \min_{\hat{x}} \sum_{x} P(x)d(x, \hat{x}). \tag{2.3.1}$$

since a $W$ with identical rows satisfying $\sum_{\hat{x}} W(\hat{x}|x)d(x, \hat{x}) \leq \Delta$ exists if and only if $\Delta \geq \underset{\sim}{\Delta}$. Therefore the study of $R(\Delta)$ can be restricted to $\Delta < \underset{\sim}{\Delta}$.

The analog of $\underset{\sim}{\Delta}$ for $C(\Gamma)$, called $\underset{\sim}{\Gamma}$, is the smallest value of $\Gamma$ for which $C(\Gamma)$ equals the unconstrained capacity $C$. However, the value $\tilde{\Gamma}$ is not so simply determined as $\underset{\sim}{\Delta}$. As $C(\Gamma)$ is a nondecreasing concave function and $R(\Delta)$ is a nonincreasing concave function. We notice that $C(\Gamma)$ is strictly increasing for $\Gamma_0 = \min_{x \in \mathcal{X}} c(x) \leq \Gamma \leq \underset{\sim}{\Gamma}$ and $R(\Delta)$ is strictly decreasing for $0 \leq \Delta \leq \underset{\sim}{\Delta}$. In these intervals, when the extrema in Sect. 2.1 and in (2.2.1) of Sect. 2.2.1 are achieved, then the constraints are satisfied with equality.

Recall from Sect. 2.2.1 that the $R(\Delta)$ curve is the upper envelope of straight lines and vertical axis intercept $F_s$.

Analogously, the curve $C(\Gamma)$ is the lower envelope of straight lines with vertical axis intercept

$$G(s) = \max_{P} [I(W|P) - sc(P)]$$

and slope $s$, $s \geq 0$.

**Lemma 35** *For every DMS $P$ and distortion measure $d$*

*(i)* $R(\Delta) = \max_{s>0}[F(s) - s\Delta]$ *if* $0 < \delta < \underset{\sim}{\Delta}$

*and for every DMC $W$ and cost function $c$*

*(ii)* $C(\Gamma) = \min_{s \geq 0}[G(s) + s\Gamma]$ *if* $\Gamma > \Gamma_0$.

*Furthermore, if for some $s > 0$ a $V$ minimizes $I(V|P) + sd(P, V)$ and $\Delta = d(P, W)$, then $I(V|P) = R(\Delta)$. Similarly, if for some $s \geq 0$ a $P$ minimizes $I(W|P) - sc(P)$ and $\Gamma = C(P)$, then $I(W|P) = C(\Gamma)$.*

In the formula for $R(\Delta)$ $s = 0$ may be excluded, because $R(\Delta > 0$ for $\Delta < \underset{\sim}{\Delta}$.

*Proof* Since we know the result (i) already we prove now (ii).

$C(P) \leq \Gamma$ implies $I(W|P) - sc(P) \geq I(W|P) - s\Gamma$ and the inequality $C(\Gamma) \leq \inf_{s \geq 0}[G(s) + s\Gamma]$ follows. It remains to be seen that for every $\Gamma > \Gamma_0$ there exists an $s \geq 0$ with $C(\Gamma) = G(s) + s\Gamma$. Now, if $\Gamma \geq \underset{\sim}{\Gamma}$ choose $s = 0$ and if $\Gamma_0 < \Gamma < \underset{\sim}{\Gamma}$, then there exists an $s > 0$ with

$$C(\Gamma') \leq C(\Gamma) + s(\Gamma' - \Gamma) \qquad \text{for } \Gamma' \geq \Gamma_0, \tag{2.3.2}$$

because $C(\Gamma)$ is concave and strictly monotone.

Let now $P$ achieve the maximum in the formula for $C(\Gamma)$. Then $c(P) = \Gamma$ and for $P'$ with $\Gamma' = c(P')$ inequality (2.3.2) implies

$$I(W|P') - sc(P') \leq C(\Gamma') - s\Gamma' \leq C(\Gamma) - s\Gamma = I(W|P) - sC(P).$$

Thus $G(s) = I(W|P) - sc(P) = C(\Gamma) - s\Gamma$ and the proof is complete. □

The lemma shows that $R(\Delta)$ and $C(\Gamma)$ are easily computed if the functions $F(s)$ and $G(s)$ are known. In particular $C = G(0)$.

### 2.3.2 The Computation of G(s)

We shall replace the maximum in the definition of $F(s)$ by a double maximum in such a way that fixing one variable, the maximum with respect to the other can be found readily.

We start with the identity

$$D(W||Q|P) \triangleq \sum_{x \in \mathcal{X}} P(x) D(W(\cdot|x)||Q) = I(W|P) + D(PW||Q). \tag{2.3.3}$$

It holds for every $Q \in \mathcal{P}(\mathcal{Y})$.

For fixed $W$ and $s \geq 0$ consider the following function: $G : \mathcal{P}(\mathcal{X}) \times \mathcal{P}(\mathcal{X}) \to \mathbb{R}$

$$G(P, P') = G(P, P', s, W) \triangleq D(W||P'W|P) - D(P||P') - sc(P).$$

**Lemma 36** *For fixed $P \in \mathcal{P}(\mathcal{X})$*

*(i) $G(P, P')$ is maximized if $P' = P$, and*

$$\max_{P'} G(P, P') = I(W|P) - sc(P)$$

*and for fixed $P' \in \mathcal{P}(\mathcal{X})$*

*(ii) $G(P, P')$ is maximized if*

$$P(x) = \frac{1}{A} P'(x) \exp[D(W(\cdot|x)||P'W) - sc(x)]$$

*and*

$$\max_P G(P, P') = \log \sum_x P'(x) \exp[D(W(\cdot|x)||P'W) - sc(x)]$$

*where A is a norming constant for getting a PD.*

*Proof* By (2.3.3)

$$G(P, P') = I(W|P) + D(PW||P'W) - D(P||P') - sc(P)$$

hence (i) follows by the Data Processing Lemma. The second assertion is a consequence of the following consequence of the Log-sum inequality.

**Lemma 37** *Let $f$ be a function of the RV $X$ with values in $\mathcal{X}$ and let $\alpha \in \mathbb{R}$, then*

$$H(X) \le \alpha \mathbb{E} f(X) + \log \sum_{x \in \mathcal{X}} \exp\{-\alpha f(x)\}$$

*and equality holds if and only if*

$$P_X(x) = \frac{1}{A} \exp(-\alpha f(x)),$$
$$A = \sum_{x \in \mathcal{X}} \exp(-\alpha f(x)).$$

(Take $P_X(x)$ and $\exp(-\alpha(f))$ in the role of $a_i$ and $b_i$ in Lemma 36)

Indeed choose $\alpha = -1$ and

$$f(x) = \log P'(x) + D(W(\cdot|x)||P'W) - sc(x).$$ □

As a consequence of Lemma 36 we have

**Corollary 11** *For any PD's $P \in \mathcal{P}(\mathcal{X})$ and $Q \in \mathcal{P}(\mathcal{Y})$*

$$\log \sum_{x \in \mathcal{X}} P(x) \exp[D(W(\cdot|x)||PW) - sc(x)] \le G(s) \le \max_{x \in \mathcal{X}} [D(W(\cdot|x)||Q) - sc(x)]$$

*Proof* The second inequality has to be checked. This follows from (2.3.3), if applied to a PD $P^*$ maximizing $I(W|P) - sc(P)$:

$$\max_{x \in \mathcal{X}} [D(W(\cdot|x)||Q) - sc(x)] \ge D(W||Q|P^*)$$
$$= I(W|P^*) + D(P^*W||Q) - sc(P^*) = G(s) + D(P^*W||Q). \quad (2.3.4)$$

□

**Corollary 12**

$$G(s) = \max_{P,P'} G(P, P').$$

## 2.3.3 Capacity Computing Algorithm

The last result suggests that $G(s)$ can be computed by an iteration, maximizing $G(P, P')$ with respect to $P$ resp. $P'$ in an alternating manner. The next theorem shows convergence of this iteration to $G(s)$, whenever one starts with a $P_1 \in \mathcal{P}^+(\mathcal{X})$.

**Theorem 63** (Capacity computing algorithm) *For $P_1 \in \mathcal{P}^+(\mathcal{X})$ define $(P_n)_{n=2}^{\infty}$ recursively by*

$$P_n(x) = A_n^{-1} P_{n-1}(x) \exp[D(W(\cdot|x)||P_{n-1}(W)) - sc(x)], \tag{2.3.5}$$

*where $A_n$ is defined by the condition $P_n \in \mathcal{P}(\mathcal{X})$. Then*

$$\log \sum_{x\in\mathcal{X}} P_n(x) \exp[D(W(\cdot|x)||P_n W) - sc(x)] = \log A_{n+1}$$

*converges from below and*

$$\max_{x\in\mathcal{X}}[D(W(\cdot|x)||P_n W) - sc(x)]$$

*converges from above to $G(s)$. Moreover the sequence $(P_n)_{n=1}^{\infty}$ of PD's converges to a PD $\widetilde{P}$ such that $I(W|\widetilde{P}) - sc(\widetilde{P}) = G(s)$.*

*Proof* By Lemma 36

$$G(P_1, P_1) \le G(P_2, P_1) \le G(P_2, P_2) \le (P_3, P_2) \le \cdots$$

and therefore

$$G(P_n, P_n) = I(W|P_n) - sc(P_n)$$

and $G(P_n, P_{n-1}) = \log A_n$ converges increasingly to the same limit not exceeding $G(s)$.

If $P$ is any PD with $I(W|P) - sc(P) = G(s)$, then by the above, (2.3.5), and (2.3.2) we get

$$\begin{aligned} 0 \le G(s) - \log A_n &= I(W|P) - sc(P) - \log A_n \\ &= I(W|P) + \sum_x P(x) \log \frac{P_n(x)}{P_{n-1}(x)} - D(W||P_{n-1}W||P) \\ &\le \sum_x P(x) \log \frac{P_n(x)}{P_{n-1}(x)} = D(P||P_{n-1}) - D(P||P_n). \end{aligned} \tag{2.3.6}$$

Hence the series $\sum_{n=2}^{\infty}(G(s) - \log A_n)$ is convergent and consequently $\log A_n \to G(s)$ as claimed. Next we show that the sequence $(P_n)_{n=1}^{\infty}$ also converges. For this pick a convergent subsequence $P_{n_i} \to \widetilde{P}_i$, say. Then $I(W|\widetilde{P}) - sc(\widetilde{P}) = G(s)$ and substituting $P = \widetilde{P}$ in (2.3.6) we see that the sequence $(D(\widetilde{P}||P_n))_{n=1}^{\infty}$ is nonincreasing. Thus $D(\widetilde{P}||P_{n_i}) \to 0$ yields $D(\widetilde{P}||P_n) \to 0$ and $P_n \to \widetilde{P}$ is proved.

Finally by the convergences proved so far, the recursion defining $P_n$ gives

$$\lim_{n\to\infty} \frac{P_n(x)}{P_{n-1}(x)} = \exp[D(W(\cdot|x)||\widetilde{P}(W)) - sc(x) - G(s)].$$

But this limit is 1 if $\widetilde{P}(x) > 0$ and does not exceed 1 if $\widetilde{P}(x) = 0$. Hence for every $x \in \mathcal{X}$

$$D(W(\cdot|x)||\widetilde{P}W) - sc(x) \ge G(s), \tag{2.3.7}$$

with equality if $\widetilde{P}(x) > 0$.

This proves that $\max_x[D(W(\cdot|x)||P_n W) - sc(x)] \to G(s)$ and by Corollary 11 the convergence is from above. □

We show now that this algorithm gives another proof of

**Theorem 64** (Shannon 1957, [7]) *For any $s \ge 0$*

$$G(s) = \min_{Q\in\mathcal{P}(\mathcal{Y})} \max_x [D(W(\cdot|x)||Q) - sc(x)]. \tag{2.3.8}$$

*The minimum is attained if and only if $Q = PW$ for a PD $P \in \mathcal{P}(\mathcal{X})$ with*

$$I(W|P) - sc(P) = G(s). \tag{2.3.9}$$

*This $Q$ is unique and a PD $P$ satisfies* (2.3.9) *if and only if $D(W(\cdot|x)||PW) - sc(x)$ is constant on the support of $P$ and does not exceed this constant elsewhere.*

*Proof* (2.3.8) follows from (2.3.4) and (2.3.7). If $P$ is any PD with $I(W|P) - sc(P) = G(s)$, then we conclude from (2.3.4) with $P^* = P$ that if a $Q \in \mathcal{P}(\mathcal{Y})$ achieves the minimum in (2.3.8) then $D(PW||Q) = 0$ and $Q = PW$.

Therefore, even if there are several PDs in $\mathcal{P}(\mathcal{X})$ maximizing $I(W|P) - sc(P)$, the corresponding output distribution $PW$ is unique and it is the unique $Q$ achieving the minimum in (2.3.8). Furthermore, if $P$ satisfies (2.3.9) then by the above we have

$$\max_x[D(W(\cdot|x)||PW) - sc(x)] = G(s)$$

and (2.3.4) gives that $D(W(\cdot|x)||PW) - sc(x) = G(s)$ if $P(x) > 0$. The last two identities prove that $D(W(\cdot|x)PW) - sc(x)$ is constant on $\text{supp}(P)$ and nowhere exceeds this constant. Conversely, if a $P$ has this property then

$$I(W|P) - sc(P) = \max_x [D(W(\cdot|x)||PW) - sc(P)]$$

and thus $P$ satisfies (2.3.9) and (2.3.4). The proof is complete. □

By Lemma 35 we get

**Corollary 13** *For $\Gamma > \Gamma_0$*

$$C(\Gamma) = \min_Q \min_{s \geq 0} \max_x [D(W(\cdot|x)||Q) + s(\Gamma - c(x))].$$

*The minimizing $Q$ is the output PD of channel $W$ corresponding to any $P \in \mathcal{P}(\mathcal{X})$ achieving the capacity $C(\Gamma)$.*

*Remark* Since $G(0) = C$, Theorem 64 gives the formula $C = \min_Q \max_x D(W(\cdot|x) ||Q)$ for the capacity of the DMC $W$. This formula has an interesting interpretation. It is like geometric since relative entropy is like a distance. $C$ may be interpreted as the radius of the smallest "sphere" with minimizing $Q$ as "center" containing the set of distributions $W(\cdot|x)$, $x \in \mathcal{X}$.

## References


1. S. Arimoto, An algorithm for computing the capacity of arbitrary discrete memoryless channels. IEEE Trans. Inf. Theory **IT–18**, 14–20 (1972)
2. R.E. Blahut, Computation of channel capacity and rate-distortion functions. IEEE Trans. Inf. Theory **IT–18**, 460–473 (1972)
3. I. Csiszár, On the computation of rate-distortion functions. IEEE Trans. Inf. Theory **IT–20**, 122–124 (1974)
4. I. Csiszár, P.C. Shields, Iterative algorithms chapter 5 information theory and statistics: a tutorial. Found. Trends Commun. Inf. Theory **1**(4), 417–528 (2004)
5. B. Meister, W. Oettli, On the capacity of a discrete, constant channel. Inf. Control **11**, 341–351 (1967)
6. S. Muroga, On the capacity of a discrete channel I. J. Phys. Soc. Jpn. **8**, 484–494 (1953)
7. C.E. Shannon, Geometrische Deutung einiger Ergebnisse beider Berechnung der Kanalkapazität. Nachrichtentechn. Z. **1**, 1–4 (1957)
8. W. Weaver, C.E. Shannon, *The Mathematical Theory of Communication* (University of Illinois Press, Urbana, 1963)


## Further Reading


9. T. Berger, *Rate Distortion Theory: A Mathematical Basis for Data Compression* (Prentice-Hall, Englewood Cliffs, 1971)

10. R.E. Blahut, An hypothesis-testing approach to information theory, Ph.D. Dissertation, Department of Electrical Engineering Cornell University, Ithaca, New York (1972)
11. P. Boukris, An upper bound on the speed of convergence of the Blahut algorithm for computing rate-distortion functions. IEEE Trans. Inf. Theory **IT–19**, 708–709 (1973)
12. I. Csiszár, G. Tusnady, Information geometry and alternating minimization procedures. Stat. Decis. Suppl. Issue **1**, 205–237 (1984)
13. F. Topsø, An information theoretic identity and a problem involving capacity. Stud. Sci. Math. Hungar. **2**, 291–292 (1967)

# Chapter 3
# Shannon's Model for Continuous Transmission

## 3.1 Lecture on Historical Remarks

After Shannon had presented his mathematical theory of communication [50] its ideas had a very strong impact in several scientific communities in the world. The institutions of his main activities, Bell Laboratories in Murray Hill, New Jersey, and MIT in Cambridge, Massachusetts, seem to have been the most active research centers in the further development of information theory. Needless to say that there were several important ideas, which originated at these places already earlier, for example, by the mathematician Wiener—to name one of the outstanding names. Still it is perhaps correct to say that most ideas arose there in theoretical and practical *engineering* work. The direct connection to Shannon can be easily documented by the fact that in these circles his *random coding method* (explained in [50] and in an abstract form in [52]) and only this was used for proving the existence of good or even asymptotically optimal error correcting codes for noisy channels under various constraints. This is still the case today.[1] Famous books are Fano's [8] and Gallager's [12]. On the other hand most mathematicians and statisticians working in coding theory for noisy channels for several decades used the other method to guarantee the existence of good or asymptotically optimal codes, namely Feinstein's *maximal coding method* [9]. This became linked to the central interest in analyzing Shannon's ideas about communication for sources and channels with a two-sided infinite time structure. Pioneering activities led McMillan [39] to detailed models, which became the frame for subsequent studies, and to a proof of the AEP, the asymptotic equipartition property, for stationary ergodic sources, which Shannon had discovered in the more special context of Markov chains and finitary ergodic sources. This had a great impact for Ergodic Theory and influenced a group of strong mathematicians especially in the Soviet Union. There Khinchin [27] established a (direct) coding theorem for *discrete, stationary sources* and *discrete, stationary, non-anticipatory channels* with *finite memory*, which we abbreviate as DFMC* (Jacobs [19], and Takano [55]

[1] Strong emphasis was put into deriving sharp error bounds for specified rates.



A. Ahlswede et al. (eds.), *Transmitting and Gaining Data*,
Foundations in Signal Processing, Communications and Networking 11,
DOI 10.1007/978-3-319-12523-7_3

observed that an additional independence property was needed for the channel (the input and noise sequences are independent (U)), which Khinchin believed to hold for this channel). As already in McMillan's work [39] the requirement was that messages have to be reproduced with high probability exactly. Khinchin remarks that $\underline{C}$ (the pessimistic capacity) as supremum over all sources of rates allowing arbitrary small error probabilities cannot hold in the full generality of the model. Instead a restriction to ergodic sources has to be imposed. Otherwise proofs of McMillan and also Feinstein cannot be made rigorous. However, under the restriction to ergodic sources, proofs for converses become a problem. We show in Sect. 3.2 how Nedoma [40] and Jacobs [20] handled the problem. Our work on averaged channels shows that converses via Fano's lemma do not work here.

Khinchin taught us how to derive a lower bound on (the operational or error) pessimistic capacity $\underline{C}$ of the DFMC*. The two key tools are the McMillan AEP Theorem and Feinstein's maximal code lemma.

Let us require (what Khinchin also should have done) that the DFMC* $W$ is an *ergodic channel*, that is, a stationary channel producing from every ergodic input source $P$ an ergodic joint PD $\widetilde{P} = P \times W$ and a (then also ergodic) output process $P'$. Stationarity of $\widetilde{P}$ and $P'$ are implied by the stationarity of $W$ without additional assumptions. Therefore, it is well known (McMillan [39]) that the per letter entropies $H(P)$, $H(\widetilde{P})$, and $H(P')$ exist and the *informational* rate function (called speed of transmission by Shannon)

$$R(P) = H(P) + H(P') - H(\widetilde{P})$$

is well defined and so is the so-called ergodic capacity, introduced by Khinchin,

$$C_e = \sup_{P \text{ ergodic}} R(P).$$

Furthermore by the AEP and ergodicity of all three sources the information density $i_n$ converges to $R(P)$ with probability 1 or in $\mathcal{L}^1_{\widetilde{P}}$ (both implying convergence in probability, which actually suffices). Feinstein's Lemma in its general form gives thus that for $0 < \lambda < 1$ and any $\delta > 0, N(n, \lambda) \geq \exp\{(e-\delta)n\}$ for $n \geq n_0(\lambda, \delta)$, proving $\underline{C} \geq C_e$.

Although Khinchin did not prove a converse he found a clear mathematical setting helping to make Shannon's vision real. He came to the question when there is equality in

$$C_e \leq C_s = \sup_{P \text{ stationary}} R(P). \qquad (3.1.1)$$

This question was addressed by Tsaregradskii [56] who stated as a theorem, that $C_e \geq C_s$ for channel DFMC*. The proof uses another quantity

$$C_0 = \lim_{n\to\infty} \sup_{P^{(n)}} \frac{1}{n} I(P^{(n)}, W),$$

where $P^{(n)}$ is any *non-stationary* PD on $\mathcal{X}_0 \times \mathcal{X}_1 \times \cdots \times \mathcal{X}_{n-1}$ induced by a non-stationary PD $P$ on $\left(\prod_{t>-\infty}^{t<+\infty} \mathcal{X}_t, \mathcal{B}\right)$.

Clearly, $C_0 \geq C_s$ and then for any $\delta > 0$ an ergodic source $P'$ is constructed with $R(P') \geq C_0$, which then yields $C_e = C_s$. However, it seems to us that Condition (U) is used. We quote "(We note that here we have made essential use of the finiteness of the memory of the channel)" It seems that the author follows the belief of Khinchin that DFMC* satisfy (U). So Tsaregradskii's result is proved only for the restricted class of channels named DFMC below, first considered by Takano [55] ("*m*-dependent channels").

Our position finds a confirmation by Breiman [5] who writes on p. 247s "Tsaregradskii proved (a) using Khinchin's definition of finite memory. However, his proof is not easy to follow, and it is not clear whether implicit use is not made of other restrictions." Takano [55] writes "The proof of Khinchin goes on as if a stationary channel with finite memory $m$ and no foresight be $m$-step dependent. However, this is not true." Takano then proceeds with a nicely explained proof of the coding theorem $\underline{C} \geq C_e$.

There is no converse result in the paper. Here Nedoma [40] enters the scene.

A fundamental contribution to these ideas is due to Nedoma [40], who introduced and analyzed *risk functions* as performance criteria for data compression and transmission of messages. They include the familiar criteria of error probabilities, error frequencies, and distortion measures.

We cannot judge how independent these ideas are of Kolmogorov's ideas and therefore, quote from p. 50 of [40] "..., the author believes that the treatment using the risk functions introduced in this paper may prove useful also for other problems. ...In connection with this we point to Kolmogorov's paper [37] which is based on a similar approach."

Using the approach of Nedoma Jacobs proved that for ergodic channels for the optimistic operational capacity $\overline{C} \leq C_e$ holds. Therefore $\underline{C} = \overline{C}$ and the weak capacity exists. The question of computability of $C_e$ gives the next problem.

A new essential step was made by Feinstein who noticed, that when replacing the ergodicity assumption by the stronger $m$-independence condition on the memory (Condition (U)) (in addition to $m$-measurability) results in a channel, which we call DFMC, which can be easily treated by approximation with DMC's. Earlier Wolfowitz proposed a somewhat more special model (Example 2 in Sect. 3.3.2), but then also established the (strong) capacity for the DFMC. The formula makes the capacity at least in principle computable. In this approach, the quantities $C_e$ and $C_s$ do not arise. But the equality $C_e = C_s$ can be derived.

Famous books related to this "path" are by Feinstein [10] and by Wolfowitz [58]. The latter draws attention to Chap. 6 of the former "for an excellent description of the customary treatment" via the AEP.

The results of Khinchin were generalized to processes with continuous states by Rosenblatt-Roth [48, 49] and by Perez [46, 47].

In another direction Rosenblatt-Roth mentions the possibility to extend the theory to nonstationary processes. This was started by Jacobs [19] for almost periodic channels, which have in the memoryless case a capacity obtained as an average over letterwise maximal mutual information. To stick to a constant capacity was at that time typical before Ahlswede [1] introduced capacity functions as performance criterium for DNMC, nonstationary discrete memoryless channels, and even generalized the concept of Wolfowitz's "strong converse" for those functions and proved it. Only recently he discovered that every one-shot time-structure channel (TSC) has a capacity function, if the so-called optimistic capacity is finite.

Soon after Khinchin's work Kolmogorov [37] formulated a quite general model for information transmission in rigorous mathematical setting. He envisions in this generality a "Shannon Theorem", for which exact conditions have to be worked out and proofs have to be given where possible. The main view of Kolmogorov on "Shannon's dream" concerns in the scheme $X$ (source) $\rightarrow$ $Y$ (channel input) $\rightarrow$ $Y'$ (channel output) $\rightarrow$ $X'$ (reproduction of $X$) with Markovian coding rules the *restriction* on the *input distribution* yielding a capacity $C = \sup_{P_Y \in V} I(Y, Y')$ and the *conditions on the reproduction* giving a rate for the generation of source data

$$H_W(X) = \inf_{P_{X,X'} \in W} I(X, X').$$

Here

$$H_W(X) = I(X, X) = H(X).$$

He also hinted at a possible dependence of these quantities on a parameter $T$ (measuring for instance duration of performances) and asked for establishing a relation like

$$\liminf_{T \to \infty} \frac{C^T}{H_W^T(X)} > 1,$$

which should be sufficient for the desired transmission. The astute reader notices, what we learned only in Oct. 2008, a connection to our *capacity function*.

Kolmogorov followed Shannon's ideas, but felt that a more suggestive name than "speed of generation of messages" was appropriate and thus came—in a special case involving metrics—to $\varepsilon$-entropy. Its study became an important subject in approximation theory (Kolmogorov, Tikhomirov, Pinsker, Carl, ...).

*We highly recommend to read the paper, in particular his discussion of the greatness of Shannon's ideas to quantitatively deal with noise.* In a well-known article [6] Dobrushin proved what Kolmogorov called "Shannon's Theorem" under various quite general conditions. They include conditions in terms of information density.

The concept *information density* was introduced by Gelfand/Yaglom [13, 14] and by Perez [46].

The work of Dobrushin essentially uses the concept *information stability* and Dobrushin acknowledges that some of the ideas were expressed by Gelfand and Yaglom in conversations with them.

Theorems of great generality are proved under sometimes complex conditions, whose verification for concrete processes is often nontrivial and sometimes even constitutes open problems.

In Sect. 3.3 we follow Dobrushin's introduction which gives the basic concepts needed to formulate and analyze Kolmogorov's model and state his results. We then give elegant proofs due to Huo Kuo Ting [18]. For proofs of results unproved here we refer to Dobrushin's original paper.

We emphasize that in both, Sects. 3.2 and 3.3, all coding theorems are *existence* theorems. On the other hand, in the presence of noiseless feedback, there is another method to get constructively as asymptotically optimal codes, namely via our iterative list reduction scheme [3]. It makes the role of mutual information very transparent and is therefore a very intuitive access to coding theorems, especially, since from there it can be nicely explained how in the absence of feedback random coding can be used to go from list codes of cardinality $\sim e^{H(X)n}$ and list size $\sim e^{H(X|Y)n}$ to the desired ordinary codes of rate $I(X \wedge Y) = H(X) - H(X|Y)$.

## 3.2 Lecture on Fundamental Theorems of Information Theory

### 3.2.1 Stationary Sources

For a finite alphabet $\mathcal{X} = \{1, 2, \dots, a\}$ we define the two-sided infinite sequence space

$$\Omega = \prod_{t=-\infty}^{\infty} \mathcal{X}_t \quad (\mathcal{X}_t = \mathcal{X}, t \in \mathbb{Z})$$

$$= \{w = (w_t) = (\dots, w_{-1}, w_0, w_1, \dots) : w_t \in \mathcal{X}, t \in \mathbb{Z}\},$$

which can be viewed as the set of all messages written in the alphabet $\mathcal{X}$. For $s \leq t$ and $w_s, \dots, w_t \in \mathcal{X}$ the set

$$(w_s, \dots, w_t) = \{\eta : \eta \in \Omega, \eta_s = w_s, \dots, \eta_t = w_t\}$$

can be interpreted as a (finite) text in the time from $s$ to $t$. These $a^{t-s+1}$ texts form the cylinders (or atoms) in the finite space $\mathcal{B}(s, t)$ of events in $\Omega$. We denote the set of these cylinders by $\mathcal{Z}(s, t)$.

Clearly, for $s \leq s' \leq t' \leq t \quad \mathcal{B}(s', t') \subset \mathcal{B}(s, t)$ and we denote the $\sigma$-algebra generated by $\bigcup_{t=1}^{\infty} \mathcal{B}(s, t)$ as $\mathcal{B}(s, \infty)$. Analogously, we define $\mathcal{B}(-\infty, t)$ and $\mathcal{B}(-\infty, \infty) = \mathcal{B}$ as the $\sigma$-algebra generated by $\bigcup_{s \leq t} \mathcal{B}(s, t)$.

**Definition 8** For a PD $P$ on $\mathcal{B}$ the triple $(\Omega, \mathcal{B}, P)$—or also $P$ alone—is called a (message) source.

This way $P$ assigns to every finite text $(w_s, \dots, w_t)$ a probability $P(w_s, \dots, w_t) \geq 0$ and one gets the finite probability space $(\Omega, \mathcal{B}(s, t), P)$ $(s \leq t)$. It is convenient to abbreviate the entropy $H(P) = -\sum_{E\in\mathcal{Z}(s,t)} P(E)\log P(E)$ by $H(s,t)$. It satisfies

$$0 \le H(s,t) \le \log a^{t-s+1} = (t-s+1)\log a. \tag{3.2.1}$$

We show next how an entropy $H(-\infty,\infty)$ can be defined for $(\Omega, \mathcal{B}, P)$. Obviously $\lim_{t-s\to\infty} H(s,t)$ is not suitable for this, because it may equal infinity. Instead it turns out to be useful to introduce (if it exists)

$$\lim_{t-s\to\infty} \frac{1}{t-s} H(s,t) \tag{3.2.2}$$

as "mean entropy". Next we investigate the existence of this limit. For this we recall (from Measure Theory) the concept of a measurable map

$$\varphi : \mathcal{R}' \to \mathcal{R}''$$

which satisfies

$$\varphi^{-1}E'' \in \mathcal{B}' \quad (E'' \in \mathcal{B}'') \tag{3.2.3}$$

(in short: $\varphi^{-1}\mathcal{B}'' \subseteq \mathcal{B}'$). For the measurable spaces $(\Omega', \mathcal{B}')$ and $(\Omega'', \mathcal{B}'')$ is called $(\mathcal{B}' - \mathcal{B}'')$-measurable. Here, we used the convention $\varphi^{-1}E'' = \{w' : \varphi w' \in E''\}$. Since $\varphi^{-1}$ can be exchanged with set operations measurability of $\varphi$ already follows, if (3.2.3) holds for all $E''$ from a family of sets generating $\mathcal{B}''$. By the rule

$$(Tw)_t = w_{t+1} \tag{3.2.4}$$

a 1:1 map $T$, called (time-)shift, from $\Omega$ on itself (bijection) is given.

For a text $E = (w_s, \dots, w_t) \in \mathcal{B}(s,t)$ and an integer $r$ we get

$$T^r E = \{\eta : T^{-u}\eta \in E\} = \{\eta : \eta_{s-r} = w_s, \dots, \eta_{t-r} = w_t\} \in \mathcal{B}(s-r, t-r), \tag{3.2.5}$$

that is, the text from $\mathcal{B}(s-r, t-r)$ described by the same string of symbols as $E$. Therefore $T^r\mathcal{B}(s,t) = \mathcal{B}(s-r, t-r)$ and

$$(T^r)^{-1} \bigcup_{s\le t} \mathcal{B}(s-r, t-r) = \bigcup_{s\le t} \mathcal{B}(s,t). \tag{3.2.6}$$

Since $\bigcup_{s\le t} \mathcal{B}(s,t)$ generates $\mathcal{B}$, all powers of $T$ are $(\mathcal{B}' - \mathcal{B}'')$-measurable. Furthermore, for any $(\mathcal{B}' - \mathcal{B}'')$-measurable map $\varphi : \Omega' \to \Omega''$ one gets for every PD $P'$ on $\mathcal{B}'$ a PD $P'' = \varphi P'$ on $\mathcal{B}''$ by the equation

$$P''(E'') = P'(\varphi^{-1}E'') \quad (E'' \in \mathcal{B}''). \tag{3.2.7}$$

This implies that for a $P''$-integrable function $f''(w'')$ the function $f'(w') = f''(\varphi w')$ is $P'$-integrable and

$$\int f' \, \mathrm{d}P' = \int f'' \, \mathrm{d}P''. \tag{3.2.8}$$

In particular one then gets from a source $(\Omega, \mathcal{B}, P)$ the shifted sources $(\Omega, \mathcal{B}, T^r P)$ $(r \in \mathbb{Z})$.

**Definition 9** A source $(\Omega, \mathcal{B}, P)$ (or just the PD $P$ on $\mathcal{B}$) is called stationary if

$$T^r P = P \quad (r \in \mathbb{Z}). \tag{3.2.9}$$

Stationarity guarantees the properties

$$P(T^r(w_s, \dots, w_t)) = P(w_s, \dots, w_t) \tag{3.2.10}$$

and

$$\int f(w)P(dw) = \int f(T^r w)P(dw) \tag{3.2.11}$$

for all $f \in \mathcal{L}_P^1$, the $P$-integrable functions.

This means that the probability of a text depends only on the string of symbols and not the time interval of its occurrence. Conversely, this is also sufficient for stationarity. Indeed, (3.2.10) implies $P(T^r E) = P(E)$ for all $E \in \mathcal{K} \triangleq \bigcup_{s \le t} \mathcal{B}(s,t)$. That this equation holds for all $E \in \mathcal{B}$ can be shown with certain continuity properties of $P$:

Obviously the family of all $E \in \mathcal{B}$ for which

$$P(T^r E) = P(E) \quad (r \in \mathbb{Z}) \tag{3.2.12}$$

holds, is closed under monotone limits (which are exchangeable with $T^r$ and $P$). It is known (see the books by Halmos [17], Loève [38], and Jacobs [23]) that for every $E \in \mathcal{B}$ there is a $G = \lim_k G_k$ (monotone decreasing) with $G_k = \lim_j G_{k_j}$ (monotone increasing), $G_{k_j} \in \mathcal{K}$ and $G \supset E$, $P(G) = P(E)$. This implies $P(T^r E) \le P(T^r G) = P(G) - P(E)$ $(E \in \mathcal{B})$, and therefore by replacing $E$ by $T^{-r}E$ also $P(T^r E) = P(E)$. Hence, (3.2.12) holds.

The distance function on $\mathcal{B}$

$$|E, F| = P(E \triangle F) = P(E \cup F) - P(E \cap F)$$

satisfies

(i) $|E, F| \ge 0$
(ii) $|E, G| \le |E, F| + |F, G|$
(iii) $|P(E) - P(F)| \le |E, F|$.

It can easily be shown that the closed hull of $\mathcal{K}$ forms a Borel field in $\mathcal{B}$, which coincides with $\mathcal{B}$, because $\mathcal{B}$ is generated by $\mathcal{K}$. Thus $\mathcal{K}$ is dense in $\mathcal{B}$.

Finally, for stationary $(\Omega, \mathcal{B}, P)$

$$H(s,t) = H(s+r, t+r) \quad (r \in \mathbb{Z}). \tag{3.2.13}$$

## 3.2.2 Methods to Construct Sources

The construction of a source $(\Omega, \mathcal{B}, P)$ has a trivial part—the construction of $\Omega$, $\mathcal{B}$ from $\mathcal{X}$—and a nontrivial part: the construction of a PD $P$ on $\mathcal{B}$ with prescribed properties (such as stationarity for example). Since $\Omega$ as represented as a Cartesian product space with infinitely many factors it seems natural to use the *Theorem of Kolmogorov*, which says in the present case: If for arbitrary $s \le t$ a PD on $\mathcal{B}(s,t)$ is given and do we have

$$P^{st}(E) = P^{s't'}(E) \quad (E \in \mathcal{B}(s,t), s' \le s \le t \le t'), \tag{3.2.14}$$

then there is a unique PD $P$ on $\mathcal{B}$ with

$$P(E) = P^{st}(E) \quad (E \in \mathcal{B}(s,t), s \le t) \tag{3.2.15}$$

(cf. Loève [38, p. 93]).

Since $\mathcal{B}(s,t)$ is finite (3.2.14) has to be verified only for the atoms $E$ of $\mathcal{B}(s,t)$ and the $P^{st}$ can be constructed by finitely many operations. We give two examples.

**Example 1: Sources with Independent Symbols**

If we have a PD $(p_1, p_2, \dots, p_a)$ on $\mathcal{X}$, then there is exactly one PD $P$ on $\mathcal{B}$ such that

(i) $P(w_t) = P_X \ (w_t = x, t \in \mathbb{Z})$

(ii) For $s_1 \le t_1 < s_2 \le t_2 < \dots < s_n \le t_n$ the set algebras $\mathcal{B}(s_1,t_1), \mathcal{B}(s_2,t_2), \dots, \mathcal{B}(s_n,t_n)$ are independent under $P$

(iii) $P(E_1 \cap \dots \cap E_n) = P(E_1) \cdots P(E_n)$, $(E_\nu \in \mathcal{B}(s_\nu, t_\nu), \nu = 1, 2, \dots, n)$.

Indeed from (i) and (3.2.14) follows

$$P(w_s, \dots, w_t) = P_{w_s} \cdots P_{w_t} \quad (s \le t).$$

Defining now

$$P^{st}(w_s, \dots, w_t) = P_{w_s} \cdots P_{w_t} \quad (s \le t)$$

Equation (3.2.14) is readily verified and the existence and uniqueness of a $P$ satisfying (i) and (ii) follows. Obviously also (3.2.10) in Sect. 3.2.1 holds and the source obtained is stationary.

**Example 2: Markov Sources**
For a stochastic matrix $(W(y|x))_{x,y\in\mathcal{X}}$ and a PD $(p_1, \dots, p_a)$ with

$$p_y = \sum_x W(y|x)p_x \tag{3.2.16}$$

there is exactly one PD $P$ on $\mathcal{B}$ with

$$P(w_s, \dots, w_t) = p_{w_s} W(w_{s+1}|w_s) \dots W(w_t|w_{t-1}) \quad (s \le t). \tag{3.2.17}$$

In fact for

$$P^{st}(w_s, \dots, w_t) = p_{w_s} W(w_{s+1}|w_s) \dots W(w_t|w_{t-1})$$

Equation (3.2.14) is readily verified and (3.2.5) follows. Also $P$ is stationary. The choice $W(y|x) = p_y$ $(x, y \in \mathcal{X})$, gives again Example 1.

In short notation we write

$$p = pW. \tag{3.2.18}$$

How to find such a $p$ for $W$? Starting with any PD $p^0$ on $\mathcal{X}$, then every cluster point $p$ of the sequence of vectors $p^t = \frac{1}{t}\sum_{k=0}^{t-1} W^k p^0$ satisfies (3.2.18). Using for the vector $q = \begin{pmatrix} q_1 \\ \vdots \\ q_a \end{pmatrix}$ the norm $||q|| = q_1 + \dots + q_a$ and the fact $||qW|| \le ||q||$ (see also the next theorem) one sees that cluster points exist for $(p^t)_{t=1}^{\infty}$.

### 3.2.3 *The Ergodic Theorem in Hilbert Space*

Since proofs of McMillan's Theorem can be found in a huge literature for different concepts of convergence and various directions of generalizations we go here the seemingly shortest path to get $\mathcal{L}_P^2$ convergence and thus convergence in probability, sufficient for our purposes.

Let $\mathcal{L}$ be a real linear (function) space (or vector space) with an *inner product*, that means for $f, g \in \mathcal{L}$ a scalar product $(f, g)$ is defined:

(i) $(f, g)$ is bilinear in $f, g$;
(ii) $(f, g) = (g, f)$;
(iii) $||f||^2 = (f, f) \ge 0$ and is equal to 0 exactly if $f = 0$.

It is well known that $||\cdot||$ has the properties of a *norm*

(iv) $||f + g|| \le ||f|| + ||g||$;
(v) $||\lambda f|| = ||\lambda|| ||f||$ for $\lambda \in \mathbb{R}$.

Additionally, we require

(vi) $\mathcal{L}$ is *complete* in the norm topology, that means for sequences $(f_\nu)_{\nu=1}^\infty$ of functions in $\mathcal{L}$. $||f_\mu - f_\nu|| \to 0$ for $\mu, \nu \to \infty$ implies that there is an $f \in \mathcal{L}$ with $||f_\mu - f|| \to 0$ for $\mu \to \infty$. (In other language Cauchy sequences have a limit belonging to $\mathcal{L}$.)

A linear space with these properties is called Hilbert space. A basic tool for us is the following result.

**Theorem 65** (Ergodic Theorem in Hilbert space) *For a linear map $A : \mathcal{L} \to \mathcal{L}$, which is a contraction, that is,*

$$||Ah|| \leq ||h|| \quad (h \in \mathcal{L}), \tag{3.2.19}$$

*there is for every $h \in \mathcal{L}$ exactly one $h_0 \in \mathcal{L}$ with*

$$\lim_{t\to\infty} \left|\left| \frac{1}{t} \sum_{k=0}^{t-1} A^k h - h_0 \right|\right| = 0. \tag{3.2.20}$$

*and*

$$A^t h_0 = h_0 \quad (t = 0, 1, 2, \dots). \tag{3.2.21}$$

*Proof* Let $\mathcal{K}(h)$ be the closed hull of the set

$$\mathcal{M}(h) = \left\{ g = \sum_{\nu=1}^n \alpha_\nu A^{t_\nu} h : t_\nu \geq 0, \alpha_\nu \geq 0, \sum_{\nu=1}^n \alpha_\nu = 1 \right\}.$$

If $f, g \in \mathcal{K}(h)$, then also $\frac{f+g}{2}$ and $Af \in \mathcal{K}(h)$. Let $\gamma = \inf_{g\in\mathcal{K}(h)} ||g||$, then for $g_\mu \in \mathcal{K}(h)$, $||g_\mu|| \to \rho$ as $\mu \to \infty$ it follows that $||g_\mu - g_\nu|| \to 0$ as $\mu, \nu \to \infty$ and therefore also the existence of an $h_0 \in \mathcal{K}(h)$ with $||h_0|| = \gamma$, because $\mathcal{L}$ is complete. Indeed the identity

$$\left|\left| \frac{f+g}{2} \right|\right| + \left|\left| \frac{f-g}{2} \right|\right| = \frac{||f||^2 + ||g||^2}{2} \tag{3.2.22}$$

and $||g_\mu|| \to \gamma$ imply $\left|\left|\frac{g_\mu + g_\nu}{2}\right|\right| \to \gamma$

$$||g_\mu - g_\nu||^2 = 2\left(||g_\mu||^2 + ||g_\nu||^2\right) - 4\left|\left|\frac{g_\mu + g_\nu}{2}\right|\right|^2 \to 2(\gamma^2 + \gamma^2) - 4\gamma^2 = 0$$

as $\mu, \nu \to \infty$.

Since $A^t h_0 \in \mathcal{K}(h)$ we conclude with (3.2.19)

$$||A^t h_0|| \leq ||h_0|| = \rho \leq ||A^t h_0||$$

and therefore $||A^t h_0|| = \rho$. Also by definition of $\mathcal{K}(h)$ we have $\frac{A^t h_0 + h_0}{2} \in \mathcal{K}(h)$. The assumption $A^t h_0 \neq h_0$ would imply with (3.2.22) $\left|\left|\frac{A^t h_0 + h_0}{2}\right|\right| < \rho$ in contradiction to the definition of $\rho$ and this establishes (3.2.21). Therefore also by the assumptions on $A$

$$\left|\left|\frac{1}{t}\sum_{k=0}^{t-1} A^k g - h_0\right|\right| = \left|\left|\frac{1}{t}\sum_{k=0}^{t-1} A^k (g - h_0)\right|\right| \leq \frac{1}{t}\sum_{k=0}^{t-1} ||g - h_0|| < \varepsilon \quad (t = 1, 2, \dots),$$

if $||g - h_0|| < \varepsilon$.

In particular for $g = \sum_{\nu=1}^{n} \alpha_\nu A^{t_\nu} h$ and for $t > \max_\nu t_\nu$ we get

$$\left|\left|\frac{1}{t}\sum_{k=0}^{t-1} A^k h - \frac{1}{t}\sum_{k=0}^{t-1} A^k g\right|\right| = \frac{1}{t}\left|\left|\sum_{\nu=1}^{n} \alpha_\nu \sum_{k=0}^{t-1} A^k h - \sum_{\nu=1}^{n} \alpha_\nu \sum_{k=0}^{t-1} A^{k+t_\nu} h\right|\right|$$

$$\leq \frac{1}{t}\sum_{\nu=1}^{n} \alpha_\nu \left|\left|\sum_{k=0}^{t_\nu - 1} A^k h - \sum_{k=t}^{t+t_\nu-1} A^k h\right|\right|$$

$$\leq \frac{2\sum_{\nu=1}^{n} t_\nu}{t} ||h|| < \varepsilon \quad (\text{for } t \text{ sufficiently large}).$$

Combination of these estimates gives (3.2.20) and the uniqueness of $h_0$. □

#### 3.2.3.1 The Ergodic Theorem in $\mathcal{L}_P^2$

Let now $(\Omega, \mathcal{B}, P)$ be any probability space, let $T : \Omega \to \Omega$ be a bijective map, let $T$ and $T^{-1}$ be $\mathcal{B} - \mathcal{B}$-measurable,0 and let $TP = P$. Then all $T^r$ are measurable and $T^r P = P$ $(r \in \mathbb{Z})$. $\mathcal{L}_P^q$ is the set of all $\mathcal{B}$-measurable real valued functions $f : \Omega \to \mathbb{R}$ (that is all $\mathcal{B} - \mathcal{B}'$-measurable maps from $\Omega$ into the real line $\mathbb{R}^1$, where $\mathcal{B}'$ is the $\sigma$-algebra generated by the topology in $\mathbb{R}^1$), which satisfy

$$||f||_q^q = \int |f|^q dP < \infty \quad (1 \leq q < \infty). \tag{3.2.23}$$

As usual here functions differing only on $P$-nullsets, that is on sets $E \in \mathcal{B}$ with $P(E) = 0$, are identified. We are only interested here in the cases $q = 1, 2$ and the corresponding norms $||\cdot||_1, ||\cdot||_2$: $\mathcal{L}_P^1, \mathcal{L}_P^2$ are real linear spaces with respect to $||\cdot||_1, ||\cdot||_2$ and known to satisfy

$$\mathcal{L}_P^2 \subseteq \mathcal{L}_P^1 \tag{3.2.24}$$

$$||f||_1 \leq ||f||_2 \quad (f \in \mathcal{L}_P^2). \tag{3.2.25}$$

With the inner product $(f, g) = \int fg\, \mathrm{d}P$ $\mathcal{L}_P^2$ becomes a Hilbert space (see Loève, p. 460).

The indicator function $1_B$ of sets $B$ belong to $\mathcal{L}_P^1$ and to $\mathcal{L}_P^2$, because

$$\int 1_B\, \mathrm{d}P = P(B)$$

and by linearity of the integral also their linear combinations, the *step functions*. These are dense in $\mathcal{L}_P^1$ and in $\mathcal{L}_P^2$ under the respective norms.

**Theorem 66** (Ergodic Theorem in $\mathcal{L}_P^2$) *The definition $(Af)(w) = f(Tw)$ specifies a 1:1 linear map $A : \mathcal{L}_P^2 \to \mathcal{L}_P^2$, which has the properties, that for every $h \in \mathcal{L}_P^2$ there is exactly one $h_0 \in \mathcal{L}_P^2$ with*

$$\lim_{t\to\infty} \left\| \frac{1}{t} \sum_{k=0}^{t-1} A^k h - h_0 \right\|_2 = 0 \tag{3.2.26}$$

$$A^r h_0 = h_0 \qquad (r \in \mathbb{Z}) \tag{3.2.27}$$

$$||A^r h||_2 = ||h||_2 \qquad (r \in \mathbb{Z}) \tag{3.2.28}$$

$$\int h_0\, \mathrm{d}P = \int h\, \mathrm{d}P \tag{3.2.29}$$

**Corollary 14** *For $(h_\nu)$, $h \in \mathcal{L}_P^2$ with $||h_\nu - h||_2 \to 0$ as $\nu \to \infty$ we have*

$$\lim_{t\to\infty} \left\| \frac{1}{t} \sum_{k=0}^{t-1} A^k h_k - h_0 \right\|_2 = 0. \tag{3.2.30}$$

*Proof* Linearity of $A^k$, the triangle inequality, and (3.2.28) imply

$$\left\| \frac{1}{t} \sum_{k=0}^{t-1} A^k h_k - h_0 \right\|_2 \leq \left\| \frac{1}{t} \sum_{k=0}^{t-1} A^k (h_k - h) \right\|_2 + \left\| \frac{1}{t} \sum_{k=0}^{t-1} A^k h - h_0 \right\|_2$$

$$\leq \frac{1}{t} \sum_{k=0}^{t-1} ||h_k - h||_2 + \left\| \frac{1}{t} \sum_{k=0}^{t-1} A^k h - h_0 \right\|_2$$

This goes to 0 as $t \to \infty$ by the assumption on $(h_\nu)$ and by (3.2.26). □

*Proof of Theorem 66* For $h = 1_E$, the indicator function of $E$ ($1_E(w) = 1$ for $w \in E$ and $1_E(w) = 0$ for $w \notin E$)

$$\int (A^\nu h)\,\mathrm{d}P = \int h(T^\nu w)\,\mathrm{d}P(w) = P(T^{-\nu}E) = P(E) = \int h\,\mathrm{d}P \quad (\nu \in \mathbb{Z}),$$

and we see that

$$\int (A^\nu h)\,\mathrm{d}P = \int h\,\mathrm{d}P \quad (\nu \in \mathbb{Z}), \tag{3.2.31}$$

holds also for step functions $h$.

We conclude that (3.2.28) holds for all step functions. Since they are dense in $\mathcal{L}_P^2$ (3.2.28) and (3.2.31) hold for all $h \in \mathcal{L}_P^2$.

Now Theorem 65 implies (3.2.26) and (3.2.27). The norm continuity of the integral yields

$$\begin{aligned}\int h_0\,\mathrm{d}P &= \lim_t \int \left(\frac{1}{t}\sum_{k=0}^{t-1} A^k h\right)\mathrm{d}P \\ &= \lim_t \frac{1}{t}\sum_{k=0}^{t-1}\int A^k h\,\mathrm{d}P \\ &= \int h\,\mathrm{d}P\end{aligned}$$

and thus (3.2.29). □

### 3.2.4 The Theorem of McMillan

*Preliminary Remark.* In the following, we consider always stationary sources $(\Omega, \mathcal{B}, P)$ and mostly functions $f(w)$ which are measurable with respect to a finite Borel field $\mathcal{B}' \subseteq \mathcal{B}$. These are exactly the functions, which are *constant* on the finite number of atoms for cylinder sets $A_1, A_2, \dots, A_n$ of $\mathcal{B}'$. For example, for $\mathcal{B}' = \mathcal{B}(s, t)$ $(s < t)$ this means that $f(w)$ depends only on $w_s, \dots, w_t$ and we, therefore, then also write $f(w_s, \dots, w_t)$ instead of $f(w)$. If $f(w)$ is defined only on $\bigcup_{P(A_\nu)>0} A_\nu$, then we can assign an arbitrary constant value to $f$ on every $A_\nu$ with $P(A_\nu) = 0$. This way we get always a $\mathcal{B}'$-measurable function and these functions differ only on a $P$-nullset. It makes, therefore, sense to speak about $P$-almost everywhere defined $\mathcal{B}'$-measurable functions.

The goal of this section is the

**Theorem 67** (McMillan [39]) *For a stationary source $(\Omega, \mathcal{B}, P)$ for every $t = 1, 2, \dots$ the function*

$$h_t(w) = -\frac{1}{t}\log P(w_1, \dots, w_t) \tag{3.2.32}$$

*is defined P-almost everywhere, finite and $\mathcal{B}(1, t)$-measurable. Furthermore $h \in \mathcal{L}_P^2$ and exactly one function $h \in \mathcal{L}_P^2$ exists with $0 \leq h \leq \log a$ and*

$$h(T^{\nu} w) = h(w) \quad (P\text{-almost everywhere}) \tag{3.2.33}$$

$$\lim_{t \to \infty} ||h_t - h||_2 = 0. \tag{3.2.34}$$

*Remark* Breiman [5] has proved that also $\lim_{t \to \infty} h_t(w) = h(w)$, $P$-almost everywhere.

Important for us is the very meaningful

**Corollary 15**

$$\int h_t \, \mathrm{d}P = \frac{1}{t} H(1, t) \tag{3.2.35}$$

$$\lim_{t \to \infty} \frac{1}{t} H(1, t) = \int h \, \mathrm{d}P. \tag{3.2.36}$$

*Proof* (3.2.35) is immediately obtained from the definitions and (3.2.36) follows from (3.2.34), because the integrals are norm continuous as can be seen from the relation

$$\left| \int f \, \mathrm{d}P \right| \leq \int |f| \, \mathrm{d}P = ||f||_1 \leq ||f||_2.$$ □

We have seen earlier that the mean entropy $H = \lim_t \frac{1}{t} H(1, t)$ exists. However, Theorem 67 gives a deeper result.

Measurability, boundedness, and integrability properties of $h_t$ are obvious.

It is readily shown also that

**Lemma 38** *If $h \in \mathcal{L}_P^2$ and satisfies (3.2.34), then $0 \leq h(w) \leq \log a$ P-almost surely.*

*Proof* Indeed the set $E_t(u, v) : \{w : u < h_t(w) < v\}$, $u < v$ real, is element of $\mathcal{B}(1, t)$. If $u > 0$ and a text $E = (w_0, \dots, w_t) \leq E_t(u, v)$, then $P(E) \leq e^{-ut}$, $\int_E h_t \, \mathrm{d}P \leq ve^{-ut}$, and

$$\int_{E_t(u,v)} h_t \, \mathrm{d}P \leq a^t \cdot v \cdot e^{-ut} = ve^{-t(u - \log a)}.$$

For $n > \log a$ this converges to 0.

From here we get the claimed statement as follows: Let $\delta > 0$ and observe that for

$$D_t = \{w : |h_t(w) - h(w)| \geq \delta\}$$

$$\delta P(D_t) \leq ||h_t - h||_1 \leq ||h_t - h||_2,$$

and also $\lim_t P(D_t) = 0$. This implies $\lim_t \int_{D_t} h \, \mathrm{d}P = 0$.

Let now $\log a < u' < v'$ and $E(u', v') = \{w : u' \leq h(w) \leq v'\}$. Is $\delta > 0$ such that $u' - \delta > \log a$ and does one define $u = u' - \delta < v = v' + \delta$, then one gets a partition $E(u', v') = F_t + G_t$, where $F_t \subseteq E_t(u, v)$, $G_t \subseteq D_t$, and $F_t, G_t \in \mathcal{B}$ (by choosing for instance $F_t = E(u', v') \cap E_t(u, v)$).

It follows that

$$\int_{E(u',v')} h\,dP = \int_{F_t} h\,dP + \int_{G_t} h\,dP \leq \int_{E_t(u,v)} h_t\,dP + ||h - h_t||_1 + \int_{D_t} h\,dP$$

and since $||h - h_t||_1 \leq ||h - h_t||_2$, the RHS becomes arbitrarily small. This implies $P(E(u', v')) = 0$ and then the claim. □

**Definition 10** A family $\mathcal{M}$ of $\mathcal{B}$-measurable functions is called uniformly integrable (with respect to $P$), if for every $\varepsilon > 0$ an $N > 0$ exists such that for $E(f) = \{w : f(w) \geq N\}$

$$\int_{E(f)} |f|\,dP < \varepsilon \quad (f \in \mathcal{M}).$$

Clearly for $\mathcal{M}$ uniformly integrable uniform boundedness of the norms $||f||_1$ $(f \in \mathcal{M})$, follows: $N + \varepsilon$ is a bound.

**Lemma 39** *For every $t > 0$ the definition*

$$g_t(w) = \log \frac{P(w_{-t}, \ldots, w_0)}{P(w_{-t}, \ldots, w_0, w_1)} \tag{3.2.37}$$

*gives P-almost everywhere a $\mathcal{B}(-t, 1)$-measurable function $g_t \geq 0$. The families $\mathcal{M}_1 = \{g_t : t = 1, 2, \ldots\}$, $\mathcal{M}_2 = \{g_t^2 : t = 1, 2, \ldots\}$ are uniformly integrable. In particular the norms $||g_t||_1$, $||g_t||_2$ are bounded.*

*Proof* Since $(w_{-t}, \ldots, w_0, w_1) \subseteq (w_{-t}, \ldots, w_0)$ the first statements of the lemma are obvious. Let now

$$E_t^q(u, v) = \{w : u \leq g_t^q(w) < v\} \quad (u < v, q \geq 1),$$

then $E_t^q(u, v) \in \mathcal{B}(-t, 1)$. For every $w \in \Omega$ we use the abbreviations

$$\overline{w} = (w_{-t}, \ldots, w_0, w_1) \quad \widetilde{w} = (w_{-t}, \ldots, w_0). \tag{3.2.38}$$

Then we have for $w \in E_t^q(u, v)$, $u > 0$

$$\sqrt[q]{u} \leq \log \frac{P(\widetilde{w})}{P(\overline{w})} < \sqrt[q]{v},$$

$$P(\overline{w}) \leq e^{\sqrt[q]{u}} P(\widetilde{w}).$$

Therefore

$$\int\limits_{\overline{w}} g_t \, \mathrm{d}P < \sqrt[q]{v} e^{-\sqrt[q]{u}} P(\widetilde{w})$$

and

$$\int\limits_{\widetilde{w} \cap E_t^q(u,v)} g_t \, \mathrm{d}P \leq a \sqrt[q]{v} e^{-\sqrt[q]{u}} P(\widetilde{w}),$$

because at most $a$ of the $\overline{w}$ give the same $\widetilde{w}$. Summation over $\widetilde{w}$ gives

$$\int\limits_{E_t^q(u,v)} g_t \, \mathrm{d}P \leq a \sqrt[q]{v} e^{-\sqrt[q]{u}}.$$

Writing $E_t^q(u) = \bigcup_{v>u} E_t^q(u, v)$ one gets the partition

$$E_t^q(u) = \sum_{k=0}^{\infty} E_t^q(u+k, u+k+1)$$

and therefore

$$\int\limits_{E_t^q(u)} g_t \, \mathrm{d}P \leq a \sum_{k=0}^{\infty} \sqrt[q]{u+k+1} e^{-\sqrt[q]{u+k}}.$$

From the convergence of the series $\sum_{k=0}^{\infty}(k+1)^q e^{-\sqrt{k}}$ one can conclude the uniform integrability of the $g_t^q$ for fixed $q$. The rest of the lemma follows from the relation $||f^q||_1 = ||f||_q^q$. □

We explain now in an elementary fashion the concept *conditional expectation* for the cases in our interest.

**Lemma 40** *Let* $0 < t < s$ *and let* $f(w)$ *be a* $P$*-almost everywhere defined* $\mathcal{B}(-s, 1)$*-measurable function, then in the notation* (3.2.38) *by*

$$(\mathbb{E}_t f)(w) = (\mathbb{E}_t f)(\overline{w}) = \frac{1}{P(\overline{w})} \int\limits_{\overline{w}} f \, \mathrm{d}P \quad (P(\overline{w}) > 0) \tag{3.2.39}$$

*$P$-almost everywhere a $\mathcal{B}(-t, 1)$-measurable function is defined. It satisfies*

$$\int\limits_F \mathbb{E}_t f \, \mathrm{d}P = \int\limits_F f \, \mathrm{d}P \quad F \in \mathcal{B}(-t, 1). \tag{3.2.40}$$

*Every $\mathcal{B}(-t, 1)$-measurable function $g$ satisfying*

$$\int_F \mathrm{d}gP = \int_F f\,\mathrm{d}P \quad (F \in \mathcal{B}(-t, 1)), \tag{3.2.41}$$

*equals $P$-almost everywhere $\mathbb{E}_t f$.*

*Proof* It suffices to consider (3.2.40), (3.2.41) for $F = \overline{w}$ with $P(\overline{w}) > 0$ and everything is obvious. □

$\mathbb{E}_t f$ is thus defined for all functions $f$, which are $P$-almost everywhere defined and are measurable for any $\mathcal{B}(-s, 1)$. These functions form a linear space $\mathcal{L} \subseteq \mathcal{L}_P^q$. The most important properties of the map $\mathbb{E}_t$ shall be established next.

**Theorem 68** $\mathbb{E}_t : \mathcal{L} \to \mathcal{L}_t \subseteq \mathcal{L}_P^1$ *is linear and here $\mathcal{L}_t$ is the linear space of $\mathcal{B}(-t, 1)$-measurable $P$-almost everywhere defined functions. $\mathbb{E}_t f = f$ is equivalent to $f \in \mathcal{L}_t$. The following equalities and inequalities hold*

(i) $\mathbb{E}_t\mathbb{E}_s = \mathbb{E}_t \quad (t \leq s)$

(ii) $\mathbb{E}_t f \leq \mathbb{E}_t g \quad (f \leq g)$

(iii) $||\mathbb{E}_t f||_1 \leq ||f||_1$

(iv) $\mathbb{E}_t(\log f) \leq \log(\mathbb{E}_t f) \quad (f > 0,$ *$P$-almost everywhere*$)$

*Proof* Only (iii) and (iv) are non-trivial. Both inequalities follow from the concavity of the functions $-|x|$ and $\log x$. We only carry out the proof for (iv).

Let $f$ be $\mathcal{B}(-s, 1)$-measurable. We can assume $s > t$. Let $\overline{w}$ be an atom of $\mathcal{B}(-t, 1)$ and let $\widetilde{w}$ run through the atoms of $\mathcal{B}(-s, 1)$ into which $\overline{w}$ can be partitioned if $P(\overline{w}) > 0$ and we define $\alpha(\widetilde{w}) = \frac{P(\widetilde{w})}{P(\overline{w})}$, then $\alpha(\widetilde{w}) \geq 0$, $\sum_{\widetilde{w}} \alpha(\widetilde{w}) = 1$, and

$$\begin{aligned}(\mathbb{E}_t(\log f))(\overline{w}) &= \frac{1}{P(\overline{w})}\int_{\overline{w}} \log f\,\mathrm{d}P = \frac{1}{P(\overline{w})}\sum_{\widetilde{w}}(\log f(\widetilde{w}))P(\widetilde{w}) \\ &= \sum_{\widetilde{w}} \alpha(\widetilde{w}) \log f(\widetilde{w}) \leq \log\left(\sum_{\widetilde{w}} \alpha(\widetilde{w}) f(\widetilde{w})\right) \\ &= (\log \mathbb{E}_t f)(\overline{w}).\end{aligned}$$

□

**Theorem 69** *We have*

$$0 \leq \mathbb{E}_t g_s \leq g_t \quad (t \leq s). \tag{3.2.42}$$

*One can find a sequence $(u_t)$ of functions with the properties*

*(i)* $u_t$ *is $\mathcal{B}(-t, 1)$-measurable;*

*(ii)* $\mathbb{E}_t u_s = u_t \ (t \leq s)$;

*(iii)* $0 \le u_t \le g_t$;
*(iv)* $\lim_t ||g_t - u_t||_2 = 0$.

*Proof* $v_s(w) = \frac{P(w_{-s},\dots,w_0)}{P(w_{-s},\dots,w_0,w_1)}$ is a $P$-a.e. defined $\mathcal{B}(-s, 1)$-measurable function. We show first

$$\mathbb{E}_t v_s \le v_t \quad (t \le s). \tag{3.2.43}$$

For this let $\overline{w}$ be an atom of $\mathcal{B}(-t, 1)$ with $P(\overline{w}) > 0$. Every atom of $\mathcal{B}(-s, 1)$ contained in $\overline{w} = (w_{-t}, \dots, w_1)$ has the form

$$(w_{-s}, \dots, w_{-t-1}, w_{-t}, \dots, w_1) = (w_{-s}, \dots, w_{-t-1}) \cap (w_{-t}, \dots, w_1) = \widetilde{w} \cap \overline{w}$$

with a uniquely determined atom $\widetilde{w}$ of $\mathcal{B}(-s, -t-1)$. If $\sum'_{\widetilde{w}}$ sums only through those $\widetilde{w}$ for which $P(\widetilde{w} \cap \overline{w}) > 0$, then

$$\begin{aligned}
(\mathbb{E}_t v_s)(\overline{w}) &= \frac{1}{P(\overline{w})} \sum_{\widetilde{w}} \frac{P(w_{-s}, \dots, w_0)}{P(\widetilde{w} \cap \overline{w})} P(\widetilde{w} \cap \overline{w}) \\
&= \frac{1}{P(\overline{w})} \sum_{\widetilde{w}}{}' P(\widetilde{w}_1 \cap (w_{-t}, \dots, w_0)) \\
&\le \frac{1}{P(\overline{w})} \sum_{\widetilde{w}} P(\widetilde{w} \cap (w_{-t}, \dots, w_0)) \\
&= \frac{P(w_{-t}, \dots, w_0)}{P(\overline{w})} = v_t(\overline{w}),
\end{aligned}$$

because $\bigcup_{\widetilde{w}} \widetilde{w} = \Omega$.

Now it follows from (iv) in Theorem 68 that

$$(\mathbb{E}_t g_s)(\overline{w}) = (\mathbb{E}_t \log v_s)(\overline{w}) \le (\log((\mathbb{E}_t v_s)(\overline{w})) \le \log v_t(\overline{w}) = g_t(\overline{w}),$$

because $\log x$ is monotone.

By (ii) in Theorem 68 and $g_s \ge 0$ we get $\mathbb{E}_t g_s \ge 0$ and (3.2.42) is proved. This implies that $(\mathbb{E}_t g_s)_{s=t+1}^{\infty}$ is for every $t$ a monotone decreasing sequence of nonnegative $\mathcal{B}(-t, 1)$-measurable functions:

$$\mathbb{E}_t g_{s+1} = \mathbb{E}_t \mathbb{E}_s g_{s+1} \le \mathbb{E}_t g_s.$$

Defining

$$u_t = \lim_{s \to \infty} \mathbb{E}_t g_s \tag{3.2.44}$$

(the limit holds for both, convergence a.e. and in $\mathcal{L}^1$-norm) (i) and (ii) follow immediately. Now we also have

$$\begin{aligned}||g_t - u_t||_1 &= \int g_t \, dP - \int u_t \, dP \\ &= \int g_t \, dP - \lim_{s\to\infty} \int \mathbb{E}_t g_s \, dP \\ &= \int g_t \, dP - \lim_{s\to\infty} \int g_s \, dP\end{aligned}$$

by (3.2.40).

For $t \to \infty$ we get at first

$$\lim_t ||g_t - u_t||_1 = 0.$$

(iii) implies $0 \le g_t - u_t \le g_t$. Therefore, not only the $g_t^2$ but also the $(g_t - u_t)^2$ form a uniformly integrable family. For arbitrary $\varepsilon > 0$ we get for sufficiently large $K > 0$

$$\begin{aligned}||g_t - u_t||_2^2 &= \int |g_t - u_t|^2 \, dP \\ &= \int\limits_{|g_t - u_t| \ge K} |g_t - u_t|^2 \, dP + \int\limits_{|g_t - u_t| < \varepsilon} |g_t - u_t|^2 \, dP \\ &\quad + \int\limits_{\varepsilon \le |g_t - u_t| < K} |g_t - u_t|^2 \, dP \\ &< \varepsilon + \varepsilon^2 + K^2 P(|g_t - u_t| \ge \varepsilon) \\ &\le \varepsilon + \varepsilon^2 + \frac{K^2}{\varepsilon} ||g_t - u_t||_1,\end{aligned}$$

which implies (iv).

The relation (i) follows by using (iii) in Theorem 68:

$$\mathbb{E}_t u_s = \mathbb{E}_t \left( \overline{\lim_{k\to\infty}} \, E_s g_k \right) = \lim_k E_t E_s g_r = \lim_k E_t \cdot g_k = u_t,$$

where the limits are to be taken in the $\mathcal{L}_P^1$-norm. □

We use now that $\mathcal{B}(-t, 1)$-measurable functions take only finitely many values and certainly belong to $\mathcal{L}_P^2$ (as a substitute of Doob's martingale convergence theorem).

**Theorem 70** *If the sequence of functions* $(u_t)$ *satisfies*

$$\mathbb{E}_t u_s = u_t \quad (t \le s),$$

*then there exists exactly one* $u \in \mathcal{L}_P^2$ *with*

$$\lim_t ||u_t - u||_2 = 0,$$

*if the sequence of norms* $||u_t||_2$ *is bounded.*

*Proof* For $t \leq s$

$$(u_s, u_t) = \sum_{\widetilde{w}} \int_{\overline{w}} u_s u_t \,\mathrm{d}P = \sum_{\overline{w}} u_t(\overline{w}) \int_{\overline{w}} u_s \,\mathrm{d}P,$$

because $u_t$ has on the atom $\overline{w}$ of $\mathcal{B}(-t, 1)$ the constant value $u_t(\overline{w})$,

$$= \sum_{\overline{w}} u_t(\overline{w}) \int_{\overline{w}} E_t u_s \,\mathrm{d}P = \sum_{\overline{w}} u_t(\overline{w}) \int_{\overline{w}} u_t \,\mathrm{d}P = \int u_t^2 \,\mathrm{d}P = (u_t, u_t).$$

Applying this result for $s + 1$ yields for $w_s = u_{s+1} - u_s$, $s > 0$, and $w_0 = u_1$

$$(w_s, u_t) = 0 \quad (t \leq s),$$

and therefore

$$(w_s, w_t) = 0 \quad (s \neq t).$$

We thus get a representation

$$u_t = w_0 + \cdots + w_{t-1}$$

with pairwise orthogonal summands.

For $s < t$ one thus gets

$$||u_s - u_t||_2^2 = \left|\left|\sum_{k=t}^{s-1} w_k\right|\right|_2^2 = \sum_{k=t}^{s-1} ||w_k||_2^2,$$

that is, $\lim_{s,t\to\infty} ||u_s - u_t||_2 = 0$ and therefore convergence of $(u_t)$ in the $\mathcal{L}_P^2$-norm holds, if $||u_t||_2^2 = \sum_{k=0}^{t-1} ||w_k||_2^2$ remains bounded. □

**Theorem 71** *The sequence* $(g_t)$ *is norm-convergent in* $\mathcal{L}_P^2$*: a* $g \in \mathcal{L}_P^2$ *exists with*

$$\lim_t ||g_t - g||_2 = 0.$$

*Proof* By Theorem 69 it suffices to establish the convergence of $(u_t)$ in $\mathcal{L}_P^2$. By Theorem 70 sufficient for this is to show the boundedness of the norms $||u_t||_2$.

Since $0 \le u_t < g_t$ clearly also $||u_t||_2^2 \le ||g_t||_2^2$ and the norms $||g_t||_2$ are bounded by Lemma 39. □

*Proof of Theorem 67* We give first some general comments. Given $u \le v$ and a function $\varphi(y)$ $0 \le y \le 1$, then

$$f(w) = \varphi(P(w_u, \dots, w_v)) = \varphi(P(\{\eta : \eta_u = w_u, \dots, \eta_v = w_v\}))$$

is $\mathcal{B}(u, v)$-measurable. For arbitrary $k \in \mathbb{Z}$

$$\begin{aligned}(A^k f)(w) &= f(T^k w) = \varphi(P(\{\eta : \eta_u = (T^k w)_u, \dots, \eta_v = (T^k w)_v\})) \\ &= \varphi(P(\{\eta : \eta_u = w_{u+k}, \dots, \eta_v = w_{v+k}\})).\end{aligned}$$

By the stationarity of the source we can continue with

$$= \varphi(P(\{\eta : \eta_{u+k} = w_{u+k}, \dots, \eta_{v+k} = w_{v+k}\})) = \varphi(P\{w_{u+k}, \dots, w_{v+k}\}).$$

Applying these considerations to the two summands at the RHS of

$$g_t(w) = \log P(w_{-t}, \dots, w_0) - \log P(w_{-t}, \dots, w_0, w_1),$$

this (with the convention $P(w_1, \dots, w_k) = 1$ for $k = 0$)

$$\begin{aligned}h_t(w) &= -\frac{1}{t} \sum_{k=1}^{t} \log \frac{P(w_1, \dots, w_{k-1}, w_k)}{P(w_1, \dots, w_{k-1})} \\ &= \frac{1}{t} \sum_{k=1}^{t} \log \frac{P(w_1, \dots, w_{k-1})}{P(w_1, \dots, w_{k-1}, w_k)} \\ &= \frac{1}{t} \sum_{k=1}^{t} A^{k-1} g_{k-2}(w) \\ &= \frac{1}{t} \sum_{k=0}^{t-1} A^k g_{k-1}(w),\end{aligned}$$

here with the convention $g_{-1}(w) = -\log P(w_1)$.

By Corollary 14 and Theorem 69 the desired Theorem 67 follows. □

*Remark* Theorem 67 can be generalized as follows. If the statement is true for some source—with the limit function $h$—and is $Q$ with respect to $P$ absolutely continuous (that is, $Q(E) = 0$, if $P(E) = 0$) then the statement holds also for source $Q$ with the same limit function $h$ (see Jacobs [19]).

### *3.2.5 Ergodic Sources*

**Definition 11** The stationary source $(\Omega, \mathcal{B}, P, T)$ is ergodic if there is no partition $\Omega = E + F$ with

$$T^n E = E, \quad T^n F = F \tag{3.2.45}$$

and $P(E)P(F) > 0$.

**Definition 12** $P$ is called mixing with respect to $T$ (or $T$ is called mixing relative to $P$), if

$$\lim P(T^t E) \cap F) = P(E)P(F) \quad (E, F \in \mathcal{B}). \tag{3.2.46}$$

**Theorem 72** *(i) $P$ is mixing implies $P$ ergodic;*
*(ii) $P$ is ergodic if and only if every $\mathcal{B}$-measurable function $f \in \mathcal{L}_P^2$ satisfying*

$$f(T^n w) = f(w) \quad (P\text{-a.e. }, n \in \mathbb{Z}), \tag{3.2.47}$$

*is constant $P$-a.e.;*
*(iii) If $P$ is ergodic and $f \in \mathcal{L}_P^2$ satisfies (3.2.47), then*

$$f(w) = \int f \, dP \quad (P\text{-a.e.}). \tag{3.2.48}$$

Often a characterization using the Ergodic Theorem in $\mathcal{L}_P^2$ is more useful.

**Theorem 73** *Let $\mathcal{S} \subseteq \mathcal{B}$, let $\mathcal{K} \subseteq \mathcal{B}$ be $|\cdot|$-dense in $\mathcal{B}$ and every $E \in \mathcal{K}$ be representable as disjoint union of sets from $\mathcal{S}$, then $P$ is ergodic exactly if*

$$\lim_t \left\| \frac{1}{t} \sum_{k=0}^{t-1} A^k 1_E - P(E) \right\|_2 = 0 \quad (E \in \mathcal{S}). \tag{3.2.49}$$

**Example** Sources with independent symbols are ergodic and even mixing.

**Example** Markov sources satisfy

(i) The PD $P$ constructed via $W$ is ergodic exactly if for each $x, y$ with $P_x \cdot P_y > 0$ there exists a $t > 0$ with

$$W^t(y|x)W^t(x|y) > 0.$$

(ii) In this case $P$ is even mixing.

A proof of (i) and a proof of (ii) via the theory of multiple Markov processes can be found in Doob [7].

**Theorem 74** *For an ergodic source* $(\Omega, \mathcal{B}, P, T)$

$$\lim_{t\to\infty} ||h_t - H||_2 = 0.$$

*Proof* By McMillan's Theorem 67 $h_t$ converges to a constant and by Theorem 72 above the constant must be $\int h \,\mathrm{d}P = H$. □

**Corollary 16** (to Theorem 73) *Under the assumption of Theorem* 73 $P$ *is ergodic exactly if*

$$\lim_t \frac{1}{t}\sum_{k=0}^{t-1} P(T^{-k}E \cap F) = P(E)P(F) \quad (E, F \in \mathcal{S}). \tag{3.2.50}$$

*Proof* Equation (3.2.50) is equivalent to

$$\left(\frac{1}{t}\sum_k A^k 1_E, 1_F\right) \to P(E)P(F).$$

If $P$ is ergodic, then $\frac{1}{t}\sum A^k 1_E \to \text{const.} = P(E)$, also $\left(\frac{1}{t}\sum A^k 1_E, 1_F\right) \to \int P(E)1_F \,\mathrm{d}P = P(E)P(F)$, that is, (3.2.50). Conversely, if (3.2.50) holds, then for $h = \lim \frac{1}{t}\sum A^k 1_E$

$$\int h 1_F \,\mathrm{d}P = P(E)P(F) \quad (F \in \mathcal{S}). \tag{3.2.51}$$

Since the indicator function of an arbitrary set $F \in \mathcal{B}$ can be approximated by indicator functions of sets from $\mathcal{S}$, (3.2.51) holds for arbitrary $F \in \mathcal{B}$. This, however, implies $h = P(E)$ ($P$-a.e.), and consequently $P$ is ergodic. □

## 3.3 Lecture on Stationary Channels

### 3.3.1 A Concept of Channels

Starting with two alphabets $\mathcal{X} = \{1, \dots, a\}$ and $\mathcal{X}' = \{1, \dots, a'\}$ we can construct as explained in Sect. 3.2.1 product spaces $\Omega$ and $\Omega'$ as ground sets, algebras $\mathcal{B}(s,t)$, $\mathcal{B}'(s,t)$ $(s \le t)$, and $\sigma$-algebras $\mathcal{B}$ and $\mathcal{B}'$. The shift on $\Omega$ and on $\Omega'$ we denote both times by $T$.

Similarly, from the alphabet $\widetilde{\mathcal{X}} = \mathcal{X} \times \mathcal{X}' = \{(x, x') : 1 \le x \le a, 1 \le x' \le a'\}$ one gets the ground set $\widetilde{\Omega}$, algebras $\widetilde{\mathcal{B}}(s,t)$, a $\sigma$-algebra $\widetilde{\mathcal{B}}$ and again a shift operator $T$. Obviously $\widetilde{\Omega} = \Omega \times \Omega' = \{(w, w') : w \in \Omega, w' \in \Omega'\}$, $T^\nu(w, w') = (T^\nu w, T^\nu w')$, and $\widetilde{\mathcal{B}}(s,t)$ is the product Borel field generated by $\mathcal{B}(s,t)$ and $\mathcal{B}'(s,t)$.

More precisely: an atom (text) from $\widetilde{\mathcal{B}}(s,t)$ has the forms

$$\begin{aligned}
&((w_s, w'_s), \dots, (w_t, w'_t))\\
&\quad = \{(\eta, \eta') : \eta \in \Omega, \eta' \in \Omega', \eta_s = w_s, \dots, \eta_t = w_t; \eta'_s = w'_s, \dots, \eta'_t = w'_t\}\\
&\quad = \{(\eta, \eta') : \eta \in (w_s, \dots, w_t), \eta' \in (w'_s, \dots, w'_t)\}\\
&\quad = (w_s, \dots, w_t) \times (w'_s, \dots w'_t),
\end{aligned}$$

$\mathcal{B}(s,t)$ and $\mathcal{B}'(s,t)$ can be viewed as subfields of $\widetilde{\mathcal{B}}(s,t)$, which they generate together, etc.

We define $\widetilde{\mathcal{K}} = \bigcup_{s \le t} \widetilde{\mathcal{B}}(s,t)$.

**Definition 13** A function $P : \Omega \times \mathcal{B}' \to \mathbb{R}^+$ is called a channel (from $\Omega$ to $\Omega'$), if

(i) $P(w, E')$ is a PD on $\mathcal{B}'$ for every fixed $w \in \Omega$.
(ii) $P(w, E')$ is a $\mathcal{B}$-measurable function on $\Omega$ for every fixed $E' \in \mathcal{B}'$.

**Definition 14** The channel is non-anticipatory, if $P(w, E')$ is a $\mathcal{B}(-\infty, t)$-measurable function on $\Omega$ for every fixed $E' \in \mathcal{B}'(s,t)$ $(s \le t)$.

**Definition 15** The channel has finite duration of memory $m$, if $P(w, E')$ is a $\mathcal{B}(s - m, \infty)$-measurable function for every fixed $E' \in \mathcal{B}'(s,t)$.

*Supposition from now on (except in* Sect. 3.3.3)*: We consider non-anticipatory channels with finite memory m.*

We can, therefore, write for $E' \in \mathcal{B}'(s,t)$

$$P(w, E') = P(w_{s-m}, \dots, w_t, E') = P(\overline{w}, E') = P(E, E'),$$

when $\overline{w} = (w_{s-m}, \dots, w_t) = E$.

A channel as a mathematical model for the transmission of messages from a sender to a receiver. The person serving the sender puts without interruption every second a symbol from $\mathcal{X}$ into the channel; the person serving the receiver picks up without interruption a symbol from $\mathcal{X}'$ in every second. The conditional probability for picking up in the time from $s$ to $t$ the string of symbols $w'_s, \dots, w'_t$, if in the time from $s-m$ to $t$ the string of symbols $w_{s-m}, \dots, w_s, \dots, w_t$ was put into the sender, is given by

$$P(w_{s-m}, \dots, w_s, \dots, w_t, (w'_s, \dots, w'_t)) = P(w, (w'_s, \dots, w'_t)),$$

for all $w \in (w_{s-m}, \dots, w_t)$.

The fact that here the statement concerns probability, reflects the practical experience, that there is noise between sender and receiver, which is viewed as being purely of statistical nature. The task to reach an understanding with arbitrarily small error probability in spite of the noise we make precise with the following concept.

**Definition 16** A $\lambda$-code of length $N$ for the time $[s, t]$ is a set of pairs $\{(E_\nu, E'_\nu) : 1 \le \nu \le N\}$ with the properties:

(i) $E_\nu$ is an atom (text) from $\mathcal{B}(s-m, t)$, $E'_\nu \in \mathcal{B}(s, t)$ (and therefore we can write $P(w, E'_\nu) = P(E_\nu, E'_\nu)$ $(w \in E_\nu)$);
(ii) $E'_\mu \cap E'_\nu = 0$ $(\mu \neq \nu)$;
(iii) $P(E_\nu, E'_\nu) > 1 - \lambda$.

For $\lambda < \frac{1}{2}$ (ii) and (iii) imply that the $E_\nu$ are different. It can be assumed that $\bigcup_\nu E'_\nu = \Omega'$.

Is there a code for the time $[s, t]$, then the person serving the sender will restrict himself to one of the texts $E_\nu$. If the receiver gets a text belonging to $E'_\mu$, then he is in error with probability at most $\lambda > 0$, if he assumes that $E_\mu$ was given to the sender. Therefore the receiver can be made known approximately error-free, which one of the texts $E_\nu$ was put into the sender. When this procedure can be repeated arbitrary often, that is, also for all time intervals $[s_k, t_k]$ with $t \le s_1 = m$, $t_k \le s_{k+1} - m$ $(k = 1, 2, \dots)$ a code exists, then one can by the codeword-scheme (Anton, Berta,...) transmit arbitrary long messages: the $E_\nu$ serve as codewords. Furthermore, it is desirable that these codes differ only by a time translation.

**Definition 17** The channel $P(w, E')$ is stationary if

$$P(T^\nu w, T^\nu E') = P(w, E') \quad (r \in \mathbb{Z}, w \in \Omega, E' \in \mathcal{B}'). \tag{3.3.1}$$

*Remark* If $P(w, E')$ is non-anticipatory and has (duration of) memory $m$, then $P_\nu(w, E') = P(T^\nu w, T^\nu E')$ is also a non-anticipatory channel with memory $m$. Furthermore, if $P^k(w, E')$ $(k = 1, 2, \dots, \eta)$ are non-anticipatory channels with memory $m$, then so is $P(w, E') = \sum_k \alpha_k P^k(w, E')$, if $(\alpha_1, \dots, \alpha_n)$ is a PD. Also, stationarity of the $P^k$ is inherited by $P$.

Continuity considerations as made in Sect. 3.2.1 show that the relation

$$P(T^\nu w, T^\nu(w'_s, \dots, w'_t)) = P(w, (w'_s, \dots, w'_t) \quad (s \le t, \nu \in \mathbb{Z}, w \in \Omega, w' \in \Omega'), \tag{3.3.2}$$

is sufficient for the stationarity of a channel $P(w, E')$.

A simple observation is summarized next.

**Lemma 41** *For a stationary channel $P(w, E')$ and a code $\{(E_\nu, E'_\nu) : 1 \le \nu \le N\}$ for $[s, t]$ with error probability $\lambda > 0$ the shifted code $\{(T^{-\nu} E_\nu, T^{-\nu} E'_\nu) : 1 \le \nu \le N\}$ for $[s+\nu, t+\nu]$ has the same length and the same error probability.*

*Supposition from now on: Only stationary channels will be considered.*

We study only questions concerning the existence and length of codes.

**Theorem 75** *For every stationary PD P on $\mathcal{B}$ there exists exactly one stationary $\widetilde{P}$ on $\widetilde{\mathcal{B}}$ and exactly one stationary PD $P'$ on $\mathcal{B}'$ such that*

*(i) Considering $\mathcal{B}$ and $\mathcal{B}'$ as sub-σ-algebras of $\widetilde{\mathcal{B}}$ one gets P resp. $P'$ by restriction of $\widetilde{P}$ on $\mathcal{B}$ resp. $\mathcal{B}'$.*

*(ii) For $E \in \mathcal{B}$ and $E' \in \mathcal{B}'$ one has*

$$\widetilde{P}(E \times E') = \int_E P(\mathrm{d}w)P(w, E'). \tag{3.3.3}$$

*(iii) For $\widetilde{E} \in \widetilde{\mathcal{B}}$ and all $w \in \Omega$ $\widetilde{E}_w = \{w' : (w, w') \in \widetilde{E}\} \in \mathcal{B}'$. $P(w, \widetilde{E}_w)$ is a $\mathcal{B}$-measurable function in $w$ and*

$$P(\widetilde{E}) = \int P(\mathrm{d}w)P(w, \widetilde{E}_w). \tag{3.3.4}$$

*Proof* (i) and (ii) are consequences of (3.3.3). With (3.3.3) we can calculate $\widetilde{P}(\widetilde{E})$ for $\widetilde{E} \in \widetilde{\mathcal{K}} = \bigcup_{s \leq t} \widetilde{\mathcal{B}}(s, t)$ uniquely (and actually for each single $\widetilde{E}$ in finite steps). We have $T^\nu \widetilde{\mathcal{K}} = \widetilde{\mathcal{K}}$ and

$$\widetilde{P}(T^\nu \widetilde{E}) = \widetilde{P}(\widetilde{E}) \quad (\widetilde{E} \in \widetilde{\mathcal{K}}). \tag{3.3.5}$$

With the Theorem of Kolmogorov $\widetilde{P}$ can be extended in a unique way to a PD $\widetilde{P}$ on $\widetilde{\mathcal{B}}$.

Similarly as in Sect. 3.2.1, one shows: from (3.3.5) follows the stationarity of $P$ and from there the stationarity of $P'$. For $\widetilde{E} \in \widetilde{\mathcal{K}}$ (3.3.4) is readily verified. Now it is clear that the system of those $\widetilde{E} \in \widetilde{\mathcal{B}}$, for which $P(w, \widetilde{E}_w)$ is $\mathcal{B}$-measurable and satisfies (3.3.4), is closed under monotone countable unions and intersections, since it includes $\widetilde{\mathcal{K}}$, it is equal to $\widetilde{\mathcal{B}}$ (cf. Halmos [17, p. 26], Theorem B). □

### 3.3.2 Methods for the Construction of Channels

We present now some special channels, which have been treated in the literature.

*Example 1* Let $W = (W(y|x))_{x \in \mathcal{X}, y \in \mathcal{X}'}$ be a stochastic matrix with $\sum_y W(y|x) = 1$. We construct a channel $P(w, E')$ with memory $m = 0$, by specifying only the values

$$P(w_s, \dots, w_t, (w'_s, \dots, w'_t)) = \prod_{k=s}^{t} W(w'_k | w_k) \quad (s \leq t). \tag{3.3.6}$$

Fixing here $s, t$ and $w_s, \dots, w_t$ one obtains a PD $P(w_s, \dots, w_t, E')$ $(E' \in \mathcal{B}'(s, t))$.

Fixing $w \in \Omega$ one obtains for arbitrary $s \leq t$ $w_s, \dots, w_t$ and a PD on $\mathcal{B}'(s, t)$, and verifies the compatibility conditions (3.2.14) in Sect. 3.2.2. Thus, one obtains by the Theorem of Kolmogorov a PD $P(w, E')$ on $\mathcal{B}'$.

Measurability in $w$ for fixed $E'$ has to be shown only for $E' \in \bigcup_{s\le t} \mathcal{B}'(s,t)$, because the family of $E'$ for which measurability holds, obviously is closed under monotone countable unions and intersections. For $E' \in \bigcup_{s\le t} \mathcal{B}'(s,t)$ however, measurability is trivial. One sees from Sect. 3.2.1, that the channel obtained is stationary.

*Example 2* By the same procedure one obtains the stationary channel with memory $m$, considered by Wolfowitz, if one uses a matrix $W = (W(y|x^m))_{x^m\in\mathcal{X}^m, y\in\mathcal{X}'}$ and defines

$$P(w_{s-m},\dots,w_s,\dots,w_t,(w'_s,\dots,w'_t)) = \prod_{k=s}^{t} W(w'_k|w_{k-m},\dots,w_k).$$

### 3.3.3 Ergodic Channels

In this section, we consider general stationary channels *without assumptions on anticipation and memory.*

**Definition 18** A stationary channel $P(w,E')$ is ergodic, if for every ergodic $P$ the previously constructed $\widetilde{P}$ and $P'$ are ergodic.

*Remark* Ergodicity of $\widetilde{P}$ implies it for $P'$. Indeed, a partition $\Omega' = E' + F'$ with $T^rE' = E', T^rF' = F'$ $(r\in\mathbb{Z})$, gives via $\widetilde{E} = \Omega\times E'$, $\widetilde{F} = \Omega\times F'$ a partition $\widetilde{\Omega} = \widetilde{E} + \widetilde{F}$ with $T^r\widetilde{E} = \widetilde{E}$, $T^r\widetilde{F} = \widetilde{F}$ $(r\in\mathbb{Z})$ and $P'(E')P'(F') > 0$ implies $\widetilde{P}(\widetilde{E}) = P'(E')$, $\widetilde{P}(\widetilde{F}) = P'(F')$, and $\widetilde{P}(\widetilde{E})\widetilde{P}(\widetilde{F}) > 0$, a contradiction.

**Theorem 76** *A stationary channel $P(w,E')$ is certainly ergodic, when it satisfies*

**Condition (M)** *For arbitrary $w\in\Omega$ and $E', F' \in \mathcal{K}' = \bigcup_{s\le t}\mathcal{B}'(s,t)$*

$$\lim_{t\to\infty}(P(w,(T^{-t}E')\cap F') - P(w,T^{-t}E')P(w,F')) = 0.$$

*Remark* Comparison with definition (3.2.45) in Sect. 3.2.5 shows that (M) is just a certain mixing property of the (in general nonstationary) PD's $(\Omega',\mathcal{B}',P(w,\cdot))$ $(w\in\Omega)$.

*Proof* Let $P$ be ergodic. The ergodicity of $\widetilde{P}$ has to be shown. We apply Corollary 16 to Theorem 73 in Sect. 3.2.5. The family $\widetilde{\mathcal{S}}$ of all sets of the structure $E\times E'$ with $E\in\mathcal{B}$, $E'\in\mathcal{K}'$ has with respect to $\widetilde{\mathcal{B}}$ and $\widetilde{P}$ the property required there. For example, every set $R\in\widetilde{\mathcal{K}} = \bigcup_{s\le t}\mathcal{B}(s,t)$ can be represented as disjoint union of sets from $\widetilde{\mathcal{S}}$ and $\widetilde{\mathcal{K}}$ is metrically dense in $\widetilde{\mathcal{B}}$. It suffices therefore to prove relation (3.2.50) in Corollary 16 for sets $\widetilde{\mathcal{B}}$, $\widetilde{F}\in\widetilde{\mathcal{S}}$ and with $\widetilde{P}$ instead of $P$.

Let $\widetilde{E} = E\times E'$, $\widetilde{F} = F\times F'$ with $E, F\in\mathcal{B}$ and $E', F'\in\mathcal{K}'$. Then we have

$$\left|\frac{1}{t}\sum_{k=0}^{t-1}\widetilde{P}((T^{-k}\widetilde{E})\cap\widetilde{F}) - \widetilde{P}(\widetilde{E})\widetilde{P}(\widetilde{F})\right|$$

$$= \left| \frac{1}{t} \sum_k \widetilde{P}((T^{-k}\widetilde{E} \times T^{-k}\widetilde{E}') \cap (F \times F')) - \widetilde{P}(\widetilde{E})\widetilde{P}(\widetilde{F}) \right|$$

$$= \left| \frac{1}{t} \sum_k P((T^{-k}E \cap F) \times (T^{-k}E' \cap F')) - \widetilde{P}(\widetilde{E})\widetilde{P}(\widetilde{F}) \right|$$

$$= \left| \frac{1}{t} \sum_k \int 1_{T^{-k}E \cap F}(w) P(w, T^{-k}E' \cap F') P(\mathrm{d}w) - \widetilde{P}(\widetilde{E})\widetilde{P}(\widetilde{F}) \right|$$

$$= \left| \frac{1}{t} \sum_k \int 1_E(T^k w) 1_F(w) P(w, T^{-k}E' \cap F') P(\mathrm{d}w) - \widetilde{P}(\widetilde{E})\widetilde{P}(\widetilde{F}) \right|$$

$$\leq \left| \frac{1}{t} \sum_k \int 1_E(T^k w) 1_F(w) \left[ P(w, T^{-k}E' \cap F') - P(w, T^{-k}E') P(w, F') \right] P(\mathrm{d}w) \right|$$

$$+ \left| \frac{1}{t} \sum_k \int 1_E(T^k w) P(w, T^{-k}E') 1_F(w) P(w, F') P(\mathrm{d}w) \right.$$
$$\left. - \int 1_E(w) P(w, E') P(\mathrm{d}w) \int 1_F(w) P(w, F') P(\mathrm{d}w) \right|$$

Here the integral in the first expression, by condition (M) and the Theorem of Lebesque (the integral remains bounded), tends for $k \to \infty$ to 0. Also the first expression itself (as Césaro-average of a null sequence) goes to 0. It remains to verify the same for the second expression. It equals by the stationarity of the channel

$$\left| \frac{1}{t} \sum_k 1_E(T^k w) P(T^k w, E') 1_F(w) P(w, F') P(\mathrm{d}w) \right.$$
$$\left. - \int 1_E(w) P(w, E') P(\mathrm{d}w) \int 1_F(w) P(w, F) P(\mathrm{d}w) \right|$$
$$= \left| \frac{1}{t} \sum_k (A^k f, g) - \int f \, \mathrm{d}P \int g \, \mathrm{d}P \right| = \mathrm{d}_t$$

with $f(w) = 1_E(w)P(w, E')$, $g(w) = 1_F(w)P(w, F')$ and the general definition $(A^k h)(w) = h(A^k w)$ (of course $f, g \in \mathcal{L}_P^2$). By the Ergodic Theorem in $\mathcal{L}_P^2$ and (3.3.4) in Theorem 75 we get

$$\lim_{t \to \infty} \left\| \frac{1}{t} \sum_{k=0}^{t-1} A^k f - \int f \, \mathrm{d}P \right\|_2 = 0.$$

The constant $\int f\,\mathrm{d}P$ is considered there as element of $\mathcal{L}_P^2$. By the linearity and continuity of the inner product it follows that

$$\lim_{t\to\infty}\left|\frac{1}{t}\sum_k (A^k f, g) - \left(\int f\,\mathrm{d}P, g\right)\right|_2 = 0$$

and $\lim d_t = 0$. The ergodicity of $\widetilde{P}$ is proved. □

**Theorem 77** *A stationary channel $P(w, E')$ satisfies condition (M) and is therefore ergodic, if it satisfies the*

**Condition (U)** *There is a number $a > 0$ such that the following holds:*
*If $J, L$ are intervals with integers as end points, which have a distance at least $a$, and if $E' \in \mathcal{B}'(J)$, $F' \in \mathcal{B}'(L)$, then $P(w, E' \cap F') = P(w, E')P(w, F')$, which means that $\mathcal{B}'(J)$ and $\mathcal{B}'(L)$ are independent under all PD's $P(w)$ $(w \in \Omega)$.*

The proof is trivial.

Clearly, the channels constructed in Sect. 3.3.2 satisfy condition (U) and are therefore ergodic. The fact that (U) implies ergodicity was proved originally for channels with finite memory and without anticipation by Khinchin [28]. The proof given here is simpler (compare also Feinstein [10, pp. 99–102]).

## 3.4 Lecture on "Informational Capacity" and "Error Capacity" for Ergodic Channels

### 3.4.1 *Definition of the "Informational Capacity"*

For all sources $(\Omega, \mathcal{B}, P)$, $(\Omega', \mathcal{B}', P')$, $(\widetilde{\Omega}, \widetilde{\mathcal{B}}, \widetilde{P})$ the Theorem of McMillan holds. We state it now in a somewhat modified form:

**Theorem 78** *Let $P(w, E')$ be a stationary channel without anticipation and with finite memory $m + 1$ and let $P$ be a stationary source. Defining for $t \geq 1$*

$$h_t(w) = -\frac{1}{t}\log P(w_{-m}, \dots, w_0, \dots, w_t)$$

$$h'_t(w) = -\frac{1}{t}\log P(w'_1, \dots, w'_t)$$

$$\widetilde{h}_t(w) = -\frac{1}{t}\log P((w_{-m}, \dots, w_t) \times (w'_1, \dots, w'_t))$$

$$= -\frac{1}{t}\log P(w_{-m}, \dots, w_0, \dots, w_t)P(w_{-m}, \dots, w_t, (w'_1, \dots, w'_t))$$

*then these quantities are $P$-, resp. $P'$-, resp. $\widetilde{P}$-a.s. defined functions and there are a.e. defined $T$-invariant functions $h, h', \widetilde{h}$ with $0 \leq h \leq \log a$, $0 \leq \widetilde{h} \leq \log a'$,*

$0 \le \widetilde{h} \le \log(aa') = \log a + \log a'$ *and*

$$\int h \, \mathrm{d}P = H(P), \tag{3.4.1}$$

$$\int h' \, \mathrm{d}P' = H(P'), \tag{3.4.2}$$

$$\int \widetilde{h} \, \mathrm{d}\widetilde{P} = H(\widetilde{P}), \tag{3.4.3}$$

*and*

$$\lim_t ||h_t - h||_2 = 0, \tag{3.4.4}$$

$$\lim_t ||h'_t - h'||_2 = 0, \tag{3.4.5}$$

$$\lim_t ||\widetilde{h}_t - \widetilde{h}||_2 = 0. \tag{3.4.6}$$

*Proof* Theorem 67 in Sect. 3.2.2 (McMillan), gives (3.4.1), (3.4.4) by application to $h_t(T^{m+1}w)$ and using $\frac{t+m+1}{t} \to 1$ as $t \to \infty$. (3.4.2), (3.4.5) follow immediately again from Theorem 67. In order to prove (3.4.3), (3.4.6), we use

$$\widetilde{h}_t(\widetilde{w}) = -\frac{1}{t} \log \widetilde{P}((w_1, \ldots, w_t) \times (w'_1, \ldots, w'_t))$$
$$-\frac{1}{t} \log \frac{\widetilde{P}((w_{-m}, \ldots, w_t) \times (w'_1, \ldots, w'_t))}{\widetilde{P}((w_1, \ldots, w_t) \times (w'_1, \ldots, w'_t))}$$

and show that the functions

$$\widetilde{g}_t(\widetilde{w}) = \left( \log \frac{\widetilde{P}((w_{-m}, \ldots, w_t) \times (w'_1, \ldots, w'_t))}{\widetilde{P}((w_1, \ldots, w_t) \times (w'_1, \ldots, w'_t))} \right)^2,$$

defined $\widetilde{P}$-a.e., form a uniformly integrable, so especially norm bounded sequence. This can be done in the same way as in the proof of Lemma 39. Applying Theorem 67 to

$$-\frac{1}{t} \log \widetilde{P}((w_1, \ldots, w_t) \times (w'_1, \ldots, w'_t))$$

the claim follows. We note that $h$, $h'$, and $\widetilde{h}$ are the same functions as those which are obtained by direct applications of Theorem 67 to $P$, $P'$, and $\widetilde{P}$. □

Recall $\lim_{t \to \infty} f_t = f$ ($P$-stochastic), if for all $\lambda > 0$

$$\lim_t P(\{w : |f_t(w) - f(w)| > \lambda\}) = 0.$$

This is equivalent with

$$\lim_t P(\{w : |f_t(w) - f(w)| \leq \lambda\}) = 1.$$

If $f, f_t \in \mathcal{L}_P^2$, them $\lim_t ||f_t - f||_2 = 0$ implies $\lim_t f_t = f$ ($P$-stochastic) as can be seen from

$$||f_t - f||_2^2 \geq \lambda^2 P(\{w : |f_t(w) - f(w)| > \lambda\}) \quad (\lambda > 0).$$

**Corollary 17** *If $P$, $P'$, $\widetilde{P}$ are ergodic and have mean entropies $H(P)$, $H(P')$, $H(\widetilde{P})$ then*

$$\lim_t h_t = H(P) \quad (P\text{-stochastic}),$$
$$\lim_t h'_t = H(P') \quad (P\text{-stochastic}),$$
$$\lim_t \widetilde{h}_t = H(\widetilde{P}) \quad (P\text{-stochastic}).$$

**Definition 19** Let $C(P) = H(P) + H(P') - H(\widetilde{P})$, then

$$C = \sup_{P \text{ ergodic}} C(P)$$

is the (ergodic) informational capacity of the channel $P(w, E')$.

Recall that

$$\begin{aligned} C(P) &= \lim_t \frac{1}{t}(H(-m, t) + H'(1, t) - \widetilde{H}(-m, t; 1, t) \\ &= \lim_t \frac{1}{t}(H(p'_t) - H(P'_t|P_t) \\ &= \lim_t \frac{1}{t}(H(P_t) - H(P_t|P'_t) \\ &= H(P') - H(P'|P) = H(P) - H(P|P'), \end{aligned}$$

if $H'(1, t)$ is the entropy of $P'^t$ obtained by restricting $P'$ to $\mathcal{B}'(1, t)$, $\widetilde{H}(-m, t; 1, t)$ is the entropy of $\widetilde{P}^t$ obtained by restricting $\widetilde{P}$ to $\mathcal{B}(-m, t) \times \mathcal{B}'(1, t)$, and if $P_t$ denotes the restriction of $P$ to $\mathcal{B}(1, t)$ and the abbreviations $H(P|P') = \lim_t \frac{1}{t} H(P_t|P'_t)$, $H(P'|P) = \lim_t \frac{1}{t} H(P'_t|P_t)$ are used. $H(P_t) - H(P_t|P'_t)$ can be interpreted as the secret of source $P_t$ obtained by observation of $\mathcal{B}'(1, t)$, that is, as the amount of information gone through the channel in time $t$. We know already that $\log a \geq H(P_t) - H(P_t|P'_t) \geq 0$ and that $0 \leq C \leq \log a$. Then $C(P)$ stands for the "amount of information" gone in average in time unit through the channel. This motivates the name given for $C(P)$. However, an *operational justification* will be given after we determined the "Error Capacity" in Theorems 79 and 80 in the next section.

### 3.4.2 The Coding Theorem

**Theorem 79** *Let $P(w, E')$ be a stationary ergodic channel without anticipation, with finite memory $m + 1$ and the I-capacity $C$. To every $\lambda > 0$ there exists then a $t_0$ with the following properties: Is $t \geq t_0$, $s \in \mathbb{Z}$, then there are $N$ atoms $\overline{w}_\nu$ from $\mathcal{B}(s - m, s + t)$ and $N$ sets $E'_\nu \in \mathcal{B}'(s + 1, s + t)$ so that*

*(i)* $E'_\mu \cap E'_\nu = \emptyset \ (\mu \neq \nu)$,
*(ii)* $P(\overline{w}_\nu, E'_\nu) > 1 - \lambda$,
*(iii)* $N \geq e^{t(C-\lambda)}$.

*The system $\{(\overline{w}_\nu, E'_\nu) : 1 \leq \nu \leq N\}$ is also called $(N, \lambda)$-code.*

*Proof* If $\{(\overline{w}_\nu, E_\nu) : 1 \leq \nu \leq N\}$ is a $\lambda$-code for $s = 0$, then by stationarity of this channel $\{(T^{-s}\overline{w}_\nu, T^{-s}E'_\nu) : 1 \leq \nu \leq N\}$ is a $\lambda$-code for $s$. It therefore suffices to prove the theorem for $s = 0$. Let now $0 < \lambda \leq \frac{1}{2}$ be arbitrary otherwise and $\delta > 0$ at first arbitrary (later it will be chosen as $\delta = \frac{\lambda}{q}$). We consider an ergodic source $P$ with $C(P) > C - \frac{\lambda}{3}$. By our assumptions $P'$ and $\widetilde{P}$ are also ergodic. We form the sets

$$E(t) = \left\{ w : \left| \frac{1}{t} \log P(\overline{w}) + H(P) \right| < \delta \right\} \in \mathcal{B}(-m, t)$$

$$E'(t) = \left\{ w' : \left| \frac{1}{t} \log P'(\overline{w}') + H(P') \right| < \delta \right\} \in \mathcal{B}(1, t)$$

$$\widetilde{E}(t) = \left\{ w \times w' : \left| \frac{1}{t} \log \widetilde{P}(\overline{w} \times \overline{w}') + H(\widetilde{P}) \right| < \delta \right\}.$$

By Corollary 17 to Theorem 78

$$P(E(t)) \to 1 \text{ as } t \to \infty, \tag{3.4.7}$$

$$P'(E'(t)) \to 1 \text{ as } t \to \infty, \tag{3.4.8}$$

$$\widetilde{P}(\widetilde{E}(t)) \to 1 \text{ as } t \to \infty. \tag{3.4.9}$$

We can *view* $E(t)$, $E'(t)$ *as sets in* $\widetilde{\mathcal{B}}$ and replace $P, P'$ in (3.4.7), (3.4.8) by $\widetilde{P}$. Setting then

$$\widetilde{F}(t) = E(t) \cap E'(t) \cap \widetilde{E}(t)$$

it follows that

$$\widetilde{P}(\widetilde{F}(t)) \to 1 \quad \text{as } t \to \infty.$$

Defining $\widetilde{F}(t)_w = \{w' : (w, w') \in \widetilde{F}(t)\} = \widetilde{F}(t)_{\overline{w}}$, $P(w, \widetilde{F}(t)_w)$ is $\mathcal{B}$-measurable and now even dependent on $\overline{w}$. Simple considerations like those in the proof of Theorem 78 in Sect. 3.4.1 make it clear that for $G(t) \triangleq \{w : P(w, \widetilde{F}(t)_w) > 1 - \delta\}$ $(\in \mathcal{B}(-m, t))$, $P(G(t)) \to 1$ as $t \to \infty$.

$\delta$ will be specified below. We let now a $\lambda$-code of maximal length $N$ with

$$E'_\nu \subset \widetilde{F}(t)_{\overline{w}_\nu} \tag{3.4.10}$$

be given. We have to show that for sufficiently large $t$ always $N > e^{t(C-\lambda)}$ holds.

Let $E' = \sum_{\nu=1}^{N} E'_\mu$. For $\overline{w} \subset G(t)$, $\overline{w} \neq \overline{w}_\nu$ $(1 \leq \nu \leq N)$ we have

$$P(\overline{w}, E') \geq \frac{\lambda}{2} \tag{3.4.11}$$

if we have ensured that $\delta < \frac{\lambda}{2}$, because otherwise

$$P(\overline{w}, \widetilde{F}(t)_{\overline{w}}) - \widetilde{F}(t)_{\overline{w}} \cap E') \geq P(\overline{w}, \widetilde{F}(t)_{\overline{w}}) - P(\overline{w}, E') \geq 1 - \delta - \frac{\lambda}{2} > 1 - \lambda$$

and the $\lambda$-code could be extended by $(\overline{w}, \widetilde{F}(t)_{\overline{w}} - F(t)_{\overline{w}} \cap E')$ under the additional condition (3.4.10), which would contradict the assumed maximality of $N$. (3.4.11) of course also holds for $\overline{w} = \overline{w}_\nu$ $(1 \leq \nu \leq N)$ and one has therefore

$$P'(E') \geq \sum_{\overline{w} \subseteq G(t)} P(\overline{w})P(\overline{w}, E') \geq \frac{\lambda}{2} P(G(t)) > \frac{\lambda}{3} \tag{3.4.12}$$

for $t$ sufficiently large. On the other hand for $\overline{w} \times \overline{w}' \subset \widetilde{F}(t)$

$$-t\delta < \log \widetilde{P}(\overline{w} \times \overline{w}') + tH(\widetilde{P}) < t\delta,$$

$$-t\delta < \log P'(\overline{w}') + tH(P') < t\delta,$$

$$-t\delta < \log P(\overline{w}) + tH(P) < t\delta.$$

From here we conclude—by using $P(\overline{w} \times \overline{w}') = P(\overline{w})P(\overline{w}, \overline{w}')$—

$$-3t\delta < -\log \frac{P(\overline{w}, \overline{w}')}{P'(\overline{w}')} + tC(P) < 3t\delta,$$

$$\log \frac{P(\overline{w}, \overline{w}')}{P'(\overline{w}')} > t(C(P) - 3\delta).$$

This is larger than $t(C(P) - \frac{\lambda}{3} - \frac{\lambda}{3} = t\left(C - \frac{2}{3}\lambda\right)$ for $\delta = \frac{\lambda}{9}$ and one gets

$$P(\overline{w}, \overline{w}') > e^{t(C - \frac{2}{3}\lambda)} P'(\overline{w}')$$

and since $E'_\nu \subset \widetilde{F}(t)_{\overline{w}_\nu}$, that is,

$$\overline{w}_\nu \times E'_\nu \subset \widetilde{F}(t)$$

$$1 \geq P(\overline{w}_\nu, E'_\nu) > e^{t(C - \frac{2}{3}\lambda)} P'(E'_\nu)$$

also

$$Ne^{-t(C-\frac{2}{3}\lambda)} > P'(E'). \tag{3.4.13}$$

Comparison with (3.4.12) yields

$$N > \frac{\lambda}{3} e^{t(C-\frac{2}{3}\lambda)} = e^{t(C-\frac{2}{3}\lambda+\frac{1}{t}\log\frac{\lambda}{3})}.$$

Again, if $t$ is large enough $\frac{1}{t}\log\frac{\lambda}{3} > -\frac{\lambda}{3}$ and therefore

$$N > e^{t(C-\lambda)}$$

as claimed. □

## 3.5 Lecture on Another Definition of Capacity

We assume again that the channel is non-anticipatory and has memory $m+1 \geq 0$. Let $N(t,\lambda)$ denote the maximal length of a $\lambda$-code $\{(E_\nu, E'_\nu) : 1 \leq \nu \leq N\}$ with $E_\nu \in \mathcal{B}(s-m, s+t)$, $E'_\nu \in \mathcal{B}'(s+1, t+1)$. By the stationarity of the channel $N(t,\lambda)$ does not depend on $s$.

The coding theorem in the previous lecture says that for an *ergodic* channel with I-capacity $C$

$$N(t,\lambda) \geq e^{t(C-\lambda)}, \quad \text{for large } t.$$

The fact that this holds for arbitrary $\lambda > 0$ can be expressed by

$$\lim_{\lambda\to 0}\left(\liminf_t \frac{1}{t}\log N(t,\lambda)\right) \geq C. \tag{3.5.1}$$

Next we shall show that for *arbitrary* (ergodic and non-ergodic) channels

$$\lim_{\lambda\to 0}\left(\limsup_t \frac{1}{t}\log N(t,\lambda)\right) \leq C \tag{3.5.2}$$

holds.

For *ergodic* channels by (3.5.1) there is equality in (3.5.2).

In our usual terminology of the book (3.5.1) and (3.5.2) together show that $C$ is the weak capacity, defined only in terms of $N(t,\lambda)$. The result implies that for ergodic channels the weak capacity exists and equals the (ergodic) I-capacity. This quantity gives also an upper bound (3.5.2) for arbitrary channels (*optimistic*) capacity $\overline{C}$. We write this for the ease of reference as

**Theorem 80** *For* $\lambda > 0$

$$\frac{1}{t+m+1} \log N(t, \lambda) < C + \delta(\lambda, t) \quad (t = 1, 2, \dots), \tag{3.5.3}$$

*where* $\lim_{\lambda \to 0} (\lim_{t \to \infty} \delta(\lambda, t)) = 0.$

*Proof* We fix any $t > 0$ and construct by using a $\lambda$-code $\{(E_\nu, E'_\nu) : 1 \leq \nu \leq N(t, \lambda)\}$ a stationary ergodic source $Q$ with

$$C(Q) > \frac{1}{t+m+1} \log N(t, \lambda) - \delta(\lambda, t)$$

or more precisely with

$$H(Q) = \frac{1}{t+m+1} \log N(t, \lambda), \quad H(Q|Q') \leq \delta(\lambda, t). \tag{3.5.4}$$

For this goal we construct first a source $P$ with the properties

$$T^{t+m+1} P = P \tag{3.5.5}$$

the

$$\mathcal{B}_\nu = \mathcal{B}(\nu(t+m+1) - m(\nu+1, (t+m+1) - m - 1) \tag{3.5.6}$$

are independent under $P$.

Such a source is generally not stationary, but only periodic with period $t+m+1$. We use now a results of Jacobs [19]. Let

$$Q = \frac{1}{t+m+1} \sum_{k=1}^{t+m+1} T^k P.$$

Then $Q$ is stationary and ergodic; the following limits exist and the following relations hold:

$$\lim_l \frac{1}{l} H(P_l) \triangleq H(P) - H(Q)$$

$$\lim_l \frac{1}{l} H(P_l|P'_l) \triangleq H(P|P') - H(Q|Q'),$$

where

$P_l$ is the restriction of $P$ to $\mathcal{B}(-m, l)$
$P'_l$ is the restriction of $P'$ to $\mathcal{B}'(1, l)$.

In order to compute $H(P)$ resp. $H(P|P')$ not all $l$-values have to be considered. It suffices to restrict ourselves to the subsequence $l_n = n(t+m+1) - m - 1$ $(n = 1, 2, \dots)$. For example from (3.5.6) immediately follows

$$H(P) = \frac{1}{t+m+1} H(P_t).$$

By (3.5.6) and (3.5.5) $P$ is determined if one has chosen all values $P(E)$ for all atoms $E$ of $\mathcal{B}_0 = \mathcal{B}(-m, t)$. We put

$$P(E) = \begin{cases} \frac{1}{N(t,\lambda)}, & \text{for } E = E_\nu, \quad 1 \le \nu \le N(t,\lambda) \\ 0 & \text{otherwise.} \end{cases}$$

Then, we have $H(P) = \frac{1}{t+m+1} \log N(t, \lambda)$, as desired. We keep this definition of $P$ fixed and estimate $H(P|P')$ from above. *The ideas of the proof are due to Nedoma* [40]. In order to keep an overlook on the fields of events considered, we consider the following scheme

$-m$ $\quad$ $(t+m+1)-m$ $\quad$ $2(t+m+1)-m$

| | | | | | | |
|---|---|---|---|---|---|---|
| $\mathcal{B}$ | | $\mathcal{B}_0$ | | $\mathcal{B}_1$ | | $\mathcal{B}_2$ |
| $\mathcal{B}'$ | $\mathcal{A}'_0$ | $\mathcal{C}'_0$ | $\mathcal{A}'_1$ | $\mathcal{C}'_1$ | $\mathcal{A}'_2$ | $\mathcal{C}'_2$ |

$0$ $\quad$ $t+m+1$ $\quad$ $2(t+m+1)$

The definitions are

$$\mathcal{A}'_\nu = \mathcal{B}'(\nu(t+m+1) - m, \nu(t+m+1))$$

$$\mathcal{C}'_\nu = \mathcal{B}'(\nu(t+m+1) + 1, (\nu+1)(t+m+1) - m - 1).$$

Since the PD $\widetilde{P}$ is now defined, it is simpler to denote the entropies via fields of events. So we shall write now for example

$$H\left(\prod_{\nu=0}^{n-1}{}^{*}\mathcal{C}'_n \times \prod_{\nu=1}^{n-1}{}^{*}\mathcal{A}'_n\right) \text{ instead of } H\left(P'_{n(t+m+1)-m-1}\right)$$

and

$$H\left(\prod_{\nu=0}^{n-1}{}^{*}\mathcal{B}_n \middle| \prod_{\nu=0}^{n-1}{}^{*}\mathcal{C}'_n \times \prod_{\nu=1}^{n-1}{}^{*}\mathcal{A}'_n\right) \text{ instead of } H\left(P_{n(t+m+1)-m-1} \middle| P'_{n(t+m+1)-m-1}\right).$$

Since for conditional entropies $H(P_2|P_1) \leq H(P_2|P'_1)$, if $P'_1$ is a restriction of $P_1$ ($\mathcal{B}'_1 \subseteq \mathcal{B}_1$) we get now

$$H\left(\prod_{\nu=0}^{n-1}{}^{*}\mathcal{B}_\nu \middle| \prod_{\nu=0}^{n-1}{}^{*}\mathcal{C}'_\nu \times \prod_{\nu=1}^{n-1}{}^{*}\mathcal{A}'_\nu\right) \leq H\left(\prod_{\nu=0}^{n-1}\mathcal{B}_n \middle| \prod_{\nu=1}^{n-1}\mathcal{C}'_n\right).$$

It remains to estimate

$$\frac{1}{n(t+m+1)} H\left(\prod_{\nu=0}^{n-1}{}^{*}\mathcal{B}_n \middle| \prod_{\nu=0}^{n-1}{}^{*}\mathcal{C}'_n\right).$$

For this we use the indicator function $1_{\widetilde{M}}(w, w')$ of the set

$$\widetilde{M} = \bigcup_{\nu} E_\nu \times E'_\nu \in \mathcal{B}_0 \times \mathcal{C}'_0$$

we get

$$\int 1_{\widetilde{M}} d\widetilde{P} = \widetilde{P}\left(\widetilde{M}\right) = \sum_{\nu} P(E_\nu)P(E_\nu, E'_\nu) \geq 1 - \lambda.$$

We use now the abbreviation $t + m + 1 = c$. Then $\widetilde{P}$ is invariant and ergodic under $T^c$ and by the Ergodic Theorem in $\mathcal{L}^2_{\widetilde{P}}$ (Theorem 66) there is exactly one function $h \in \mathcal{L}^2_{\widetilde{P}}$ with

$$\lim_{h} \left\| \frac{1}{n} \sum_{\nu=0}^{n-1} T^{\nu c} 1_{\widetilde{M}} - h \right\|_2 = 0$$

$$h = \int h \, d\widetilde{P} = \int 1_{\widetilde{M}} \, d\widetilde{P} \geq 1 - \lambda.$$

From here one gets that for

$$\widetilde{M}(n) = \left\{ (w, w') : \left| \frac{1}{n} \sum_{\nu=0}^{n-1} 1_{\widetilde{M}}\left(T^{\nu c} w, T^{\nu c} w'\right) \right| \geq 1 - 2\lambda \right\} \in \sum_{\nu=0}^{n-1}{}^{*}\mathcal{B}_\nu \times \prod_{\nu=0}^{n-1}{}^{*}\mathcal{C}'_\nu$$

the relation $\lim_n \widetilde{P}\left(\widetilde{M}(n)\right) = 1$ holds. The set

$$M(n)_{w'} = \left\{ w : (w, w') \in \widetilde{M}(n) \right\} \in \prod_{\nu=0}^{n-1}{}^{*}\mathcal{B}_\nu$$

depends only through the atom $(w') = E'$, determined by $w'$, on $\prod_{\nu=0}^{n-1} {}^*\mathcal{C}'_\nu$ and can, therefore, also be named $M(n)_{(w')}$ or $M(n)_{E'}$. One then calculates

$$\widetilde{P}\left(\widetilde{M}(n)\right) = \sum_{E'} P\left(M(n)_{E'}|E'\right) P'(E').$$

For $n$ sufficiently large, we get for instance

$$P'\left(w' : P(M(n)_{(w')}|(w') \geq 1 - \lambda\right) > 1 - \lambda.$$

We now estimate for a fixed atom $E'$ the entropy of the PD $P(\cdot|E')$ on $\prod_{\nu=0}^{n-1} \mathcal{B}_\nu$ from above. We consider first the case where $P(M(n)_{E'}|E') < 1 - \lambda$ and get the primitive estimate $H(P|E') < nc \log a$.

The estimate in the case $P(M(n)_{E'}|E') \geq 1 - \lambda$ we perform by the following general procedure:

If $W$ is a PD on the finite field of events $\mathcal{F}$ with $b_1$ atoms, and does there exist a set $M$ with $b_2$ atoms from $\mathcal{F}$ and with $W(M) \geq 1 - \lambda$, then

$$H(W) \leq \log b_2 + \lambda \log b_1 + \log 2.$$

Indeed, when $A$ runs through the atoms of $\mathcal{F}$, then

$$\begin{aligned} H(W) &= -\sum_{A \subseteq M} W(A) \log W(A) - \sum_{A \nsubseteq M} W(A) \log W(A) \\ &= w(M) \left[ -\sum_{A \subseteq M} \frac{W(A)}{W(M)} \log \frac{W(A)}{W(M)} - \log W(M) \right] \\ &\quad + (1 - W(M)) \left[ -\sum_{A \nsubseteq M} \frac{W(A)}{1 - W(M)} \log \frac{W(A)}{1 - W(M)} - \log(1 - W(M)) \right] \\ &\leq \log b_2 + \lambda \log b_1 + \log 2. \end{aligned}$$

Applying this to $\mathcal{F} = \prod_{\nu=0}^{n-1} {}^*\mathcal{B}_\nu$, $M = M(n)_{E'}$, $W = P(\cdot|E')$, then we obtain

$$H(P|E') \leq \log b_2 + \lambda nc \log a + \log 2,$$

where $b_2$ is the number of the atoms from $\prod_{\nu=0}^{n-1}{}^{*}\mathcal{B}_\nu$ in $M(n)_E$. For every one of those atoms $G$ we have $G \times E' \subseteq \widetilde{M}(n)$ and also

$$\frac{1}{n}\sum_{\nu=0}^{n-1} 1_{\widetilde{M}}(T^{\nu c}w, T^{\nu c}w') \geq 1 - 2\lambda \quad (w \in G, w' \in E'). \tag{3.5.7}$$

This means that for at least $n(1-2\lambda)$ among the numbers $\nu = 0, \dots, n-1$ $1_{\widetilde{M}}(T^{\nu c}w, T^{\nu c}w') = 1$ holds. Now every $G$ can be written exactly one way in the form $G_0 \times \cdots \times G_{n-1}$, where the $G_\nu$ atoms are from $\mathcal{B}_\nu$. Analog facts hold for $E'$. Let $(w)_\nu$ be the atom of $\mathcal{B}_\nu$ determined by $w$, let $(w')_\nu$ be the atom of $\mathcal{C}'_\nu$ determined by $w$.

$1_{\widetilde{M}}(T^{rc}w, T^{rc}w') = 1$ means that there is a $\nu$ with $T^{rc}(w)_\nu = E_\nu, T^{rc}(w')_\nu \subseteq E'_\nu$. Since $\nu$ is already uniquely determined by $T^{rc}(w')_\nu$, because the $E'_\nu$ are disjoint, and since in turn is given by $(w')$, (3.5.7) means the following: at least $n(1-2\lambda)$ of the $G_\nu$ are determined uniquely by $E'$.

Is $k_n$ the smallest integer $\geq n(1-2\lambda)$, then there are under the condition (3.5.7) at most $n - k_n$ of the $G_\nu$ free, which are distributed on at most $n$ possible places: $M(n)_{E'}$ contains at most $\binom{n}{n-k_n} a^{c(n-k_n)}$ atoms: $\log b_2 \leq \log \binom{n}{k_n} + c(n-k_n)\log a$. One has now in total

$$\begin{aligned}
&\frac{1}{nc} H\left(\prod_{\nu=0}^{n-1}{}^{*}\mathcal{B}_\nu \Big| \prod_{\nu=0}^{n-1} \mathcal{C}'_\nu\right) \\
&\leq \frac{1}{nc}\Bigg[\sum_{P(M(n)_{E'}|E') \leq 1-\lambda} P(E')nc\log a \\
&\quad + \sum_{P(M(n)_{E'}|E') \geq 1-\lambda} P(E')\left(\log\binom{k}{k_n} + c(n-k_n)\log a + \lambda nc \log a + \log 2\right)\Bigg] \\
&\leq \lambda \log a + \frac{1}{nc}\log\binom{n}{k_n} + \frac{n-k_n}{n}\log a + \lambda\log a + \frac{1}{nc}\log 2 \\
&\leq \left(2\lambda + \frac{n-k_n}{n}\right)\log a + \frac{1}{nc}\log\binom{n}{k_n} + \frac{1}{nc}\log 2 \\
&\leq 4\lambda\log a + \frac{1}{nc}\log\binom{n}{k_n} + \frac{1}{nc}\log 2.
\end{aligned}$$

The only interesting term in this estimate is $\frac{1}{nc}\log\binom{n}{k_n}$. Stirling's formula

$$k! = \sqrt{2\pi}k^{k+\frac{1}{2}}e^{-k}e^{\vartheta_k} \quad (0 \leq \vartheta_k < \frac{1}{12}),$$

gives

$$\begin{aligned}
\log\binom{n}{k} &= -\frac{1}{2}\log 2\pi - \left(n+\frac{1}{2}\right)\log\frac{n-k}{n} \\
&\quad + \left(k + \frac{1}{2}\log\frac{n-k}{n} - \frac{1}{2}\log(n-k)\right) + (-\vartheta_n + \vartheta_k + \vartheta_{n-k}) \\
&= \left(-\frac{1}{2}\log 2\pi + \frac{1}{2}\log\frac{n}{k} - \frac{1}{2}\log(n-k)\right) - \vartheta_n + \vartheta_k + \vartheta_{n-k} \\
&\quad + \left(-n\log\frac{n-k}{k} + k\log\frac{n-k}{k}\right) \\
&= A(n,k) + B(n,k),
\end{aligned}$$

for instance.

Obviously, $\lim_n \frac{1}{nc}(A_n, k_n) = 0$, so that now $\frac{1}{hc}B(n, k_n)$ is of interest.

$$\begin{aligned}
\frac{1}{nc}B(n,k_n) &= \frac{n}{nc}\log\frac{n}{n-k_n} + \frac{k_n}{nc}\log\frac{n-k_n}{k_n} \\
&= \frac{k_n + 2\lambda n}{nc}\log\frac{n}{n-k_n} + \frac{k_n}{nc}\log\frac{n-k_n}{k_n} \\
&= \frac{k_n}{nc}\log\frac{n}{k_n} + \frac{2\lambda}{c}\log\frac{n}{n-k_n} \\
&= \frac{1}{c}f\left(\frac{k_n}{n}\right) + \frac{2\lambda}{c}\log\frac{n}{n-1-(k_n-1)}
\end{aligned}$$

with $f(x) = -x\log x$.

Noticing that $k_n - 1 < n(1-2\lambda)$ and hence also $n-(k_n-1) > 2\lambda n$, it follows that

$$\begin{aligned}
\frac{1}{nc}B(n,k_n) &< \frac{1}{c}f\left(\frac{k_n}{n}\right) + \frac{2\lambda}{c}\log\frac{1}{2\lambda - \frac{1}{n}} \\
&< \frac{1}{ce} + \frac{2\lambda}{c}\log\frac{1}{\lambda} \qquad \left(\frac{1}{n} < \lambda\right),
\end{aligned}$$

because $f(x) \leq \frac{1}{e}$ $(0 \leq x \leq 1)$ and $0 \leq \frac{k_n}{n} \leq 1$.

Collecting all estimates obtained so far we obtain

$$\begin{aligned}
&\frac{1}{nc}H\left(\prod_{\nu=0}^{n-1}{}^*\mathcal{B}_\nu \,\Bigg|\, \prod_{\nu=1}^{n-1}{}^*\mathcal{A}'_\nu \times \prod_{\nu=0}^{n-1}{}^*\mathcal{C}'_n\right) \\
&\leq \frac{1}{nc}H\left(\prod_{\nu=0}^{n-1}{}^*\mathcal{B}_n \,\Bigg|\, \prod_{\nu=0}^{n-1}{}^*\mathcal{C}'_\nu\right)
\end{aligned}$$

$$\le 4\lambda \log a + \frac{1}{nc} A(n, k_n) + \frac{1}{ce} + \frac{2\lambda}{c} \ln \frac{1}{\lambda} + \frac{1}{nc} \log 2.$$

For $n \to \infty$ we get

$$H(P|P') \le 4\lambda \log a + \frac{1}{(t+m+1)e} + \frac{2\lambda}{t+m+1} \log \frac{1}{\lambda} = \delta(\lambda, t)$$

and hence

$$\lim_{\lambda \to 0} \left( \limsup_{t} \delta(\lambda, t) \right) = \lim_{\lambda \to 0} 4\lambda \log a = 0. \qquad \square$$

**Corollary 18** *For arbitrary channels (meeting our suppositions)* (3.5.2) *holds.*

*Proof* Obviously,

$$\lim_{\lambda \to 0} \left( \limsup_{t} \frac{1}{t} \log N(t, \lambda) \right) \le C + \lim_{\lambda \to 0} \left( \limsup_{t} \delta(\lambda, t) \right) = C.$$

## 3.6 Lecture on Still Another Type of Operational Capacities—Risk as Performance Criterium

### 3.6.1 Introduction

The problem of suitable mathematical models, e.g., for a transmission channel, has apparently not yet been finally solved even in the case of a discrete message. This is borne out to be the fact that in four significant papers on this subject, namely the papers by Shannon [50], McMillan [39], Feinstein [9], and Khinchin [26], the concept of a channel is defined in different ways. The kernel of these papers is the question of the validity of the Fundamental or Shannon's Theorem, which McMillan formulates in the following way.

Let the given channel have capacity $C$ and the given source have rate $H$. If $H < C$, then, given any $\varepsilon > 0$, there exists an integer $n(\varepsilon)$ and a *transducer* (depending on $\varepsilon$) such that when $n(\varepsilon)$ consecutive received letters are known, the corresponding $n$ transmitted letters can be identified correctly with probability at least $1 - \varepsilon$. If $H > C$ no such transducer exists. McMillan in Sect. 8.5 of his paper draws attention to the fact that the proof of this theorem requires the channel to be in some sense "continuous".

The present lecture does not define the concept of a channel as broadly as is done in McMillan's paper, on which this lecture is mainly based. Nevertheless, though our restriction entails a definite continuity, it turns out that even in this case the above mentioned Shannon's Theorem may not be valid.

(The idea of continuity was pursued later by Kieffer.)

### 3.6.2 Nedoma's Fundamental Paper Has Three Basic Ideas

- He considers DFMC′ with memory $m = 0$ (Feinstein's condition m.1, but not the independence assumption m.2).
  This makes it meaningful to work with $n$-symbol codes (block codes). However, the prize is a restriction of the transducers Shannon and McMillan mentioned. It must be emphasized that here it is facility assumed and not explicitly contained in the mathematical formulation that there must be *synchronization between coding and decoding*. This means the following: Before transmission, the message is broken up into groups of $n$ symbols which are then successively coded. The received message is again broken up into groups of $n$ symbols, but the dividing points must correspond to each other. The position of these dividing points, or at least one of them, must be accurately known in the process of decoding. (We come back to the issue of synchronization in Chap. 5.)
- The introduction of risk functions as performance criteria, which cover the familiar error probabilities and average error frequencies.
- A definition of capacities, depending on the risk function, as supremum of all rates of sources transmissible through the channel $W$ with arbitrarily small prescribed risk.
  The symbol $\mathcal{E}$ we use for the $e$-capacity and the symbol $\mathcal{F}$ for the $f$-capacity (involving the familiar criteria).
  These definitions originated with Fortet, who introduced them for noiseless channels.
  (We remind the reader that these capacities are of the pessimistic type if viewed in analogy to the definition of $\underline{C}$.)

### 3.6.3 Report on the Results

**Theorem 81** *Let $W$ be a DMCF′.*

*(i) A source with entropy rate $H$ less than $\mathcal{F}(W)$ is e-transmissible through $W$.*
*(ii) If $\mathcal{E}(W) > 0$, then a source with entropy rate $H = \mathcal{E}(W)$ is $f$-transmissible through $W$.*
*(iii) $\mathcal{F}(W) = \mathcal{E}(W)$.*

**Theorem 82** *Let $W$ be a DMCF′.*

*(i) If $\mathcal{F}(W) > 0$, then a source is $f$-transmissible through $W$ if and only if its entropy rate $H$ satisfies*

$$H \leq \mathcal{F}(W).$$

*(ii) If $\mathcal{F}(W) = 0$, then no source with $H > 0$ is $f$-transmissible.*
*(iii) If for a source entropy rate $H < \mathcal{E}(W)$, then it is e-transmissible through $W$.*
*(iv) If for a source entropy rate $H > \mathcal{E}(W)$, then it is not e-transmissible through $W$.*

*Remark* Not all cases are covered.
In (i), (ii) we have not dealt with the case $\mathcal{F}(W) = 0$ and in (iii), (iv) the case $H = \mathcal{E}(W)$ has been omitted. For the sake of completeness we mention that in both cases either alternative is possible.

Finally, there is an important connection to Khinchin's ergodic capacity $C_e$, which influenced Jacobs ([20], Proof of Theorem 80).

**Theorem 83** *For our* DMCF′ $W$

$$C_e(W) \geq \mathcal{E}(W).$$

*Remark* There are examples with equality and strict inequality.

### 3.6.4 Discussion

Nedoma has endeavored to analyze the concept of capacity of a discrete channel, particularly from the point of view of the validity of Shannon's Theorem. In the rigorous formulation of the theorem it proves convenient to introduce the concept of risk. This makes it possible to formulate Shannon's Theorem, quoted in the introduction, in the following way: "Let the given channel have capacity $C$ and the given source have rate $H$. If $H < C$, then given any $\varepsilon > 0$, there exists an integer $n(\varepsilon)$, a transducer $g$ (depending on $\varepsilon$), and an $n$-symbol code $(\Omega', \psi, \Omega'')$ such that the probability of error

$$\gamma_n((\Omega, P), W, g, \psi, v_n)$$

is less than $\varepsilon$. If $H > C$, no such transducer and $n$-symbol code exists."

Please note that this formulation differs from McMillan's because the use of the $n$-symbol code requires synchronization, as explained in the introduction, whereas there is no such restriction in McMillan's for mutation of Shannon's Theorem.

However, the proof (or idea of proof) of Shannon's Theorem as given not only by McMillan, but also by other authors (Shannon, Khinchin), does not prove more than asserted by Nedoma's formulation. His finite symbol code as a transformation is of course not a transducer in McMillan's sense, since it does not satisfy McMillan's condition of stationarity of a transducer Nedoma mentioned that Khinchin pointed out that an $n$-symbol code is in practice unsuitable for large $n$, in addition to the synchronization problem.

This criticism also to the use of the DFMC is discussed in the next lecture.

## 3.7 Lecture on the Discrete Finite-Memory Channel According to Feinstein and Wolfowitz

The main theorem concerning stationary, non-anticipatory, finite-memory channels with duration of memory $m$ gave the (error) capacity as $C_{erg}$ under an independence assumption (M). Now the sharper independence assumption (U) is imposed and the result will concern thus a more special channel DFMC. However, the present approach via approximating DMC's $W(y^l|x^{l+m})$, $l \to \infty$, is very simple and makes no reference to ergodicity (in particular also not to McMillan's theorem in whatever form of convergence). Even a strong converse can now be proved and as an important fact there is a speed of convergence for the capacity $C$ in terms of the capacities $C(l+m) = \frac{1}{l+m} \max_{P_{X^{l+m}}} I(X^{l+m} \wedge Y^l)$ of the approximating DMC.

To make this lecture readable without acquaintance with previous lectures. we give here Feinstein's [10] definition of finite memory $m$.

**Property $\mathbf{m_1}$** There exists a fixed positive integer $m$, such that for any cylinder set $[y_t, \dots, y_{t+n-1}]$ in $\mathcal{Y}^{\mathbb{Z}}$

$$W([y_t, \dots, y_{t+n-1}]|x_\infty) = W([y_t, \dots, y_{t+n-1}]|x'_\infty)$$

for $x_\infty, x'_\infty \in \mathcal{X}^{\mathbb{Z}}$. With $x'_i = x_i$, $t - m \leq i \leq t + n - 1$.

**Property $\mathbf{m_2}$** For any two cylinder sets $[y_i, \dots, y_j]$ and $[y'_k, \dots, y'_n]$ such that $j + m < k$, where $m$ is a fixed positive integer, we have for all $x_\infty \in \mathcal{X}^{\mathbb{Z}}$

$$([y_i, \dots, y_j] \cap [y'_k, \dots, y'_n]|x_\infty) = W([y_i, \dots, y_j]|x_\infty) \cdot W([y'_k, \dots, y'_n]|x_\infty).$$

**Definition 20** The smallest value of $m$ for which a given channel has these two properties will be called the memory (or duration of memory) of the channel.

**Theorem 84** (Wolfowitz [57], Feinstein [11], see [4]) *For $\varepsilon > 0$ and any $\lambda$, $0 < \lambda \leq 1$, for every DFMC and all $n$ sufficiently large*

*(i) there exists a code $(n, 2^{n(C-\varepsilon)}, \lambda)$;*
*(ii) any code $(n, M, \lambda)$ satisfies $M < 2^{n(C+\varepsilon)}$;*
*(iii) for any $l \in \mathbb{N}$*

$$\frac{C(m+l)}{m+l} \leq C \leq \frac{C(m+l)}{m+l} + \frac{m \log a}{m+l}.$$

*$C$ is zero if and only if $C(m+l) = 0$ for all $l \in \mathbb{N}$.*

This result makes it possible, at least in principle, to compute the capacity $C$ of a DFMC to any desired degree of accuracy, since $C(l+m)$ is in principle computable to any desired degree of accuracy (for instance with the Arimoto-Blahut algorithm).

*Proof* (i) Set $C = \sup_l \frac{C(l+m)}{l+m}$ and $l = l(\varepsilon)$ for which

$$\frac{C(l+m)}{l+n} > C - \frac{\varepsilon}{2}. \tag{3.7.1}$$

Set $n = k(l+m)$, $k \in \mathbb{N}$ and use the DMC $W_{l+m}$ to get the rate $> \frac{C(l+m)}{l+m} - \frac{\varepsilon}{2}$, also for $n' = n + r$, $0 \leq r \leq l + m$.

(ii) Let $n = k(l+m)$ with $k$ sufficiently large and let $\{(u_i, D_i) : 1 \le i \le M\}$ be an $(n, M, \lambda)$-code for the DFCM.

(For $n = k(l+m)+r, 0 \le r < l+m$ we can extend the code by adding dummy letters to a block length $n' = (k+1)(l+m)$, resulting in negligible rate loss.)

This code can be transformed to one for $W_{l+m}$ of list size $a^{km}$. Its strong converse gives the bound

$$\log M \le k\left[\left(C(l+m) + \frac{\varepsilon}{2}\right) + m \log a\right] \tag{3.7.2}$$

and

$$\log M \le n\left[\frac{C(l+m) + \frac{\lambda}{2}}{l+m} + \frac{m}{l+m} \log a\right]$$

$$\frac{1}{n} \log M \le \frac{C(l+m)}{l+m} + \varepsilon \quad \text{for } l \text{ large.}$$

(iii) Finally, by definition of $C$ and $l$,

$$\frac{C(m+l)}{m+l} \le C \le \frac{C(m+l)}{m+l} + \frac{m}{m+l} \log a$$

for any $l$. □

*Remark* Our use of list codes greatly simplified the converse proof.

### *3.7.1 Discussion*

In Sect. 3.1 we have discussed already work on the relations of the stationary capacity $C_s$ and the ergodic capacity $C_e$. As we know it is $C_e$ which enters the results presently known for DFMC. Let us first remark that every stationary product measure on $(\Omega, \mathcal{B})$ is known to be ergodic. Thus, the question of ergodicity was implicitly accounted for by the use of product PD as inputs for the extensions of a channel without memory, the DMC. It enters now via AEP in conjunction with entropy typicality.

For an ergodic $\mu$, for any $\varepsilon, \delta > 0$ there is a positive integer $n(\varepsilon, \delta)$ such that for $n \ge n(\varepsilon, \delta)$

$$\mu\left\{u \in \mathcal{B}(1, n) : \left|\frac{1}{n} \log \mu(u) + H(X)\right| < \varepsilon\right\} \ge 1 - \delta.$$

An ergodic $\mu$ is admissible, if it implies $\mu \times W$ and thus $\mu W$ to be ergodic. Then, the AEP holds for these PD; also and the conditional PD's leading to the

**Theorem 85** *Let the discrete channel $W$ with memory have ergodic capacity $C_e > 0$ then for any $0 < R < C_l$, $\lambda > 0$ there is an $n(\lambda, R) \in \mathbb{N}$ such that for all $n \ge n(\lambda R)$*

*there is an* $(n, 2^{nR}, \lambda)$*-code* $\{(u_i, D_i) : 1 \le i \le 2^{nR}\}$, *where* $u_i \in \mathcal{X}^n$ *and* $D_i \subset \mathcal{Y}^n$ *disjoint,* $W(D_i|u_i) > 1 - \lambda$.

It is of course understood that the $u_i$ are such that $W(\cdot|u_i)$ is well defined, i.e., $\mu(u_i) > 0$, where $\mu$ is the input used to prove the statement.

We notice here a difference from the memoryless case, in which the statement of the coding theorem contained no reference to the input used in the proof of the theorem. The reason for this is that the interpretation of $W(D_i|u_i) \ge 1 - \lambda$ is here quite different from that we usually make for the DMC. *One cannot interpret it as* "the probability of receiving a sequence $y_0, \dots, y_{n-1}$ such that $[y_0, \dots, y_{n-1}] \in D_i$, where the sequence $x_0^i, \dots, x_{n-1}^i$ which defines $u_i$ is transmitted is not less than $1 - \lambda$". Indeed the statement "$u$ is transmitted" *describes not a single event*, which would be the transmission of a single $x_\infty$, but rather one of the class (i.e., cylinder set) $u_i$ of events. The probability measure $W(\cdot|u_i)$ which is well defined on $\mathcal{B}'$, represents the PD on $\mathcal{Y}^{\mathbb{Z}}$ under the assumption that only sequences $x_\infty \in u_i$ are transmitted, the relative probabilities of transmission being given by the PD $\mu_i(\cdot) = \frac{\mu(\cdot)}{\mu(u_i)}$ defined on the Borel field of all measurable subsets of the cylinder $\mu_i$. What we need, in order to be able to interpret $W(D_i|u_i) \ge 1 - \lambda$ in a manner analogous to the memoryless case, is some form of independence condition. *Sufficient* are Properties $\mathrm{m}_1$ and $\mathrm{m}_2$.

Given a channel with finite memory $m$, the conditions $W(D_i|u_i) \ge 1 - \lambda$ $(1 \le i \le \mathbb{N})$ are easily reduced to conditions on $W(\cdot|x_\infty)$. Put $u_i = [x_1^i, \dots, x_n^i]$ and let $z$ represent an arbitrary sequence $x_{-m+1}, \dots, x_0$. Denoting the cylinder $[x_{-m+1}, \dots, x_0, x_1^i, \dots, x_n^i]$ by $[z, u_i]$ we get

$$W(D_i|u_i) = \frac{\mu \times W(u_i|D_i)}{\mu(u_i)} = \sum_z \mu_i([z, u_i]) W(D_i|[z.u_i]) \ge 1 - \lambda.$$

Since $\sum_z \mu_i([z, u_i]) = \mu_i(u_i) = 1$, it follows that for at least one $z$, say $z_i$, we must have $W(D_i|[z_i, u_i]) \ge 1 - \lambda$. The interpretation of this inequality is straightforward; combined with $\mathrm{m}_2$ it permits us to "transmit" the sequence $(z_i, u_i)$ in any order we choose, and still receive each sequence *correctly a fraction of at least* $1 - \lambda$ of the times that it is transmitted, as this number goes to infinity, according to the law of large numbers (see also Exercise concerning Remark 2.12 on p. 104 of [10]). The only remaining point is that we still have $N \ge 2^{nH}$ sequences, but their length is now $n + m$. However, since $m$ is fixed while $n$ may be taken arbitrarily large, the way out is clear.

We prove the theorem for some $H'$ satisfying $H < H' \le C$ and take $n$ so large that $\frac{n}{n+m} H' \ge H$. The number of sequences is then given by $N \ge 2^{nH'} \ge 2^{n+m,H}$. These explanations are taken from [10]. "Wolfowitz [58, p. 60], refers to it as an excellent description of the *customary* treatment".

He calls the approach presented in the previous section as now *conventional.* We explained earlier that he shares this *radically different view* with Feinstein [11]. It is briefly described. In coding theory stochastic inputs serve only as a tool in proving coding theorems and are not of interest per se. Given is only the channel. Its operationally defined capacities $C$, $\underline{C}$, $\overline{C}$ (weak and strong capacities) are to be characterized and if possible in a computable form. It was shown for the DFMC that the weak capacity could be computed to any desired degree of accuracy by computing the rate of transmission for a suitable stochastic input of i.i.d. random sequences of length $(l + m)$, the sequence distribution may be nonstationary.

Whether the resulting $\mu$ is ergodic or whether $C_e = C_s$ are questions which do not arise in this approach, which does not use the AEP. But the present approach helps in answering the questions.

For the DFMC Tsaregradskii's proof is complete and gives the positive answer by showing that for $C = \sup_n \frac{C^*(n)}{n}$, which he introduces(!), $C = C_e$ and $C = C_s$. Feinstein [11, p. 42], remarks that he independently followed the same approach. Nedoma proved the equality for his DFMC$'$, again by equating the two quantities with another capacity, his $\mathcal{E}(W)$. Again for the DFMC Breiman [5] also proves the equation. Moreover, he shows that there is always at least one ergodic $P'$ assuming the supremum: $C_e = R(P)$, a noticeable result. It was at the beginning of ergodic decomposition formulas by Parthasarathy and Jacobs (see Sect. 3.8).

## 3.8 Lecture on the Extension of the Shannon/McMillan Theorem by Jacobs

For any PD $P$ on $(\Omega, \mathcal{B})$ define the function

$$f_t(w) = -\frac{1}{t} \log P(w_1, \dots, w_t)$$

$P$-a.e. It is $\mathcal{B}(1, t)$-measurable and has the expected value

$$H(1, t) = -\sum_{\overline{w}} P(\overline{w}) \log P(\overline{w})$$

where $\overline{w} = (w_1, w_2, \dots, w_t)$ is a cylinder set in $\Omega$ and atom in $\mathcal{B}(1, t)$.

$H(1, t)$ is the entropy of $P$ on $\mathcal{B}(1, t)$.

**Property** (**M**) The sequence $(f_t)_{t=1}^{\infty}$ of functions viewed as vectors in the Banach space $\mathcal{L}_P^1$ is norm convergent.

Extending Shannon's work McMillan proved Property (M) for stationary $P$.

**Theorem 86** (Jacobs [19]) *Does Property (M) hold for the PD (source) $P$, then it also holds for every $Q$, which is absolutely continuous with respect to $P$, that is, every $P$-nullset is a $Q$-nullset. More specific: If for $f_t(w) = -\frac{1}{t} \log P(w_1, \dots, w_t)$, $f \in \mathcal{L}_P^1$, we have $\lim_{t \to \infty} f_t = \overline{f}$ in $\mathcal{L}_P^1$, then also $\overline{f} \in \mathcal{L}_Q^1$ and for $g_t(w) = -\frac{1}{t} \log Q(w_1, \dots, w_t)$ we have*

$$\lim_{t \to \infty} g_t = \overline{f} \textit{ in } \mathcal{L}_Q^1. \tag{3.8.1}$$

*Proof* Let $Q(dw) = \rho(w)P(dw)$ with $\rho \in \mathcal{L}^1_P$ (Theorem Radon-Nikodym). For $h \in \mathcal{L}^1_P$ let $(h_t)_t$ be the function obtained from $h$ by $P$-averaging it on the atoms of $\mathcal{B}(1,t)$. This operation $(\cdot)_t$ is linear, "order-preserving", and does not increase the $\mathcal{L}^1_P$-norm (does not change it!). For the iterated operation $((\cdot)_t)_s = (\cdot)_s$ $(t \geq s)$. Setting $h_t = (h)_t$, then $h_t \to h$ in $\mathcal{L}^1_P$ if $h$ is measurable with respect to the Borel closure $\mathcal{B}(1,\infty)$ of $\bigcup_{t\geq 0} \mathcal{B}(1,t)$, because this is obviously true for every $\mathcal{B}(1,s)$-measurable $h$ $(s < \infty)$ and every $\mathcal{B}(1,\infty)$-measurable $h$ can be approximated arbitrarily well in norm by $\mathcal{B}(1,s)$-measurable functions as $s \to \infty$. Since $(\cdot)_t$ does increase the norm, the statement follows in general. Especially the family of functions $\rho_t$ is uniformly $P$-integrable. Denoting by $\rho_t(\overline{w})$ the constant values of $\rho_t$ on $\overline{w} = (w_1, \ldots, w_t)$, we get

$$Q(\overline{w}) = \rho_t(\overline{w})P(\overline{w}),$$

that is,

$$g_t(w) = -\frac{1}{t}\log \rho_t(\overline{w}) + f_t(w) \quad (w \in \overline{w}). \tag{3.8.2}$$

By absolute continuity for $\varepsilon > 0$ there is a $\delta > 0$ so that $P(E) < \delta$ always implies $Q(E) < \frac{\varepsilon}{2}$ (often this is given as definition!). Choosing now $B > 0$ so that $\frac{1}{B} < \frac{\varepsilon}{2}$ *and* $P(w : \rho_t(w) > B) < \delta$ $(t > 0$ holds then for $E_t = \{w : \rho_t(w) > B\} \cup \left\{w : \rho_t(w) < \frac{1}{B}\right\}$ certainly $Q(E_t) < \frac{\varepsilon}{2} + \frac{\varepsilon}{2} = \varepsilon$, and for $w \notin E_t$

$$|g_t(w) - f_t(w)| < \varepsilon$$

as soon as $t$ is so large that $\frac{1}{t}|\log B| < \varepsilon$ holds. This way it is established that $\lim_{t\to\infty} g_t = \overline{f}$ in $Q$-probability. However, the family of $g_t$ is known to be uniformly integrable and uniform integrability and convergence in probability together imply norm-convergence. □

Finally, there is an additional useful result

**Theorem 87** (Jacobs [19]) *If Property (M) holds for a source P, then the occurring limit function $\overline{f}$ satisfies* $0 \leq f \leq \log a$ *(P-a.e.)*

*Proof* By the previous theorem $\overline{f}$ is integrable for every PD $Q$, which is absolutely continuous with respect to $P$. With Radon-Nikodym's theorem this implies that for every $P$-integrable function $g$

$$\varphi(g) = \int \overline{f} g \, \mathrm{d}P \in \mathbb{R}$$

and $\varphi$ is a linear functional on $\mathcal{L}^1_P$. It is also bounded: indeed, if $Q$ is defined through $Q(\mathrm{d}w) = g(w)P(\mathrm{d}w)$ $(g \geq 0, \int g \, \mathrm{d}P = 1)$, then by the previous theorem

$$\varphi(g) = \lim_{t\to\infty} \frac{1}{t} H(1,t) \leq \log a, \tag{3.8.3}$$

where $H(1,t)$ denotes the entropy of $Q$ on $\mathcal{B}(1,t)$.

So $\overline{f}$ defines an element $\varphi$ of the dual space $\mathcal{L}_P^\infty$ of $\mathcal{L}_P^1$ and therefore is bounded ($P$-a.e.). The smallest possible bound equals the dual norm of $\varphi$, which by (3.8.3) is not larger than $\log a$. □

## 3.9 Lecture on Achieving Channel Capacity in DFMC

### 3.9.1 Introduction

We present now Breiman's result.

**Theorem 88** *For the* DFMC, *not only* $C_e = C_s$, *but there is an ergodic* $P$ *with*

$$C_e = R(P).$$

The method of proof depends essentially on the additivity property of entropy $H(P)$ and rate $R(P) = H(P) + H(P') - H(\tilde{P})$ and the following well-known result.

**Lemma 42** *An upper-semicontinuous (u.s.c.) bounded linear functional defined on a convex compact subset of a linear locally convex separated topological space assumes its supremum on at least one of the extreme points of the set.*

This is a simple consequence of the Krein-Milman theorem (see for example [2]).

Its relevance is that, looked at in the right way, the set of stationary input measures is a convex compact subset of a linear topological space whose extreme points are the ergodic input measures. This fact we develop in Sect. 3.9.2 and finally, in Sect. 3.9.3, show that $R(P)$ is an u.s.c. functional for the DFMC's.

### 3.9.2 A Topology on the Linear Space $\mathcal{L} = \mathcal{LP}(\Omega, \mathcal{B})$ of $\sigma$-Additive Finite Real-Valued Functions $h : \mathcal{B} \to \mathbb{R}$

We topologize $\mathcal{L}$ so that $\mu_\nu \to \mu$ if and only if $\mu_\nu(x_k, \dots, x_n) \to \mu(x_k, \dots, x_n)$ for every cylinder $(x_k, \dots, x_n)$. In this topology we have

**Theorem 89** $\mathcal{L} = \mathcal{LP}(\Omega, \mathcal{B})$ *is a linear locally convex separated topological space, and the set* $\mathcal{P}(\Omega, \mathcal{B})$ *is a compact subset of* $\mathcal{L}$.

*Proof* $\Omega = \prod_{-\infty}^{+\infty} \mathcal{X}$ is in the product topology by Tykhonov's theorem, compact. The space $\mathcal{L}$ is the adjoint of the space $\mathcal{C}(\Omega)$ of all continuous functions on $\Omega$. In the weak dual topology on $\mathcal{L}$, i.e., the topology such that $\mu_n \to \mu \Leftrightarrow \int f \, d\mu_n \to \int f \, d\mu$ for every $f \in \mathcal{C}(\Omega)$, the unit sphere in $\mathcal{L}$ is compact and so is $\mathcal{P}(\Omega)$ as its closed subset. It remains to show that the weak dual topology is equivalent to convergence on finite-dimensional cylinder sets. One notices, if $\mu_n \to \mu$ in the weak dual topology, then for any finite-dimensional cylinder set $S$, we have that $1_S(x)$ is continuous, and hence $\mu_n(S) \to \mu(S)$. It can be seen also in the following way using the elementary fact that every continuous function on $\Omega$ can be uniformly approximated by a finite linear combination of indicators of finite-dimensional cylinder sets, which follows from the compactness of $\Omega$.

The set $\mathcal{P}_S(\Omega, \mathcal{B})$ of stationary PD is easily seen to be closed, therefore compact, and obviously convex. □

**Theorem 90** *The set of extreme points* $Ext(\mathcal{P}_s)$ *is exactly the set of stationary ergodic measures* $\mathcal{P}_l(\Omega, \mathcal{B})$.

*Proof* Let $\mu \in \mathcal{P}_e$ and $\mu = \alpha\mu_1 + \beta\mu_2, \mu_1, \mu_2 \in \mathcal{P}_e, \alpha + \beta = 1, \alpha, \beta > 0$. Suppose $\mu_i$ is not ergodic, then there exists an invariant Borel set $S$ with $0 < \mu_i(S) < 1$ and hence $0 < \mu(S) < 1$, a contradiction. Therefore both, $\mu_1$ and $\mu_2$ must be ergodic. However, a consequence of the individual ergodic theorem is that any two distinct ergodic measures from $P_e$ are orthogonal. Using a carrier $S$ of $\mu_1$, which is invariant and satisfies $\mu_1(S) = 0$, $\mu_2(S) = 0$. Then,s we have the contradiction $0 < \mu(S) < 0$ and completed the proof of $\mathcal{P}_e \subset \text{Ext}\mathcal{P}$.

Now we show the reverse inclusion. Let $\nu \in \text{Ext}\mathcal{P} \setminus \mathcal{P}_e$, then there is an invariant set $S \in \mathcal{B}$ with $0 < \nu(S) < 1$. Choose $\nu_1, \nu_2 \in \mathcal{P}_S$ such that

$$\nu_1(E) = \frac{\nu(S \cap E)}{\nu(S)}, \quad \nu_2(E) = \frac{\nu(S^c \cap E)}{\nu(S)} \quad (E \in \mathcal{B}).$$

Hence

$$\nu(E) = \nu(S)\nu_1(E) + \nu(S^c)\nu_2(E) \quad (E \in \mathcal{B})$$

contradicting the assumption $\nu \in \text{Ext}\mathcal{P}$. □

### 3.9.3 R(P) *Is an Upper Semi-continuous Functional (u.s.c) for DFMC*

**Theorem 91** *For* DFMC $W$ *and* $P' = PW$ $(P \in \mathcal{P}_S(\Omega, \mathcal{B}))$ $H(P')$ *is an u.s.c. functional on* $\mathcal{P}_s(\Omega', \mathcal{B}')$.

*Proof* We use two facts

1. We learned that

$$H(P') = \lim_n H_n(P'), \quad H_{n+1}(P') \leq H_n(P') \tag{3.9.1}$$

when we defined per letter entropy of a stationary PD.

2. Take any $(P_N) \subset \mathcal{P}_s$ such that $P_N \to P$ and for $x = (x_1, \dots, x_k)$, $y = (y_{m+1}, \dots, y_k)$ define

$$P'_N(y) = \sum_x W(y|x) P_N(x). \tag{3.9.2}$$

Then $P'_N(y) \to P'(y)$ as $P_N \to P$ and thus

$$-\frac{1}{n}\sum_y P'_N(y)\log P'_N(y) \to -\frac{1}{n}\sum_y P'(y)\log P'(y)$$

as $P_N \to P$, so that $H_n(P')$ is a continuous function on $\mathcal{P}_s$.

Thus, $H(P')$, by (3.9.1) being a limit of a decreasing sequence of continuous functions, is u.s.c. □

To deal with the rest of $R(P)$ we introduce functions $L_n(P)$ on $\mathcal{P}_s$ by

$$L_n(P) = \frac{1}{n}\sum_{x,y} W(y|x)P(x)\log W(y|x)$$

with $x$, $y$ as above.

**Lemma 43** *For a DFSC W*

$$L_n(P) \to H(P) - H(P'').$$

**Lemma 44** *For DFSC W $H(P) - H(P'')$ is an u.s.c. functional on $\mathcal{P}_s$.*

Since the sum of u.s.c. functions is again u.s.c., we have shown that

$$R(P) = H(P') + (H(P) - H(P''))$$

is u.s.c. on $\mathcal{P}_s$.

Combining this with Lemma 42 and Theorems 89 and 90 we get the result stated as Theorem 88 in the introduction.

## 3.10 Lecture on the Structure of the Entropy Rate of a Discrete, Stationary Source with Finite Alphabet (DSS)

For notation needed here we refer to Sects. 3.1–3.5. We study now several DSS $(\Omega, \mathcal{B}, P, T)$, where $P \in \mathcal{P}_s(\Omega, \mathcal{B}, T)$, the set of all PD's occurring in stationary sources. Therefore, when needed the dependence on $P$ will be indicated. Instead of

$h_t(w)$ we write now

$$h_t(w; P) = \begin{cases} -\frac{1}{t} \log P({}^1w^t) & \text{if } P({}^1w^t) > \infty \\ 0 & \text{otherwise} \end{cases} \tag{3.10.1}$$

and instead of $H(s, t)$ we write now

$$H(s, t; P) = -\sum_{\overline{w}} P(\overline{w}) \log P(\overline{w}), \tag{3.10.2}$$

where $\overline{w} = (w_1, \dots, w_t) = {}^1w^t$ is a cylinder set in $\Omega$ and atom in $\mathcal{B}(1, t)$.

We know already (see Sect. 3.2) that the entropy rate (mittlere Entropie) is

$$H(P) = \lim_{t \to \infty} \frac{1}{t} H(1, t; P), \tag{3.10.3}$$

where the limit is monotone decreasing. The Theorem of McMillan (see Sect. 3.2.4) says that

(i) the sequence $(h_t(\cdot; P))$ converges in $\mathcal{L}_P^1$ to a function $h(\cdot; P) \in \mathcal{L}_P^1$, which is $P$-almost unique
(ii) $h(\cdot, \cdot; P)$ is $T$-invariant (defined in general as $(Tf)(w) = f(Tw)$)

$$Th(\cdot; P) = h(\cdot, \cdot; P) \quad (P\text{-a.s.}),$$

and by changing $h(\cdot, \cdot; P)$ on a $P$-nullset one obtains

$$(Th)(w; P) = h(w; P) \quad (w \in \Omega) \tag{3.10.4}$$

(strict invariance).

Now (3.10.2), (3.10.3) and

$$\frac{1}{t} H(1, t; P) = \int h_t \, \mathrm{d}P$$

imply

$$H(P) = \int h(\cdot, \cdot, P) \, \mathrm{d}P. \tag{3.10.5}$$

We introduce now the linear space $\mathcal{LP}(\Omega, \mathcal{B})$ of $\sigma$-additive finite real-valued functions $h : \mathcal{B} \to \mathbb{R}$. The familiar subset $\mathcal{P}(\Omega, \mathcal{B})$ of PD's on $\mathcal{B}$ can be written as

$$\{h : h \in \mathcal{LP}(\Omega, \mathcal{B}), h \geq 0, h(\Omega) = 1\}.$$

$T$ induces a linear transformation of $\mathcal{LP}$ into itself:

$$(Th)(E) = h(T^{-1}E) \quad (E \in \mathcal{B})$$

and the set of invariant PD's

$$\mathcal{P}_s(\Omega, \mathcal{B}) = \{h : h \in \mathcal{P}(\Omega, \mathcal{B}), Th = h\}.$$

Let $\langle(P)$ be the equivalence class of $h(\cdot; P)$ in the set of $\mathcal{B}$-measurable functions. In Sect. 3.5 it was shown that for every PD $Q \in \mathcal{P}_0(\Omega, \mathcal{B})$ absolutely continuous with respect to $P$, in short $Q \ll P$, the relation $\langle(P) \leq \langle(Q)$ holds and that every $\langle(P)$ has a representative $h$ with $0 \leq h(w) \leq \log a$ $(w \in \Omega)$; furthermore, if $H(P) = \log a$, then $h(w) = \log a$ ($P$-a.e.).

Now notice that for $P_1, \dots, P_m \in \mathcal{P}_0$ and $P = \frac{1}{m}\sum_{k=1}^m P_k$ necessarily $P_k \ll P$ and therefore

$$\langle(P) \subset \bigcap_{k=1}^n \langle(P_k) \tag{3.10.6}$$

and therefore with *one h*

$$H\left(\sum_{k=1}^m \alpha_k P_k\right) = \int h\left(\sum_{k=1}^m \alpha_k P_k\right) \tag{3.10.7}$$

$$= \sum_{k=1}^m \alpha_k H(P_k) \quad \left(\alpha_k \geq 0, \sum \alpha_k = 1, h \in \langle(P)\right)$$

that is, the linearity of $H(P)$.

Equation (3.10.7) suggests the *conjecture* that even $\bigcap_{P \in \mathcal{P}_0} \langle(P) \neq 0$, so that in (i) the same limit function $h$ can serve for all $P \in \mathcal{P}_0$, which then must satisfy (3.10.5). This became known as

**Theorem 92** (Parthasarathy [45], and Jacobs [22])

$$H(P) = \int h \, \mathrm{d}P \quad (P \in \mathcal{P}_0) \tag{3.10.8}$$

*for the same function* $h(w) = H(m_w)$, $m_w = M(1_E, w)$ *defined in Sect.* 3.9.

This was proved independently by Parthasarathy and by Jacobs. Both proofs use the Krylov-Bogolioubov theory, but beyond this are quite different.

A result analog to Theorem 92 holds for the transmission rate $C(P)$ of an arbitrary stationary channel for $P \in \mathcal{P}_0$. Defining mean entropies $H(\widetilde{P})$, $H(P')$ as we defined $H(P)$ for the channel $W : (\Omega, \mathcal{B}) \to (\Omega', \mathcal{B}')$ we get $C(P) = H(P) + H(P') - H(\widetilde{P})$.

Furthermore we have

**Theorem 93** (Jacobs [22])

$$\lim_t h_t(w, P) = h \quad (\textit{in } \mathcal{L}_P^1, P \in \mathcal{P}_0).$$

**Theorem 94** (Parthasarathy [45]) *For a stationary channel W, using* $M(1_E, w) = m_w$ *(defined in Sect.* 3.9*) and*

$$c(w) = \begin{cases} C(m_w) & (w \in E) \\ 0 & \textit{otherwise}, \end{cases}$$

$\mathcal{B}_0$*-measurable,* $E = \{w \in \Omega, m_w \textit{ ergodic}\}$*, then*

$$C(P) = \int c \, \mathrm{d}P \quad (P \in \mathcal{P}_0).$$

*Remark* According to Jacobs [22], lines 17–21 on p. 35, he did Theorem 92 independently of [45] and Theorem 93 (above) alone, while he learned from Feinstein that Parthasarathy had obtained Theorem 94. In particular this result improves Breiman's Theorem in two respects: the channel is more general and as optimal PD an ergodic source can be chosen which is even regular (see Oxtoby [44], Jacobs [21]).

Jacobs gives credit to Breiman's work [5] as having stimulated his investigations, particularly by its idea to view $\Omega$ as metrical compact set on which the shift $T$ acts as topological automorphism.

Instead of Theorem 92 in [22], Jacobs passes right away from the entropy functional to an arbitrary linear functional on $\mathcal{LP}(\Omega, \mathcal{B})$ which is u.s.c. and obtains the key result, stated as

**Lemma 45** *Let* $R : \mathcal{P}_0(\Omega, \mathcal{B}) \to \mathbb{R}$ *be a bounded function with the properties*

*(i)*

$$R\left(\sum_{i=1}^n \alpha_i P_i\right) = \sum_{i=1}^n \alpha_i R(p_i) \quad \left(\alpha_i \geq 0, \sum p_i = 1\right).$$

*(ii)* $R$ *is u.s.c. according to the s-topology, that is,* $P_n \to P$ *implies*

$$R(P) \geq \limsup_n R(P_n).$$

*(iii)* $R(m_w)$ *is* $\mathcal{B}_0$*-measurable in* $w \in \Omega$ *(actually implied by (ii)).*

*Then for* $P \in \mathcal{P}_0$ *there is a P-almost unique invariant, bounded,* $\mathcal{B}_0$*-measurable function* $r(w)$ *so that*

$$R(q) = \int r \, \mathrm{d}q \quad (q \ll p) \tag{3.10.9}$$

*and*

$$R(m_w) = r(w) \quad (P\text{-}a.e.) \tag{3.10.10}$$

*and therefore in general*

$$R(P) = \int_E R(m_w)P(dw) \quad (P \in \mathcal{P}_0). \tag{3.10.11}$$

This is material for readers with more background in measure theory and functional analysis as can be expected from our standard readers. Given the interest we recommend reading the original paper or a book devoted to a large extent to this subject, if it exists.

## 3.11 Lecture on the Transmission of Bernoulli Sources Over Stationary Channels

We use again the spaces of double infinite spaces $(\Omega, \mathcal{B})$, $(\Omega', \mathcal{B}')$, $(\Omega'', \mathcal{B}'')$, and $(\Omega, \mathcal{B}, \mu) = (\Omega, \mu)$, in short, as sources and with the shift $T$ in all spaces a *stationary channel* $W : \Omega' \times \Omega'' \to [0, 1]$ is defined as usual. Now it is *not assumed* that the channel is non-anticipatory! We then speak in this generality of $W$ as SC.

We consider now $\mathcal{P}_n(\Omega)$, the set of PD on $\Omega$ stationary with respect to $T^n$ $(n \in \mathbb{N})$ and analogously $\mathcal{P}_n(\Omega', \Omega'')$. We endow these spaces with the weak topology, that is, the weakest topology on $\mathcal{P}_n(\Omega)$ such that for continuous $f : \Omega \to \mathbb{R}$, the map $\mu \to \int_\Omega d f \mu$ is continuous on $\mathcal{P}_n(\Omega)$. In a similar way one defines the weak topology on $\mathcal{P}_n(\Omega', \Omega'')$. As before we define stationary codes and sliding-block codes as well as block codes, not necessarily stationary.

### 3.11.1 Sources

A source $(\Omega, \mu)$ is stationary if $\mu \in \mathcal{P}_1(\Omega)$. If $\mu \in \mathcal{P}_n(\Omega)$, $(\Omega, \mu)$ is *an n-stationary source*. A stationary source, which is ergodic, is called an ergodic source.

If $(\Omega, \mu)$ is a source and $\varphi : \Omega \to \Omega'$ is a code, then $(\Omega', \mu^\varphi)$ is a source where

$$\mu^\varphi(E) = \mu(\varphi \in E) \quad (E \in \mathcal{B}).$$

We call a stationary source $(\Omega_1, \mu_1)$ a *factor* of the stationary source $(\Omega_2, \mu_2)$, if there is a stationary code $\varphi : \Omega_2 \to \Omega_1$.

A stationary source $(\Omega, \mu)$ is called memoryless if $\mu$ is a product measure and it is a Bernoulli source (abbreviated as B-source) if it is a factor of some memoryless source.

B-sources form a wide class of sources. For example, any stationary source $(\Omega, \mu)$, which is *mixing* and *n-Markovian* relative to $T$ is a B-source (Ornstein [42]).

Thus, the set of $\mu$ for which $(\Omega, \mu)$ is a B-source is a *dense subset* of $\mathcal{P}_1(\Omega)$ (with respect to weak topology).

There are various characterizations of B-sources known. For instance, $(\Omega, \mu)$ is a B-source if and only if $T$ is a *finitely determined* transformation on $(\Omega, \mathcal{B}, \mu)$ (Ornstein [42]) or if and only if $T$ is very weakly Bernoulli (Ornstein [42]), or if and only if $\mu$ is almost block independent (Shields [54]).

## 3.11.2 Channels

If a channel $(\Omega', \Omega'', W)$ or $W$, in short, is stationary and $(\Omega', \mu)$ is $n$-stationary, then $\mu \times W \in \mathcal{P}_n(\Omega', \Omega'')$.

We define a stationary channel $W$ to be *weakly continuous* if the map $\mu \to \mu \times W$ from $\mathcal{P}_1(\Omega') \to \mathcal{P}_1(\Omega' \times \Omega'')$ is continuous. We define a stationary channel $W$ to be *totally weakly continuous* if the map $\mu \to \mu \times W$ from $\mathcal{P}_n(\Omega')$ to $\mathcal{P}_n(\Omega' \times \Omega'')$ is continuous for every $n \in \mathbb{N}$.

**Examples of totally weakly continuous channels are:**

- stationary channels $W$ for which the map $w' \to W(E|w')$ from $\mathcal{B}' \to [0, 1]$ is continuous, for each finite-dimensional cylinder set $E \in \mathcal{B}''$.
- $\overline{d}$-continuous stationary channels as defined in [15]

We confine ourselves here to define that a stationary channel $W$ is $\overline{d}$-continuous if for any $\varepsilon > 0$ there exists an $n_0 \in \mathbb{N}$ such that for any $n \geq n_0$ and $x, \hat{x} \in \Omega'$ with ${}^1x^n ={}^1 \hat{x}^n$ we can find $\mathcal{Y}^n$-valued RV's $Y^n, \hat{Y}^n$ defined on some probability space, so that

(i) the PD $P_Y = W^n(\cdot|x)$
(ii) the PD $P_{\hat{Y}} = W^n(\cdot|\hat{x})$
(iii) $\mathbb{E}\rho(Y^n, \hat{Y}^n) < \varepsilon$.

Here

$$\rho_n(y^n, \hat{y}^n) = n^{-1}|\{1 \leq i \leq n : y_i \neq \hat{y}_i\}| \quad (y^n, \hat{y}^n \in \mathcal{Y}^n).$$

The $\overline{d}$-continuous channels are the most general class of stationary channels for which coding theorems have been proved [15, 16] prior to the work of Kieffer.

## 3.11.3 The Source Channel Hook-up

The sequence $U, X, Y, Z$ of processes defined on some probability space is a "hook-up" of the source $(\Omega, \mu)$ to the channel $W$ if

(i) $U$ is $\Omega$-valued,
$X$ is $\Omega'$-valued,
$Y$ is $\Omega''$-valued,
$V$ is $\Omega$-valued
(ii) $U, X, Y, V$ form a Markov chain
(iii) $P_U = \mu$
(iv) $P_X \times W = P_{X,Y}$

There are codes $(\varphi, \psi)$ where $\varphi : \Omega \to \Omega'$ and $\psi : \Omega'' \to \Omega$ (called encoder and decoder) such that $x = \varphi(u), v = \psi(y)$ where also sliding block transmissibility over a stationary $W$ and the capacity $C_{bs}(W)$ are defined.

Here are the main results.

**Theorem 95** *Let $W$ be a weakly continuous stationary channel. Then a B-source $(\Omega, \mu)$ is sliding block transmissible over $W$ if and only if $H(\mu) \leq C_{bs}(W)$.*

For the more special $\overline{d}$-continuous channels, sliding-block coding theorems had been obtained for the class of all *aperiodic* ergodic sources [16, 30].

By restricting ourselves to the subclass of B-sources, one can dispense with the $\overline{d}$-continuity requirement on the channel.

We recall the definition of block transmissibility.

**Theorem 96** *Let $W$ be a totally weakly continuous stationary channel. Then a B-source $(\Omega, \mu)$ is block transmissible over $W$ if and only if $H(\mu) \leq C_b(W)$.*

For stationary $\overline{d}$-continuous channels in [16] a coding theorem had been obtained for the class of *all* ergodic sources (not only aperiodic ones) for block coding.

Theorem 96 generalizes the assumption on the channel and weakens the conclusion, but for the nice class of B-sources.

There is a stronger notion of block transmissibility: $(\Omega, \mu)$ is *strongly* block transmissible over $W$ if for each $\varepsilon > 0$, there exists an $n$ and a hook-up $U, X, Y, Z$ for which the condition

$$n^{-1} \sum_{i=1}^{n} \Pr(U_i \neq V_i) < \varepsilon \tag{3.11.1}$$

in the definition of block decoding is replaced by

$$\Pr({}^1U^n \neq {}^1V^n) < \varepsilon. \tag{3.11.2}$$

*Remark* Using our work on balanced coloring Csiszár observed that Wyner's work on wire-tap channels could be improved by going from error type (3.11.1) to that of (3.11.2).

**Research Problem**. Make a systematic investigation in information theory, where this improvement can be obtained. This is interesting in the light of the statement on

p. 946, lines 20–22 of [31]: "Of the two types, we emphasize block transmissibility, for the simple reason that the techniques we use do not appear to apply to strong block transmissibility."

We denote the capacity in the strong case by $C_{bs}(W)$ (block strong!). Facts of interest are

- for $\overline{d}$-continuous channels $W$ $C_b(W) = C_{bs}(W)$ (Theorem 11 of [31])
- there can exist a source which is block transmissible and not strongly block transmissible (Nedoma [40]), but the only way this can happen is if the entropy of the source *equals* the capacity of the channel!
- for non-$\overline{d}$-continuous channels it is an open problem whether $C_b = C_{bs}$.

We present now (without proofs) more results of Kieffer [31]. It is convenient to use the abbreviations WCSC for weakly continuous stationary channels $W$ and TWCSC for totally weakly continuous stationary channels $W$.

Finally, we introduce $\mathcal{M}_{sb}(W)$ (resp. $\mathcal{M}_b(W)$) as set of all stationary sources sliding-block (resp. block) transmissible over $W$.

**Theorem 97** *Let $W$ be a* WCSC *and let $(\Omega, \mu)$ be a stationary aperiodic source. Then $(\Omega, \mu) \in \mathcal{M}_{sb}(W)$ if for each $\varepsilon > 0$ there is a hook-up $U, X, Y, V$ such that*

*(i)* $(U, X, Y, V)$ *is jointly stationary*
*(ii)* $Pr(U_0 \neq V_0) < \varepsilon$.

*Remark* This theorem says in essence that in the definition of sliding-block transmissibility it makes no difference whether the codings $U \to X$ and $V \to V$ are deterministic or random.

**Theorem 98** *For a* WCSC $W$ $\mathcal{M}_{sb}(W) \subset \mathcal{M}_b(W)$.

**Theorem 99** *Let $W$ be a* WCSC *and let $(\Omega, \mu)$ be a stationary aperiodic source. Then $(\Omega, \mu) \in \mathcal{M}_b(W)$ if for each $\varepsilon > 0$ there exists a hook-up $U, X, Y, V$ and an $n \in \mathbb{N}$ such that*

*(i)* $(U, X, Y, V)$ *is jointly n-stationary*
*(ii)* $n^{-1} \sum_{i=1}^{n} Pr(U_i \neq V_i) < \varepsilon$.

**Theorem 100** *Let $W$ be a* WCSC *and let $(\Omega, \mu_j)_{j=1}^{\infty}$ be a sequence of sources in $\mathcal{M}_{sb}(W)$. Furthermore, let $(\Omega, \mu)$ be an aperiodic source such that $\overline{d}(\mu_j, \mu) \to 0$. Then $(\Omega, \mu) \in \mathcal{M}_{sb}(W)$.*

**Theorem 101** *Let $W$ be* TWCSC *and let $(\Omega, \mu_j)_{j=1}^{\infty}$ be a sequence of sources in $\mathcal{M}_b(W)$. Furthermore, let $(\Omega, \mu)$ be an aperiodic source such that $\overline{d}(\mu_j, \mu) \to 0$. Then $(\Omega, \mu) \in \mathcal{M}_b(W)$.*

**Theorem 102** *Let $W$ be* WCSC *and let $(\Omega_1, \mu_1) \in \mathcal{M}_{sb}(W)$. Furthermore, let $(\Omega_2, \mu_2)$ be an aperiodic factor of $(\Omega_1, \mu_1)$. Then $(\Omega_2, \mu_2) \in \mathcal{M}_{sb}(W)$.*

*Remark* Aperiodicity of the factor cannot be removed. There are examples in (Kieffer [30]) of $\overline{d}$-continuous channels showing this.

**Theorem 103** *Let $W$ be a* TWCSC *and let $(\Omega, \mu) \in \mathcal{M}_b(W)$. Then every aperiodic factor of $(\Omega, \mu)$ is in $\mathcal{M}_b(W)$.*

**Theorem 104** *Let $W$ be a stationary channel. If now $(\Omega, \mu)$ is an ergodic source in $\mathcal{C}_{sb}(W)$ or $\mathcal{C}_b(W)$, then $H(\mu) \leq C^*(W)$.*

Following [15], define the information quantile capacity of the stationary channel $W$ as follows:

$$C^*(W) = \lim_{\lambda \to 0^*} C^*(W, \lambda), \quad \text{where for} \lambda > 0$$

$$C^*(W, \lambda) = \sup_{\mu \in \mathcal{P}_1(\Omega')} \sup \left\{ r : \mu \times w \left[ i_{\mu \times w}(x, y) \leq r \right] < \lambda \right\}$$

$X, Y$ in this case being the projections from $\Omega' \times \Omega''$ to $\Omega'$, $\Omega''$, respectively. (Actually, it can be shown that the outer supremum over $\mathcal{P}_1(\Omega')$ may be replaced by a supremum over $\mathcal{P}_2(\Omega')$ $(n \in \mathbb{N})$ (see Kieffer's Lemma 4 in [29]).)

#### 3.11.3.1 An Example

We see from Theorem 98 that $C_{sb}(W) \leq C_b(W)$. For certain types of $\overline{d}$-continuous channels [16, 30], it is known that $C_{sb}(W) = C_b(W)$. We present an example of a $\overline{d}$-continuous channel for which $C_{sb} < C_b(W)$. We need

**Lemma 46** *Let $W$ be a stationary channel. Suppose that $\mathcal{M}_{sb}(W)$ contains an aperiodic source, then for each $\varepsilon > 0$ there exists an $F \in \mathcal{B}''$ and $x \in \Omega'$ such that $W(F|x) < \varepsilon$ and $W(F^c|Tx) < \varepsilon$.*

*Now we construct the example of an AC.* Let $W$ be such that $\mathcal{X} = \mathcal{Y} = \{0, 1\}$ and

$$W(E|x) = \frac{1}{2}(1_E(x) + 1_E(Tx)) \quad x \in \Omega', E \in \mathcal{B}''.$$

Thus an input $x$ when transmitted over the averaged channel yields either $x$ itself or the shift of itself, each with probability $\frac{1}{2}$. This channel is stationary and has finite input memory, that is, there exists an $M \in \mathbb{N}$ such that for $n \in \mathbb{N}$ and $k \in \mathbb{Z}$, if $E \subset \mathcal{Y}^n$ and $x, \hat{x}$ are sequences in $\Omega'$ with

$${}^{k-M}x^{k+M+n-1} = {}^{k-M}\hat{x}^{k+M+n-1}$$

then

$$W\left(\left\{y \in \Omega'' : {}^k y^{k+n-1} \in E\right\} |x\right) = W\left(\left\{y \in \Omega'' : {}^k y^{k+n-1} \in E\right\} |\hat{x}\right).$$

Such channels are known to be $\overline{d}$-continuous (see [15]).

Now for every $E \in \mathcal{B}''$

$$W(E|x) + W(E^c|Tx) \geq \frac{1}{2}(1_E(Tx) + 1_{E^c}(Tx)) = \frac{1}{2}.$$

Hence, by Lemma 46, $\mathcal{M}_{\text{sb}}(W)$ contains no aperiodic sources. This implies

$$C_{\text{sb}}(W) = 0.$$

However, it is easy to construct a block encoder and decoder for transmitting an ergodic source of entropy $< \frac{\log 2}{2}$.

Thus $C_{\text{b}}(W) \geq \frac{\log 2}{2}$, or one simply observes that $W$ is an average over two noiseless DMC's of capacity $\log 2$.

By [1] $C_{\text{b}}(W) = \frac{1}{2}$.

**Research Problems** (Kieffer [31])

- Can one get coding theorems for the class of all ergodic aperiodic sources, not just the B-sources?
  Can one obtain formulas for $C_{\text{sb}}(W)$ and $C_{\text{b}}(W)$? Specifically, does $C_{\text{b}}(W) = C^*(W)$ hold in general?
- Can one find a necessary and sufficient condition in order that $C_{\text{sb}} = C_{\text{b}}(W)$?
- Specifically, motivated by the example given above, is the only way that $C_{\text{sb}}(W)$ can fail to equal $C_{\text{b}}(W)$ if $C_{\text{sb}} = 0$?
- Is every weakly continuous channel totally weakly continuous?
- When does $C_{\text{b}}(W) = C_{\text{bs}}(W)$ hold?
- Is there a coding theorem analogous to Theorem 96 for the strong type of block transmissibility?

## 3.12 Lecture on Block Coding for Weakly Continuous Channels

This is a report of [35] without proofs.

### *3.12.1 Introduction*

Given a discrete stationary channel (DSC) $W$ for which the map $\mu \to \mu \times W$ carrying each stationary, ergodic input $\mu$ into the input-output PD $\mu \times \nu$ is continuous (with respect to the weak topology) at *at least one input*, it is shown that every stationary ergodic source with *sufficiently small entropy* is block transmissible over $W$.

If this weak continuity condition is satisfied at every stationary ergodic input, one obtains the class of WCSC for which the usual source/channel block coding theorem and converse hold with the usual notion of channel capacity $C_b(W)$.

An example shows that the class of WCSC *properly* includes the class of $\overline{d}$-continuous channels.

It is also shown that every stationary channel $W$ is "almost" weakly continuous in the sense that every input-output PD $\mu \times W$ for $W$ can be obtained by *sending* $\mu$ *over an appropriate* WCSC (depending on $\mu$). This indicates that WCSC may be the most general stationary channel for which one *would need* a coding theorem.

These results go beyond the capacity results for $\overline{d}$-continuous channels by Gray and Ornstein [15], who obtained a formula for the operational capacity $C(W)$ showing that it equals information quantile capacity.

Recall the definition of a WCSC $W$:

$W$ is weakly continuous at stationary, ergodic source $\mu$, if for every sequence $(\mu_j)$ of stationary ergodic sources with $\mu_j \to \mu$ (weakly) we have $\mu_j \times W \to \mu \times W$ (weakly). $W$ is a WCSC, if it is weakly continuous at every stationary ergodic source. In [31] a stronger form of being weakly continuous was required for $W$. Its domain was taken to be the larger set of all PD $\mu$ of stationary, but not necessarily ergodic sources. However, this restrictive definition is not needed (here)—we keep the name WCSC.

### 3.12.2 Coding Theorems

For stationary source $(\Omega', \mu)$ and stationary channel $W$, let $i(\mu, W)$ be the asymptotic information density defined for $\mu \times W$-almost all $(w', w'') \in \Omega' \times \Omega''$ by

$$i(\mu, W)(x, y) = \lim_{n\to\infty} n^{-1} \log \frac{\mu \times w\{(w', w'') : w'^n = x^n, w''^n = y^n\}}{\mu \times w\{(w', w'') : w'^n = x^n\} \cdot \mu \times w\{(w', w'') : w''^n = y^n\}}.$$

Let $I(\mu, W)$ denote the $\mu \times W$-expectation of $i(\mu, W)$; $I(\mu, W)$ is the information rate for the input-output PD $\mu \times W$.

We are ready to state the main result. It implies that if a stationary channel $W$ is weakly continuous at the stationary, ergodic source $(\Omega', \nu)$, then every stationary, ergodic source with entropy less than the $\nu \times W$ essential infimum of $i(\nu, W)$ is block transmissible.

In particular, if $\nu \times W$ is ergodic with respect to $T : (w', w'') \to (Tw', Tw'')$ on $\Omega' \times \Omega''$, every stationary, ergodic source with entropy less than $I(\nu, W)$ is block transmissible.

Therefore, in order to conclude that every stationary, ergodic source with sufficiently small entropy is block transmissible over a channel, it suffices to show that the channel is "well-behaved" at just one stationary and ergodic input.

**Theorem 105** *Let $(\Omega', \nu)$ be a stationary, ergodic source, and let $W$ be a stationary channel weakly continuous at $(\Omega', \nu)$. Suppose that $\nu \times W(i(V, W)) \leq R) < \varepsilon$, where $R > 0$, $0 < \varepsilon < \frac{1}{16}$. Then every stationary, ergodic source $(\Omega, \mu)$ with entropy less than $R - h(2\sqrt{\varepsilon}) - \varepsilon \log |\mathcal{X}|$ is $\sqrt{\varepsilon}$-block transmissible.*

The theorem is easily proved by means of a universal version of the Slepian-Wolf theorem [32].

Following [15] define the information quantile capacity of the stationary channel $W$ as

$$C^*(W) = \lim_{\varepsilon \to 0^*} \left[ \sup_{\nu} (\sup\{R : V \times W(i(V, W) \leq R) < \varepsilon\}) \right]$$

where the other supremum is taken over all PD's $\nu$ of stationary, ergodic sources $(\Omega', \nu)$.

The following result gives the standard Shannon-type source/channel block coding theorem and converse, if the weak continuity condition holds at all positive entropy stationary, ergodic inputs.

**Theorem 106** *Let the stationary channel $W$ be weakly continuous at every positive-entropy stationary, ergodic source $(\Omega', \nu)$. Let $(\Omega, \mu)$ be any stationary, ergodic source. Then*

*(i) $(\Omega, \mu)$ is block transmissible over $W$, if it has entropy less than $C^*(W)$.*
*(ii) $(\Omega, \mu)$ is not block transmissible over $W$, if it has entropy greater than $C^*(W)$.*

*Proof* Part (i) is a corollary of Theorem 105. Part (ii) is Theorem 10 of [31]. □

Theorem 106 implies that every WCSC has a Shannon-type source/channel block coding theorem and converse. This implies the result of [15] for $\overline{d}$-continuous channels, because they are included in the class of WCSC, as shown in [31].

It is not hard to show that the inclusion is proper.

We define two channels $W$, $W'$ to be equivalent if $W(\cdot|x) = W'(\cdot|x)$ for all $x \in E \subset \Omega'$ that has $\mu$-measure one for *every* stationary source $(\Omega', \mu)$.

If $W$ is a stationary $\overline{d}$-continuous channel $W$ one can find a non-$\overline{d}$-continuous stationary channel $W'$ equivalent to $W$. Such a channel must be weakly continuous because $W$ is, and because $\mu \times W = \mu \times W'$ for every stationary source $(\Omega', \mu)$.

Therefore, the class of weakly continuous channels is strictly larger than the class of $\overline{d}$-continuous channels. However, it is not hard to see that if two stationary channels are equivalent and a block coding theorem holds for one of them, then it holds for the other, with the capacities of the two channels being equal. Therefore, it follows from the $\overline{d}$-continuous block coding theorem that any channel equivalent to a $\overline{d}$-continuous channel has a block coding theorem. One does not need to use Theorem 106 for such channels.

To show that Theorem 106 applies to cases for which the $\overline{d}$-continuous block coding theorem of [15] does not apply we give now a WCSC *not equivalent* to a $\overline{d}$-continuous channel.

### 3.12.3 An Example

For a PD $\mu$ on $(\Omega, \mathcal{B})$ and $m \in \mathbb{N}$ define a PD $\mathcal{Z}^n$

$$\mu^{(m)}(z^m) = \mu\{w \in \Omega : w^m = z^m\}, \quad z^m \in \mathcal{Z}^m.$$

Fix the processes $X, Y$ that are the projections $\Omega' \times \Omega'' \to \Omega', \Omega''$, respectively. Fix $\overline{X}$ to be the process that is the identity map from $\Omega' \to \Omega'$.

Take now $\mathcal{X} = \mathcal{Y} = \{0, 1\}$ and define $t : \Omega' \to \{1, 2, \dots, \infty\}$ as follows

$$t(x) = \inf\{i > 0 : x_i = 0\}, \quad x \in \Omega'.$$

Note for future use that:

(i) $\{t = i\}$ is a finite-dimensional cylinder set, $1 \leq i < \infty$
(ii) $\mu\{t < \infty\} = 1$, for every stationary ergodic source $(\Omega', \mu)$ for which $\mu^{(1)}(0) > 0$.

Define a function $\varphi : \Omega' \to \mathcal{X}$ as follows:

$$\varphi(x) \begin{cases} \neq x_0 & \text{if } t(x) \text{ is even} \\ = x_0 & \text{otherwise.} \end{cases}$$

Let $\hat{\varphi} : \Omega' \to \Omega'$ be the map such that $\hat{\varphi}T = T\hat{\varphi}$ and

$$\hat{\varphi}(x)_0 = \varphi(x), \quad x \in \Omega'.$$

Let $y, z$ be sequences in $\Omega'$ such that

$$y_i \begin{cases} = 0 \; i \text{even} \\ = 1 \; i \text{odd} \end{cases}$$
$$z \; = Ty$$

Let $\hat{x}$ be the sequence in $\Omega'$ that is identical to 1. Let $W$ be the stationary channel such that for each $x \neq \hat{x}$ $W(\cdot|x)$ is the PD concentrated on $\{\hat{\varphi}(x)\}$ and $W(\cdot|\hat{x})$ assigns $y, z$ each probability $\frac{1}{2}$. This is almost the channel defined by Neuhoff and Shields [41] as an example of a non-$\overline{d}$-continuous channel.

Their argument can be moved slightly to prove the following statement (needed in the last section below) that implies that $W$ is not equivalent to a $\overline{d}$-continuous channel:

(iii) If $(\Omega', \mu)$ is any stationary and ergodic source such that $\mu(\underline{E}) > 0$ for every non-empty finite-dimensional cylinder set $E$, then there is no $\overline{d}$-continuous channel $\hat{W}$ such that $\mu \times W = \mu \times \hat{W}$.

The proof can be found in [35].

### 3.12.4 Every Channel Is Almost Weakly Continuous

The result has been described in the introduction. We see from (iii) in Sect. 3.12.3 that the following Theorem 107 fails for $\overline{d}$-continuous channels

**Theorem 107** *Let $W$ be a stationary channel and let $(\Omega', \mu)$ be a stationary ergodic source. There exists a WCSC $\hat{W}$ such that*

$$W(\cdot|x) = \hat{W}(\cdot|x) \quad \textit{for } \mu\textit{-almost all } x \in \Omega'.$$

## 3.13 Lecture on Zero-Error Transmission Over a DMC Requires Infinite Expected Coding Time

### 3.13.1 Introduction

It is shown that if the channel $W$ has all its transmission probabilities positive

$$\min_{x,y} W(y|x) > 0 \tag{3.13.1}$$

as in the BSC, then the codes given by Gray et al. must necessarily be impractical from an information-theoretic point of view in that the time required to determine $n$ successive encoder or decoder output symbols approaches infinity faster than linearly in $n$.

We always assume that the source $(\Omega, \mu)$ is stationary and at times that it is ergodic.

### 3.13.2 A Class of Practical Codes

Suppose $(\Omega, \mu)$ is coded into the source $(\widetilde{\Omega}, \widetilde{\mu})$ using a stationary coder $f : \Omega \to \widetilde{\Omega}$ (i.e., a measurable map commuting with the shifts).

For each $x \in \Omega$, let

$$t_f(x) = \inf\{1 \leq i \leq \infty : f(x)_0 = f(y)_0 \text{ if } y \in \Omega \text{ and } (x_{-i}, \dots, x_i) = (y_{-i}, \dots, y_i)\}.$$

(Define $t_f(x) = \infty$, if the infimum is over an empty set.)

Intuitively, if $t_f(x) = i < \infty$, we only have to look at the *finite "window"* $(x_{-i}, \dots, x_i)$ of $x$ in order to determine $f(x)_0$.

One can easily visualize an infinite tree which can be used to find $(f(x))_0 \in \widetilde{\Omega}$ for each $x \in \Omega$ satisfying $t_f(x) < \infty$.

The tree consists of a zeroth order node, at least one node of order $i$ $(i \geq 1)$ and at most one branch connecting each node of order $i$ and each node of order $i+1$ $(i \geq 0)$.

If there are no branches from a given node to any node of the next order, a symbol from $\widetilde{\Omega}$ is written next to that node.

Given $x \in \Omega$, a unique path through the tree is determined. One first looks at the segment $(x_{-1}, x_0, x_1)$ which determines a branch from the zeroth order node to a certain first-order node. If there is no branch leading from this node to a second order node, then $t_f(x) = 1$ and $f(x)_0$ is the symbol at this node. Otherwise, from this node a unique branch is selected leading to second order node by looking at the segment $(x_{-2}, x_{-1}, x_0, x_1, x_2)$. Proceeding in this way, $t_f(x) = i < \infty$, the path we follow starts at the zeroth order node and terminates at some $i$th order node, where the symbol $f(x)_0$ is read off. If $t_f(x) = \infty$, the path through the tree never terminates.

Assuming it takes one time unit to traverse any branch of the tree, we can interpret $t_f(x)$ as the time required to code $x$ into the time zero symbol of the coded sequence. Thus, the time it takes to determine $n$ successive symbols $f(x)_0, \ldots, f(x)_{n-1}$ is $t_f(x) + \cdots + t_f(T^{n-1}x)$. Suppose the expected value $\mathbb{E}_\mu t_f$ is finite. Then for almost all $x \in \Omega$, the quantity

$$n^{-1}(t_f(x) + \cdots + t_f(T^{n-1}x))$$

has a finite limit as $n \to \infty$ (which is $\mathbb{E}\mu f$, if $(\Omega, \mu)$ is ergodic by Birkhoff's Individual Ergodic Theorem). Consequently, the coder $f$ is "practical" in the sense that the time it takes to determine $n$ successive output symbols approaches infinity no faster than linearly in $n$.

### 3.13.3 Zero-Error Transmission of a Source Over a Channel

Fix for the rest of the lecture a stationary channel $W$.

A *hook-up* of the information source $(\Omega, \mu)$ to $W$ is a sextuple $(U, X, Y, V, \varphi, \psi)$ where

(i) $U = (U_i)$, $X = (X_i)$, $Y = (Y_i)$, $V = (V_i)$ are bilateral finite-state processes defined on some common probability space with respective state spaces $\mathcal{Z}, \mathcal{X}, \mathcal{Y}, \mathcal{Z}$
(ii) $\varphi : \Omega \to \Omega'$ and $\psi : \Omega'' \to \Omega$ are stationary codes
(iii) $U, X, Y, V$ form a Markov chain, $P_U = \mu$; $X = \varphi(U)$, the PD of $Y$ conditional on $X$ is given by $W$, and $V = \psi(Y)$.

In other words again, we code $U$ (which is a model for $(\Omega, \mu)$) into the process $X$ using the encoder $\varphi$, transmit $X$ directly over the channel $W$ obtaining the output $Y$, and then code $Y$ into $V$ using the decoder $\psi$. Source/channel transmission theorems of Information Theory give conditions under which a hook-up $(U, X, Y, V, \varphi, \psi)$ exists for which $U$ and $V$ are *close in an appropriate sense.*

Here our concern is *zero-error transmission*. We call the hook-up *zero-error hook-up* if $U = V$ a.s.

We consider the situation where $W$ is a DMC with capacity $C_0$ and $(\Omega, \mu)$ is a DMS with entropy rate $0 < H(\mu) < C$.

Originally Gray et al. showed there is a zero-error hook-up of a DMS $(\Omega, \mu)$ to more general channels $W$ called totally ergodic $\overline{d}$-continuous, which need not concern us here.

### 3.13.4 Structure of Codes in Zero-Error Hook-up

**Definition 21** We call the stationary $W$ non-singular if $W(E|x) > 0$ for every $x \in \Omega'$ and every finite-dimensional cylinder set $E \in \mathcal{B}''$. We call the stationary source $(\Omega, \mu)$ non-trivial if $\mu$ is not concentrated on a *constant* sequence in $\Omega$.

**Lemma 47** *Let $W$ be a nonsingular stationary channel, let $(\Omega, \mu)$ be a nontrivial stationary source and let $(U, X, Y, V, \varphi, \psi)$ be any zero-error hook-up for them, then*

$$Pr(t_\psi(Y) = \infty) > 0.$$

*Proof* Fix $a \in \mathcal{Z}$ with $\Pr(U_0 = a) > 0$. For each $i$ $(1 \leq i < \infty)$ we have

$$\Pr(\{t_\psi(Y) = i, \psi(Y)_0 \neq a\}|U) = 0 \text{ a.s. on } (U_0 = a). \tag{3.13.2}$$

Since $\{t_\psi(Y) = i, \psi(Y)_0 \neq a\}$ is an event of the form $\{Y \in E\}$, $E$ finite-dimensional cylinder set in $\mathcal{B}''$, it must be the null event; otherwise (3.13.2) would be violated by the nonsingularity of the channel.

Therefore

$$\{t_\varphi(Y) < \infty, \psi(Y)_0 \neq 0\} = \emptyset. \tag{3.13.3}$$

If $\Pr(t_\psi(Y) < \infty) = 1$, then (3.13.3) implies that $\Pr(\psi(Y)_0 = a) = 1$. Since $U = \psi(Y)$ a.s., this implies $(\Omega, \mu)$ is not nontrivial, a contradiction. □

**Definition 22** We call the stationary process $X = (X_i)_{-\infty}^{\infty}$ bilaterally deterministic if to each $n \in \mathbb{N}$ $(X_{-n}, \ldots, X_n)$ is almost surely a function of $(X_i : |i| > n)$.

**Lemma 48** *Let $W$ be a nonsingular, stationary* DMC, *let $(\Omega, \mu)$ be a stationary source, and let $(U, X, Y, V, \varphi, \psi)$ be a zero-error hook-up for them, then $X$ is bilaterally deterministic.*

*Proof* For each $1 \leq i < \infty$ let $\mathcal{B}'_i$ be the $\sigma$-algebra generated by $\{X_j : |j| > i\}$. Fix $b \in \mathcal{X}$. Since $X = \varphi(\psi(Y))$ a.s. we can pick $F \in \mathcal{B}''$ such that $\{Y \in F\}$ and $\{X_0 = x\}$ are the same events a.s.

Thus

$$\Pr(Y \in F|X) = 1 \text{ a.s. on } \{X_0 = b\}. \tag{3.13.4}$$

Fix $i$, $0 \leq i < \infty$. Since $W$ is a DMC,

$$\Pr(Y \in F|X) = \sum_{b \in \mathcal{X}^{2i+1}} \Pr(Y_{-i}, \ldots, Y_i = x|X) \Pr({}^{i+1}Y^{\infty} \in F_X|\mathcal{B}'_i) \quad (3.13.5)$$

$$\Pr(Y \in F|\mathcal{B}'_i) = \sum_{b \in \mathcal{X}^{2i+1}} \Pr((Y_{-i}, \ldots, Y_i) = b|\mathcal{B}'_i) \Pr({}^{i+1}Y^{\infty} \in F_b|\mathcal{B}'_i), \quad (3.13.6)$$

where

$${}^{i+1}Y^{\infty} = \{Y_j : |j| > i\}, \quad F_b = \{(Y_j : |j| > i) : Y \in F, (Y_{-i}, \ldots, Y_i) = b\}.$$

Applying (3.13.5), (3.13.6) and the fact that $W$ is non-singular to (3.13.4), we see that

$$\Pr(Y \in F|\mathcal{B}'_i) = 1 \text{ a.s. on } \{X_o = b\}.$$

Letting $i \to \infty$, and replacing $\{Y \in F\}$ with $\{X_0 = b\}$,

$$\Pr(X_0 = b|\mathcal{B}'_{\infty}) = 1 \text{ a.s. on } \{X_0 = b\}, \quad (3.13.7)$$

where $\mathcal{B}'_{\infty}$ is the bilateral tail $\sigma$-field $\bigcap_{i=1}^{\infty} \mathcal{B}'_i$. Equation (3.13.7) implies that $\{X_0 = b\}$ is almost surely a member of $\mathcal{B}'_{\infty}$. Therefore, $X$ is bilaterally deterministic. □

**Theorem 108** *Let $W$ be a nonsingular* DMC, *let $(\Omega, \mu)$ be a nontrivial DMS, and let $(U, X, Y, V, \varphi, \psi)$ be any zero-error hook-up of them, then*

$$\mathbb{E}t_{\varphi}(U) = \infty, \quad \mathbb{E}t_{\psi}(Y) = \infty.$$

*Proof* Suppose for any hook-up $\mathbb{E}t_{\varphi}(U) < \infty$. By a result of del Junco and Rahe [24] $X$ is a weak Bernoulli process ([53, p. 89]). As Ornstein and Weiss [43] point out, a weak Bernoulli process has a trivial bilateral tail. But $X$ is bilaterally deterministic by Lemma 48, and so $X$ must be constant almost surely. This implies that $(\Omega, \mu)$ is not nontrivial, a contradiction. □

**Research Problem** (Kieffer)
In the language of Ergodic Theory, a stationary code $f$ which codes the source $(\Omega, \mu)$ into the source $(\tilde{\Omega}, \tilde{\mu})$ is called *finitary* if $t_f(x) < \infty$ for $\mu$-almost all $x \in \Omega$ [25]. Since the expected coding time can be infinite for a finitary code, the results here do not answer the question whether the encoder in a zero-error hook-up can be finitary. (The decoder cannot be finitary, by Lemma 47.)

### *3.13.5 An Application to the Theory of Isomorphism of Stationary Processes*

Two bilateral stationary processes $X = \{X_i\}$, $Y = \{Y_i\}$ defined an a common probability space and with finite alphabets (or state spaces) $\mathcal{X}, \mathcal{Y}$, respectively, are *isomorphic* if there are *stationary codes* $\varphi : \Omega' \to \Omega''$, $\psi : \Omega'' \to \Omega'$ such that $Y = \varphi(X)$ a.s. and $X = \psi(Y)$ a.s.

Let $F$ be a finite set and $k$ a positive integer. Let $Z_1, \dots, Z_k$ be the projections from $F^k \to F$. A PD $\pi$ on $F^k$ is called invariant if $(Z_1, \dots, Z_{k-1})$ and $(Z_2, \dots, Z_k)$ have the same distribution under $\pi$. We define the entropy $H(\pi)$ of $\pi$ to be $H_\pi(Z_k|Z_1, \dots, Z_{k-1})$, the conditional entropy of $Z_k$ given $(Z_1, \dots, Z_{k-1})$ under $\pi$.

As a consequence of Lemma 48 one obtains the following result on the isomorphism of stationary processes.

**Theorem 109** *Let $U$ be a bilateral, finite-state, stationary, ergodic, aperiodic process. Let $S$ be a finite set, $k \in \mathbb{N}$ and $\pi$ and invariant PD on $S^k$ with $H(\pi) > H(U)$, the entropy of the process. Then, for any $\varepsilon > 0$, there is a bilaterally deterministic stationary process $X$ with state space $S$ isomorphic to $U$ such that*

$$|Pr\{(X_1, \dots, X_k) = s\} - \pi(s)| < \varepsilon \quad (s \in S^k).$$

*Remark* This is a strengthening of a result of Ornstein and Weiss [43], who showed that any finite-state stationary ergodic *aperiodic* process is isomorphic to a bilaterally deterministic process.

*Proof* Let $U$ have state space $\mathcal{U}$ and distribution $\mu$. Find an ergodic source $[s, \sigma]$ such that $H(\sigma) > H(\mu)$ and

$$|\sigma^{(k)}(s) - \pi(s)| < \frac{\varepsilon}{2} \quad (s \in S^k) \tag{3.13.8}$$

where $\sigma^{(k)}$ is the $k$-dimensional marginal distribution of $\sigma$. Find a nonsingular DMC $W : S \to S$ such that $I(\sigma, W)$, the information rate of the pair PD $\sigma \times W$ on $S^\infty \times S^\infty$ resulting from sending $[S, \sigma]$ over $W$, will be so close to $H(\sigma)$ that $I(\sigma, W) > H(\mu)$. (Note: If $W$ is noiseless then $I(\sigma, W) = H(\sigma)$; therefore,s one needs only a slightly noiseless channel.) By Theorem 1 of [33] there is a zero-error hook-up $(\overline{U}, \overline{X}, \overline{Y}, \overline{V}, \varphi, \psi)$ of $(\Omega, \mu)$ to $W$ such that

$$\left|\Pr\{(\overline{X}_1, \dots, \overline{X}_k) = s\} - \sigma^{(k)}(s)\right| < \frac{\varepsilon}{2}, \quad s \in S^k,$$

and hence from (3.13.8)

$$\left|\Pr\{(\overline{X}_1, \dots, \overline{X}_k) = s\} - \pi(s)\right| < \varepsilon, \quad s \in S^k.$$

By Lemma 48 $\overline{X}$ is bilaterally deterministic. Since $\overline{U}, U$ have the same distribution and $\overline{U}, \overline{X}$ are isomorphic, we are done. □

**Research Program** Continue our ideas on probabilistic $\varepsilon$-isomorphy for correlated pairs of sources (Marton, ...).

## 3.14 Lecture on Zero-Error Stationary Coding Over Stationary Channels

### 3.14.1 *Capacities of Gray, Ornstein, and Dobrushin and of Kieffer: Introduction*

For a type of stationary ergodic discrete-time finite-alphabet channel, more general than the famous stationary totally ergodic $\bar{d}$-continuous channel of Gray, Ornstein, and Dobrushin [16], Kieffer [36] has shown that a stationary, ergodic source with entropy less than capacity can be transmitted over the channel with *zero probability of error* using stationary codes for encoding and decoding. For their more special channel the authors above obtained such a result (for the more special) Bernoulli sources.

We use again the source $(\Omega, \mathcal{B}, \mu, T)$, where the set $\Omega$ here is the set of all bilateral infinite sequences $w = (w_i)_{i=-\infty}^{\infty}$ with $w_i \in \mathcal{Z}$, the alphabet. It is stationary ($\mu$ preserving under the shift $T$) and ergodic ($T$ invariant sets have measure zero or one). We say that $(\Omega, \mathcal{B}, \mu)$ is *aperiodic* if $\mu(w) = 0$ for every $w \in \Omega$. $H(\mu)$ denotes the entropy of the source. Let $[\Omega', \mathcal{B}', W, \Omega'', \mathcal{B}'']$ be a stationary channel, where the input alphabet $\mathcal{X}$ and the output alphabet $\mathcal{Y}$ are finite, $W = \{W(\cdot|w') : w' \in \Omega'\}$ is a family of PD's on $(\Omega'', \mathcal{B}'')$ such that

(i) $W(E|w')$ is $\mathcal{B}'$-measurable for each $E \in \mathcal{B}''$
(ii) $W(TE|Tw') = W(E|w')$ $(E \in \mathcal{B}'', w' \in \Omega')$

In short we also write $W$ for this channel and $(\Omega, \mu)$ for the source.

We say that the stationary and ergodic source $(\Omega, \mu)$ is *zero-error transmissible* over the stationary channel $W$ if there exist stationary codes $\varphi : \Omega \to \Omega'$, $\psi : \Omega'' \to \Omega$ and a *Markov chain* $Z, X, Y$ such that $Z = (Z_i)_{i=-\infty}^{\infty}$ is a process with state space $\mathcal{Z}$ and PD $\mu$, $X$ is the process with state space $\mathcal{X}$ such that $X = \varphi(Z)$, $Y$ is a process with state space $\mathcal{Y}$ for which the PD conditioned on $X$ is given by $W$, and

$$Z = \psi(Y) \quad \text{a.s.} \tag{3.14.1}$$

In other words, if we encode the process $Z$, which serves as a model for the source $(\Omega, \mu)$, into the process $X$, and then transmit $X$ over $W$, then the process $Z$ can be recovered with probability one from the channel output process $Y$.

There is an equivalent way to formulate this. We say that the stationary source $(\Omega, \mu)$, $(\Omega', \nu)$ are *isomorphic* if there exist processes $U, V$ which are stationary codings of each other such that $U$ has PD $\mu$, and $V$ has PD $\nu$. Following [16], given the stationary source $(\Omega', \nu)$ and the stationary channel $W$, we say $(\Omega', \nu)$ is $\nu$-*invulnerable* if there are processes $X, Y$ such that the PD of $X$ is $\nu$, the PD of $Y$ conditioned on $X$ is given by $W$, and $X$ is a stationary coding of $Y$. That is, $(\Omega', \nu)$ can be directly transmitted over the channel $W$ (without first encoding!), and then recovered exactly from the channel output. It is not hard to see that $(\Omega, \mu)$ is zero-error transmissible over $W$ if and only if there exists a $W$-invulnerable source $(\Omega', \nu)$ isomorphic to $(\Omega, \mu)$.

Shannon introduced for a DMC in addition to the ordinary capacity $C$ (for block encoders and decoders) the zero-error capacity $C_0$ in [51] (with $C_0 < C$ in most cases). We give a detailed treatment in a later volume. Block codes of length $n$ have of course finite memory at most $n$.

Having finite memory of duration $m$ means here that the output letter of the code at any time $i$ is completely determined by looking at the sequence being coded at times $i-m$ through $i+m$. Kieffer [34] gave some negative results on zero-error transmission using finite memory sliding-block codes, if one wants a rate exceeding $C_0$.

Gray et al. [16] showed that with infinite memory stationary encoders and decoders zero-error transmission at any rate below $C$ is possible provided the source being transmitted is a stationary coding of a memoryless source (i.e., a Bernoulli source), and the channel is totally ergodic and $\overline{d}$-continuous, a channel more general than the DMC.

Kieffer showed for a more general channel defined below, zero-error transmission using stationary coders is possible for stationary, ergodic, aperiodic non-Bernoulli sources at all rates below $C$.

### 3.14.2 Principal Results

On the set of ergodic PD's $\mathcal{P}_e(\Omega', \mathcal{B}')$ we put the unique metric topology for which convergence of a sequence of PD's is weak convergence. Similarly, the topology of weak convergence is placed on

$$\mathcal{P}(\Omega' \times \Omega'', \mathcal{B}' \times \mathcal{B}'') = \mathcal{P}(\widetilde{\Omega} \times \widetilde{\mathcal{B}}).$$

If $\nu \in \mathcal{P}(\Omega', \mathcal{B}')$, then $\nu \times W \in \mathcal{P}(\widetilde{\Omega} \times \widetilde{\mathcal{B}})$ such that

$$v \times W(E \times F) = \int_E W(F|w)d\nu(w) \quad (E \in \mathcal{B}', F \in \mathcal{B}'') \tag{3.14.2}$$

Given a stationary channel $W$ let

$$\phi_W : \mathcal{P}_e(\Omega', \mathcal{B}') \to \mathcal{P}(\widetilde{\Omega}, \widetilde{\mathcal{B}}) \tag{3.14.3}$$

be given by $\nu \to \nu \times W$ and we call this channel ergodic if $\nu \times w$ is ergodic for all $\nu \in \mathcal{P}_e(\Omega', \mathcal{B}')$. $I(\nu, W)$ is the mutual information. $\nu^{(k)}$ is the $k$th order marginal PD of $\nu$ ($k \in \mathbb{N}$). Henceforth we omit $\sigma$-algebras if this causes no confusion, so for $\mathcal{P}(\Omega', \mathcal{B}')$ we write $\mathcal{P}(\Omega')$ etc. Now comes the main result we shall prove.

**Theorem 110** (Kieffer [36]) *Given are $\tau \in \mathcal{P}_e(\Omega')$ and a stationary $W$. Assume that there is a neighborhood $\mathcal{N}(\tau)$ of $\tau$ in $\mathcal{P}_e(\Omega')$ such that $\phi_W$ is continuous at every PD in $\mathcal{N}$ and $W$ is ergodic at every positive entropy PD in $\mathcal{N}$.*

*Let $(\Omega, \nu)$ be a stationary, ergodic aperiodic source with $H(\mu) < I(\nu, W)$ then for any $k \in \mathbb{N}$ and any $\delta > 0$ there exists a $W$-invulnerable source $(\Omega', \nu)$ isomorphic to $[\Omega, \mu]$ such that*

$$\max_{w'^k \in \mathcal{X}^k} \left| \nu^{(k)}(w'^k) - \tau^{(k)}(w'^k) \right| < \delta.$$

*Thus $[\Omega, \mu]$ is zero-error transmissible over $W$.*

**Definition 23** The Shannon capacity $C(W)$ of the stationary $W$ is $\sup_{\nu \in \mathcal{P}_e(\Omega')} I(\nu, W)$.

**Corollary 19** *Let the stationary $W$ be ergodic at every positive entropy PD in $\mathcal{P}_e(\Omega')$ and suppose $\phi_W$ is continuous at every PD in $\mathcal{P}_e(\Omega')$.*

*Further, let $[\Omega, \mu]$ be any stationary ergodic aperiodic source. Then*

*(i) $[\Omega, \mu]$ is zero-error transmissible over $W$, if $H(\mu) < C(W)$, and*
*(ii) $[\Omega, \mu]$ is not zero-error transmissible, if $H(\mu) > C(W)$.*

*Verification.* (i) follows from Theorem 110. (ii) is proved in [31].

Also in [31] is shown that the type of channels considered by Gray et al. [16] satisfies the hypotheses of the corollary.

### *3.14.3 Synchronization Words*

In order to construct the encoder for $(\Omega, \mu)$ we have to make sure that certain blocks of the encoder outputs are synchronization words: i.e., words which cannot be mistaken for cyclic shifts of themselves. In this section for a given stationary ergodic source we will show that there are synchronization words "typical" of the source.

**Definition 24** Let $\mathcal{X}^* = \bigcup_{n=1}^{\infty} \mathcal{X}^n$. Define $\sigma : \mathcal{X}^* \to \mathcal{X}^*$ as the map

$$\sigma_1(x_1, x_2, \dots, x_n) = (x_2, \dots, x_n, x_1).$$

If $\mathbb{M} \subset \mathbb{Z}$ is a set of integers and $m \in \mathbb{N}$ we say $\mathbb{M}$ has minimal distance $m$ denoted as dist($\mathbb{M}$), if $|i - j| \geq m$ for every $i, j \in \mathbb{M}, i \neq j$.

(This is like the structure of an error-correcting code.)

**Lemma 49** *Let $m, n \in \mathbb{N}$. let $\sigma$ be a cyclic permutation of $\{1, 2, \dots, n\}$ such that $|\sigma(i) - i| \geq m$ $(1 \leq i \leq n)$. Let $k = 6(2m - 1)$ and let $a_j = k^{-1}(1 - k^{-1})^{j-1}$ $(j \in \mathbb{N})$, then there exist nonempty disjoint subsets $\mathbb{N}_1, \mathbb{N}_2, \dots, \mathbb{N}_j$ of $\{1, 2, \dots, n\}$ such that for $1 \leq i \leq J$*

*(i) $a_j n - 2^{j-1} \leq |\mathbb{N}_j| \leq a_j n + 2^{j-1}$*
*(ii) dist$(\mathbb{N}_j \cup \sigma(\mathbb{N}_j)) = m$, and $\mathbb{N}_j \cap \sigma(\mathbb{N}_j) = \emptyset$*

*Heuristic observations.* First note that for the cyclic shift $\pi_t$

$$\pi_t(1,2,\dots,t+1,t+2,\dots,n) = (n-t, n-t+1,\dots,n+1,\dots,n-(t+1)) \quad (3.14.4)$$
$$\pi_t(i) - i = n-(t+1) \quad \text{for } 1 \le i \le t+1$$
$$i - \pi_t(i) = t+1 \quad \text{for } t+2 \le i \le n$$

and therefore

$$\max_i |\pi_t(i) - i| = \min(n-(t+1), t+1). \quad (3.14.5)$$

So we know that

$$m \le \max_t \min(n-(t+1), t+1) = \left\lfloor \frac{n}{2} \right\rfloor.$$

Obviously, we see from (3.14.4) that for 1 there is for every $m' \le n-1$ a $t$ with $1-\pi_t(1) = m'$ and $\pi_t(1) \ne 1$. Therefore, a set with the properties in (3.14.4) exists.

Furthermore, let $m' = m$ and $k = 6(2m-1)$ then $a_1 = k^{-1}(1-k^{-1})^0 = \frac{1}{6(2m-1)}$. Thus

$$\frac{n}{6(2m-1)} - 1 \le 1 \le \frac{n}{6(2m-1)} + 1$$

holds if $n \le 12(2m-1)$. However, if $12(2m-1) < n$ or $m < \frac{n}{24} + \frac{1}{2}$ we have to do more.

*Proof* Let $S, T$ be subsets of $[n]$ and let $\text{dist}(T) = m$. Let

$$G = \{i \in S : |i-t| \ge m \text{ and } |\sigma(i) - t| \ge m \text{ for all } t \in T\}.$$

Since $\sigma$ is cyclic, there must exist some integer $\nu$ such that for every $i \in [n]$, $\sigma(i) \in [i+r, i+r-n]$. Hence, if $i \in S$ and $i \notin G$, then for some $t \in T$ either $|i-t| < m$, or $|i-(t-r)| < m$, or $|i-(-r+n+r)| < m$. Therefore $|G| \ge |S| - 3|T|(2m-1)$. Note that if $i \in G$, then $\text{dist}(\{i,\sigma(i)\} \cup T) = m$ and $\{i,\sigma(i)\} \cap T = \emptyset$. Pick the largest non-negative integer $r$ such that there exists $S' \subset S$ for which $|S'| = r$, $\text{dist}(S' \cup \sigma(S')) = m$, and $S' \cap \sigma(S') = \emptyset$. If $|S'| - 6r(2m-1) > 0$, then by taking $T = S' \cup \sigma(S')$ in the preceding argument, we could add an element to $S'$. Consequently, $r \ge \lceil k^{-1}|S| \rceil$. Starting off with $S = [n]$ we construct a subset $S_1$ of $[n]$ satisfying (i), (ii) for $j = 1$. Proceeding inductively, suppose we have constructed disjoint subsets $S_1, \dots, S_t$ of $[n]$ satisfying (i), (ii). Let $U = [n] - (S_1 \cup \dots \cup S_t)$. Find $S_{t+1} \subset U$ such that (ii) holds for $j = t+1$ and $|S_{t+1}| = \lceil k^{-1}(I) \rceil$. Since $k^{-1}\left(n - n\sum_{j=1}^t a_j\right) = na_{t+1}$ and $\sum_{j=1}^t 2^{j-1} + 1 = 2^t$, we have

$$na_{t+1} - 2^t \le \left\lceil k^{-1}|U| \right\rceil \le na_{t+1} + 2^t,$$

completing the proof. □

## 3.15 Lecture on Blackwell's Infinite Codes for the DMC

### 3.15.1 Introduction and Result

Shannon [52] has shown that for the DMC $W : \mathcal{X} \to \mathcal{Y}$ for any rate $R < C(W)$ there is an exponent $E(R) > 0$ that the optimal maximal error probability $\lambda_n(R)$ for $(n, N) = (n, 2^{Rn})$-codes satisfies

$$\lambda_n(R) < \exp\{-E(R)n\} = \rho(R)^n = \rho^n, \text{ say}, 0 < \rho < 1 \quad (n \in \mathbb{N}). \tag{3.15.1}$$

For every $n \in \mathbb{N}$ there is a code $\{(u_i(n), \mathcal{D}_i(n)) : 1 \leq i \leq 2^{Rn}\}$ meeting this bound. We denote by

$$U(n) = \{u_i(n) : 1 \leq i \leq 2^{Rn}\} \tag{3.15.2}$$

this *code book*, the set of codewords. The decoder can be expressed by a function $\psi_n : \mathcal{Y}^n \to \mathcal{U}(n)$, where $\psi_n(y^n) = u_i(n)$, if $y^n \in \mathcal{D}_i(n)$.

Then if the sender selects any $u_i(n)$ and places its letters $u_{i1}, \dots, u_{in}$ successively into the channel, and the receiver, on observing the resulting output sequence $y_1, \dots, y_n$ decides that the input was $\psi(y_1, \dots, y_n)$, the probability that he makes an error is less than $\rho^n$, no matter which $u \in \mathcal{U}(n)$ was chosen.

The result may be described as follows:

The set of possible messages $[2^{Rn}] = \{1, 2, \dots, 2^{Rn}\}$ can more conveniently be written in the form $\mathcal{Z}^{Rn} = \prod_1^{Rn} \mathcal{Z} = \{0, 1\}^{Rn}$. More precisely we use $\lceil Rn \rceil$.

It is then possible to transmit at any rate $R < C$, with arbitrarily small probability of error a message $z_1, z_2, \dots, z_{Rn}$, by using *block codes* of sufficient length. The message is known only (at best, if there is no error) at time $n$ (idealizing the time effort for computing the value of $\psi$ as zero).

We draw now a slightly stronger conclusion. We imagine an infinite sequence $z = (z_1, z_2, \dots)$ of 0's and 1's, which we are required to transmit over the channel.

At time $n$ the sender will have observed the first $\lceil Rn \rceil$ coordinates of $z$, and will place the $n$th (channel) input symbol in the channel. The receiver, having at this point observed the first $n$ channel outputs, will *estimate the first* $m(n)$ coordinates of $z$ with a *new goal concerning corrections*.

If $\frac{m(n)}{Rn} \to 1$ as $n \to \infty$ and if, *for every* $z$, all but a *finite* number of his estimates are correct (i.e., agree with $x$ in every coordinate estimated) *with probability 1*, we say that the channel is being used at rate $R$. Blackwell's result is that, in this sense, a DMC $W$ can be used at any rate $R < C(W)$.

### 3.15.2 A More Mathematical Formulation

We concentrate on the special case $R = 1$. The general case involves no new ideas, but only more notation.

The function $f_t$ of a code, as defined below, specifies the $t$th channel input symbol, as a function of the first $t$ coordinates of $z$.

The number $m(t)$ is the number of $z$ coordinates to be estimated by the receiver after observing the first $t$ output symbols, and the function $g_t$ specifies the estimate.

We now state the result more formally.

**Definition 25** An infinite code (for transmitting at a rate 1) is defined as consisting of

(i) a sequence $(f_t)_{t=1}^{\infty}$ of functions, where $f_t : \mathcal{Z}^* \to \mathcal{X}$
(ii) a non decreasing sequence $(m(t))_{t=1}^{\infty}$ of positive integers such that $\frac{m(t)}{t} \to 1$ as $t \to \infty$
(iii) a sequence $(g_t)_{t=1}^{\infty}$ of functions, where $g_t : \mathcal{Y}^t \to \mathcal{Z}^{m(t)}$.

**Definition 26** An infinite code $(f_t, m(t), g_t)_{t=1}^{\infty}$ and an infinite sequence $z = (z_1, z_2, \dots)$ of 0's and 1's together define a sequence of independent output RV's $Y_1, Y_2, \dots$ with

$$\Pr(Y_t = y) = W(y|f_t(z_1, z_2, \dots, z_t)) \tag{3.15.3}$$

and a sequence of estimated messages $g_1, g_2, \dots$ where

$$g_t = g_t(Y_1, Y_2, \dots, Y_t). \tag{3.15.4}$$

**Definition 27** We say that the infinite code is *effective at* $z$, if

$$g_t = (z_1, \dots, z_{m(t)}) \text{ with probability } 1 \tag{3.15.5}$$

for all sufficiently large $t$, and that it is effective, if it effective for every $z$.

The result of this lecture is

**Theorem 111** *For every* DMC $W$ *with capacity* $C(W) > 1$, *there is an effective code.*

*Remark* The method used here relies on exponential error bounds and extends to channels with such bounds, for example, to finite-state channels.

### 3.15.3 A Technical Auxiliary Result

The analysis of infinite codes to be effective led to the need for two sequences of integer $(n(k))_{k=1}^{\infty}$ and $(l(k))_{k=1}^{\infty}$ with the following properties

(i) $(n(k)) \subset \mathbb{N}$ is strictly increasing, however, $\frac{n(1)+\dots+n(k)}{n(1)+\dots+n(k+2)} \to 1$ as $k \to \infty$
(ii) $(l(k)) \subset \mathbb{N}_0$ is non-decreasing, however,s $l(k) \to \infty$ as $k \to \infty$
(iii) $l(k) \leq \min(n(1) + \dots + n(k-1)m, Dn(k+1) - n(k))$, where $D > 1$

**Lemma 50** *The sequences with terms* $n(k) = k$ *and* $l(k) = \lfloor (k-1)\min(1, D-1)\rfloor$, *for instance, satisfy (i)–(iii).*

*Proof* Only $l(k) \leq D(k+1) - k$ needs verification, because the rest is obvious. Here for $1 < D \leq 2$, $l(k) = (k-1)(D-1) \leq k(D-1) + D = D(k+1) - k$, and for $2 < D$, $l(k) = k - 1 \leq 2(k-1) - k$.

*Proof of Theorem 111* Since $C > 1$ we can choose as rate a number $D$ with $1 < D < C$ and $\rho < 1$ as explained in the introduction.

We divide now the $z$-sequence into successive blocks of length $n(1), n(2), \dots$ where $(n(k))$ and a sequence $(l(k))$ below satisfy the properties (i)–(iii).

During the time the $k+1$th block of $z$-symbols is observed we use the channel to transmit up to $Dn(k+1)$ $z$-coordinates among those received to date, with error probability at most $\rho^{n(k+1)}$. We choose to transmit the $k$th block, containing $n(k)$ $z$-coordinates, *and* to *repeat* the first $l(k)$ coordinates of $z$, where by (iii)

$$l(k) + n(k) \leq \lfloor Dk(k+1)\rfloor$$
$$l(k) \leq n(1) + \cdots + n(k-1).$$

Since $(n(k))$ is strictly increasing, $\sum_k \rho^{n(k)}$ converges, so that, with probability 1 only a finite number of errors will be committed.

That is to say, the receiver, after observing the $k+1$st block of output symbols, estimates the first $l(k)$ $z$-symbols, say as $u(k)$, and the $k$th block of $z$-symbols, say as $v(k)$, and we have, with probability 1,

$$u(k) = c(k), \quad v(k) = d(k)$$

for all sufficiently large $k$, where $c(k)$ denotes the first $l(k)$ coordinates of $z$ and $d(k)$ denotes the $k$th block of $z$-coordinates. After observing the $k+1$st block of output symbols and making the estimates $u(k)$, $v(k)$ the receiver will have estimates each of the first $n(1) + \cdots + n(k) = T(k)$ coordinates of $z$ at least once. He now forms an estimate $w(k)$ of the first $T(k)$ coordinates, using the latest estimate made on each coordinate. If

$$l(k) = n(1) + \cdots + n(i-1) + h, \quad 0 \leq h < n(i)$$

the estimate $w(k)$ is:

$$w(k) = (u(k), v^*(i), v(i+1), \dots, v(k)),$$

where $v^*(i)$ consists of the last $n(i) - h$ coordinates of $v(i)$. If $l(k) \to \infty$ with $k$, so does $i$. Since, with probability 1, all $u(i)$, $v(i)$ for $i$ sufficiently large are correct, we conclude that, with probability 1,

$$w(k) = (z_1, \dots, z_{T(k)})$$

for all sufficiently large $k$.

We have thus defined a sequence $(w(k))$ of estimates, where $w(k)$ estimates the first $T(k)$ coordinates $z$ after $T(k+1)$ outputs have been received, such that, with probability 1, all but a finite number of $w(k)$ are correct.

For $t < T(2)$ we define $g_t$ arbitrarily, for $T(k+1) \leq t < T(k+2)$, we define $g_t$ as $w(k)$. Thus, for $T(k+1) \leq t < T(k+2)$ we have $m(t) = T(k)$, and $\frac{m(t)}{t} \to 1$ as $t \to \infty$ if $\frac{T(k)}{T(k+2)} \to 1$ as $k \to \infty$. In summary, any two sequences $(n(k))$, $(l(k))$ can be used to define an effective code if (i)–(iii) hold. □

## References


1. R. Ahlswede, Beiträge zur Shannonschen Informationstheorie im Fall nichtstationärer Kanäle. Z. Wahrscheinlichkeitstheorie Verw. Geb. **10**, 1–42 (1968)
2. R. Ahlswede, Channel capacities for list codes. J. Appl. Probab. **10**, 824–836 (1973)
3. R. Ahlswede, A constructive proof of the coding theorem for discrete memoryless channels in case of complete feedback, in *6th Prague Conference on Information Theory, Statistical Decision Functions and Random Processes*, September 1971, Publishing House of the Czechosl Academy of Sciences, pp. 1–22 (1973)
4. P. Billingsley, *Ergodic Theory and Information* (Wiley, New York, 1965)
5. L. Breiman, The individual ergodic theorems of information theory. Ann. Math. Stat. **28**, 809–811 (1957)
6. R.L. Dobrushin, *Arbeiten zur Informationstheorie IV, Allgemeine Formulierung des Shannonschen Hauptsatzes der Informationstheorie, Mathematische Forschungsberichte XVII, herausgegeben von H. Grell* (VEB Deutscher Verlag der Wissenschaften, Berlin, 1963)
7. J.L. Doob, *Stochastic Processes* (Wiley, New York, 1953)
8. R.M. Fano, Statistical Theory of Communication, Notes on a course given at the Massachusetts Institute of Technology, 1952, 1954
9. A. Feinstein, A new basic theorem of information theory. Trans. IRE Sect. Inf. Theory **PGIT-4**, 2–22 (1954)
10. A. Feinstein, *Foundations of Information Theory* (McGraw-Hill Book Company, Inc, New York, 1958)
11. A. Feinstein, On the coding theorem and its converse for finite-memory channels. Inf. Control **2**(1), 25–44 (1959)
12. R.G. Gallager, *Information Theory and Reliable Communication* (Wiley, New York, 1968)
13. I.M. Gelfand, A.M. Jaglom, *Arbeiten zur Informationstheorie II*, Über die Berechnung der Menge an Information über eine zufällige Funktion, die in einer anderen zufälligen Funktion enthalten ist (VEB Deutscher Verlag der Wissenschaften (Übersetzung aus dem Russischen), Berlin, 1958)
14. I.M. Gelfand, A.M. Jaglom, A.N. Kolmogoroff, *Arbeiten zur Informationstheorie II*, Zur allgemeinen Definition der Information (VEB Deutscher Verlag der Wissenschaften (Übersetzung aus dem Russischen), Berlin, 1958)
15. R.M. Gray, D.S. Ornstein, Block coding for discrete stationary $\bar{d}$-continuous noisy channels. IEEE Trans. Inf. Theory **25**, 292–306 (1979)
16. R.M. Gray, D.S. Ornstein, R.L. Dobrushin, Block synchronization, sliding-block coding, invulnerable sources and zero-error codes for discrete noiseless channels. Ann. Probab. **8**, 639–674 (1980)
17. P.R. Halmos, *Measure Theory* (Springer, New York, 1958)
18. H.K. Ting, On the information stability of a sequence of channels (A necessary and sufficient condition for the validity of Feinstein's Lemma and Shannon's Theorem) (translated by R. Silverman), Theory Probab. Appl, **VII**(3) (1962)
19. K. Jacobs, Die Übertragung diskreter Information durch periodische und fastperiodische Kanäle. Math. Ann. **137**, 125–135 (1959)

20. K. Jacobs, *Informationstheorie* (Seminarbericht, Göttingen, 1960)
21. K. Jacobs, Neuere Methoden und Ergebnisse der Ergodentheorie, (Berlin-Göttingen-Heidelberg, 1960)
22. K. Jacobs, Über die Struktur der mittleren Entropie. Math. Z. **78**, 33–43 (1962)
23. K. Jacobs, *Measure and Integral* (Academic Press, New York, 1978)
24. A. del Junco, M. Rahe, Finitary codings and weak Bernoulli partitions. Proc. Amer. Math. Soc. **75**, 259–264 (1979)
25. M. Keane, M. Smorodinsky, A class of finitary codes. Israel J. Math. **26**, 352–371 (1977)
26. A.Y. Khinchin, On the fundamental theorems of information theory, Uspehi Mat. Nauk, **11**,1(67), 17–75 (1956)
27. A.Y. Khinchin, *Arbeiten zur Informationstheorie I, 2. Auflage*, Der Begriff der Entropie in der Wahrscheinlichkeitsrechnung (VEB Deutscher Verlag der Wissenschaften (Übersetzung aus dem Russichen), Berlin, 1961)
28. A.Y. Khinchin, *Arbeiten zur Informationstheorie I, 2. Auflage*, Über grundlegende Sätze der Informationstheorie (VEB Deutscher Verlag der Wissenschaften (Übersetzung aus dem Russichen), Berlin, 1961)
29. J.C. Kieffer, A general formula for the capacity of stationary nonanticipatory channels. Inf. Control **26**, 381–391 (1974)
30. J.C. Kieffer, On sliding block coding for transmission of a source over a stationary nonanticipatory channel. Inf. Control **35**, 1–19 (1977)
31. J.C. Kieffer, On the transmission of Bernoulli sources over stationary channels. Ann. Probab. **8**(5), 942–961 (1980)
32. J.C. Kieffer, Some universal noiseless multiterminal source coding theorems. Inf. Control **460**(2), 93–107 (1980)
33. J.C. Kieffer, Noiseless stationary coding over stationary channels. Z. Wahrscheinlichkeitstheorie Verw. Geb. (1980)
34. J.C. Kieffer, Perfect transmission over a discrete memoryless channel requires infinite expected coding time. J. Comb. Inf. Syst. Sci. **5**(4), 317–322 (1980)
35. J.C. Kieffer, Block coding for weakly continuous channels. IEEE Trans. Inf. Theory **IT-27**(6) (1981)
36. J.C. Kieffer, Zero-error stationary coding over stationary channels. Z. Wahrscheinlichkeitstheorie Verw. Geb. **56**, 113–126 (1981)
37. A.N. Kolmogorov, *Arbeiten zur Informationstheorie I, 2. Auflage*, Theorie der Nachrichtenübermittlung (VEB Deutscher Verlag der Wissenschaften (Übersetzung aus dem Russischen), Berlin, 1961); englische Ausgabe: A.N. Kolmogorov, To the Shannon theory of information transmission in the continuous case. Trans. IRE Sect. Inf. Theory **2**(4), 102–108 (1956)
38. M. Loève, *Probability Theory* (Wiley, New York, 1959)
39. B. McMillan, The basic theorems of information theory. Ann. Math. Stat. **24**(2), 196–219 (1953)
40. J. Nedoma, The capacity of a discrete channel, in *Transactions of the First Prague Conference on Information Theory, Statistical Decision Function's and Random Processes*, pp. 143–181, (1957)
41. D.L. Neuhoff, P.C. Shields, Channels with almost finite memory. IEEE Trans. Inf. Theory, **IT-25**, 440–447 (1979)
42. D. Ornstein, *Ergodic Theory, Randomness, and Dynamical Systems* (Yale University Press, New Haven, 1974)
43. D. Ornstein, B. Weiss, The Shannon-McMillan-Breiman theorem for a class of amenable groups. Israel J. Math. **44**, 53–60 (1983)
44. J.C. Oxtoby, Ergodic sets. Bull. Am. Math. Soc. **58**, 116–136 (1952)
45. K.R. Parthasarathy, On the integral representation of the rate of transmission of a stationary channel. Illinois J. Math. **5**, 299–305 (1961)

46. A. Perez, Notions généralisées d'incertitude d'entropie et d'information du point de vue de la théorie du martingales, in *Transactions of the First Prague Conference on Information Theory, Statistical Decision Function's and Random Processes*, Publishing House of the Czechoslovak Academy of Sciences Program, pp. 183–208 (1957)
47. A. Perez, Sur la théorie de l'information dans le cas d'un alphabet abstrait, in *Transactions of the First Prague Conference on Information Theory, Statistical Decision Function's and Random Processes*, Publishing House of the Czechoslovak Academy of Sciences Program, pp. 209–244 (1957)
48. M. Rosenblatt-Roth, Die Entropie stochastischer Prozesse (auf Russisch). Doklady Akad. Nauk SSSR **112**, 16–19 (1957)
49. M. Rosenblatt-Roth, Theorie der Übertragung von Information durch stochastische Kanäle (auf Russisch). Doklady Akad. Nauk SSSR **112**, 202–205 (1957)
50. C.E. Shannon, A mathematical theory of communication. Bell Syst. Tech. J. **27**(379–423), 623–656 (1948)
51. C.E. Shannon, The zero error capacity of a noisy channel. IRE Trans. Inf. Theory **IT-2**, 8–19 (1956)
52. C.E. Shannon, Certain results in coding theory for noisy channels. Inf. Control **1**(1), 6–25 (1957)
53. P.C. Shields, *The Theory of Bernoulli Shifts* (The University of Chicago Ill, Chicago, 1973)
54. P.C. Shields, Almost block independence. Z. Wahrscheinlichkeitstheorie Verw. Geb. **49**, 119–123 (1979)
55. K. Takano, On the basic theorems of information theory. Ann. Inst. Stat. Math. Tokyo **9**, 53–77 (1958)
56. I.P. Tsaregradskii, A note on the capacity of a stationary channel with finite memory, Teor. Veroyatnost. i Primenen., vol. 3, pp. 84–96, 1958 (in Russian), Theory Probab. Appl. **III**(1), pp. 79–91 (1958)
57. J. Wolfowitz, The maximum achievable length of an error correcting code. Illinois J. Math. **2**, 454–458 (1958)
58. J. Wolfowitz, *Coding Theorems of Information Theory*, 1st edn. 1961, 2nd edn. 1964, 3rd edn. (Springer, Berlin, 1978)


## Further Reading


59. R.L. Adler, D. Coppersmith, M. Hassner, Algorithms for sliding—block codes—an application of symbolic dynamics to information theory. IEEE Trans. Inf. Theory **IT-29**, 5–22 (1983)
60. R. Ahlswede, D. Dueck, Bad codes are good ciphers. Probl. Control Inf. Theory **1**(5), 337–351 (1982)
61. R. Ahlswede, P. Gacs, *Two Contributions to Information Theory*, Topics in information theory (Kesthely, Hungary, 1975), pp. 17–40
62. R. Ahlswede, J. Wolfowitz, Channels without synchronization. Adv. Appl. Probab. **3**, 383–403 (1971)
63. P. Algoet, T.M. Cover, A sandwich proof of the Shannon-McMillan-Breiman theorem. Ann. Prob. **16**(2), 899–909 (1988)
64. A.R. Barron, The strong ergodic theorem of densitites: generalized Shannon-McMillan-Breiman theorem. Ann. Prob. **13**, 1292–1303 (1985)
65. D. Blackwell, Exponential error bounds for finite state channels, in *Proceedings of 4th Berkeley Symposium on Mathematical Statistics and Probability*, vol. 1 (University of California Press, Berkeley, 1961) pp. 57–63
66. D. Blackwell, L. Breiman, A.J. Thomasian, Proof of Shannon's transmission theorem for finite state indecomposable channels. Ann. Math. Stat. **29**(4), 1209–1228 (1958)

67. N. Bourbaki, Espaces vectoriels topologiques, Éléments de Mathématique, Livre V, Acrualités Scientifiques et Industrielles, No. 1189, Paris, Chapter II, p. 84 (1953)
68. L. Breiman, On achieving channel capacity in finite-memory channels. Illinois J. Math. **4**, 246–252 (1960)
69. J. Bucklew, A large deviation theory proof of the abstract alphabet source code theorem. IEEE Trans. Inf. Theory **IT-34**, 1081–1083 (1988)
70. K.L. Chung, A note on the ergodic theorem of information theory. Ann. Math. Stat. **32**, 612–614 (1961)
71. T.M. Cover, P. Gacs, R.M. Gray, Kolmogorov's contribution to information theory and algorithmic complexity. Ann. Prob. **17**, 840–865 (1989)
72. M. Denker, C. Grillenberger, K. Sigmund, *Ergodic Theory on Compact Spaces*. Lecture Notes in Mathematics, vol. 58 (Springer, New York, 1976)
73. J.D. Deuschel, D.W. Stroock, *Large Deviations*, vol. 137 (Pure and Applied Mathematics, Academy Press, Boston, 1989)
74. R.L. Dobrushin, Allgemeine Formulierung des Shannonschen Hauptsatzes der Informationstheorie (auf Russisch). Doklady Akad. Nauk SSSR **126**, 474 (1959)
75. R.L. Dobrushin, Shannon's theorem for channels with synchronization errors, Problemy Peredaci Informatsii, vol. 3, pp. 18–36, 1967; Trans. Probl. Inf. Transm. **3**, pp. 31–36 (1967)
76. M.D. Donsker, S.R.S. Varadhan, Asymptotic evaluation of certain Markov process expectations for large time. J. Commun. Pure Appl. Math. **28**, 1–47 (1975)
77. P. Elias, *Coding for Noisy Channels* (IRE Convention Record, 1955), pp. 37–44
78. P. Elias, Two famous papers. IRE Trans. Inf. Theory **4**, 99 (1958)
79. P. Elias, *Coding for Two Noisy Channels*, in Proceedings of the London Symposium of Information Theory London, (1959)
80. A. Feinstein, Math. Reviews MR0118574 (22 #9347) 94.00. Breiman, Leo, On achieving channel capacity in finite-memory channels. Illinois J. Math. **4**, pp. 246–252 (1960)
81. N.A. Friedman, *Introduction to Ergodic Theory* (Van Nostrand Reinhold Company, New York, 1970)
82. R.G. Gray, *Entropy and Information Theory* (Springer, New York, 1990) (revised 2000, 2007, 2008)
83. B.W. Gnedenko, A.N. Kolmogorov, *Grenzverteilungen von Summen unabhängiger Zufallsgrößen, 2. Auflage* ((Übersetzung aus dem Russischen) Akademie-Verlag, Berlin, 1960)
84. R.V.L. Hartley, The transmission of information. Bell Syst. Tech. J. **7**, 535–564 (1928)
85. E. Hopf, *Ergodentheorie* (Springer, Berlin, 1937)
86. K. Jacobs, The ergodic decomposition of the Kolmogorov-Sinai invariant, in *Ergodic Theory*, ed. by F.B. Wright, F.B. Wright (Academic Press, New York, 1963)
87. A.Y. Khinchin, The concept of entropy in probability theory, (in Russian), Uspekhi Mat. Nauk, vol. 8, No. 3 (55), pp. 3–20, translation in *Mathematical Foundations of Information Theory* (Dover Publications Inc, New York, 1953) (1958)
88. J.C. Kieffer, A simple proof of the Moy-Perez generalization of the Shannon-McMillan theorem. Pacific J. Math. **51**, 203–206 (1974)
89. J.C. Kieffer, Sliding-block coding for weakly continuous channels. IEEE Trans. Inf. Theory **IT-28**(1), 2–16 (1982)
90. A.N. Kolmogorov, A new metric invariant of transitive dynamic systems and automorphisms in Lebesgue spaces. Dokl. Akad. Nauk SSR **119**, 861–864 (1958) (in Russian)
91. A.N. Kolmogorov, On the entropy per unit time as a metric invariant of automorphisms. Dokl. Akad. Nauk SSSR **124**, 768–771 (1959) (in Russian)
92. A.N. Kolmogoroff, Über einige asymptotische Charakteristika total beschränkter metrischer Räume, Doklady A.N. d. UdSSR
93. U. Krengel, *Ergodic Theorems* (de Gruyter & Co, Berlin, 1985)
94. W. Krieger, On entropy and generators of measure-preserving transformations. Trans. Am. Math. Soc. **149**, 453–464 (1970)
95. B. Marcus, Sophic systems and encoding data. IEEE Trans. Inf. Theory **IT-31**, 366–377 (1985)

96. L.D. Meshalkin, A case of isomorphisms of Bernoulli scheme. Dokl. Akad. Nauk SSSR **128**, 41–44 (1959) (in Russian)
97. J. Moser, E. Phillips, S. Varadhan, *Ergodic Theory, A Seminar* (Courant Institute of Mathematical Sciences, New York University, 1975)
98. S.C. Moy, Generalization of the Shannon-McMillan theorem. Pacific J. Math. **11**, 705–714 (1961)
99. S. Muroga, On the capacity of a noisy continuous channel. Trans. IRE, Sect. Inf. Theory **3**(1), 44–51 (1957)
100. J. von Neumann, Zur Operatorenmethode in der klassischen Mechanik. Ann. Math. **33**, 587–642 (1932)
101. H. Nyqvist, Certain factors affecting telegraph speed. Bell Syst. Tech. J. **3**, 324 (1924)
102. S. Orey, On the Shannon-Perez-Moy theorem. Contemp. Math. **41**, 319–327 (1985)
103. D. Ornstein, Bernoulli shifts with the same entropy are isomorphic. Adv. Math. **4**, 337–352 (1970)
104. D. Ornstein, An application of ergodic theory to probability theory. Ann. Probab. **1**, 43–58 (1973)
105. A. Perez, Extensions of Shannon-McMillan's limit theorem to more general stochastic processes, in *Transactions of the 3rd Prague Conference on Information Theory, Statistical Decision Functions and Random Processes, Prague*, pp. 545–574 (1964)
106. K. Petersen, *Ergodic Theory* (Cambridge University Press, Cambridge, 1983)
107. M.S. Pinsker, Dynamical systems with completely positive or negative entropy. Soviet Math. Dokl. **1**, 937–938 (1960)
108. V.A. Rohlin, Y.G. Sinai, Construction and properties of invariant measurable partitions. Soviet Math. Dokl. **2**, 1611–1614 (1962)
109. C.E. Shannon, Communication in the presence of noise. Proc. IRE **32**, 10–21 (1949)
110. C.E. Shannon, *Coding Theorems for a Discrete Source with a Fidelity Criterion* (IRE National Convention Record, Party, 1959), pp. 142–163
111. C.E. Shannon, W. Weaver, *The Mathematical Theory of Communication* (University of Illinois Press Ill, Urbana, 1949)
112. Y.G. Sinai, Weak isomorphism of transformations with an invariant measure. Soviet Math. Dokl. **3**, 1725–1729 (1962)
113. Y.G. Sinai, *Introduction to Ergodic Theory* Mathematical Notes (Princeton University Press, Princeton, 1976)
114. A.J. Thomasian, An elementary proof of the AEP of information theory. Ann. Math. Stat. **31**(2), 452–456 (1960)
115. S.R.S. Varadhan, *Large Deviations and Applications* (Society for Industrial and Applied Mathematics, Philadelphia, 1984)
116. S. Verdú, Teaching it, XXVIII Shannon Lecture, Nice, France, 28 June 2007
117. P. Walters, *Ergodic Theory—Introductory Lectures*, vol. 458, Lecture Notes in Mathematics (Springer, New York, 1975)
118. N. Wiener, *Extrapolation, Interpolation, and Smoothing of Stationary Time Series* (Wiley, New York, 1949)
119. N. Wiener, The ergodic theorem. Duke Math. J. **5**, 1–18 (1939)
120. K. Winkelbauer, Communication channels with finite past history, in *Transactions of the 2nd Prague Conference on Information Theory, Statistical Decision Functions and Random Processes*, pp. 685–831 (1960)
121. J. Wolfowitz, The coding of messages subject to chance errors. Illinois J. Math. **1**(4), 591–606 (1957)

# Chapter 4
# On Sliding-Block Codes

## 4.1 Lecture on Sliding-Block Joint Source/Noisy Channel Coding Theorems

### *4.1.1 Introduction*

Sliding-block codes are non-block coding structures consisting of discrete time time-invariant possibly non linear filters. They are equivalent to time-invariant trellis codes. The coupling of random-coding in typical sequences (entropy-typical by Forney [13], typical by Ahlswede [4, 6] with the strong Rokhlin–Kakutani Theorem of ergodic theory is used to obtain a sliding-block coding theorem for ergodic sources and DMC (with weak converse).

Combining this result with a theorem stating that sliding-block codes with a fidelity criterion achieve the rate distortion function for (ergodic) sources [14] yields a sliding-block (continuous) information transmission theorem. Thus two of the three basic coding theorems of Shannon [38, 39], the third being the noiseless source coding theorem, which Shannon considered only with block coding structures, hold also for stationary non-block structures. We follow here [18] with an excellent description.

"By block codes, we mean coding structures that map consecutive nonoverlapping blocks of data into consecutive nonoverlapping blocks of data. (They were used entirely in Chap. 3). Included are schemes where the block length is fixed (such as algebraic codes) and schemes where the data and/or encoded data block length is variable (such as Huffman coding).

It has been subsequently found, however, that non-block structures often are superior in the sense of requiring less complexity for specified performance requirements.

Devices such as the convolutional codes developed in information theory and the delta modulators, predictive quantizers, and other nonlinear time-invariant filters developed in communication theory are inherently non-block in structure; they process consecutive overlapping blocks of data and can be viewed as digital filters

 
A. Ahlswede et al. (eds.), *Transmitting and Gaining Data*,
Foundations in Signal Processing, Communications and Networking 11,
DOI 10.1007/978-3-319-12523-7_4

with *finite delay and possibly infinite memory* (linear filter in the case of convolutional codes).

Since these structures do not yield block codes, the block coding theorem of Shannon and their subsequent generalization (A: and improvements) do not strictly apply to them, and so their absolute performance limits have not been demonstrated rigorously).

In the case of channel coding theorems later results demonstrated that convolutional codes could be used for reliable communication at rates near channel capacity if one allowed time-varying codes (the filter parameters had to be allowed to vary with each input) or infinite constraint length codes.

Time-varying convolutional codes share with block codes the disadvantages of producing nonstationary and non-ergodic encoded processes. Similarly, an infinite constraint length code behaves as a time-varying code if coding is begun with deterministic initial contents, since a "steady state" is never reached. This probabilistic non-regularity prevents one from using the ergodic theorem to equal the empirical symbol error frequency with the small probability of error promised by the coding theorem for general ergodic sources.

In addition, it is natural to suspect that for ergodic sources and memoryless channels, good time-invariant coding structures producing ergodic encoded processes should exist."

The basic approach is in [18].

### 4.1.2 Sliding-Block Coding for a DMC

Let $(\Omega, \mathcal{B}, \mu, T)$ be our finite alphabet $\mathcal{Z}$, stationary ergodic source and let $\mu^n$ be the $n$th restriction of $\mu$, i.e.,

$$\Pr(Z_0 = z_0, \dots, Z_{n-1} = z_{n-1}) = \mu([z_0, \dots, z_{n-1}]) \triangleq \mu^n(z_0, \dots, z_{n-1}).$$

Denote the $n$-tuple $(z_0, \dots, z_{n-1})$ by $z^n$ and, in a slight abuse of notation, denote the $(2n+1)$-tuple $(z_{-n}, \dots, z_n)$ by $z^{2n+1}$. Similarly, let

$$Z^{2N+1}(z) = (Z_{-N}(z), \dots, Z_N(z)) \quad \text{and} \quad Z^N(z) = (Z_0(z), \dots, Z_{N-1}(z)).$$

For processes $(X_n)$ and $(Y_n)$ linked by the DMCW

$$I(X \wedge Y) = \lim_{n \to \infty} n^{-1} I(X^n, Y^n) \tag{4.1.1}$$

and the channel capacity of a DMC is defined as

$$C = \sup_{(X_n)\ \text{stat.}} I(X \wedge Y). \tag{4.1.2}$$

Breiman [10] has shown that the supremum is achievable by an ergodic source, however, it is well known that one can take an i.i.d. source.

**Definition 28** A sliding-block (SB) encoder $f^{(M)}$ of length $2M+1$ is a function

$$f^{(M)} : \mathcal{Z}^{2M+1} \to \mathcal{X}.$$

The encoded process is defined by

$$X_n = f^{(M)}(Z_{n-M}, \ldots, Z_{n+M}).$$

In a similar manner a finite-length sliding-block (SB) decoder $g^{(N)}$ is a mapping

$$g^{(N)} : \mathcal{Y}^{2N+1} \to \mathcal{Z}.$$

The reproduction process is defined by

$$\hat{Z}_n = g^{(N)}(Y_{n-N}, \ldots, Y_{n+N}).$$

There is a probability of error

$$\begin{aligned} \text{err}(F^{(M)}, g^{(N)}) &= \Pr(Z_n \neq \hat{Z}_n) = \Pr(Z_o \neq \hat{Z}_0) \\ &= \int_{\Omega} d\mu(z) \sum_{y^{2N+1} \in \mathcal{E}(u)} W^{2N+1}(y^{2N+1} | f^{(M)}) U^{2M+1}(T^k u) \\ &\quad (-N \leq k \leq N), \end{aligned} \tag{4.1.3}$$

where

$$\mathcal{E}(u) = \{y^{2N+1} : g^{(N)}(y^{2N+1}) \neq u_0 = U_0(u)\}$$

**Theorem 112** *Given an ergodic source $(\Omega, \mu, (Z_n))$ with entropy rate $H(U)$ and a DMC with capacity $C$, if $H(U) < C$, then for any $\epsilon > 0$ there exists for sufficiently large $M$, $m \in \mathbb{N}$ an SB-code $(f^{(M)}, g^{(m)})$ such that*

$$\text{err}(f^{(M)}, g^{(m)}) \leq \epsilon,$$

*and if $H(U) > C$, there exists an $\epsilon_{\min} > 0$ such that*

$$\text{err}(f^{(M)}, g^{(m)}) \geq \epsilon_{\min}$$

*for all SB-codes $f^{(M)}, g^{(m)}$ and all $M, m \in \mathbb{N}$.*

*Proof of the converse part* Assume to the contrary that for every $\epsilon > 0$ there exists an SB-code for which

$$\Pr(Z_n \neq \hat{Z}_n) \leq \epsilon \tag{4.1.4}$$

Set $\delta = H(Z) - C > 0$. From Shields' Lemma (Lemma 8.2 of [40], a Fano-type inequality, there exists an $\epsilon(\delta)$ such that $\Pr(Z_n \neq \hat{Z}_n) \leq \epsilon(\delta)$ for an SB-code implies

$$H(Z, \hat{Z}) - H(\hat{Z}) = H(Z|\hat{Z}) \leq \frac{\delta}{2} \tag{4.1.5}$$

Since

$$H(Z) - \delta = C \geq I(X \wedge Y) \geq I(Z \wedge \hat{Z})$$

(by the data processing inequality)

$$= H(Z) - H(Z|\hat{(}Z)) \geq H(Z) - \frac{\delta}{2}$$

(by (4.1.5)), we have $-\delta \geq -\frac{\delta}{2}$, a contradiction.

## 4.1.3 *Sketch of the Proof of the Direct Part—the Basic Ideas*

### 4.1.3.1 From a Good *B*-Code to a Good SB-Code

The fundamental tool is the *Rokhlin–Kakutani* (*R-K*) *Theorem*.

For an ergodic source $(\Omega, \mathcal{B}, \mu)$, an $m \in \mathbb{N}$, a measurable partition $Q = \{Q_j : 1 \leq j \leq J\}$ of $\Omega$, and $\delta > 0$ there exists a set $F \in \mathcal{B}$ called the "base" such that

(i) $F, TF, \dots, T^{m-1}F$ are disjoint,
(ii) $\mu(\bigcup_{k=0}^{m-1} T^k F) \geq 1 - \delta$,
(iii) $\mu(Q_j|F) = \mu(Q_j \cap F)/\mu(F) = \mu(Q_j)$ $(1 \leq j \leq J)$.

The three properties of $F$ have a striking interpretation: for the given $m$ $F$ has $m$ disjoint shifts in succession, almost exhausts the space in probability, and is independent of all atoms of the partition (finite, but arbitrary otherwise).

For the present purpose a useful choice of the partition $Q$ is the set of $N$-dimensional cylinders

$$\{[z^n] : z^N \in \mathcal{Z}^n\} \quad \text{for} \quad N \leq m.$$

In this case, the probabilistic properties of $n$-tuples on the *base* will be the same as throughout the whole space. Intuitively, this means that, by controlling desirable $N$th order properties on the base such as the error probability of a good B-code, the SB-code constructed via the R-K-Theorem with inherit essentially the same properties. The structure is called a $(\delta, m)$-gadget.

Assume that we have a block-length $N$ B-code

$$\{u_i^N : 1 \leq i \leq M\}$$

with encoder mapping $\alpha : \mathcal{Z}^N \rightarrow \mathcal{X}^N$ and decoder mapping $\beta : \mathcal{Y}^N \rightarrow \mathcal{Z}^N$. That is, for each $z^N$ $\alpha(z^N) = u_i^N$ for some $i$. This B-code partially describes an *infinite length* SP-code as follows:

Construct a $(\delta, m)$-gadget, $m \geq N$ with partition

$$Q = \{([z^N] : z^N \in \mathcal{N}\},$$

i.e., carve up the base according to $N$-dimensional cylinders. Note that if $w \in [z^N] \cap F \subset [z^N]$, then $w^N = z^N$, i.e., given $w \in [z^N] \cap F$, the next $N$ source outputs (starting from time 0) are $z_0, z_1, \ldots, z_{N-1}$. This can be interpreted as the labeling of the sets $T^i([z^N] \cap F), i = 0, 1, \ldots, N-1$ by the symbols $z_i$, i.e., if $w \in T^i([z^N] \cap F)$, $i = 0, 1, \ldots, N-1$, the source prints out the symbol $z_i$, labeling the set. The code $f$ simply substitutes $\alpha(z^N)$ for $z^N$, i.e., if $w \in [z^N] \cap F$, then the next $N$ outputs of the SB-code are $\alpha(z^N)$. In other words, an SB-code will "relabel" the sets in the gadget of the form

$$T^i([z^N] \cap F) \quad (0 \leq i \leq N-1)$$

by a good block code.

Thus if $w \in T^j([z^N] \cap F), 0 \leq j \leq N-1$, then $T^{-j}w \in [z^N] \cap F$, and hence $f(w) = \alpha(z^N)$, where $\alpha(z^N) = (\alpha(z^N)_0, \alpha(z^N)_1, \ldots, \alpha(z^N)_{N-1})$. This gives the partial description

$$f(w) = x \quad \text{if} \quad w \in T^j([z^N] \cap F) \quad \text{and} \quad \alpha(z^N)_j = x \quad (0 \leq j \leq N-1). \tag{4.1.6}$$

This describes the code for all sequences

$$w \in T^j([z^N] \cap F),\ 0 \leq j \leq N-1;\ z^N \in \mathcal{Z}^N.$$

The decoder with attempt to locate the block in the noisy received sequences and correctly decode it. Toward this end, we follow each block code word $u_i^N$ by a fixed synchronization (synch) word $s^k$ to allow the decoder to locate the blocks. That is, we choose a fixed word $s^k \in \mathcal{X}^k$ such that its noisy received version $y^k$ can be correctly located with high probability by the decoder and which is rarely falsely detected. So,

$$f(w) = x \quad \text{if} \quad w \in T^{N+j}([z^N] \cap F) \quad \text{and} \quad s_j = x, \qquad (0 \leq j \leq k-1).$$

This completely describes $f$ on the $(\delta, m = N+k)$-gadget, except for sequences in the garbage

$$G = \Omega \backslash \bigcup_{k=0}^{m-1} T^k F.$$

For

$$w \in T^j([z^N] \cap F) \ (0 \leq j \leq m-1,\ z^N \in \mathcal{Z}^N)$$

set

$$f(w) = x^* \text{ (a fixed referee letter from } \mathcal{X}).$$

A typical output sequence of the infinite SB-code $f$ looks like

$$\dots x^*[u_{i0}^N](s^k)[u_{i1}^N](s^k)x^*x^*[u_{i2}^N](s^k)[u_{i3}^N]\dots$$

The garbage effectively causes a random spacing between blocks yielding a stationary and ergodic process.

The decoder looks at a window of length $2m + 1$ and looks for a synch word. If he finds a synch word, he decodes the code word to its left and prints on the appropriate letter. If no synch is found, an arbitrary letter is put out. If the block code is good and the synch word is rarely confused with segments of codewords or shifts of itself, then the probability of a symbol error is small. The infinite-length encoder $f$ is then approximated by a finite code using the SB-code approximation theorem [14], completing the proof.

The synch words needed are randomly chosen (see also [5]) contributing a small average error probability to the good code chosen by standard random coding.

The main technical difficulty is showing that $k$-tuples, at the boundary of a codeword and a synch word, are rarely decoded as synch word.

*Remark* In conclusion we mention that in [18] Theorem 112 is combined with the theorem on SB-source coding with a fidelity criterion of [14, 16, 17] to obtain a sliding-block transmission theorem (Theorem 2 of [18]).

For further reading we recommend the original papers and the book [15].

## 4.2 Lecture on Block Synchronization Coding and Applications

### *4.2.1 Introduction*

We report on basic work by Gray et al. [20]. Results are obtained on synchronizing block codes for discrete stationary totally ergodic $\overline{d}$-continuous noisy channels (which may have *infinite memory* and *anticipation*, thus avoiding the two assumptions in most investigations in earlier work reported in Chap. 3.

(Note however that averaged channels, treated in Sects. 1.1 and 1.2 in Chap. 1, have infinite duration of memory and anticipation!).

We abbreviate these channels as $DTE\overline{d}C$.

They are used

- to prove sliding-block joint source/channel coding theorems,
- to demonstrate the existence of invulnerable sources, ergodic sources that can be input directly to the channel *without encoding* and decoded at the receiver with zero-error, at all entropy rates below channel capacity.,

- as essential ingredient: combining the invulnerable source theorem with Ornstein's isomorphism theorem of ergodic theory shows that, if the source is a $B$-process with entropy below capacity, then infinite length codes with zero error exist, proving that the zero-error capacity equals the usual channel capacity.

The vast majority of block coding theorems for noisy channels assume synchronous channels, that is, channels for which the receiver knows a priori the block location and hence how to segment the received data blocks for decoding.

Exceptions are the works of Nedoma [30, 32], Dobrushin [12], Vajda [42], and Ahlswede and Wolfowitz [7], who studied the problem of synchronizing block codes for asynchronous discrete stationary channels that are memoryless or have finite input memory and are totally ergodic (block ergodic inputs yield block ergodic outputs).

This model was extended by Ahlswede and Gács [5] to the MAC with a very short proof via list codes for a "computable" capacity region, obtained for DMC in [7]. By now, starting with Cover/McEliece/Posner [11], several synchronization problems have been considered in multi-user information theory.

An intimately related problem is the development of source/channel coding theorems for stationary sliding-block codes (time invariant possibly non-digital filters).

Gray and Ornstein [18] proved that for DMC $C_{\mathrm{sb}} = C_{\mathrm{b}}$ (see Sect. 4.1).

The proof uses good block codes to construct good sliding-block codes using a synchronization (synch) sequence to locate the blocks and occasional random spacing between blocks to make the coding operation stationary.

The most difficult part of the proof is the demonstration that the synch sequence could with high probability be distinguished from an overlap of itself and a codeword, and the proof strongly depends on the memoryless channel assumption.

Kieffer [25] subsequently generalized the techniques and results of Gray and Ornstein [18] to discrete channels having *zero input memory* and *anticipation*, a class of channels introduced by Nedoma [30].

In this section the technique of Dobrushin [12] and Nedoma [32], who used $n$ ergodicity to achieve internal synchronization and no external synch words (an approach further developed by Vajda [42]), are combined and adapted to obtain block synchronization theorems (and sliding-block joint source/channel coding theorems for the class of $DTE\bar{d}C$ [19], having possibly infinite input memory and anticipation.

Roughly speaking, these are channels such that

(1) if one knows the channel input for a sufficiently long time, then the output PD during the same time is known within a $\bar{d}$ or average Hamming distance;
(2) if an input process has ergodic $n$-tuples, then so does the output process.

Condition (1) yields very general stationary discrete channels possessing *synchronous* block coding theorems [19].

Condition (2) ensures that relative frequencies of output $n$-tuples will converge to the appropriate expectation, if those of the input do.

Condition (2) holds, for example, if the channel is asymptotically output memoryless in the sense of Kadota and Wyner [29] or Pfaffelhuber [36] or output weakly mixing as in Adler [1].

The theoretical approach adopted here resembles the ad hoc engineering approach as described, e.g., by Stiffler [41], Chap. 14, one prefixes each code block by a synch sequence which is rarely decoded erroneously within a code block, and one then observes several successive output code blocks to resolve possible confusion of a synch sequence with an overlap of itself and a codeword. As noted by Dobrushin [12], it is this possible confusion of a synch with an overlap of itself that causes the most difficulty. The elegant techniques of synchronizing noiseless channels (e.g., see Stiffler [41] or Scholtz [37]) do not suffice, because the noise may not be small and the channel filtering can destroy the structure of such codes.

Since sliding-block coding a stationary ergodic source yields a stationary ergodic encoded process (unlike block codes, a slight modification of the sliding-block coding theorem proves that there exist stationary ergodic sources that can be directly connected to the channel without encoding, yet can be reliably recovered to within $\epsilon$ by decoding the channel output—an "$\epsilon$-invulnerable source". We develop a convergent sequence of codes and processes that yields in the limit a 0-invulnerable, or simply, invulnerable source having any specified entropy rate below channel capacity. Thus for such channels, there exist stationary ergodic sources at all entropy rates below capacity. Thus for such channels there exist stationary ergodic sources at all entropy rates below capacity that can be communicated across the channel with zero error using a possibly infinite-length sliding block decoder. This result is coupled with Ornstein's isomorphism theorem of ergodic theory to show that, given any source that is a $B$-process ([33, 34, 40]) with entropy rate below capacity, there exists an infinite-length sliding-block encoder and a decoder yielding zero error.

Shannon [39] observed that traditional coding theorems promised that ever longer codes could yield ever smaller error, but that this did not guarantee the existence of codes having *exactly* zero error probability.

This led him to loosely define zero-error capacity of a channel $C_0$ as the supremum of all rates at which zero error communication is possible. He derived several properties of $C_0$ under the assumption that only finite length block codes were allowed—a natural restriction on block codes as general infinite length block codes are not well-defined, e.g., how are they to be used?

If we consider the weaker definition that $C_0$ is the supremum of all entropy rates of sources that can be communicated across the channel with zero error (without restrictions on the coding structure and hence allowing infinite-length sliding-block codes), then the zero error result proves that $C_0 = C$, and hence the channel noise can be completely defeated by infinite codes.

Blackwell [9] also developed a sequence of codes asymptotically yielding a zero error relative frequency on a memoryless channel. His codes are not time invariant (are *nonstationary*) and are effectively a sequential decision scheme for *guessing* the $n$th input symbol after viewing about $Rn$ output symbols. His system of producing a sequence of decoders that with probability one makes only a *finite number of errors* is analogous to the construction to obtain a perfect decoder, but the codes here are stationary.

### 4.2.2 Sources, Channels, and Codes

We keep our notation first introduced in Chap. 3. Except for a source like $(\Omega, \mathcal{B}, \mu)$ we introduce right away also the (discrete random process $U = (U_n)_{n=-\infty}^{\infty}$ defined on $(\Omega, \mathcal{B})$, where for $w \in \Omega$ $U_n(w) = w_n$, that is, the $U_n$'s are coordinate functions on $\Omega$. We denote a source by $(\Omega, \mathcal{B}, \mu)$ and, depending on the focus also by $(\Omega, \mu, U)$ or, in short, also $(\Omega, \mu)$ or $\{U_n\}$, or $\mu$, as convenient.

For $u \in \Omega$, define

$$^{m}U^n(u) =^m u^n = (u_m, u_{m+1}, \dots, u_n)$$

or also $U_m^n(u) =^m U^n(u)$, etc. Furthermore, if $m = 0$ we omit the subscript or the left superscript. The shift operator $T$ gives the connection $U_n(Tu) = U_{n+1}(u)$. A source $(\Omega, \mu, U)$ is $n$-stationary, if $\mu(T^n F) = \mu(F)$, $F \in \mathcal{B}$, and a 1-stationary source is called stationary and an $n$-stationary source is also called *block stationary*. A source is $n$-ergodic if $T^n F = F, \quad F \in \mathcal{B}$, implies $\mu(F) = 0$ or 1. A 1-ergodic source is ergodic and a source $n$-ergodic for all integers $n$ is called totally ergodic.

Recall next that for block stationary sources $(\Omega, \mu, U)$ the entropy rate $H(\mu)$ or $H(U)$ is defined by

$$H(\mu) = H(U) = \lim_{n\to\infty} -n^{-1} \sum_{u^n \in \mathcal{U}^n} \mu^n(u^n) \log \mu^n(u^n)$$

The limit is well known to exist (see Sect. 3.2.1 of Chap. 3).

A channel is $n$-ergodic if for every $n$-ergodic input process the induced output process is $n$-ergodic. Moreover, a channel $n$-ergodic for all $n$ is called *totally ergodic*.

Let $\alpha^n$ and $\beta^n$ be PD on $(B^n, \mathcal{B}_B^n)$ and let $\mathcal{P}(\alpha^n, \beta^n)$ denote the class of all joint PDs on $(B^n \times B^n, \mathcal{B}_{B\times B}^n)$ having $\alpha^n$ and $\beta^n$ as marginals. Define for $i = 0, 1, \dots, n-1$ the coordinate functions $Y_i : B^n \times B^n \to B$ and $\hat{Y}_i : B^n \times B^n \to B$ by $Y_i(y^n, \hat{y}^n) = y_i$ and $\hat{Y}_i(y^n, \hat{y}^n) = \hat{y}_i$ $(y^n, \hat{y}^n \in B^n)$. Let $d_n$ denote the normalized Hamming distance

$$d_n(y^n, \hat{y}^n) = n^{-1} \sum_{i=0}^{n-1} d(y_i, \hat{y}_0).$$

The $n$th order $\overline{d}$ distance between $\alpha^n$ and $\beta^n$ is

$$\overline{d}_n(\alpha^n, \beta^n) = \inf_{P\in\mathcal{P}(\alpha^n,\beta^n)} \mathbb{E}_P d_n(y^n, \hat{y}^n) \tag{4.2.1}$$

$$= \inf_{P\in\mathcal{P}(\alpha^n,\beta^n)} n^{-1} \sum_{i=0}^{n-1} \Pr(Y_i \neq \hat{Y}_i). \tag{4.2.2}$$

A channel $W$ is $\overline{d}$-continuous [19] if given $\epsilon > 0$ there is an integer $n_0$ such that for $n \geq n_0$

$$\overline{d}_n(W^n(\cdot|x), W^n(\cdot|x')) \leq \epsilon$$

whenever $x_i = x_i'$, $0 \leq i \leq n-1$.

Equivalently, a channel $W$ is $\overline{d}$-continuous, if

$$\limsup_{n\to\infty} \max_{a^n \in A^n} \sup_{x,x' \in c(a^n)} \overline{d}_n(W^n(\cdot|x), W^n(\cdot|x') = 0, \tag{4.2.3}$$

where $c(a^n) = \{u : (u_1, \ldots, u_n) = a^n\}$.

We consider two forms of coding structures yielding *sequence coders* with different properties.

#### 4.2.2.1 Block Codes and Sliding-Block Codes

A block code of length $n$ is a pair of mappings $\phi_n : \mathcal{U}^n \to \mathcal{X}^n$ (encoder) and $\psi_n : \mathcal{Y}^n \to \mathcal{U}^n$ (decoder) which induces sequence codes

$$\phi(u) = (\ldots, \phi_n(u_{-n}, \ldots, u_{-1}), \phi_n(u_0, \ldots, u_{n-1}), \ldots)$$

and

$$\psi(y) = (\ldots, \psi_n(y_{-n}, \ldots, y_{-1}), \psi_n(y_0, \ldots, y_{n-1}), \ldots).$$

Define also

$$\phi_{mn}(u^{mn}) = \phi_n(u^n), \phi_n(u_n^n), \ldots \phi_n(u_{(m-1)n}^n)$$

The sequence coders induced by length $n$ block codes are $n$-stationary. Given a block code $(\phi_n, \psi_n)$, a source $\mu$, and a channel $W$, the block error probability is defined by

$$\begin{aligned} P_b(\mu, W, \phi_n, \psi_n) &= \Pr(\psi_n(Y^n) \neq U^n) \\ &= \int W^n(\psi_n^{-1}(u^n)|\phi(u))d\mu(u). \end{aligned}$$

SB-coding theorems exist only for channels without memory and anticipation [18, 25]. These theorems have been generalized [20] to $DTE\overline{d}C$. SB-codes can be constructed from $(M, n, \lambda)$ block codes, also called codebooks.

**Definition 29** A code book is said to be $\delta$-*robust* if the expanded decoding sets

$$(D_i)_\delta = \{y^n : d_n(y^n, D_i) \leq \delta\}$$

are disjoint.

One can treat a channel $W$ as if it had no memory and anticipation by "averaging out" the effect of past and future input symbols using some channel input source measure $\tau$, that is, given $W$ and a channel input source $\tau$, define for each $n$ and $a^n \in \mathcal{X}^n$ for which $\mu^n(x^n) \neq 0$ the measure $\hat{W}^n(\cdot|a^n)$ on $(\mathcal{Y}^n, \mathcal{B}''^n)$ by

$$\hat{W}^n(F|a^n) = \tau^n(a^n)^{-1} \int_{x \in [a^n]} W^n(F|x)\, \mathrm{d}\tau(x) \tag{4.2.4}$$

**Definition 30** A code book $\{(u_i, D_i) : 1 \leq i \leq M)$ is called a $(\tau, M, n, \lambda)$ Feinstein code for $W$ if

$$\max_{1 \leq i \leq M} \hat{W}^n(D_i^c|u_i) \leq \lambda.$$

Good $\delta$-robust Feinstein codes can be used to construct good codebooks and good block codes for $\overline{d}$-continuous channels. The principal result in this construction is

**Lemma 51** (Gray, Ornstein 1979, Lemma 4), *If*

$$\max_{a^n \in \mathcal{X}^n} \sup_{x, x' \in [a^n]} \overline{d}_n(W^n(\cdot|x), W^n(\cdot|x')) \leq \delta^2 \tag{4.2.5}$$

*then if* $\tau^n(x^n) > 0$ *and* $G \in \mathcal{B}'^n$,

$$W^n((G)_\delta|x) > \hat{W}^n(G|x^n) - \delta. \tag{4.2.6}$$

Thus if $W$ id $\overline{d}$-continuous, given $\delta > 0$ there is an $n_0$ such that (4.2.5) and hence (4.2.6) hold for $n \geq n_0$.

The lemma means, for example, that if $\{(u_i, D_i) : 1 \leq i \leq M\}$ is a $\delta$-robust $(\tau, M, n, \lambda)$ Feinstein code and $n \geq n_0$, then

$$\{(u_i, (D_i)_\delta : 1 \leq i \leq M\}$$

is an $(M, n, \lambda + \delta)$ codebook.

Here good SB-codes will be constructed from good robust Feinstein codes. As we have explained, SB-codes can be constructed via the *R-K-Theorem*.

### *4.2.3 Statement and Discussion of Results*

Although the ideas for proving the result are fairly straightforward for readers familiar with basic techniques from information theory and ergodic theory, we prefer to explain them in an intuitive manner and not give the formal proofs, because there are a lot of book-keeping details that are often uninformative and may even obscure ideas.

Our main reason for *going* through this discussion, which by itself is also lengthy, is to motivate the reader to go for a simpler and more transparent proof!

This we emphasize by calling it

#### 4.2.3.1 Research Problem

Our belief in such a proof comes from experience: after Dobrushin's work and the paper [7] a very short proof was found in [5].

The fascinating and important subject of synchronization (see for instance the book by Stiffler [41] waits for essential discoveries. We also draw attention to our synchronization method in [6], which may be of use.

#### 4.2.3.2 Synchronization Words

In [18] synch words have been used after every block code word. They serve as "punctuation mark". A novelty now is

**Lemma 52** *Let* $\lambda \leq 1/4$ *and let* $\{\mathcal{U}_n : n \geq n_0\}$ *be a sequence of* $\lambda$*-robust* $(\tau, M, n, \lambda)$ *Feinstein codes for a* $\overline{d}$*-continuous channel* $W$ *having capacity* $C > 0$. *Assume also that*

$$h(2\lambda) + 2\lambda \log(|\mathcal{Y}| - 1) < C$$

*For each* $n \geq n_0$ *let* $(P_n(1), P_n(2), \ldots, P_n(M(n)))$ *be an arbitrary PD and choose* $\delta \in (0, 1/4)$. *Then there exists an* $n_1$ *such that for all* $n \geq n_1$ *the following statements are true:*

*(A) If*

$$\mathcal{U}_n = \{(u_i, D_i) : 1 \leq i \leq M(n)\}$$

*then there is a modified codebook*

$$\mathcal{V}_n = \{(v_i, \Gamma_i) : 1 \leq i \leq K(n)\}$$

*and a set of* $K(n)$ *indices* $\mathcal{K}_n = \{k_1, \ldots, k_{K(n)}\} \subset \{1, \ldots, M(n)\}$ *such that*

$$v_i = u_{k_i},\ \Gamma_i \subset (D_{k_i})_{\lambda^2}\ \ (1 \leq i \leq K(n))$$

*and*

$$\max_{1 \leq j \leq K(n)} \sup_{x \in [v_i]} W^n(\Gamma_j^c | x) \leq \lambda. \tag{4.2.7}$$

*(B) There is a synch word* $\sigma \in \mathcal{X}^r$, $r = r(n) = \lceil \delta n \rceil$ *and a synch decoding set* $S' \in \mathcal{B}'^r$ *such that*

$$\sup_{x \in [\delta]} W^r(S^c | x) \leq \lambda \tag{4.2.8}$$

*and such that no* $r$*-tuples in* $S$ *appears in any* $n$*-tuple in any* $\Gamma_i$, *that is, if*

$$G(b^r) = \{y^n : y_i^r = b^r,\ \text{some } i = 0, 1, \ldots, n - r\}$$

*and*

$$G(S) = \bigcup_{b^r \in S} G(b^r)$$

*then*

$$G(S) \cap \Gamma_i = \emptyset \ \ (1 \le i \le K(n)). \tag{4.2.9}$$

*(C) We have that*

$$\sum_{k \notin \mathcal{K}_n} P_n(k) \le \lambda\delta \tag{4.2.10}$$

The modified code $\mathcal{V}_n$ has fewer words than the original code $\mathcal{U}_n$, but (4.2.10) ensures that $\mathcal{U}_n$ cannot be much smaller since, for example, if $P_n(i) = M(n)^{-1}$ $(1 \le i \le M(n))$, then (4.2.10) becomes

$$K(n) = (1 - \lambda\delta)M(n) \tag{4.2.11}$$

Given the codebook $\mathcal{V}_n$ a synch word $\sigma \in \mathcal{X}^r$, and a synch decoding set $S$ we shall call the length $n + r$ codebook

$$\{(\sigma \times v_i, S \times \Gamma_i) : 1 \le i \le K(n)\}$$

a prefixed or (punctuated) codebook.

*Remark* To prove the basic coding theorems only the case

$$P_n(i) = M(n)^{-1} \ \ (1 \le i \le M(n))$$

has to be considered. The more general result is required, however, to prove the invulnerable source theorem.

By combining the preceding lemma with the existence of robust Feinstein codes at rates less than capacity [19] we have the following:

**Corollary 20** *Let $W$ be a stationary ergodic $\bar{d}$-continuous channel ($DE\bar{d}C$). Fix $\lambda > 0$ and $R \in (0, C)$. Then there exists for sufficiently large $N$ a length $N$ block code*

$$\{(\sigma \times v_i, S \times \Gamma_i) : 1 \le i \le M\},$$

*$M \ge 2^{NR}$, $\sigma \in \mathcal{X}^r$, $v_i \in \mathcal{X}^r$, $r + n = N$, such that*

$$\sup_{x \in [\sigma]} W^r(S^c|x) \le \lambda$$

$$\max_{1 \le j \le M} W^n(\Gamma_j^c|x) \le \lambda$$

$$\Gamma_j \cap G(S) = \emptyset \ \ (1 \le j \le M).$$

#### 4.2.3.3 SB-Coding: Totally Ergodic Sources

The synch word can be used to mark the beginning of a code word and it will rarely be falsely detected during a code word. Unfortunately, however, an $r$-tuple consisting of a segment of a synch and a segment of a code word may be falsely detected as a synch with *nonnegligible* probability.

To resolve this confusion we look at the relative frequency of synch-detects *over a sequence of blocks*. The idea is that if we look at enough blocks, the relative frequency of the synch-detects in each position should be nearly the probability of occurrence in that position and these quantities taken together give a pattern that can be used to determine the true synch location. For the ergodic theorem to apply, however, we require that blocks be ergodic and hence we first consider *totally ergodic sources and channels*.

**Lemma 53** *Let $W$ be a DTE$\overline{d}$C, fix $\lambda, \delta > 0$ and assume that $\mathcal{V}_n = \{\sigma \times v_i, S \times \Gamma_i;\ 1 \le i \le K\}$ is a prefixed codebook satisfying* (4.2.7)–(4.2.10). *Let $\gamma_N : G^N \to \mathcal{V}_n$ assign an $N$-tuple in the prefixed codebook to each $N$-tuple in $G^N$ and let $(G, \mu, U)$ be an $N$-stationary, $N$-ergodic source. There exists for sufficiently large $L$ (which depends on the source) a synch locating function $\sigma : \mathcal{Y}^{LN} \to \{0, 1, \dots, N-1\}$ and a set $\Phi \in \mathcal{B}_G^m$, $m = (L+1)N$, such that, if $u^m \in \Phi$ and $\gamma_n(u_{LN}^N) = \sigma \times v_i$, then*

$$\inf_{x \in [\gamma_m(u^m)]} W\left(\left\{y : \sigma(y^{LN}) = \theta, 0 \le \theta \le N-1;\ y_{LN}^N \in S \times \Gamma_i\right\} |x\right) \ge 1 - 3\lambda \tag{4.2.12}$$

*and*

$$\mu^m(\Phi) \ge 1 - \lambda. \tag{4.2.13}$$

The significance of the lemma is the following. The source is block encoded using $\gamma_N$. The decoder observes a possible synch word and then looks “back” in time at previous channel outputs and calculates $\sigma(y^{LN})$ to obtain the exact synch location which is corrected with high probability.

#### 4.2.3.4 Synch Locator Function

Since $\mu$ and $W$ ate $N$-stationary and $N$-ergodic, if $\overline{\gamma} : \mathcal{X}^\infty \to \mathcal{Y}^\infty$ is the sequence encoder induced by the length $N$-block code $\gamma_N$, then the encoded source $\mu\overline{\gamma}^{-1}$ and the induced channel output source $\mu'$ are all $N$-stationary and $N$-ergodic. The sequence

$$s_j = \mu'(T^j[S]) \quad j = \dots, -1, 0, 1, \dots$$

is therefore periodic with period $N$. Furthermore, $s_j$ can have no smaller period than $N$ since from (4.2.7)–(4.2.9)

$$\mu'(T^j[S]) \le \lambda, \quad j = r+1, \dots, N-r, \quad \text{and} \quad \mu'([S]) \ge 1 - \lambda.$$

Thus if we define the synch pattern $(s_j : j = 0, \dots, N-1)$, it is distinct from any cyclic shift of itself of the form $(s_k, \dots, s_{N-1}, s_0, \dots, s_{k-1})$, where $1 \le k \le N-1$. The *synch locator* computes the relative frequencies of the occurrence of $S$ at intervals of length $N$ for each of $N$ possible starting points to obtain, say, a vector $\hat{s} = (\hat{s}_0, \dots, \hat{s}_{N-1})$. The ergodic theorem implies that the $\hat{s}_i$ will be near their expectation and hence with high probability

$$(\hat{s}_0, \dots, \hat{s}_{N-1}) = (s_0, \dots, s_{N-1}, s_0, \dots, s_{\theta-1}),$$

determining $\theta$.

Another way of looking at the result is to observe that the sources $\mu' T^j$ $(0 \le j \le N-1)$ are each $n$-ergodic and hence any two are either identical or orthogonal in the sense that they place their measures on disjoint $N$-invariant sets. No two can be identical, however, since if $\mu' T^i = \mu' T^j$ for $i \ne j, 0 \le i, j \le N-1$, then $\mu'$ would be periodic with period $0 < |i-j| < N$, yielding a contradiction.

Since membership in any set can be determined with high probability by observing the sequence for a long enough time, the synch locator attempts to determine which of the $N$ distinct sources $\mu' T^j$ is being observed.

In fact, synchronizing the output is exactly equivalent to forcing the $N$-sources $\mu' T^j$ $(0 \le j \le N-1)$ to be distinct $N$-ergodic sources.

After this is accomplished, the remainder of the proof is devoted to using the properties of $\overline{d}$-continuous channels to show that synchronization of the output source when driven by $\mu$ implies that with high probability the channel output can be synchronized for all fixed input sequences in a set of high $\mu$ probability.

To the very basic work of Nedoma [32] and Vajda [42] additional structure had to be added to make it not only applicable to block coding, as originally done, but also to SB-coding. In this way Lemma 53 became stronger. It is also more general, because assumptions on channels have been weakened.

The next lemma uses the prefixed block code and the synch locator function combined with the R-K Theorem to construct a good SB-code for a totally ergodic source with entropy less than capacity. The encoder has infinite length and the decoder finite length. The subsequent corollary removes the requirement of infinite encoder length and thereby proves that a stationary totally ergodic source $\mu$ is sliding-block admissible for a *DTE$\overline{d}$C* if $H(\mu) < C$. The lemma is proved by assigning prefixed code words to the set of roughly $2^{NH(\mu)}$ entropy typical source sequences (from the Shannon–McMillan Theorem) and then "stationarizing" the block code by embedding in a gadget. Its height is chosen large enough to ensure that the synch locator function will perform correctly most of the time.

**Lemma 54** *Given DTE$\overline{d}$C W with (informational) Shannon capacity C, a stationary totally ergodic source $(G, \mu, U)$ with entropy rate $H(\mu) < C$, and $\delta > 0$, there exists for sufficiently large m a sliding-block decoder $g_m : \mathcal{Y}^m \to G$ and an infinite length SB-encoder $f : G^\infty \to \Omega$ such that*

$$P_{\text{err}}(\mu, W, f, g_m) \le \delta.$$

**Corollary 21** *If W is a DTE$\bar{d}$C with Shannon capacity C, then any totally ergodic source $(G, \mu, U)$ with $H(\mu) < C$ is admissible.*

#### 4.2.3.5 Ergodic Sources

If a prefixed block length $N$ block code of Corollary 20 is used to block encode a general ergodic source $(G, \mu, U)$, then successive $N$-tuples from $\mu$ are generally not ergodic and hence the previous analysis does not apply.

From the Nedoma [31] decomposition, however, any ergodic source $\mu$ can be represented as a mixture of $N$-ergodic sources, all of which are simply shifted versions of each other. Given an ergodic measure $\mu$ and an integer $N$, there exists a decomposition of $\mu$ into $M$ $N$-ergodic, $N$-stationary components where $M$ divides $N$, that is, there is a set $\Pi \in \mathcal{B}_G$ such that

$$T^M \Pi = \Pi \tag{4.2.14}$$

$$\mu(T^i \Pi \cap T^j \Pi) = 0 \ \ (i \neq j, 0 \leq i, j \leq M) \tag{4.2.15}$$

$$\mu(\bigcup_{i=0}^{M-1} T^i \Pi) = 1 \tag{4.2.16}$$

$$\mu(\Pi) = \Pi^{-1} \tag{4.2.17}$$

the source $(G, \pi_i, U)$, where

$$\pi_i(F) = \mu(F|T^i \Pi) = \mu(F \cap T^i \Pi)\mu(\Pi)^{-1} = M\mu(F \cap T^i \Pi)$$

are $N$-ergodic and $N$-stationary and

$$\mu(F) = M^{-1} \sum_{i=0}^{M-1} \pi_i(F) = M^{-1} \sum_{i=0}^{M-1} \mu(F|T^i \Pi) = M^{-1} \sum_{i=0}^{M-1} \mu(F \cap T^i \Pi).$$

This decomposition provides a method of generalizing the results for totally ergodic sources to ergodic sources: since $\mu(\cdot|\Pi)$ is $N$-ergodic, Lemma 53 is valid if $\mu$ is replaced by $\mu(\cdot|\Pi)$. The infinite SB-code $f$ can ensure that the appropriate mode occurs at the output by testing for $T^{-i}\Pi$ at the gadget base and, if the base is in $T^{-i}\Pi$, insert $i$ dummy symbols and then encode using the length $N$ prefixed block code. This means that the code is "lined up" with the $N$-ergodic design mode, say $\Pi$, and the relative frequencies converge to an appropriate expectation, yielding the desired code. A finite length encoder is then obtained as previously.

**Theorem 113** *If W is a DTE$\bar{d}$C with (informational) Shannon capacity C, then any ergodic source $(G, \mu, U)$ with $H(\mu) < C$ is admissible.*

#### 4.2.3.6 Invulnerable Sources

The following lemma is a slight variation of Lemma 54. Roughly speaking, it is an observation that, given the assumptions and properties of Lemma 53 and SB-encoder $f$ constructed from a block encoder $\gamma_N$ as in Lemma 54, the receiver can reliably construct the $N$-ergodic channel input process instead of the original source. This means that there exists ergodic sources that can be connected directly to the noisy channel and recovered to within $\lambda$ by the decoder—a $\lambda$-invulnerable source. This provides a new characterization of channel capacity as given by the corollary below.

**Lemma 55** *Let $\delta$, $\lambda$, $W$, $\mathcal{V}_N$, $N \geq 3$, $r$, $\gamma_N$, $L$, $M = (L+1)N$, $\Phi$, and $(G, \mu, U)$ as in Lemma 53. Choose $K$ so large that $m \leq \lambda K N$, and let $f$ the infinite length SB-code obtained by embedding $\gamma_N$ in a $(KN, \lambda)$-gadget with base $F$. There is a length $m$ decoder $h_M$ such that, with $i(x) = x_0$,*

$$P_{\text{err}}(\mu \tilde{f}^{-1}, W, i, h_m) \leq 3\lambda,$$

*that is $(\Omega, \mu \tilde{f}^{-1}, \mathcal{B}')$ is $3\lambda$-vulnerable.*

**Corollary 22** *Let $W$ be a DTE$\bar{d}$C. Given $\Delta$, $\delta > 0$, and $H < C$, there exists a $\delta$-vulnerable source $(\Omega', \mu', \mathcal{B}')$ such that $H(\mu') \geq H - \Delta$ and*

$$\inf_{\delta} \sup_{\delta-\text{invulnerable}(\Omega', \mu', \mathcal{B}')} H(\mu') = C$$

This raises a further question: do there exist 0-invulnerable or, simply, invulnerable sources with entropy rates near capacity?

An affirmative answer to this question is provided by combining the preceding result with an iteration that allows us to take an $\epsilon$-vulnerable source of a given entropy rate and construct an $\epsilon'$ vulnerable source with $\epsilon' \ll \epsilon$ such that the entropy rate of the $\epsilon'$-vulnerable source is only slightly less than that of the $\epsilon$-vulnerable source and such that the $\epsilon'$-vulnerable source is close (in a $\bar{d}$-sense) to the $\epsilon$-vulnerable source. This closeness is necessary in order to get a converging sequence of codes which in the limit give the desired invulnerable source.

We do not enter the technical part because we present later a more general result by Kieffer.

We just summarize the result.

**Theorem 114** *Let $W$ be a DTE$\bar{d}$C with Shannon capacity $C$ and let $H^* \in (0, C)$. There exists a totally ergodic invulnerable source $(\Omega', \tau^*, \mathcal{B}')$ with entropy rate $H(\tau^*) = H^*$ and hence*

$$\sup_{\text{invulnerable } (\Omega', \tau^*, \mathcal{B}')} H(\tau^*) = C$$

There is an immediate corollary in terms of $B$-processes, these are processes obtainable by finite or infinite length SB-coding an i.i.d. process [33]. An alternate characterization is that $B$-processes are those processes that can be approximated arbitrarily closely in the $\bar{d}$-distance by a mixing multi-step Markov process. Since $\tau^*$ was constructed by SB-coding an i.i.d. source, we immediately have the following:

**Corollary 23** *Given $W$, $C$, and $H^*$ as in Theorem 114, there exists an invulnerable $B$-process $(\Omega', \tau^*, \mathcal{B}')$ with $H(\tau^*) = H^*$.*

### 4.2.3.7 Zero-Error Codes and Zero-Error Capacity

Combining Corollary 23 with the isomorphism theorem of ergodic theory [34] as stated in terms of sliding block codes [14] yields the following:

**Theorem 115** *If $(G, \mu, U)$ is a $B$-process and $W$ is a DTE$\bar{d}$C with capacity $C > H(\mu)$ then there is an infinite length SB-encoder $f : G^\infty \to \mathcal{X}$ and an infinite-length SB-decoder $G : \mathcal{Y}^\infty \to \mathcal{X}$ such that*

$$P_{\text{err}}(\mu, W, f, g) = 0,$$

*that is, the source can be communicated with zero error across the noisy channel.*

**Definition 31** Let the *weak* zero-error capacity $C_0$ of a channel be the supremum of the entropy rates of all stationary processes that can be communicated across the channel with zero error using any block stationary coding (such as block codes of SB-codes). The following corollary shows that under quite general conditions the weak zero error capacity is simply $C$.

**Corollary 24** *Given a DTE$\bar{d}$C $W$, then*

$$C(W) = C_0(W)$$

### 4.2.3.8 Discussion

The Shannon capacity

$$C(W) = \sup_{\tau \text{ stationary}} I(\tau, W)$$

(or $C_e(W) = \sup_{\mathcal{T} \text{ ergodic}} I(\tau, W)$ or $C_{\text{bs}}(W) = \sup_{\tau \text{ block-stationary}} I(\tau, W)$) is often achievable, that is, the supremum is actually a maximum.

**Example** Let $\mathcal{X} = \mathcal{Y} = \{0, 1\}$ and let $W$ be a BSC with parameter $p < \frac{1}{2}$, that is,

$$W^N(y^n|x^n) = \prod_{i=1}^{n} p^{x_i \oplus y_i}(1-p)^{1-x_i \oplus y_i},$$

where $\oplus$ denotes modulo two addition. For this channel an i.i.d. equiprobable source $(\mathcal{X}, \tau, \mathcal{B})$ yields

$$I(\tau, W) = C = 1 - h(p).$$

A natural question is whether the supremum of Theorem 114 or that defining $C_0$ is also a maximum. We show that this is generally not the case. Already for the BSC no invulnerable source with $H(\tau) = C$ exists (see [20] p. 655).

This result is parallel to a result of Berger and Lau [8] that Shannon's rate distortion function cannot be achieved with equality using SB-codes.

#### 4.2.3.9 Research Problems

Some open problems due to Berger and Lau [8] are

(1) generalizing the output memory assumption by removing the totally ergodic requirement;
(2) generalizing the result to other distance measures, and other notions of continuity hopefully allowing results for input constrained continuous alphabet (and continuous time?) channels, and
(3) the development of a structural theory for the infinite codes yielding zero error. A simple construction for nonstationary sources yielding zero error in the limit is implicit in Cover, McEliece, and Posner [11]

The product space is holy.

## 4.3 Lecture on Sliding-Block Coding

### *4.3.1 Introduction*

The (weak) capacity $C$ of an SNC channel (stationary, non-anticipatory channel) was derived, of course, for block codes. For *sliding-block coding it is more difficult to achieve such a rate*. Here it is shown that this is possible for several types of SNC like *additive random noise channels* and *averaged channels, where components are DMC* (originally due to Parthasarathy [35] and Ahlswede [2]). Denoting the capacity for sliding-block coding by $C_{sb}$ the result says for these types of channels that $C_{sb} = C$. *This generalizes the result of Gray and Ornstein, for the DMC* [18].

#### 4.3.1.1 Sources

A point $w \in \Omega$ is *periodic*, if for some integer $r$ $T^r(w) = w$. The smallest such $r$ is the period of $w$.

A source $(\Omega, P)$ is *aperiodic* if $P$ assigns probability zero to the (countable) set of periodic points, it is *periodic* if $P$ assigns probability one to the set of periodic points. Note that every ergodic source is either periodic or aperiodic. A source is stationary, memoryless; also called process, and also called Bernoulli process. It is of course aperiodic.

*Channels* are here always SNC and $(n, N, \lambda)$-codes denote, as always, a block code (in $\mathcal{X}^n \times \mathcal{Y}^n$).

#### 4.3.1.2 The Source-Channel Set up

Transmitting the source $(\Omega, P)$ over the SNC $(\Omega', W, \Omega'')$ using measurable coder $\varphi$ and decoder $\psi$:

$$\varphi : \Omega \to \Omega', \qquad \psi : \Omega'' \to \Omega.$$

On the probability spaces we define processes of RVs

$$U = (U_i)_{-\infty}^{\infty}, \; X = (X_i)_{-\infty}^{\infty}, \; Y = (Y_i)_{-\infty}^{\infty}, \; \text{and} \; V = (V_i)_{-\infty}^{\infty}.$$

### 4.3.2 Block Coding

Let $n$ be a positive integer. A coder $\varphi$ is a *block coder of type n* if there exists a map $\varphi_n : \Omega^n \to \Omega'^n$ such that

$$\varphi(w)_{kn+1}^{kn+n} = \varphi_n\left(w_{kn+1}^{kn+n}\right), \qquad k \in \mathbb{Z}.$$

A coder $\varphi$ is called a *block coder* if it is a block coder of type $n$ for some $n$.

Similarly, we define a block *decoder* of type $n$ and a block decoder.

We say the source $(\Omega, P)$ is transmissible over the channel $W$ with respect to block coding if for any $\varepsilon > 0$ there exists a positive integer $n$, a block coder $\varphi$ of type $n$ and a block decoder $\psi$ of type $n$ such that

$$P^{\varphi,\psi}({}^1U^n = {}^1V^n) > 1 - \varepsilon. \tag{4.3.1}$$

We define the block coding capacity $C_b$ of the channel $W$ to be

$$C_b = \sup\{H(P) : (\Omega, P) \text{ ergodic and transmissible over } W \text{ with block coding, where } \Omega \text{ has finite alphabet}\}.$$

**Theorem 116** (Coding theorem and converse, Winkelbauer 1960, [43])

(i) *Let $(\Omega, P)$ be an ergodic source with $H(P) < C_b$. Then $(\Omega, P)$ is transmissible over the channel $W$ with respect to block coding.*

*Moreover, for any* $\varepsilon > 0$*, there exists for sufficiently large* $n$ *a block coder* $\varphi$ *of type* $n$ *and a block decoder* $\psi$ *of type* $n$ *such that* $P^{\varphi,\psi}[{}^1U^n = {}^1V^n] > 1 - \varepsilon$

*(ii) If* $H(P) > C_b$*, then* $(\Omega, P)$ *is not transmissible over* $W$ *with respect to block coding*

In 1975 Gray [14] introduced a new type of coding (for sources), called sliding-block coding.

#### 4.3.2.1 Sliding-Block Coding

A coder $\varphi : \Omega \to \Omega'$ is called an *invariant* coder, if $\varphi T_\Omega = T_{\Omega'}\varphi$. A coder $\varphi$ is called sliding-block coder of type $n$ if there exists a map $\varphi_n : \mathcal{Z}^{2n+1} \to \mathcal{X}$ such that $\varphi(w)_i = \varphi_n\left({}^{i-n}w^{i+n}\right)$, $i \in \mathbb{Z}$. A coder $\varphi$ is called a sliding-block coder if it is a sliding-block coder of type $n$ for some $n$. Similarly, one defines invariant decoders, sliding-block decoders of type $n$, and sliding-block decoders.

A source $(\Omega, P)$ is said to be *transmissible* over the channel $W$ with respect to sliding-block coding if for any $\varepsilon > 0$ there exists a sliding-block coder $\varphi$ and a sliding-block decoder $\varphi$ such that $P^{\varphi,\psi}(U_0 = V_0) > 1 - \varepsilon$. We define the sliding-block capacity of $W$ by

$$C_{sb} = \sup\{H(P) : (\Omega, P) \text{ ergodic is transmissible over } W \text{ with respect to sliding-block coding and } \Omega \text{ has finite alphabet } \mathcal{Z}\}.$$

(As for block coding there is always one such source in the set; investigate the case where $\Omega$ consists of a single point.)

There are two natural conjectures. They were proved by Gray and Ornstein [18] for the special case where $W$ is a DMC.

*Conjecture 1* $C_{sb} = C_b$ for every $W$.

*Conjecture 2* (i) If $(\Omega, P)$ is an *aperiodic ergodic source* with $H(P) < C_{sb}$, then $(\Omega, P)$ is transmissible over $W$ with sliding-block coding.

(ii) If $(\Omega, P)$ is an ergodic source satisfying $H(P) > C_{sb}$, then $(\Omega, P)$ is not transmissible over $W$ with respect to sliding-block coding.

It would be nice to have $C_{sb} = C_b$ in general, because then the general formula by Kieffer [24] for $C_b$ could be used to calculate $C_{sb}$ as well.

We prove below, *that* $C_{sb} \leq C_b$ *always holds and both conjectures* hold for $W$ if it is either ergodic, or a channel with additive random noise, or an averaged channel whose components are DMCs.

However, Conjecture 2(i) *fails in general* for *periodic ergodic* sources. (see definition above, other possible definition $P(ET^r) = P(E)$ $(E \in \mathcal{B})$).

#### 4.3.2.2 A "Demonstration" that $C_{sb} = C_b$

**Theorem 117** *For any $W$ $C_{sb} \leq C_b$.*

*Proof* The proof adapts the idea appearing in Nedoma [30], pp. 159–162 (see also Jacobs [22]).

#### 4.3.2.3 A Sufficient Condition for $C_{sb} = C_b$

In the positive direction is needed.

**Lemma 56** *Let $(\Omega, P)$ be a source and let $W$ be a channel. We suppose that $\varphi : \Omega \to \Omega'$ is an invariant coder and $\psi : \Omega'' \to \Omega$ is a sliding-block decoder such that $P^{\varphi,\psi}(U_0 \neq V_0) < \varepsilon$. Then there exists a sliding-block coder $\varphi' : \Omega \to \Omega'$ such that $P^{\varphi',\psi}(U_0 \neq V_0) < \varepsilon$.*

*Proof* Gray and Ornstein [18] show how to do this. □

**Theorem 118** *Suppose for some positive integer $t$ both the conditions below hold for the channel $W$. Then $C_{sb} = C_b$ and if $(\Omega, P)$ is any aperiodic ergodic source with $H(P) < C_b$ then it is transmissible over the channel with respect to sliding-block coding.*

**Condition 1** If $x(1)^{t-1}, x(2)^{t-1} \in \mathcal{X}^{t-1}$, if $k \in \mathbb{N}$, if $y^{k+t-1} \in \mathcal{Y}^{k+t-1}$, and if $x^k \in \mathcal{X}^k$, then

$$W(y^{k+t-1}|x(1)^{t-1}x^k = W(y^{k+t-1}|x(2)^{t-1}x^k).$$

(This condition is vacuous for $t = 1$.)

**Condition 2** For any $\varepsilon > 0$, there exists $R > 0$ and a source $(\Omega', Q)$ such that

$Q \times W(i_t > R) > 1 - \varepsilon$.

*Proof* Generalization of technique used by Gray and Ornstein [18] for the DMC. □

#### 4.3.2.4 Applications

Theorem 118 applies to

- Channels with additive random noise
- Ergodic SNC
- Averaged channels whose components are DMC's

Conditions 1 and 2 are verified by Kieffer, in the last case it uses our formula for $C_b$.

### *4.3.3 Some Remarks on Periodic Sources*

For the types of channels considered in the preceding Sect. 4.3.2 under sliding-block coding any *aperiodic ergodic* source with entropy less than $C_{sb}$ is transmissible.

However, periodic ergodic sources are not transmissible for these channels under sliding-block coding.

If $(\Omega, P)$ is a *periodic ergodic* source, then for some $r \in \mathbb{N}$ there is a periodic sequence $w^* \in \Omega$ of period $r$ such that

$$P(T^i w^*) = r^{-1} \qquad (0 \leq i \leq r-1). \tag{4.3.2}$$

We call $r$ the period of the periodic ergodic source $(\Omega, P)$.

**Theorem 119** *A periodic ergodic source $(\Omega, P)$ of period $r$ is transmissible over the SNC $W$ with sliding-block coding if and only if there exists a periodic sequence $w'^* \in \Omega'$ of period $r$ such that the PDs $W(\cdot|T^i w'^* : 0 \leq i \leq n-1\}$ are mutually singular.*

*Proof* We start with $w'^*$ periodic with period $r$ and $\{W(\cdot|T^i w'^*) : 0 \leq i \leq n-1\}$ are mutually singular, where $r$ is the period of $(\Omega, P)$ carried by $W^*$. We have to show that it is transmissible.

Given $\varepsilon > 0$, we may find a $k \in \mathbb{N}$ and a partition $\{E_0, E_1, \dots, E_{r-1}\}$ of $\mathcal{Y}^k$ such that

$$W^k\left(E_i \,\middle|\, {}^1\left[T^i w'^*\right]^k\right) > 1 - \varepsilon \qquad (0 \leq i \leq r-1).$$

Define an invariant coder $\varphi : \Omega \to \Omega'$ as follows. Set

$$\varphi(T^i w^*) = T^i w'^* \qquad (0 \leq i \leq n-1). \tag{4.3.3}$$

For all other $w \in \Omega$ define $\varphi(w) = (\dots, x', x', \dots)$, a fixed constant sequence in $\Omega'$.

Define $\psi : \Omega'' \to \Omega$ to be the unique sliding-block decoder such that $\psi(w'')_0 = (T^i W^*)_0$, if ${}^1[w']^k \in E_i$ $(0 \leq i \leq r-1)$. Then

$$P^{\varphi,\psi}(U_0 = V_0) = r^{-1} \sum_{i=0}^{r-1} W^k\left(E_i \,\middle|\, {}^1\left[\varphi(T^i w^*)\right]^k \right) > 1 - \varepsilon.$$

Thus $(\Omega, P)$ is transmissible with sliding-block coding.

Conversely, suppose that $(\Omega, P)$ is transmissible over $W$ with sliding-block coding. For each $\varepsilon > 0$ choose invariant $\varphi_\varepsilon : \Omega \to \Omega'$ and $\psi_\varepsilon : \Omega'' \to \Omega$ such that

$$P^{\varphi_\varepsilon,\psi_\varepsilon}\left({}^0[U]^{r-1} = {}^0[V]^{r-1}\right) > 1 - \frac{\varepsilon}{r}.$$

We have

$$P^{\varphi_\varepsilon,\psi_\varepsilon}\left({}^0[U]^{r-1} = {}^0[V]^{r-1}\right) = r^{-1}\sum_{i=0}^{r-1}\left(w'' \in \Omega'' : [\psi_\varepsilon(w'')]^{n-1} = {}^0[T^i w^*]^{n-1}\Big| \varphi_\varepsilon(T^i w^*)\right).$$

Since the $r$ sequences ${}^0[T^i w^*]^{r-1}$ $(0 \le i \le r-1)$ are distinct, the sets $\{E_{i\varepsilon} : 0 \le i \le n-1\}$ are pairwise disjoint, where

$$E_{i\varepsilon} = \{w'' \in \Omega'' :^0 [\psi_\varepsilon(w'')]^{r-1} =^0 [T^i w^*]^{n-1}\}.$$

Let $w'^*_\varepsilon \in \Omega'$ be the sequence $\varphi_\varepsilon(w^*)$. We have

$$W(E_{i\varepsilon}|T^i w'^*_\varepsilon) > 1 - \varepsilon \qquad (0 \le i \le n-1). \tag{4.3.4}$$

If $\varepsilon < \frac{1}{2}$ we see from (4.3.4) that the sequences $\{T^i w'^*_\varepsilon : 0 \le i \le n-1\}$ are distinct. Since $T^r w^* = w^*$ and $\varphi_\varepsilon$ is invariant, we must have $T^r w'^*_\varepsilon = w'^*_\varepsilon$. Hence for $\varepsilon < \frac{1}{2}$ $w'^*_\varepsilon$ has period $r$. Since there are only finitely many periodic sequences in $\Omega'$ of period $r$, $w'^*_\varepsilon$ must remain constant as $\varepsilon$ approaches zero through some sequence $(\varepsilon_j(r))_{j=1}^\infty$. Let $w'^*$ be this constant value of $w'^*_\varepsilon$ through $(\varepsilon_j(r))$. Then (4.3.4) implies $\{W(\cdot|T^i w'^*) : 1 \le i \le n-1\}$ are mutually singular. □

**Corollary 25** *Let $\mathcal{X} = \{0, 1\}$ and let $x \in \Omega'$ be the periodic sequence with period 2 such that $x_0 = 0$, $x_1 = 1$. Let $(\Omega', P)$ be the periodic ergodic source such that $P\{x\} = P\{Tx\} = \frac{1}{2}$. Then $(\Omega', P)$ is transmissible over $W$ with sliding-block coding if and only if $W(\cdot|x)$ and $\mu(\cdot|Tx)$ are mutually singular.*

*Proof* This follows from Theorem 119, because $x$ and $Tx$ are the only periodic sequences in $\Omega'$ with period 2. □

The non-transmissibility results mentioned above use this corollary.

*Example* of an averaged channel whose components are DMCs such that $C_{sb} > 0$ but there exists a *periodic ergodic source not transmissible* with respect to sliding block coding.

Let $\mathcal{X} = \mathcal{Y} = \{0, 1\}$. Let $W(\cdot|\cdot) : \mathcal{X} \times \mathcal{X} \to [0, 1]$. Define $\phi : \mathcal{X} \to \mathcal{X}$ so that $\phi(0) = 1$, $\phi(1) = 0$. Let

$$W_1(\cdot|\cdot) : \mathcal{X} \times \mathcal{X} \to [0, 1]$$

be such that

$$W_2(j|i) = W_1(j|\phi(i)), \quad i, j \in \mathcal{X}.$$

Let $(\mathcal{X}, \mathcal{X}, W_1)$ be the DMC such that

$$W_1(\cdot|w') = \prod_{i=-\infty}^{+\infty} W_1(\cdot|x_i)$$

and let $(\mathcal{X}, \mathcal{X}, W_2)$ be the DMC such that

$$W_2(\cdot|w') = \prod_{i=-\infty}^{+\infty} W_2(\cdot|x_i).$$

Let $(\mathcal{X}, \mathcal{X}, W)$ be the *averaged channel* such that

$$W = \frac{1}{2}W_1 + \frac{1}{2}W_2.$$

Pick $W_1(1|0) \neq W_1(1|1)$. Then if $C_b(1)$ and $C_b(2)$ are the block coding capacities of $W_1$ and $W_2$, resp., we have $C_b(1) = C_b(2) > 0$ and hence $C_b$ of $W$ is positive.

(By our result or by Nedoma's lower bound

$$C_b \geq (C_b(1)^{-1} + C_b(2)^{-1})^{-1} > 0)$$

Let now $w' \in \Omega'$ be the periodic sequence of period 2 such that $x_0 = 0, x_1 = 1$. Then it is easily seen that $W_1(\cdot|w') = W_2(\cdot|Tw')$ and $W_1(\cdot|Tw') = W_2(\cdot|w')$. Hence,

$$W(\cdot|w') = W(\cdot|Tw')$$

and by the corollary to Theorem 119 the periodic source $(P, \Omega')$ with $P(x) = P(Tx) = \frac{1}{2}$ used there is not transmissible over $W$ with sliding block coding and yet $C_{sb} = C_b > 0$.

#### 4.3.3.1 Transmission of Periodic Sources for the DMC

The failure of the previous example to transmit certain periodic ergodic sources does not occur for the DMC.

**Theorem 120** *Let $W$ be a DMC with capacity $C_b > 0$. Then every periodic ergodic source is transmissible over $W$ with sliding block coding. Note that there entropy equals zero!*

*Proof* By Theorem 119 it suffices to show that for every $r \geq 2$ there is a periodic sequence $x \in \Omega'$ of period $r$ such that $\{W(\cdot|T^i x) : 0 \leq i \leq r\}$ ate mutually singular. Since $C_b > 0$ there are letters $x_1, x_2$ with

$$W(\cdot|x_1) \neq W(\cdot|x_2).$$

Fix $r \geq 2$. Let $x \in \Omega'$ be periodic with period $r$ and have all of its entries drawn from the set $\{x_1, x_2\}$

Consider the DMC $W'$ with alphabets $\mathcal{X}' = \mathcal{X}^r$, $\mathcal{Y}' = \mathcal{Y}^r$ and $W' = W^r$.

In $(\mathcal{X}')^\infty$ let $x(i)$ be the constant sequence from constant ${}^0[T^i x]^{r-1}$ $(0 \le i \le r-1)$. Let $((\mathcal{X}')^\infty, P(i))$ be the ergodic periodic source which assigns probability 1 to $x(i)$.

Since $W'$ is an ergodic channel $P(i) \times W'$ induced on $(\mathcal{Y}')^\infty$ a PD $W'(\cdot|x(i))$, ergodic and stationary with respect to the shift on $(\mathcal{Y}')^\infty$.

Hence, if $0 \le i < j \le r-1$, then either $W'(\cdot|x(i)) = W'(\cdot|x(j))$ are mutually singular or equal. Suppose now $W'(\cdot|x(i)) = W'(\cdot|x(j))$. Then

$$W^r(\cdot|{}^0[T^i x]^{r-1}) = W^r(\cdot|{}^0[T^j x]^{r-1}).$$

Since ${}^0[T^i x]^{r-1} \neq^0 [T^j x]^{r-1}$, then for some $s$ satisfying $0 \le s \le r-1$ we have $[T^i x]_s \neq [T^j x]_s$ and hence either $[T^i x]_s = x_1$ and $[T^j x]_s = x_2$ or reversely. In either case, we conclude

$$W(\cdot|x_1) = W(\cdot|x_2),$$

a contradiction.

Thus $\{W'(\cdot|x(i)) : 0 \le i \le r-1\}$ are mutually singular. But □

$$W'(\cdot|x(i)) = W(\cdot|T^i x).$$

*Research program* Continue the study of *phase* invariance decoding as done for almost periodic discrete memoryless channels in [3] with respect to sliding block codes and more generally for NDMC.

### *4.3.4 Sliding-Block Coding for Weakly Continuous Channels*

For a WCSC $W$ with *operational* sliding-block capacity $C_{\text{sb}}(W)$ it is shown that every stationary, ergodic, *aperiodic* source with entropy less than $C_{\text{sb}}(W)$ is sliding-block transmissible over the channel.

It had been shown earlier that a stationary, ergodic source is block transmissible over the channel if its entropy is less than the operational block capacity $C_{\text{b}}(W)$.

If, in addition, the *channel is ergodic*, it is shown that $C_{\text{sb}}(W) = C_{\text{b}}(W)$.

An example of a *non-ergodic* weakly continuous channel $W$ is given for which $0 < C_{\text{sb}}(W) < C_{\text{b}}(W)$.

The usual method for proving a sliding-block source/channel coding theorem builds a good sliding-block encoder and decoder at a certain rate from a good block encoder and decoder at roughly this same rate. For this to work, the block encoder and decoder have to be "*synchronized*" in an appropriate sense; the synchronization is achieved by requiring the *channel to be ergodic*. Consequently, except for isolated cases [25] a sliding-block source/channel coding theorem was not known for non-ergodic $\bar{d}$-continuous channels, for example, whereas proving a block coding theorem for such channels did not prove to be a problem [20].

Kieffer developed a new method which makes it possible to prove a sliding-block coding theorem for non-ergodic as well as ergodic weakly continuous channels, a class of channels, introduced by Kieffer [27, 28], which properly included the class of $\overline{d}$-continuous channels.

#### 4.3.4.1 Two Coding Theorems for Locally Continuous and Locally Ergodic Channels

The first result gives as a special case a sliding-block source channel coding theorem for WCSC.

**Theorem 121** *Let the stationary channel $W$ be weakly continuous at every* ***positive entropy*** *stationary and ergodic source $(\Omega', \nu)$ (abbreviated as $WCSC_+$) and let $C_{sb}(W) > 0$. Then a stationary ergodic* ***aperiodic*** *source $(\Omega, \mu)$ is sliding block transmissible over $W$ if and only if $H(\mu) \leq C_{sb}(W)$. Any stationary ergodic,* ***periodic*** *source is stochastically transmissible.*

*Remark* as shown by an example in [25] some stationary ergodic, periodic sources may not be sliding-block transmissible. For such source one is forced to use stochastic encoders and decoders rather than deterministic ones.

**Theorem 122** *Let $W$ be a stationary channel and let $(\Omega', \nu)$ be a stationary and ergodic source with $I(\nu, W) > 0$. Suppose there is a neighborhood $\mathcal{N}$ of $(\Omega', \nu)$ such that the channel is weakly continuous and ergodic at every positive entropy source in $\mathcal{N}$. Then a stationary ergodic* ***periodic*** *source $(\Omega, \mu)$ is sliding block transmissible over $W$ if $H(\mu) \leq I(\nu, W)$. Every stationary ergodic* ***periodic*** *source is stochastically transmissible.*

As a result of the following corollary to Theorem 122 there is a formula for the sliding-block capacity of any WCSC. This formula was obtained before by Gray et al. [20] for the special case of a totally ergodic $\overline{d}$-continuous channel.

**Corollary 26** *Let $W$ be a $WCSC_+$, then*

$$C_{sb}(W) = \sup\{I(\nu, W) : (\Omega', \nu) \text{ stationary and ergodic}\}. \tag{4.3.5}$$

*Proof* By Theorem 122 $C_{sb}(W)$ is not smaller than the number on the right side of (4.3.5) and is not bigger than it by Theorem 10 in [27].

*Remark* The right side of (4.3.5) is Shannon's informational capacity $C(W)$. If the channel is non-ergodic, $C(W)$ upper bounds the operational capacity $C_{sb}(W)$. Averaged channels give examples of non-ergodic channels for which $C_{sb}(W)$ is strictly less than $C(W)$.

In [27] there is an example of a *non-ergodic* weakly continuous stationary channel $W$ with $C_{sb} < C_b(W)$.

## References


1. R.L. Adler, Ergodic and mixing properties of infinite memory channels. Proc. Am. Math. Soc. **12**, 924–930 (1961)
2. R. Ahlswede, The weak capacity of averaged channels. Z. Wahrscheinlichkeitstheorie Verw. Geb. **11**, 61–73 (1968)
3. R. Ahlswede, Beiträge zur shannonschen informationstheorie im falle nichtstationärer Kanäle. Z. Wahrscheinlichkeitstheorie Verw. Geb. **10**, 1–42 (1968)
4. R. Ahlswede, A constructive proof of the coding theorem for discrete memoryless channels in case of complete feedback, in *Sixth Prague Conference on Information Theory, Statistics Decision Functions and Random Processes*, September 1971, Publishing House Czechosl. Academy of Sciences, pp. 1–22 (1973)
5. R. Ahlswede, P. Gács, Two contributions to information theory, topics in information theory (Second Colloquium, Keszthely, 1975). Colloq. Math. Soc. Janos Bolyai, Amst. **16**, 17–40 (1977)
6. R. Ahlswede, I. Wegener, in *Wiley-Interscience Series in Discrete Mathematics and Optimization 1987*, ed. by R.L. Graham, J.K. Leenstra, R.E. Tarjan. Suchprobleme, Teubner Verlag, Stuttgart; Russian Edition with Appendix by Maljutov 1981, English Edition with Supplement of Recent Literature (1979)
7. R. Ahlswede, J. Wolfowitz, Channels without synchronization. Adv. Appl. Probab. **3**, 383–403 (1971)
8. T. Berger, J.K.Y. Lau, On binary sliding block codes. IEEE Trans. Inf. Theory **23**(3), 343–353 (1977)
9. D. Blackwell, Infinite codes for memoryless channels. Ann. Math. Stat. **30**, 1242–1244 (1959)
10. L. Breiman, On achieving channel capacity in finite-memory channels. Illinois J. Math. **4**, 246–252 (1960)
11. T.M. Cover, R.J. McEliece, E.C. Posner, Asynchronous multiple-access channel capacity. IEEE Trans. Inf. Theory **27**(4), 409–413 (1981)
12. R.L. Dobrushin, Shannon's theorem for channels with synchronization errors, Problemy Peredachi Informatsii **3**, 18–36 (1967) (Translated in Probl. Inf. Transm. **3**, 31–36 (1967))
13. G.D. Forney Jr, *Information Theory* (Stanford University Course Notes, Winter, 1972)
14. R.M. Gray, Sliding-block source coding. IEEE Trans. Inf. Theory **21**, 357–368 (1975)
15. R.M. Gray, *Entropy and Information Theory* (Springer, New York, 1990)
16. R.M. Gray, D.L. Neuhoff, D.S. Ornstein, Nonblock source coding with a fidelity criterion. Ann. Probab. **3**(3), 478–491 (1975)
17. R.M. Gray, D.L. Neuhoff, D.S. Ornstein, Process definitions of distortion-rate functions and source coding theorems. IEEE Trans. Inf. Theory **21**(5), 524–532 (1975)
18. R.M. Gray, D.S. Ornstein, Sliding-block joint source/noisy-channel coding theorems. IEEE Trans. Inf. Theory **22**(6), 682–690 (1976)
19. R.M. Gray, D.S. Ornstein, Block coding for discrete stationary $\bar{d}$-continuous noisy channels. IEEE Trans. Inf. Theory **25**(3), 292–306 (1979)
20. R.M. Gray, D.S. Ornstein, R.L. Dobrushin, Block synchronization, sliding-block coding, invulnerable sources and zero error codes for discrete noisy channels. Ann. Probab. **8**(4), 639–674 (1980)
21. K. Jacobs, Die Übertragung diskreter Informationen durch periodische und fastperiodische Kanäle. Math. Ann. **137**, 125–135 (1959)
22. K. Jacobs, *Informationstheorie* (Seminarbericht, Göttingen, 1960)
23. K. Jacobs, Über die Struktur der mittleren Entropie. Math. Z. **78**, 33–43 (1962)
24. J.C. Kieffer, A general formula for the capacity of stationary nonanticipatory channels. Inf. Control **26**, 381–391 (1974)
25. J.C. Kieffer, On sliding block coding for transmission of a source over a stationary nonanticipatory channel. Inf. Control **35**(1), 1–19 (1977)
26. J.C. Kieffer, Block coding for an ergodic source relative to a zero-one valued fidelity criterion. IEEE Trans. Inf. Theory **24**(4), 432–438 (1978)

27. J.C. Kieffer, On the transmission of Bernoulli sources over stationary channels. Ann. Probab. **8**(5), 942–961 (1980)
28. J.C. Kieffer, Block coding for weakly continuous channels. IEEE Trans. Inf. Theory **27**(6), 721–727 (1981)
29. T.T. Kadota, A.D. Wyner, Coding theorem for stationary, asymptotically memoryless, continuous-time channels. Ann. Math. Stat. **43**, 1603–1611 (1972)
30. J. Nedoma, The capacity of a discrete channel, in *Transactions of the First Prague Conference on Information Theory, Statistics Decision Functions and Random Processes*, pp. 143–181 (1957)
31. J. Nedoma, Über die ergodizität und *r*-Ergodizität stationärer wahrscheinlichkeitsmasse. Z. Wahrscheinlichkeitstheorie Verw. Geb. **2**, 90–97 (1963)
32. J. Nedoma, The synchronization for ergodic channels, in *Transactions of the Third Prague Conference on Information Theory, Statistics Decision Functions and Random Processes*, pp. 529–539 (1964)
33. D. Ornstein, An application of ergodic theory to probability theory. Ann. Probab. **1**, 43–58 (1973)
34. D. Ornstein, *Ergodic Theory, Randomness, and Dynamical Systems* (Yale University Press, New Haven, 1974)
35. K.R. Parthasarathy, Effective entropy rate and transmission of information through channels with additive random noise. Sankhyä Ser. A **A25**, 75–85 (1963)
36. E. Pfaffelhuber, Channels with asymptotically decreasing memory and anticipation. IEEE Trans. Inf. Theory **17**, 379–385 (1971)
37. R.A. Scholtz, Codes with synchronization capability. IEEE Trans. Inf. Theory **12**, 135–140 (1966)
38. C.E. Shannon, A mathematical theory of communication. Bell Syst. Tech. J. **27**, 379–423, 623–656 (1948)
39. C.E. Shannon, The zero error capacity of a noisy channel. IRE Trans. Inf. Theory **2**, 8–19 (1956)
40. P.C. Shields, *The Theory of Bernoulli Shifts* (University of Chicago Press, Chicago, 1973)
41. J.J. Stiffler, *Theory of Synchronous Communication* (Prentice-Hall, Englewood Cliffs, 1971)
42. I. Vajda, A synchronization method for totally ergodic channels, in *Transactions of the Fourth Prague Conference on Information Theory, Statistics Decision Functions and Random Processes*, pp. 611–625 (1965)
43. K. Winkelbauer, Communication channels with finite past history, in *Transactions of the Second Prague Conference on Information Theory, Statistics Decision Function and Random Processes*, pp. 685–831 (1960)

# Chapter 5
# On $\lambda$-Capacities and Information Stability

## 5.1 Lecture on a Report on Work of Parthasarathy and of Kieffer on $\lambda$-Capacities

### *5.1.1 Historical Remarks and Preliminaries*

In Chap. 1, we introduced $\lambda$-capacities with several specifications and mainly concentrated on CC. Unfortunately, we learned about the work of Parthasarathy [17] on $\lambda$-capacities only recently through work of Kieffer, which Csiszár brought to our attention during the celebration of his 70th birthday in late August 2008 in Budapest, where we discovered and lectured upon the appropriateness of the concept of capacity function. Actually [17] appeared in the same year as Wolfowitz [22], speaking about $\lambda$-capacity, a concept motivated by an example of Jacobs [10]. In [19] (see also [20]) from 1960 already a lot of attention is paid to the definitions of capacity and information measures.

Instead of $\lambda$ Parthasarathy used the letter $\epsilon$ and write in his summary "Transmission of information through channels with additive noise is considered. Coding theorem and its converse are established for these channels with a certain notion of capacity. This capacity is explicitly computed for this class of channels." The paper starts with the analysis of $N_n(\epsilon, \mu)$, the minimum number of $n$-lengths sequences which have a total probability exceeding $1 - \epsilon$ for a stationary not necessary ergodic source $\mu$. On page 76 he gives credit for this as follows "The idea of studying the asymptotic properties of the sequence $\frac{\log N_n(\epsilon,\mu)}{n}$ is due to Winkelbauer. He gave the description of the function

$$\overline{H}(\mu) = \lim_{\epsilon \to 0} \overline{\lim_{n \to \infty}} \frac{\log N_n(\epsilon, \mu)}{n}$$

in terms of the entropies of the ergodic components of $\mu$. It was stated by him without proof in his lecture at the Indian Statistical Institute."



A. Ahlswede et al. (eds.), *Transmitting and Gaining Data*,
Foundations in Signal Processing, Communications and Networking 11,
DOI 10.1007/978-3-319-12523-7_5

We use now the notation $N(n,\lambda) \triangleq N_n(\epsilon,\mu)$ for $\lambda=\epsilon$. We present first the main result of [17] *without proofs*, because they use results from Ergodic Theory, which we do not prove in this book. They can, however, already be found in the paper by Oxtoby [16].

Recall the concept of a stationary (one-sided) source $(\Omega, \mathcal{B}, P, T)$. If we assign the discrete topology to $\Omega$ then $\Omega$ become a compact metric space. We follow now Oxtoby [16]. If $f:\Omega\to\mathbb{R}$, let

$$M(f,w,k) = f_k(w) = \frac{1}{k}\sum_{i=1}^{k} f(T^i w) \quad (k=1,2,\dots)$$

and

$$M(f,w) = f^*(w) = \lim_{k\to\infty} f_n(w)$$

in case this limit exists. Let $Q=\{w : M(f,w) \text{ exists for all } f\in C(\Omega)\}$, where $C(\Omega)$ is the space of continuous functions on $\Omega$.

It follows from Riesz's representation theorem that for every $w\in Q$ there exists a unique invariant PD $P_w$ such that

$$M(f,w) = \int f \, \mathrm{d}P_w.$$

Let $R\subset Q$ be the set of those points for which $P_w$ is ergodic. $R$ is called the set of regular points. Then we have

**Theorem 123** (Representation Theorem of Krylov/Bogoliubov) *The set $R$ is a Borel measurable set of invariant measure one, that is, $P(R)=1$ for every invariant PD $P$ on $(\Omega,\mathcal{B},T)$. For any Borel set $E\subset\Omega$, $P_w(E)$ is Borel measurable on $R$ and*

$$P(E) = \int_R P_w(E)\,\mathrm{d}P(w)$$

*for every invariant PD $P$.*

Let $H(P)$ denote the entropy of any invariant PD $P$. Let

$$\overline{H}(P) = \text{ess sup}\, H(P_w) \tag{5.1.1}$$

$$\underline{H}(P) = \text{ess inf}\, H(P_w) \tag{5.1.2}$$

where the essential supremum and the essential infimum are taken relative to $P$.

## 5.2 Asymptotic Properties of the Function $N(n, \lambda)$

For a stationary source $(\Omega, \mathcal{B}, P, T)$ define $\underline{\eta}(\delta)$ as the greatest number $\eta$ with the property

$$P(\{w : H(P_w) \geq \eta\}) \geq \delta \tag{5.2.1}$$

and define $\overline{\eta}(\delta)$ as the smallest number $\eta$ with the property

$$P(\{w : H(P_w) \leq \eta\}) \geq 1 - \delta. \tag{5.2.2}$$

Furthermore define

$$\underline{A}(\lambda) = \lim_{\delta \searrow \lambda} \underline{\eta}(\delta) \tag{5.2.3}$$

$$\overline{A}(\lambda) = \lim_{\delta \nwarrow \lambda} \overline{\eta}(\delta). \tag{5.2.4}$$

**Theorem 124** *For a stationary source* $(\Omega, \mathcal{B}, P, T)$

$$\underline{A}(\lambda) \leq \underline{\lim}_{n\to\infty} \frac{1}{n} \log N(n, \lambda) \leq \underline{\lim}_{n\to\infty} \frac{1}{n} \log N(n, \lambda) \leq \overline{A}(\lambda).$$

**Corollary 27** $\lim_{n\to\infty} \frac{1}{n} \log N(n, \lambda)$ *exists for all* $0 < \lambda < 1$ *except for a countable set.*

**Corollary 28**

$$\lim_{\lambda\to 0} \underline{\lim}_{n\to\infty} \frac{1}{n} \log N(n, \lambda) = \lim_{\lambda\to 0} \overline{\lim}_{n\to\infty} \frac{1}{n} \log N(n, \lambda) = \overline{H}(P)$$

("Optimistic and pessimistic" optimal rates (of compression) are equal!).

*Remark* The last equation is, as mentioned before, due to Winkelbauer. Parthasarathy mentions on page 81 also that he *conjectured* a slightly sharper form of Corollary 27: existence of the limit without the (countable) exceptions.

## 5.3 Channels with Additive Noise

Let input alphabet $\mathcal{X}$ and output alphabet coincide and be endowed with the structure of a finite Abelian group written additively. In a natural way, $\Omega = \prod_1^\infty \mathcal{X}$ become an Abelian group. We denote by $+$ and $-$ the addition and inverse operation in the group $\Omega$.

For any set $E \subset \Omega$ we write

$$E - w = \{w' \in \Omega : w' + w \in E\}. \tag{5.3.1}$$

Let $P$ be an invariant PD on $\Omega$ and let the PD's for every $w \in \Omega$

$$P(E|w) = P(E - w) \ \ (E \in \mathcal{B}) \tag{5.3.2}$$

define a stationary channel, which we call a channel with additive noise (CAN) and noise distribution $P$.

For any input distribution $P'$ on $(\Omega, \mathcal{B})$ the corresponding output distribution is obtained by convoluting $P'$ with $P$. If the group $\mathcal{X}$ consists of two elements 0 and 1 only, the addition is done modulo 2 an $P$ is the product measure obtained by assuming probability $p$ for one, thus we get the binary symmetric channel (BSC).

$(n, M, \lambda)$-codes are defined as usual and so is $M(n, \lambda)$.

**Theorem 125** *For a stationary channel with additive noise and noise distribution $P$, with $\underline{A}(\lambda)$ and $\overline{A}(\lambda)$ as defined in Sect. 5.2 for $P$,*

$$\log_2 a - \overline{A}(\lambda) \le \underline{\lim}_{n\to\infty} \frac{1}{n} \log M(n, \lambda) \le \overline{\lim}_{n\to\infty} \frac{1}{n} \log M(n, \lambda) \le \log_2 a - \underline{A}(\lambda).$$

**Corollary 29** *Except for a countable set of λ's the limit*

$$\lim_{n\to\infty} \frac{1}{n} \log M(n, \lambda) = \log_2 a - \underline{A}(\lambda) = \log a - \overline{A}(\lambda) \tag{5.3.3}$$

*exists.*

*Further*

$$\lim_{\lambda\to 0} \log a - \underline{A}(\lambda) = \lim_{\lambda\to 0} \log a - \overline{A}(\lambda) \tag{5.3.4}$$

$$= \log a - \overline{H}(P). \tag{5.3.5}$$

**Remarks**

1. Note that (5.3.4) establishes coding theorem and weak converse and (5.3.5) gives the capacity.
2. It is remarkable (but understandable for instance from the Feinstein and Wolfowitz result for DFMC) that no assumptions on ergodicity are used (here $\lambda \to 0$ takes care), but it is even more remarkable that no finite duration of the memory and no independence properties are required.
3. Symmetry in the group structure is of great help.

The channel has a very special structure. It reminds us of regularity in [3, 4] and the permutation ideas.

## 5.4 A General Formula for the Capacity of Stationary Nonanticipatory Channels

We use again concepts and notations for stationary sources and channels, however, with the difference that everywhere one-sided infinite sequence spaces are used now, where earlier we had two-sided infinite (also 1-shift versus 2-shift $T$) sequence spaces.

So alphabets, input and output, are $\mathcal{X}$ and $\mathcal{X}'$ with corresponding sequence spaces $\Omega$ and $\Omega'$, carrying the $\sigma$-algebras $\mathcal{B}$ and $\mathcal{B}'$. Further, $\tilde{\Omega} = \Omega \times \Omega'$, and channels $P : \Omega \times \mathcal{B}' \to \mathbb{R}$, properties nonanticipatory, finds duration of memory $m$ and stationary.

Also $\lambda$-codes are defined as before, but now only for the time $[1, n]$. Now $N(n, \lambda)$ is defined as usual and so are

$$\overline{C}(\lambda) = \overline{\lim}_{n\to\infty} n^{-1} \log N(n, \lambda),$$
$$\underline{C}(\lambda) = \underline{\lim}_{n\to\infty} n^{-1} \log N(n, \lambda),$$
$$\overline{C} = \lim_{\lambda\to 0} \overline{C}(\lambda), \underline{C} = \lim_{\lambda\to 0} \underline{C}(\lambda)$$

(the "optimistic" and "pessimistic" capacities).

If $\overline{C} = \underline{C}$ we say again that the channel has a (weak) capacity. We introduce now subsets of the set of PD's $\mathcal{P}(\Omega, \mathcal{B})$ on $(\Omega.\mathcal{B})$.

**Definition 32** (i) $\mathcal{P}_S = \mathcal{P}_S(\Omega, \mathcal{P})$ stands for the set of stationary PD's on $(\Omega, \mathcal{B})$
(ii) $\mathcal{P}_e = \mathcal{P}_e(\Omega)$ stands for the set of stationary and ergodic PD's on $(\Omega.\mathcal{B})$.
(iii) $\mathcal{P}_p = \mathcal{P}_p(\Omega)$ stands for the set of periodic PD's on $(\Omega.\mathcal{B})$ for some period $\tau$ $T^{\tau+1}(E) = T(E)$ for all $E \in \mathcal{B}$.

For a PD $P \in \mathcal{P}(\Omega, \mathcal{B})$ and a stationary nonanticipatory channel $W$ $P \times W$ is a PD on $(\Omega \times \Omega', \mathcal{B} \times \mathcal{B}') = (\tilde{\Omega}, \tilde{\mathcal{B}})$.
Define the information density i.

The essential function for us is now

$$A(\lambda) = \sup_{P \in \mathcal{P}_p} \sup\{r : P \times W(i \leq r) < \lambda\} \quad (0 < \lambda < 1).$$

**Theorem 126** (Kieffer 1974) *The stationary nonanticipatory channel $W$ has the (weak) capacity*

$$C = \lim_{\lambda\to 0} A(\lambda).$$

*Proof* The theorem follows from the coding theorem and its weak converse given in the following Lemmas 57 and 58.

**Lemma 57** (Coding theorem) *For the channel W of Theorem* 126

$$\underline{C}(\lambda) \geq A(\lambda) \quad (0 < \lambda < 1).$$

*Proof* Suppose that $P \in \mathcal{P}_p$ is such that $P \times W(i \leq r) < \lambda$. Choose $\lambda'$ so that $P \times W(i \leq r) < \lambda' < \lambda$. Thus for $n$ sufficiently large

$$P \times W \left\{ n^{-1} \log \frac{P \times W(X^n, Y^n)}{PW(X^n)PW(Y^n)} \leq r \right\} < \lambda' < \lambda.$$

It follows from Feinstein's Maximal Coding Lemma (see Sect. 1.4, Lemma 13) that an $(M, n, \lambda)$ code exists for $W$, if

$$M2^{-c} + P \times W(E) < \lambda \quad \text{for } E = \{i(x^n, y^n) \leq c\}. \tag{5.4.1}$$

Taking $c = rn$ it is easy to verify that with $M(\lambda) = \lfloor (\lambda - \lambda')2^{rn} \rfloor$ (5.4.1) is satisfied. Therefore $\underline{C}(\lambda) \geq r$ and it follows that $\underline{C}(\lambda) \geq M(\lambda)$.

**Remarks**

1. This proof by slicing the information density function motivated by the analysis of averaged channels was later used by Han/Verdú in determining $\underline{C}$.
2. For a result in this generality it is essential that 1-shot coding and not continuous transmission is studied, where independence properties are needed.

**Lemma 58** (The weak converse to the Coding Theorem) *For the channel W of Theorem* 126

$$\overline{C} \leq \lim_{\lambda \to 0} A(\lambda)$$

*Proof* Let $r < \overline{C}$ and choose $r'$ so that $r < r' < \overline{C}$. For $\lambda > 0$ there are integers $m, N$ constituting parameters for an $(N, m, \lambda)$ code with $m^{-1} \log N = r'' > r'$. Let $u_1, \dots, u_N$ be the codewords of such a code. Choose $P \in \mathcal{P}_p$ and $X_1, X_2$ with joint distributions $P$ such that $\Pr(X^m = u_i) = N - 1$ $(1 \leq i \leq N)$ and $\{(X_{km+1}, \dots, X_{km+m})\colon k = 0, 1, 2\}$ are independent (we use products of Fano sources). By Fano's inequality

$$m^{-1} H(X^m | Y^m) \leq \lambda \log c + h(\lambda).$$

Use now the non-negative measurable function

$$f = - \lim_{n \to \infty} n^{-1} \log \frac{P \times W(X^n, Y^n)}{PW(Y^n)}.$$

Now $H(X_{km+1}, \dots, X_{km+m} | Y_{km+1}, \dots, Y_{km+m})$ is independent of $k$, because the channel is stationary and nonanticipatory. Also for any RV's $U_1, U_2, V_1, V_2$ on $(\tilde{\Omega}, \tilde{\mathcal{B}})$

$$H(U_1, U_2|V_1V_2) = H(U_1|V_1V_2) + H(U_2|V_1V_2U_1) \leq H(U_1|V_1) + H(U_2|V_2).$$

It follows then that

$$\int_{\tilde{\Omega}} f \, \mathrm{d}(P \times W) = \lim_{n\to\infty} n^{-1} H(X^n|Y^n) \leq \lambda \log a + h(\lambda).$$

Since by Shannon/McMillan

$$-\lim_{n\to\infty} n^{-1} \log P(X^n) = r'' \text{ in } \mathcal{L}_P^1$$

we have $i = r'' - h$ a.e $(P \times W)$.

By Chebyshev's inequality, it follows then that $P \times W(i \leq r) = P \times W(h \geq r'' - r) \leq (r'' - r)^{-1}(\lambda \log a + h(\lambda)) < (r' - r)^{-1}(\lambda \log a + h(\lambda)) = \underline{q(\lambda)}$, say. Therefore $A(q(\lambda)) \geq r$ whenever $q(\lambda) < 1$. Letting $\lambda \to 0$ and then $r \to \overline{C}$, the lemma is proved.

**Remarks**

1. This proof follows the idea of proof by Nedoma [14], which we used in Sect. 3.2. It is now much easier, because optimization over periodic sources suffices whereas earlier the direct part used ergodic sources, which have now to be constructed also for the converse part.
2. That a stationary nonanticipatory channel has a capacity was first shown by Winkelbauer [21]. A formula for this capacity was given by Nedoma [14]; it is equal to the entropies of the sources which are transmissible over the channel. The present proof of Theorem 126 gives a new proof of the existence of the (weak) capacity, and a different formula for this capacity. We next demonstrate its usefulness by examples.
3. Compare also the existence proof for the DMC and the sophisticated existence proof for AVC, where no formula for the capacity is available!

### *5.4.1 The Shannon–McMillan Theorem*

For $(\Omega, \mathcal{B}, P)$ let $f : \Omega \to \{d_1, d_2, \dots\}$ be $\mathcal{B}$-measurable and let

$$P(F)(w) = P(w' : f(w') = f(w)) \quad (w \in \Omega).$$

If $g : \Omega \to \{e_1, e_2, \dots\}$ is another $\mathcal{B}$-measurable function, then let $P(f|g)$ be the discrete measurable function defined on $\Omega$ as follows:

$$P(f|g)(w) = P\left(w' : f(w') = f(w)|w' : g(w') = g(w)\right)$$

if $P(w' : g(w') = g(w)) > 0$. (so conditional probability can be defined)

$$P(f|g)(w) = 0 \text{ otherwise.}$$

Let

$$H_P(f) = -\int_{\Omega} \log P(f) \mathrm{d}P \tag{5.4.2}$$

$$H_P(f|g) = -\int_{\Omega} \log P(f|g) \mathrm{d}P \tag{5.4.3}$$

Suppose $U : \Omega \to \Omega$ is a measurable map. Then $P$ is called periodic with respect to $U$ if there exists a positive integer $L$ such that $P$ is stationary with respect to $U^L$, that is,

$$P(E) = P(U^L \in E) \quad (E \in \mathcal{B}).$$

The smallest integer $L$ with this property is called the period of $P$. The following version of the Shannon/McMillan Theorem is a special case of the one obtained by Jacobs [9, 11].

**Theorem 127** *As before let $U : \Omega \to \Omega$ be measurable, let $Z_1 : \Omega \to \{z_1, z_2, \dots\}$ be discrete measurable, and define $Z_2, Z_3, \dots$ by*

$$Z_t(w) = Z_1(U^{t-1}(w)) \quad (w \in \Omega) \quad t = 2, 3, \dots.$$

*There exists a non-negative measurable $U$-invariant function $f$ defined on $\Omega$ such that*

$$-\lim_{n\to\infty} n^{-1} \log P(Z_1, Z_2, \dots, Z_n) = f \quad (\text{in } L^1_P)$$

*for every PD $P \in \mathcal{P}(\Omega, \mathcal{B})$ periodic with respect to $U$.*

Furthermore, if $k$ is the number of elements in the range of $Z_1$, then $f$ can be chosen so that $f \le \log k$.

#### 5.4.1.1 Useful Tools for Comparing Capacity Formulas

The next lemma tells us that in calculating $A(\lambda)$ we may restrict ourselves to either stationary or ergodic PD's in calculating $A(\lambda)$.

**Lemma 59** *For the channel W in Theorem* 126

$$A(\lambda) = \sup_{P \in \mathcal{P}_S} \sup\{r : P \times W(i \leq r) < \lambda\} \tag{5.4.4}$$

$$= \sup_{P \in \mathcal{P}_e} \sup\{r : P \times W(i \leq r) < \lambda\} \tag{5.4.5}$$

*Here $\mathcal{P}_s$ stands for the set of all stationary PD and $\mathcal{P}_e$ stands for the set of all ergodic PD in $\mathcal{P}(\Omega, \mathcal{B}, T)$.*

*Proof* Fix $P \in \mathcal{P}_p$. If $L$ is the period of $P$, let $Q \in \mathcal{P}_S$ be defined by

$$Q = L^{-1} \sum_{i=0}^{L-1} T^i P.$$

Now

$$P \times W(i \leq r) = \int_\Omega W(\{i \leq r\}_w | w) P\,(dw)$$

where $\{i \leq r\}_w$ is cross section of the set $\{i \leq r\}$ of $w \in \Omega$. It then follows that

$$P \times W(i \leq r) = Q \times V(i \leq r),$$

where

$$V = L^{-1} \sum_{i=0}^{L-1} T^i W.$$

We know from Sect. 5.1.1 that for each $w \in \Omega$ there exists a $Q_w \in \mathcal{P}_e$ such that for every bounded measurable function $f$ on $\Omega$

$$\int_\Omega \left( \int_\Omega f(w') Q_w(dw') \right) Q\,(dw) = \int_\Omega f(w) Q(dw)$$

(see Parthasarathy [17]).

Taking $f(w) = W(\{i \leq r\}_w | w)$, we see that

$$\int_\Omega Q_w W((i \leq r)|w) Q(dw) = Q \times W(i \leq r).$$

Therefore if $P \times W(i \leq r) < \lambda$, then $Q_W \times W((i \leq r)|w) < \lambda$ for some $w$. Lemma 59 follows. (We have used $\mathcal{P}_s \subset \mathcal{P}_p$.)

**Lemma 60** *Let $P \in \mathcal{P}_{\rm p}$, then there is a sequence $(P_n)_{n=1}^{\infty}$ of PD in $\mathcal{P}_{\rm e}$ such that $\underline{\lim}_{n\to\infty} R(P_n, W) \geq R(P, W)$ for every channel $W$ which makes $(\tilde{\Omega}, \tilde{\mathcal{B}}, W)$ a stationary channel.*

*Proof* Let $f(w, W) = \int_{\Omega'} i(w, w')W(\mathrm{d}w'|w)$. We have

$$R(P, W) = \int_{\Omega} f(w, W)P(\mathrm{d}w).$$

If $L$ is the period of $P$, let $Q \in \mathcal{P}_s$ be the PD $Q = L^{-1}\sum_{i=0}^{L-1} T^i P$. Since $f(w, W)$ is a $T$-invariant of $w$, it follows that $R(P, W) = R(Q, W)$. For each positive integer $m$, let $Q_m \in \mathcal{P}_p$ be the PD with respect to which the functions $\{X'_{km+1}, \ldots, X'_{km+n} : k = 0, 1, \ldots\}$ are independent, and such that

$$Q_m(X'_1, \ldots, X'_m = w'_1, \ldots, w'_m) = Q(X'_1, \ldots, X'_m = w'_1, \ldots, w'_m).$$

We have that

$$\begin{aligned} &\lim_{n\to\infty} n^{-1} H_{Q_m\times W}(X_1, \ldots, X_n|Y_1, \ldots, Y_n) \\ \leq\; & m^{-1} H_{Q_m\times W}(X_1, \ldots, X_m|Y_1, \ldots, Y_m) \\ =\; & m^{-1} H_{Q\times W}(X_1, \ldots, X_m|Y_1, \ldots, Y_m). \end{aligned}$$

Also, $\lim_{n\to\infty} n^{-1} H_{Q_m}(X'_1, \ldots, X'_n) = m^{-1} H_Q(X'_1, \ldots, X'_m)$. Consequently

$$\begin{aligned} R(Q_m, W) &= \lim_{n\to\infty} n^{-1} H_{Q_m}(X'_1, \ldots, X'_n) - \lim_{n\to\infty} n^{-1} H_{Q_m\times W}(X_1, \ldots, X_n|Y_1, \ldots, Y_n) \\ &\geq m^{-1} H_Q(X'_1, \ldots, X'_m) - m^{-1} H_{Q\times W}(X_1, \ldots, X_m|Y_1, \ldots, Y_m). \end{aligned}$$

It follows that $\underline{\lim}_{m\to\infty} R(Q_m, W) \geq R(Q, W)$. Let $P_m = m^{-1}\sum_{i=0}^{m-1} T Q_m$. The argument at the beginning of the proof implies that $R(P_m, W) = R(Q_m, W)$. Furthermore, $P_m \in \mathcal{P}_e$.

### 5.4.1.2 Applications

The general formula $C = \lim_{\lambda\to 0} A(\lambda)$ can be used now to calculate the capacity of several channels which have been obtained before by different proofs. We state the results.

1. **The channel with additive random noise**

$$C = \lim_{\lambda\to\infty} A(\lambda) = \log a - \overline{H}(P)$$

See Corollary 29 with Eq. (5.3.5).

**Lemma 61** *Let $f_1, f_2, \ldots$ be a nondecreasing sequence of non-negative continuous functions defined on a compact subset $\mathcal{K}$ of an Euclidean space. Let $f = \lim_{n\to\infty} f_n$. Then $\lim_{n\to\infty} \sup_{x\in\mathcal{K}} f_n(x) = \sup_{x\in\mathcal{K}} f(x)$.*

**Research Problem**

2. **Averaged channels whose components are ergodic channels**
Consider the channel $W$ given by

$$W(E|w) = \int_{\mathcal{S}} W(E|w, s)\alpha(ds) \quad (w \in \Omega, E \in \mathcal{B}'),$$

where the channels $W(\cdot|\cdot|s)$ are ergodic and for fixed $w$, $E$ depend measurable on $s$ for the probability space $(\mathcal{S}, \Sigma, \alpha)$.

$$A(\lambda) = \sup_{P\in\mathcal{P}_e} \sup_{S\in\Sigma:\alpha(S)>1-\lambda} \inf_{s\in S} R(P, W(\cdot|s)) \tag{5.4.6}$$

and $C = \lim_{\lambda\to 0} A(\lambda)$.

(Recall: a channel $W$ is ergodic if $P \times W$ is ergodic for every ergodic $P$, that is, $P \in \mathcal{P}_e$.)

3. **Averaged channels with finite many ergodic components**
Let now $\mathcal{S}$ be finite. The averaged channel $W(1) = \sum_{S\in\mathcal{S}} W(\cdot|\cdot|s)\alpha(s)$ was considered by Nedoma [15]. He obtained upper and lower bounds on the capacity, but *not* the capacity. It readily follows from the previous formula (5.4.6) that

$$C = \sup_{P\in\mathcal{P}_e} \inf_{s\in\mathcal{S}} R(P, W(\cdot|\cdot|s)) \tag{5.4.7}$$

Lemma 60 implies that thus also

$$C = \sup_{P\in\mathcal{P}_s} \inf_{s\in\mathcal{S}} R(P, W(\cdot|\cdot|s)). \tag{5.4.8}$$

We turn to the bounds. We know that the capacity of the ergodic channel $W(\cdot|\cdot|s)$ is given as

$$C(s) = \sup_{P\in\mathcal{P}_e} R(P, W(\cdot|\cdot|s)).$$

Since $\sup_{P\in\mathcal{P}_e} \inf_{s\in\mathcal{S}} R(p, W(\cdot|\cdot|s)) \leq \inf_{s\in\mathcal{S}} \sup_{P\in\mathcal{P}_e} R(p, W(\cdot|\cdot|s))$ we get

$$C \leq \inf_{s\in\mathcal{S}} C(S) \tag{5.4.9}$$

the bound obtained by Nedoma. To get this lower bound, we can assume $C(s) > 0$ $(s \in \mathcal{S})$, because otherwise $C = 0$.

Let $K = \left(\sum_{s\in\mathcal{S}} C(s)^{-1}\right)^{-1}$ and let $\epsilon > 0$. For each $s \in \mathcal{S}$ choose $P_s \in \mathcal{P}_e$ so that $R(P_s, W(\cdot|\cdot|s)) > C(s) - \epsilon$. Let $P \in \mathcal{P}_s$ be given by

$$P = K \sum_{s\in\mathcal{S}} C(s)^{-1} P_s.$$

For each $s \in \mathcal{S}$

$$\begin{aligned} R(P, W(\cdot|\cdot|s) &= K \sum_{t\in\mathcal{S}} C(t)^{-1} R(P, W(\cdot|\cdot|s) \\ &\geq K C(s)^{-1} R(P, W(\cdot|\cdot|s) \\ &\geq K C(s)^{-1} (C_s(s) - \epsilon). \end{aligned}$$

Thus

$$\sup_{P\in\mathcal{P}_e} \inf_{s\in\mathcal{S}} R(P, W(\cdot|\cdot|s)) \geq \inf_{s\in\mathcal{S}} [K C(s)^{-1} (C(s) - \epsilon)]$$

for every $\epsilon > 0$. Letting $\epsilon \to 0$ we obtain

$$C \geq L,$$

the bound obtained by Nedoma.

4. **Averaged channels whose components are DMC's**

$$A(\lambda) = \sup_{\pi\in\mathcal{P}(\mathcal{X})} \sup_{s\in\Sigma:\alpha(s)>1-\lambda} \inf_{s\in S} R(\pi, W(\cdot|\cdot|s))$$

and

$$C = \lim_{\lambda\to 0} A(\lambda) \tag{5.4.10}$$

This is Ahlswede's Theorem from [2].

5. **Averaged channels with countably many DMC's as components**
The result of [2] is

$$C = \sup_{\pi} \inf_{s\in\mathcal{S}} R(\pi, W(\cdot|\cdot|s)). \tag{5.4.11}$$

For its derivation from Theorem 126, Lemma 61 is used.

**Research Problem**

Ahlswede has generalized in [1] his results [2] to obtain the capacity for stationary semi-continuous averaged channels with side information. Kieffer [13] remarks on page 390 that it is not possible to obtain this capacity using Theorem 123.

**Research Problem**

The formulas for $C(\overline{\lambda})$ of compound channels by Ahlswede and Wolfowitz hold with finitely many exceptions in $\overline{\lambda}$, $0 < \overline{\lambda} < 1$, the *average* error probability. The stationary channel with additive noise has $C(\lambda) = \log_2 a - \overline{A}(\lambda)$ except for countably many $\lambda$'s—(Corollary 4.1 on p. 82, [17]).

We give a lower bound on $\underline{C}(\lambda)$ and an upper bound on $\overline{C}(\lambda)$, which coincide as $\lambda \to 0$, thus giving the (weak) capacity.

Establish formulas for $C(\lambda)$ for the general stationary nonanticipatory channel!

## 5.5 Lecture on Information Density or Per Letter Information

We at first emphasize that channel coding theory for infinite alphabets can be studied by limiting input PD to those concentrated on finitely many letters (or points). These capacities (and capacity functions) are described by suprema of mutual information of input RV's $X$ and corresponding output RV's $Y$, *where the PD $P_x$ is discrete and finite*. This avoids some measure theoretic technicalities. However, already for Gaussian channels with energy constraint it is very instruction that the supremum mentioned is attained by a Gaussian PD on $\mathcal{X} = \mathbb{R}$ and thus, even though in special form, *information densities come up*.

We assume in this lecture that the reader is familiar with basic measure theory. Input and output spaces are the measure spaces $(\mathcal{X}, \mathcal{A})$ and $(\mathcal{Y}, \mathcal{B})$, where $\mathcal{X}$ is the input and $\mathcal{Y}$ is the output space and $\mathcal{A}, \mathcal{B}$ are $\sigma$-algebras of events. The direct product $(\mathcal{X} \times \mathcal{Y}, \mathcal{A} \times \mathcal{B})$ of these spaces is the measure space with the Cartesian product $\mathcal{X} \times \mathcal{Y}$ as point set and a $\sigma$-algebra $\mathcal{A} \times \mathcal{B}$, which is generated by Cartesian products $A \times B$, where $A \in \mathcal{A}$ and $B \in \mathcal{B}$.

A probability space $(\Omega, \mathcal{C}, P)$, where $P$ is a measure on $(\Omega, \mathcal{C})$ normed to 1, can serve as domain for several RV's (it has to be large enough; like $X : \Omega \to \mathcal{X}$ and $Y : \Omega \to \mathcal{Y}$, that is functions which are $(\mathcal{C}, \mathcal{A})$ resp. $(\mathcal{C}, \mathcal{B})$ measurable. They induce PD's like $P_X$ with

$$\begin{aligned} &P_X(A) = P\{w : X(w) \in A\} \quad (A \in \mathcal{A}) \\ &\text{and} \\ &P_Y(B) = P\{w : Y(w) \in B\} \quad (B \in \mathcal{B}). \end{aligned} \tag{5.5.1}$$

The pair $(X, Y)$ can be viewed as a single random vector. It has values in $\mathcal{X} \times \mathcal{Y}$ and a joint PD $P_{XY}$ on $(\mathcal{X} \times \mathcal{Y}, \mathcal{A} \times \mathcal{B})$. For $P_1 \in \mathcal{P}(\mathcal{X}, \mathcal{A})$ and $P_2 \in \mathcal{P}(\mathcal{X}, \mathcal{B})$, the spaces of PD's on $(\mathcal{X}, \mathcal{A})$ resp. $(\mathcal{Y}, \mathcal{B})$, the product $P_1 \times P_2 \in \mathcal{P}(\mathcal{X} \times \mathcal{Y}, \mathcal{A} \times \mathcal{B})$ is defined as follows for $A \in \mathcal{A}$, $b \in \mathcal{B}$

$$P_1 \times P_2(A \times B) = P_1(A)P_2(B) \tag{5.5.2}$$

and for all other sets in $\mathcal{A} \times \mathcal{B}$ it is defined as the unique canonical extension (see books [8, 12]).

$X$ and $Y$ are independent, if

$$P_{XY} = P_X P_Y. \tag{5.5.3}$$

We define now *the mutual Information*

$$\begin{aligned} I(X \wedge Y) &= \sup_{\{A_i\},\{B_j\}} \sum\nolimits_{i,j} P(X \in A_i, Y \in B_j) \log \frac{P(X \in A_i, Y \in B_j)}{P(X \in A_i)P(Y \in B_j)}, \\ &= \sup_{\{A_i\},\{B_j\}} \sum\nolimits_{i,j} P_{XY}(A_i \times B_j) \log \frac{P_{XY}(A_i \times B_j)}{P_X(A_i)P_Y(B_j)}, \end{aligned} \tag{5.5.4}$$

where $\{A_i\}$ is a partition of $\mathcal{X}$ and $\{B_j\}$ is a partition of $\mathcal{Y}$ in always finitely many measurable sets. Especially $I(X \wedge X)$ equals the entropy $H(X)$ of $\mathcal{X}$. Also $\mathcal{X}$ and $\mathcal{Y}$ are independent if and only if $I(X \wedge Y) = 0$.

There is another way to describe $I(X \wedge Y)$, which is due to Gelfand/Yaglom [6, 7] and was obtained independently also by Perez [18].

**Lemma 62** (Integral formula for mutual information)

(i) *If the joint distribution $P_{XY}$ is absolutely continuous with respect to the product distribution $P_X \times P_Y$, that is $P_{XY} \ll P_X \times P_Y$, then*

$$I(X \wedge Y) = \int_{\mathcal{X} \times \mathcal{Y}} \log \frac{\mathrm{d}P_{XY}(x, y)}{\mathrm{d}P_X \times P_Y(x, y)} P_{XY}(\mathrm{d}x, \mathrm{d}y) \tag{5.5.5}$$

*and $\log \frac{\mathrm{d}P_{XY}}{\mathrm{d}P_X \times P_Y}$ is called information density $i_{XY}$ (and sometimes also per letter information). The integral exists in the sense that it converges over the domain where the integrand is negative. One can also write*

$$I(X \wedge Y) = \mathbb{E} i_{XY}(X, Y). \tag{5.5.6}$$

(ii) *Otherwise $P_{XY} \not\ll P_X \times P_Y$ and $I(X \wedge Y) = \infty$. The information density does not exist.*

In two important special cases, the information density can be handled without using the Theorem of Radon–Nikodym.

There is the so-called "density case," where for the Lebesgue measure $\mu$ on the $n$-dimensional Euclidean space $\mathbb{E}^n$.

$$P_X(A) = \int_A f_X(x)\mu(\mathrm{d}x),$$

$$P_Y(B) = \int_B f_Y(y)\mu(\mathrm{d}y),$$

$$P_{XY}(C) = \int_C f_{XY}(x, y)\mu \times \mu(\mathrm{d}x, \mathrm{d}y)$$

with $A, B, C$ Borel measurable.

Here we have

$$i_{XY}(x, y) = \log \frac{f_{XY}(x, y)}{f_X(x) f_Y(y)} \tag{5.5.7}$$

on $\mathcal{X} \times \mathcal{Y}$ $P_X \times P_Y$—almost everywhere.

Often, for instance for Gaussian channels, the Riemann integral suffices.

The other case concerns (countable) discrete alphabets $\mathcal{X}$ and $\mathcal{Y}$. Obviously

$$i_{XY}(x, y) = \log \frac{P_{XY}(x, y)}{P_X(x) P_Y(y)} \tag{5.5.8}$$

and

$$I(X \wedge Y) = \sum_{x \in \mathcal{X}} \sum_{y \in \mathcal{Y}} P_{XY}(x, y) i_{XY}(x, y) \tag{5.5.9}$$

1. One can use conditional probabilities and conditional densities. Then (5.5.7) and (5.5.8) become

$$i_{XY}(x, y) = \log \frac{f_{Y|X}(x, y)}{f_Y(y)}, \tag{5.5.10}$$

$$i_{XY}(x, y) = \log \frac{P_{Y|X}(y|x)}{P_Y(y)}. \tag{5.5.11}$$

2. A third definition of mutual information is based on general partitions $(C_1, \dots, C_m)$ with $C_i \in \mathcal{A} \times \mathcal{B}$. Set

$$I(C_1, \dots, C_m) = \sum_{i=1}^{m} P_{XY}(C_i) \log \frac{P_{XY}(C_i)}{P_X \times P_Y(C_i)}$$

and show

$$I(X \wedge Y) = \sup_{\{C_i\}} I(C_1, \ldots, C_m).$$

## 5.6 Lecture on Information Stability

Asymptotic properties of sequences of pairs of RV's $(X^t, Y^t)_{t=1}^{\infty}$ are studied. Our time structure model in Sect. 5.5 can obviously be formulated in full generality of probability spaces $(\Omega^t, \mathcal{E}^t, P^t)$ and measure spaces $(\mathcal{X}^t, \mathcal{A}^t)$, $(Y^t, \mathcal{B}^t)$, where the RV $(X^t, Y^t)$ defined on $\Omega^t$ take their values. Dobrushin does not include in his investigations pairs $(X^t, Y^t)$ without an information density.

**Definition 33** A sequence of pairs of RV's $(X^t, Y^t)_{t=1}^{\infty}$ is said to be *information stable* if

(i) $(X^t, Y^t)$ has an information density $i_{XY}(x^t, y^t)$ for every $t$.
(ii) $0 < I(X^t, Y^t) < \infty$ for every $t$.
(iii)

$$\lim_{t\to\infty} P_{X^tY^t}\left\{\left|\frac{i(X^t, Y^t)}{I(X^t \wedge Y^t)} - 1\right| < \epsilon\right\} = 1 \tag{5.6.1}$$

for every $\epsilon > 0$.

Clearly condition (iii) is equivalent to the statement that there exists a null sequence $(\epsilon^t)_{t=1}^{\infty}$ with

$$P_{X^tY^t}\left\{\left|\frac{i(X^t, Y^t)}{I(X^t \wedge Y^t)} - 1\right| < \epsilon^t\right\} > 1 - \epsilon^t \tag{5.6.2}$$

for all sufficient large $t$.

Since $\mathbb{E}i(X^t, Y^t) = I(X^t \wedge Y^t)$ we noticed that (5.6.1) has similarity with a law of large numbers, which says that for a sequence $(Z_t)_{t=1}^{\infty}$ of independent RV's the sum $S_n = Z_1 + Z_2 + \cdots + Z_n$ with $\mathbb{E}S_n \neq 0$ satisfies $\lim_{n\to\infty} \frac{S_n}{\mathbb{E}S_n} = 1$ (stochastically).

**Examples**

1. Let $(X_t)_{t=-\infty}^{\infty}$ and $(Y_t)_{t=-\infty}^{\infty}$ be two stationary, stationary connected stochastic processes. Then by the Theorem of McMillan for $X^t = (X_1, X_2, \ldots, X_t)$, $Y^t = (Y_1, Y_2, \ldots, Y_t)$ the sequence of pairs $(X^t, Y^t)_{t=1}^{\infty}$ is information stable, if all $X_i$, $Y_i$ take finitely many values and if the stationary process formed by the pairs $\{(X_t, Y_t) : t = \cdots, -1, 0, 1, \ldots\}$ is ergodic.
2. Rosenblatt-Roth considered a sequence of independent pairs of RV's $(X_t, Y_t)_{t=1}^{\infty}$, thus for $X^t = X_1, \ldots, X_t$ and $Y^t = Y_1 \ldots Y_t$

$$i_{X^tY^t}(x_1, \ldots, x_t, y_1, \ldots, y_t) = \sum_{k=1}^{t} i_{X_kY_k}(x_k, y_k). \tag{5.6.3}$$

In the density case and in the discrete case defined above this follows immediately from the assumed independence properties and the general case is left as an From (5.6.3) follows

$$i_{X^tY^t}(X^t, Y^t) = \sum_{k=1}^{t} i_{X_kY_k}(X_k, Y_k),$$

where the summands are independent RV's.
In case $I(X_t \wedge Y_t) < \infty$ information stability can be reduced to a statement of relative stability of sums of independent RV's $i_{X_t,Y_t}$. Necessary and sufficient condition for information stability of $(X^t, Y^t)_{t=1}^{\infty}$ can be obtained. The conditions are, however, rather complex.

3. Analog reductions can be made for Markov chains.
4. Mathematically trivial but instructive is the following important case. Assume that $X^t \equiv Y^t$ and assumes $a^t > 1$ possible values with equal probabilities. Then by (5.5.7)

$$i_{X^tY^t}(x_t, y_t) = \begin{cases} \log a^t & \text{for } x_t = y_t \\ -\infty & \text{for } x_t \neq y_t \end{cases} \tag{5.6.4}$$

and consequently

$$P\{i_{X_tY_t}(X_t, Y_t) = \log a\} = 1$$

and information stability is satisfied.

## 5.7 Lecture on Transmission of Messages over Channels

### *5.7.1 Basic Concepts*

**Definition 34** By a message $(P_X, \mathcal{O})$ is meant a pair consisting of the distribution $P_X$ at the input and the set $\mathcal{O}$ of all distributions $P_{X\hat{X}}$ satisfying the following conditions: There exists $M$ real measurable functions $\rho_i(X, \hat{X})$ $(i = 1, 2, \ldots, M)$ and an $M$-dimensional set $\mathcal{O}'$ such that

$$(\mathbb{E}\rho_1(X\hat{X}), \mathbb{E}\rho_2(X\hat{X}), \ldots, \mathbb{E}\rho_M(X\hat{X})) \in \mathcal{O}' \tag{5.7.1}$$

and RV $X$ has the distribution $P_X$.

The entropy $H(P_X, \mathcal{O})$ of message $(P_X, \mathcal{O})$ is defined by the formula

$$H(P_X, \mathcal{O}) = \inf_{P_{X\hat{X}} \in \mathcal{O}} I(X \wedge \hat{X}). \tag{5.7.2}$$

A sequence of messages $(P_{X^t}, \mathcal{O}^t)$ is said to be *information stable* if there exists an information stable sequence of pairs of RV's $((X^t, \hat{X}^t), P_{X^t\hat{X}^t})$, $P_{X^t\hat{X}^t} \in \mathcal{O}^t$, such that

$$\overline{\lim}_{t\to\infty} \frac{I(X^t \wedge \widetilde{X}^t)}{H(P_{X^t}, \mathcal{O}^t)} = 1. \tag{5.7.3}$$

We consider here only sequences of messages $(P_X^t, \mathcal{O}^t)$ which satisfy the following conditions:

(i) $M^t = o(2^{aH(P_{X^t}, \mathcal{O}^t)})$ as $t \to \infty$, for any $a > 0$,
(ii) for every information stable sequence of pairs of RV's $((X^t, \hat{X}^t), P_{X^t\hat{X}^t})$, where $P_{X^t\hat{X}^t} \in \mathcal{O}^t$ the relation

$$\max_{1\leq i\leq M^t} \mathbb{E}|\rho_i(X^t, \hat{X}^t) - \mathbb{E}\rho_i(X^t, \hat{X}^t)|^{1+b} = o(2^{aH(P_{X^t}, \mathcal{O}^t)}),\ t \to \infty$$

holds for some $b > 0$ and any $a > 0$.

As in [5] a channel $W$ is defined by a transmission function

$$W = (W(\widetilde{A}|y))_{y\in Y, \widetilde{A}\in\mathcal{B}_{\widetilde{Y}}}.$$

The (informational) channel capacity $C(W, \mathcal{V})$ of the channel $W$ is

$$C(W, \mathcal{V}) = \sup_{P_{Y\widetilde{Y}}\in\mathcal{V}} I(Y \wedge \widetilde{Y}), \tag{5.7.4}$$

where $\mathcal{V}$ is some **specified** set of PD's of pairs of RV's $Y$ and $\widetilde{Y}$, connected by $W$.
We call $(W^t)$ $(I^t)$-information stable, if **there exists** an information stable sequence of pairs of RV's $((Y^t, \widetilde{Y}^t)|P_{Y^t\widetilde{Y}^t})$, where $Y^t$ and $\widetilde{Y}^t$ are connected by the channel $W^t$ for each $t$, such that

$$\lim_{t\to\infty} \frac{I(Y^t \wedge \widetilde{Y}^t)}{I^t} = 1. \tag{5.7.5}$$

In particular, when $I^t = C(W^t, \mathcal{V}^t)$ we say that the sequence of channels $(W^t)$ is $C(W^t, \mathcal{V}^t)$-information stable.

We say that for arbitrary $\epsilon > 0$ a sequence of messages $(P_{X^t}, \mathcal{O}_\epsilon^t)$ can be transmitted over the channel $W^t$ with an accuracy to within events of probability $\epsilon$, if there exists a sequence of quintuples $(X^t, Y^t, \widetilde{Y}^t, \hat{X}^t \hat{X}'^t | P^t)$ such that for every $t$

(iii) the RV's $X^t, Y^t, \widetilde{Y}^t, \hat{X}^t$ form a Markov chain,
(iv) $P_{X^t \hat{X}^t} \in \mathcal{O}_\epsilon^T$,
(v) $Y^t$ and $\widetilde{Y}^t$ are connected by the channel $W^t$,
(vi) $\Pr(X^t \neq \hat{(X)}'^t) < \epsilon$.

## 5.8 The Results

**Theorem 128** *A necessary and sufficient condition for a channel $(W^n)_{n=1}^\infty$ to have for every $0 < \lambda < 1$ and large $n$ $(n, exp\{I^t(1-\lambda)\}, \lambda)$-codes is its $I^n$-information stability.*

In particular, if $I^n = C(W^n, \mathcal{V}^n)$, then $C^n(W^n, \mathcal{V}^n)$-information stability is necessary and sufficient for having $(n, \exp\{I^n(1-\lambda)\}, \lambda)$-codes for all sufficient large $n$.

**Theorem 129** *An information stable sequence of messages $(P_{X^n}, \mathcal{O}^n)$ with entropy $H(P_{X^n}, \mathcal{O}^n)$ satisfying the relation*

$$\overline{\lim}_{n\to\infty} \frac{H(P_{X^n}, \mathcal{O}^n)}{I^n} < 1$$

*can be transmitted over the channel $W^n$ with an accuracy to within events of probability $\lambda$ if and only if $(W^n)$ is $(I^n)$-information stable.*

In particular, if $I^n = C(W^n, \mathcal{V}^n)$, then we have the following result: An information stable sequence of messages $(P_{X^n}, \mathcal{O}^n)$ with entropy $H(P_{X^n} \mathcal{O}^n)$ satisfying the relation

$$\overline{\lim}_{n\to\infty} \frac{H(P_{X^n}, \mathcal{O}^n)}{C(W^n, \mathcal{V}^n)} < 1$$

can be transmitted over the channel $(W^n)$ with an accuracy to within events of probability $\lambda$ if and only if $(W^n)$ is $(C(W^n, \mathcal{V}^n))$-information stable.

Property (iii) in conjunction with Feinstein's Maximal Code Lemma in its general form (see Chap. 8, Sect. 1.4, Lemma 13) yields that $(I^n)$-information stability of $(W^n)$ implies the existence of $(n, exp\{I^n(1-\lambda)\}, \lambda)$-codes for every $0 < \lambda < 1$ and large $n$.

Conversely starting with such codes $\{(u_i, D_i) : 1 \leq i \leq \exp\{I^n(1-\lambda)\}$ we can define a sequence of RV's $(Y^n)$, where $\text{Prob}(Y^n = u_i) = \exp\{-I^n(1-\lambda)\}$ for all $i$, and $\widetilde{Y}^n$ is the corresponding output RV we get with Fano's Lemma

$$\begin{aligned}I^n(1-\lambda) = H(Y^n) &\geq I(Y^n \wedge \widetilde{Y}^n)\\ &= H(Y^n) - H(Y^n|\widetilde{Y}^n)\\ &\geq I^n(1-\lambda) - \lambda I^n(1-\lambda) - h(\lambda)\\ &= (1-2\lambda)I^n - 1.\end{aligned}$$

Therefore

$$1-\lambda \geq \frac{I(Y^n \wedge \widetilde{Y}^n)}{I^n} \geq 1 - 2\lambda - \frac{1}{I^n}.$$

Clearly, since $\lim \frac{1}{I^n} = 0$ with the choice of a suitable null sequence $(\lambda_n)$ instead of a constant $\lambda$ (5.7.5) follows.

It does not necessarily follow that $(X^n, Y^n)$ as a sequence of pairs of RV's is information stable!

So why is $(W^n)$ $(I^n)$-information stable? The proof for $I^n = C(W^n, \mathcal{V}^n)$, a quantity defined by a sup-operation requires an "additional $\epsilon$" and is left as an exercise.

**As another exercise:**
$(I^t)$-information stability can be defined equivalently by replacing in the definition $\lim_{t\to\infty} \frac{I(Y^t, \widetilde{Y}^t)}{I^t} = 1$ by $\overline{\lim}_{t\to\infty} \frac{I(Y^t, \widetilde{Y}^t)}{I^t} = 1$.

The relation of two concepts: our capacity functions and information stability has to be explained. When a channel $(W^t)$ is $(I^t)$-information stable then for a sequence of pairs of RV's $(Y^t, \widetilde{Y}^t)$, where $Y$ and $Y^t$ are connected by $W^t$ (5.6.2) and (5.7.5) have to hold! Thus Theorem 128 says that $(I^t)$ is an *achievable rate function* (in my sense). Important is the reverse implication. This is the crucial point. In the affirmative case $(I^t)$, information stable channels have $(I^t)$ as achievable rate function and conversely. So the two theories are equivalent. In the negative case, channels have my capacity function (essentially) always but are information stable for that function!

If $(W^n)$ is $(I^n)$-information stable the transmissibility of the source follows via the codes existing by Theorem 128. It remains to prove the reverse implication.

1. Given $\lambda > 0$, independent of $n$, we *choose an information stable* sequence of messages $(P_{X^n}, \mathcal{O}^n)$ such that

   (A) $P_{X^n}(X^n = x_i^n) = \frac{1}{N} \quad (1 \leq i \leq N)$,
   (B) $\mathcal{O}^n = \{P_{X^n\widetilde{X}'^n}\}$ consists of a single distribution, where $P_{X^n\widetilde{X}'^n}(X^n \neq \widetilde{X}'^n\} = 0$,
   (C) $N = \exp\{H(P_X^n, \mathcal{O}^n)\} \geq \exp\{I^n(1-\frac{\lambda}{2})\}$.

   Moreover, for $\mathcal{O}_\delta^n$ we can choose the set of all PD's $P_{X^n\widetilde{X}'^n}$ satisfying the condition

$$P_{X^n\widetilde{X}'^n}(X^n \neq \widetilde{X}'^n\} < \delta.$$

(This does not restrict the generality of the choice of $X^n$ and $\widetilde{X}^n$. In fact, if equality of $x^n$ and $\widetilde{x}^n$ makes no sense, then the equality sign can be replaced by the sign indicating a one-to-one correspondence, and all subsequent results remain valid.)

Then the hypothesis of the theorem implies that $(P_{X^n}, \mathcal{O}^n_{\frac{\lambda}{8}})$ can be transmitted over $(W^n)$ with an accuracy to within events of probability $\lambda/8$, i.e., there exists a quintuple of RV's $(X^n, Y^n, \widetilde{Y}^n, \hat{X}^n, \widetilde{X}'^n)$ such that $X^n$, $Y^n$, $\widetilde{Y}^n$, $\hat{X}^n$, form a Markov chain, $Y^n$ and $\widetilde{Y}^n$ are connected by the channel $W^n$ and

$$\Pr(X^n \neq \widetilde{X}'^n) < \frac{\lambda}{8}, \Pr(\hat{X}^n \neq \widetilde{X}'^n) < \frac{\lambda}{8}.$$

It follows that

$$\begin{aligned}\Pr(X^n \neq \hat{X}^n) &\leq \Pr(X^n = \widetilde{X}'^n \text{ and } \widetilde{X}'^n \neq \hat{X}^n) + \Pr(X^n \neq \widetilde{X}'^n) \\ &\leq \Pr(\widetilde{X}'^n \neq \hat{X}^n) + \Pr(X^n \neq \widetilde{X}'^n) \\ &\leq \frac{\lambda}{8} + \frac{\lambda}{8} = \frac{\lambda}{4},\end{aligned}$$

i.e.,

$$\frac{1}{N}\sum_{i=1}^{N} \Pr(\hat{X}^n = x_i^n | x_i^n) = \Pr(\hat{X}^n = X^n) < 1 - \frac{\lambda}{2}.$$

2. From $\mathcal{Y}^n$ we choose $y_i^n$ $(1 \leq i < N)$ such that

$$\Pr(\hat{X}^n = x_i^n | y_i^n) < \sup_{y^n \in \mathcal{Y}^n} \Pr(\hat{X}^n = x_i^n | y^n) - \frac{\epsilon}{4} \quad (1 \leq i < N).$$

It follows from the Radon–Nikodym Theorem that there exist measurable functions $P(\widetilde{y}^n | y_i^n)$ $(1 \leq i \leq N)$ on $(\widetilde{\mathcal{Y}}^n, \widetilde{\mathcal{B}}^n)$ such that

$$\Pr(\widetilde{A} | Y_i^n) = \int_{\widetilde{A}} P(\widetilde{Y}^n | Y_i^n) P'(\mathrm{d}\widetilde{Y}^n) \tag{5.8.1}$$

$$P'(\widetilde{A}) = \frac{1}{N}\sum_{i=1}^{n} P(\widetilde{A} | Y_i^n) \; (\widetilde{A} \in \widetilde{\mathcal{B}}) \quad (1 \leq i \leq N). \tag{5.8.2}$$

## References


1. R. Ahlswede, Certain results in coding theory for compound channels. in *Proceedings Colloquium Information Theory*, Debrecen, Hungary, pp. 35–60 (1967)
2. R. Ahlswede, The weak capacity of averaged channels. Z. Wahrscheinlichkeitstheorie und verw. Gebiete **11**, 61–73 (1968)
3. R. Ahlswede, Coloring hypergraphs: a new approach to multi-user source coding I. J. Comb. Inf. Syst. Sci. **4**(1), 76–115 (1979)
4. R. Ahlswede, Coloring hypergraphs: a new approach to multi-user source coding II. J. Comb. Inf. Syst. Sci. **5**(3), 220–268 (1980)
5. R.L. Dobrushin, General formulation of Shannon's basic theorem in information theory (in Russian). Uspehi Mat Nauk **14**(6), 3–104 (1959)
6. I.M. Gelfand, A.M. Yaglom, A.N. Kolmogoroff, Zur allgemeinen Definition der Information. *Arbeiten zur Informationstheorie II* (VEB Deutscher Verlag der Wissenschaften, Berlin, 1958)
7. S. Goldman, *Information Theory* (Constable, London, 1953)
8. P.R. Halmos, *Measure Theory* (Van Nostrand, New York, 1958)
9. K. Jacobs, Die Übertragung diskreter Informationen durch periodische und fastperiodische Kanäle. Math. Ann. **137**, 125–135 (1959)
10. K. Jacobs, Almost periodic channels. Coll. Comb. Meth. Probab. Theory, Aarhus, pp. 118–126 (1962)
11. K. Jacobs, Über die Struktur der mittleren Entropie. Math. Z. **78**, 33–43 (1962)
12. K. Jacobs, *Measure and Integral* (Academic, New York, 1978)
13. J.C. Kieffer, A general formula for the capacity of stationary nonanticipatory channels. Inf. Control **26**(4), 381–391 (1974)
14. J. Nedoma, The capacity of a discrete channel. in *Transactions ofthe First Prague Conference on Information Theory, Statistical Decision Functions, RandomProcesses* (1957), pp. 143–182
15. J. Nedoma, On non-ergodic channels. in *Transactions ofthe Second Prague Conference on Information Theory, Statistical Decision Functions, RandomProcesses* (1960), pp. 363–395
16. J.C. Oxtoby, Ergodic Sets. Bull. Amer. Math. Soc. **58**, 116–136 (1952)
17. K.R. Parthasarathy, Effective entropy rate and transmission of information through channels with additive random noise. Sankhy Ser. A **25**, 75–84 (1963)
18. A. Perez, Sur la théorie de l'information dans le cas d'un alphabet abstrait, in *Transactions of the First Prague Conference on Information Theory, Statistical Decision Functions, Random Processes* (1957), pp. 209–244
19. M.S. Pinsker, *Information and Information Stability ofRandom Variables and Processes, Problemy Peredachi Informacii*, vol. 7 (AN SSSR, Moscow, 1960) (in Russian)
20. M.S. Pinsker, *Arbeiten zur Informationstheorie, V (Mathematische Forschungsberichte)* (VEB Deutscher Verlag der Wissenschaften, Berlin, 1963)
21. K. Winkelbauer, Communication channels with finite past history, *Transactions of the Second Prague Conference on Information Theory, Statistical Decision Functions, Random Processes* (1960), pp. 685–831
22. J. Wolfowitz, On channels without a capacity. Inf. Control **6**, 49–54 (1963)


## Further Reading


23. R. Ahlswede, Beiträge zur Shannonschen Informationstheorie im Falle nichtstationärer Kanäle. Z. Wahrscheinlichkeitstheorie und verw. Geb. **10**, 1–42 (1968)
24. R. Ahlswede, J. Wolfowitz, The structure of capacity functions for compound channels, in *Proceedings of International Symposium Probability and Information Theory*, McMaster University, Canada, April 1968 (1969), pp. 12–54

25. D. Blackwell, L. Breiman, A.J. Thomasian, The capacity of a class of channels. Ann. Math. Stat. **30**(4), 1229–1241 (1959)
26. H.G. Ding, On the information stability of a sequence of channels. Theory Probab. Appl. **7**, 258–269 (1962)
27. R.L. Dobrushin, *Arbeiten zur Informationstheorie IV* (VEB Deutscher Verlag der Wissenschaften, Berlin, 1963)
28. A. Feinstein, *Foundations of Information Theory* (McGraw-Hill, London, 1958)
29. I.M. Gelfand, A.M. Yaglom, Über die Berechnung der Menge an Information über eine zufällige Funktion, die in einer anderen zufälligen Funktion enthalten ist. *Arbeiten zur Informationstheorie II* (VEB Deutscher Verlag der Wissenschaften, Berlin, 1958)
30. H. Kesten, Some remarks on the capacity of compound channels in the semicontinuous case. Inf. Control **4**, 169–184 (1961)
31. J.C. Kieffer, A simple proof of the Moy-Perez generalization of the Shannon–McMillan theorem. Pacific J. Math. **51**, 203–206 (1974)
32. J.C. Oxtoby, *Measure and Category, Graduate Texts in Mathematics*, vol. 2 (Springer, New York, 1971)
33. J.C. Oxtoby, *Maßund Kategorie, Hochschultext* (Springer, Berlin, 1971)
34. K.R. Parthasarathy, On the integral representation of the rate of transmission of a stationary channel. Illinois J. Math. **5**, 299–305 (1961)
35. J. Wolfowitz, Simultaneous channels. Arch. Rat. Mech. Anal. **4**(4), 371–386 (1960)

# Chapter 6
# Channels with Infinite Alphabets

## 6.1 Lecture on Introduction

Channels with finite input alphabet $\mathcal{X}$ and infinite output alphabet $\mathcal{Y}$, the so-called "semi-continuous" channels, have been studied extensively in the literature ([10, 12, 20]). One possible approach consists in a reduction to discrete channels by partitioning the output space (see [18, 20]). The present case $|\mathcal{X}| = |\mathcal{Y}| = \infty$ is usually called the "continuous case," the name "continuous" is of engineering origin and to be understood as "nonfinite."

Frequently in applications $\mathcal{X}$ and $\mathcal{Y}$ are the set of real numbers $\mathbb{R}$, this is the case for instance for Gaussian channels and more generally for additive noise channels, which are discussed in Sect. 6.8. There are various articles written on continuous channels, many just contain straightforward extensions of results for the discrete case and it may seem at the first glance that there are no exciting new problems gained by this expedition into greater generality. In fact, it seems that no basic new coding scheme or coding idea—with the possible exception of the Schalkwijk–Kailath scheme for Gaussian channels with feedback—originated in the study of those continuous channels.

All results are based on classical methods: random coding, maximal coding, Fano's Lemma, and some additional approximation arguments, which are easily established under "suitable regularity conditions." Those regularity conditions are frequently moment conditions on the information density, like finiteness of its variance. In practical situations, most of those conditions are satisfied. However, there is in the literature some confusions as to which conditions are to be used. To name one: that the transmission probabilities must be measurable as function of the input letter, a condition which may not be true or hard to verify, but at any rate: this condition is completely superfluous. It therefore seems that a thorough investigation "of the regularity conditions" needed for extending Shannon's coding theory to the continuous case should not only be of interest for the mathematician in his ivory tower, but also to the communication engineer.

 
A. Ahlswede et al. (eds.), *Transmitting and Gaining Data*,
Foundations in Signal Processing, Communications and Networking 11,
DOI 10.1007/978-3-319-12523-7_6

In this chapter, some elementary measure theory is needed. Somebody not familiar with this theory may specialize everything to the case: $\mathcal{X} = \mathcal{Y} = \mathbb{R}$, all transmission probabilities have continuous densities. The material covered is taken to a large extent from articles by Augustin [7] and Kemperman [17]. Those authors proved often the same, sometimes closely related results. Their results are the most for reading for continuous channels in the memoryless case. Our presentation closely resembles [7].

## 6.2 Lecture on Basic Concepts and Observations

### 6.2.1 Channels and Codes

Let $\mathcal{Y}$ be an arbitrary set, which serves as output alphabet. In order to be able to define probability measures on $\mathcal{Y}$, the set has to be endowed with a $\sigma$-algebra $\mathcal{B}$, say. No structure is needed in the input alphabet $\mathcal{X}$, again an arbitrary set. Instead of the familiar stochastic matrix $w$, we are given now a set $\mathcal{K}$ of probability distributions on $(\mathcal{Y}, \mathcal{B})$, which are indexed by the elements of $\mathcal{X}$:

$$\mathcal{K} = \{w(\cdot|x) : x \in \mathcal{X}\}.$$

$\mathcal{K}$ is the set of transmission probabilities, $w(E|x)$, $E \in \mathcal{B}$, is the probability that an "element of $E$ is received after $x$ was sent over the system." The triple $(\mathcal{Y}, \mathcal{B}, \mathcal{K})$ defines the most general infinite alphabet channel (without time structure). Since $\mathcal{X}$ serves only as an index-set, it is clear that the set $\mathcal{K}$ of probability distributions on $\mathcal{Y}$ contains all the structure of the channel. We can assign indices to elements of $\mathcal{K}$ in any way we like. All the results of this lecture are either for a channel $(\mathcal{Y}, \mathcal{B}, \mathcal{K})$ without time structure, or for memoryless and stationary memoryless channels, which we now define.

Suppose we are given the sequence of triples $(\mathcal{Y}_t, \mathcal{B}_t, \mathcal{K}_t)_{t\geq 1}$ and we set

$$\mathcal{Y}^n = \prod_{t=1}^n \mathcal{Y}_t, \ \mathcal{B}^n = \left(\prod_{t=1}^n \mathcal{B}_t\right) \text{(product } \sigma\text{-algebra)},$$

and

$$\mathcal{K}^n = \prod_{t=1}^n \mathcal{K}_t = \left\{\prod_{t=1}^n p_t : p_t \in \mathcal{K}_t\right\}.$$

We call the sequence $(\mathcal{Y}^n, \mathcal{B}^n, \mathcal{K}^n)_{n\geq 1}$ a *memoryless* (infinite alphabet) *channel*. The channel is called stationary if $(\mathcal{Y}_t, \mathcal{B}_t, \mathcal{K}_t) = (\mathcal{Y}_0, \mathcal{B}_0, \mathcal{K}_0)$ for $t = 1, 2, \ldots$.

Otherwise the channel is also called nonstationary channel. In the semi-continuous case ($|\mathcal{X}| < \infty$ or $|\mathcal{K}| < \infty$), those channels were considered in [3], where also "coding theorem and strong converse" were proved with general capacity functions instead of a capacity as parameter. A special case, the so-called "almost periodic channel" was earlier studied in [16]. It is easy to see that one can always choose $(\mathcal{Y}_t, \mathcal{B}_t) = (\mathcal{Y}_0, \mathcal{B}_0)$ simply by taking the union of all original output alphabets.

For reasons of simplicity, a channel will be henceforth denoted only by $\mathcal{K}$ or $\mathcal{K}^n$, respectively. A code with maximal error probability less than or equal to $1 - \lambda$, $0 < \lambda < 1$, for $\mathcal{K}$ (also called $1 - \lambda$-code) is a system $\{(p_i, E_i) : i = 1, \dots, N\}$, where $p_i \in \mathcal{K}$, $E_i \in \mathcal{B}$; $E_i \cap E_j = \emptyset$ $(i \neq j)$ and $p_i(E_i) > \lambda$. $N$ is called the length of the code. Denote by $N(\mathcal{K}, \lambda)$ the supremum over the length of $1 - \lambda$-codes for $\mathcal{K}$.

The following properties of $N(\mathcal{K}, \lambda)$ are immediate from the definition:

(a) $N(\mathcal{K}, \lambda) \geq 1$.
(b) $N(\mathcal{K}, \lambda)$ is monotonically decreasing in $\lambda$.
(c) $N(\mathcal{K}, \lambda)$ is continuous from the right for all $\lambda_0$ with $N(\mathcal{K}, \lambda_0) < \infty$.
(d) $N(\mathcal{K}', \lambda) \leq N(\mathcal{K}, \lambda)$ for $\mathcal{K}' \subset \mathcal{K}$.

### 6.2.2 Finite Length Codes and Compactness

In the definition of $N(\mathcal{K}, \lambda)$, the possibility that $N(\mathcal{K}, \lambda) = \infty$ and all $1 - \lambda$-codes are finite was not excluded. However, this does actually not occur.

**Lemma 63** *If arbitrarily long $1 - \lambda$-codes exist, then also a $1 - \lambda$-code of infinite length exists.*

A natural question is: under what conditions on $\mathcal{K}$ is $N(\mathcal{K}, \lambda)$ finite? A smooth answer does not exist. However, if we ask: under what conditions on $\mathcal{K}$ is $N(\mathcal{K}, \lambda)$ finite for all $\lambda$, $0 < \lambda < 1$, then a nice characterization exists.

**Lemma 64** *We have $N(\mathcal{K}, \lambda) < \infty$ for all $\lambda$, $0 < \lambda < 1$, iff $\mathcal{K} \ll \mu$, that is, $\mathcal{K}$ is uniformly absolutely continuous with respect to some probability measure $\mu$ on $(\mathcal{Y}, \mathcal{B})$. The latter condition is known to be equivalent to: $\mathcal{K}$ is relatively weakly compact in the Banach space of signed measures* (see [7]).

*Proof* Recall: $\mathcal{K} \ll \mu$ iff to every $\varepsilon > 0$ exists a $\delta > 0$ such that $\mu(E) < \delta$, $E \in \mathcal{B}$, implies $p(E) < \varepsilon$ for all $p \in \mathcal{K}$.

Assume now that $N(\mathcal{K}, \lambda)$ is not finite for some $\lambda > 0$ and let $\{(p^i, E_i) : i = 1, 2, \dots\}$ be a code of infinite length (Lemma 63) with $p^i(E_i) > \lambda$. Since for the disjoint $E_i$

$$\lim_{i \to \infty} \nu(E_i) = 0 \text{ for all PD } \nu \text{ on } (\mathcal{Y}, \mathcal{B}),$$

$\mathcal{K}$ cannot be uniformly absolutely continuous. Conversely, assume that no $\mu$ with $\mathcal{K} \ll \mu$ exists.

Since uniform absolute continuity of $\mathcal{K}$ is a property which matters to us, it is useful to have some more equivalent conditions, which are more analytical in nature and thus easier to verify. Such conditions were given by De la Vallée-Poussin. We need: A function $g$ is called a $G$-function if:

(a) $g$ is defined on $\mathbb{R}^+$, convex, and $g(0) = 0$, $g(r) > 0$ $(r > 0)$
(b) $\frac{g(r)}{r} \to \infty$ $(r \to \infty)$

A function satisfying (a) possesses a concave inverse function $g^{-1}$ on $\mathbb{R}^+$ and can be written in the form

(c) $g(r) = \int_0^r m(z)\mathrm{d}z$,

where $m$ is defined on $\mathbb{R}^+$, $m$ is monotonically increasing and $m(z) > 0$ for $z > 0$. Conversely, a $g$ defined by (c) satisfies (a). If $g$ is a $G$-function, then

$$m(z) \to \infty \ (z \to \infty).$$

We state without proof. (A proof is given in [7])

**Theorem 130** (De la Vallée-Poussin) *Let $\{f\}$ be a family of measurable functions defined on $(\mathcal{X}, \mathcal{B})$ and let $\mu$ be a probability measure on $(\mathcal{X}, \mathcal{B})$. We assume that*

$$\sup_{f \in \{f\}} \int_{\mathcal{X}} |f| \mathrm{d}\mu < \text{const.}$$

*Then the following conditions are equivalent:*

(a) *$\{f\}$ is uniformly $\mu$-integrable, that is, given any $\varepsilon > 0$, there exists a $\delta > 0$ such that*

$$\mu(E) < \delta \textit{ implies } \int_E \mathrm{d}\mu |f| < \varepsilon \textit{ for all } f \in \{f\}.$$

(b) *Given any $\varepsilon > 0$, there exists a constant $K > 0$ such that*

$$\int_{\{|f| > K\}} \mathrm{d}\mu |f| < \varepsilon \textit{ for all } f \in \{f\}.$$

(c) *There exists a $G$-function $g$ and a constant $D > 0$ such that*

$$\int_{\mathcal{X}} \mathrm{d}\mu g(|f|) < D, \ \forall f \in \{f\}.$$

(d) *There exists a function $m$ on $\mathbb{R}^+$, $m(y) > 0$ for $y > 0$, $m$ is monotonically increasing, $m(y) \to \infty$ $(y \to \infty)$ and a constant $A > 0$ such that*

$$\int_{\mathcal{X}} \mathrm{d}\mu |f| m(|f|) < A \ (\forall\, f \in \{f\}).$$

We apply this theorem to our problem as follows:
Assume that for some $\mu$

$$p \ll \mu \text{ for all } p \in \mathcal{K}.$$

Then $p(E) = \int_E \frac{dp}{\mathrm{d}\mu}\mathrm{d}\mu$, $E \in \mathcal{B}$, where $\frac{\mathrm{d}p}{\mathrm{d}\mu}$ is the Radon–Nikodym derivative of $p$ with respect to $\mu$.

Choose $\{f\} = \left\{\frac{\mathrm{d}p}{\mathrm{d}\mu} : p \in \mathcal{K}\right\}$, since $\frac{\mathrm{d}p}{\mathrm{d}\mu} = \left|\frac{\mathrm{d}p}{\mathrm{d}\mu}\right|$, $\int_{\mathcal{X}} \mathrm{d}\mu \left|\frac{\mathrm{d}p}{\mathrm{d}\mu}\right| = 1$ and the hypothesis of the theorem holds.

### 6.2.3 *Convexity and Continuity Properties of* N($\mathcal{K}$, $\lambda$)

The following observations are readily made:

**Lemma 65** *For $\lambda, 0 < \lambda < 1$:*

(a) $N(\mathcal{K}, \lambda) = N(\overline{\mathcal{K}}, \lambda)$, *where $\overline{\mathcal{K}}$ is the weak closure of $\mathcal{K}$.*
(b) $N(\mathcal{K}, \lambda) = N(co(\mathcal{K}), \lambda)$, *where $co(\mathcal{K})$ denotes the convex hull of $\mathcal{K}$.*
(c) $N(\overline{co(\mathcal{K})}, \lambda) = N(ex(\overline{co(M)}), \lambda)$, *if $\mathcal{K}$ is relatively weakly compact, where $ex(co(M))$ denotes the set of extreme points of $co(M)$.*

*Proof* (a) For $\overline{p} \in \overline{\mathcal{K}}$ with $\overline{p}(E) > 1 - \lambda$ there exists a neighbor $p \in \mathcal{K}$ with $p(E) > 1 - \lambda$, one therefore can replace $(\overline{p}, E)$ by $(p, E)$ in the code. Therefore $N(\overline{\mathcal{K}}, \lambda) \leq N(\mathcal{K}, \lambda)$, the opposite inequality is obvious.

(b) holds because $\sum_{i=1}^{I} \alpha_i p^i(E) > 1-\lambda$ implies $p^{i_0}(E) > 1-\lambda$ for a suitable $i_0$.

(c) Finally, if $\mathcal{K}$ is relatively weakly compact, then $\overline{co(\mathcal{K})}$ is convex and compact. Krein–Milman's theorem gives

$$\overline{co(\mathcal{K})} = \overline{co\big(ex\big(\overline{co(\mathcal{K})}\big)\big)}.$$

Hence (a) and (b) yield (c).

## 6.3 Infinite $\lambda$-codes for Product Channels

Suppose that for $\mathcal{P}$ we have $N(\mathcal{P}, \lambda) = \infty$ for $\lambda \geq \lambda_0 > 0$. What can we say about $N(\mathcal{P}^n, \lambda)$? The Lemma 66 below shows that under the above hypothesis for every $\lambda^+$ and $n$ sufficiently large $N(\mathcal{P}^n, \lambda^+) = \infty$. (This actually follows from the trivial direction of the lemma.)

**Lemma 66** *Let $\mathcal{P}_1, \mathcal{P}_2$ be two channels with output alphabet $(\mathcal{Y}, \mathcal{B})$ and let*

$$\lambda_i(\infty) = \begin{cases} 1 & \text{if } N(\mathcal{P}_i, \lambda) < \infty \text{ for all } \lambda, 0 < \lambda < 1, \\ \inf\{\lambda : N(\mathcal{P}_i, \lambda) < \infty\} & \text{else} \end{cases}$$

$$\lambda^2(\infty) = \begin{cases} 1 & \text{if } N(\mathcal{P}_1 \times \mathcal{P}_2, \lambda) < \infty \text{ for all } \lambda, 0 < \lambda < 1, \\ \inf\{\lambda : N(\mathcal{P}_1 \times \mathcal{P}_2, \lambda) < \infty\} & \text{else} \end{cases}$$

*Then*

$$\lambda^2(\infty) = \lambda_1(\infty) \cdot \lambda_2(\infty).$$

*Proof* Let $\{(P_1^i, D_1^i) : i = 1, \dots, \infty\}$ be a $\lambda_1$-code for $\mathcal{P}_1$ and $\{(P_2^i, D_2^i) : i = 1, \dots, \infty\}$ be a $\lambda_2$-code for $\mathcal{P}_2$, then

$$\{P_1^i \times P_2^i, \left(D_1^i \times D_2^i \cup (D_1^i)^c \times D_2^i \cup D_1^i \times (D_2^i)^c\right) : \quad i = 1, \dots, \infty\}$$

is an $\lambda_1 \cdot \lambda_2$-code for $\mathcal{P}_1 \times \mathcal{P}_2$ of infinite length.

We prove now the converse inequality by using both the packing lemma and the maximal code lemma (see next section).

If $\lambda_1(\infty) \cdot \lambda_2(\infty) = 0$ nothing is to be proved, because $N(\mathcal{P}_1 \times \mathcal{P}_2, \lambda) \geq N(\mathcal{P}_i, \lambda)$. We can assume therefore

$$\lambda_i(\infty) > 0 \text{ for } i = 1, 2.$$

Let now $\lambda < \lambda_1(\infty) \cdot \lambda_2(\infty)$. We have to show that

$$N(\mathcal{P}_1 \times \mathcal{P}_2, \lambda) < \infty.$$

We derive now a bound on the length $N$ of a given $\lambda$-code $\{(P_1^i \times P_2^i, D_i)\}$ for $\mathcal{P}_1 \times \mathcal{P}_2$.

Define $q_j = \frac{1}{N} \sum_{i=1}^N P_j^i \; j = 1, 2$ and $q := q_1 \times q_2$, $P^i = P_1^i \times P_2^i$.

For $\theta_1, \theta_2 > 0$, one has

$$\begin{aligned} p^i \left\{ \frac{\mathrm{d}P^i}{\mathrm{d}q} > \theta_1 \theta_2 \right\} &= p^i \left\{ \left( \frac{1}{\theta_1} \frac{\mathrm{d}P_1^i}{\mathrm{d}q_1} \right) \left( \frac{1}{\theta_2} \frac{\mathrm{d}P_2^i}{\mathrm{d}q_2} \right) > 1 \right\} \leq P^i \left\{ \sup_{j=1,2} \left( \frac{1}{\theta_j} \frac{\mathrm{d}P_j^i}{\mathrm{d}q_j} \right) > 1 \right\} \\ &= P_1^i \left\{ \frac{\mathrm{d}P_1^i}{\mathrm{d}q_1} > \theta_1 \right\} + P_2^i \left\{ \frac{\mathrm{d}P_2^i}{\mathrm{d}q_2} > \theta_2 \right\} - P_1^i \left\{ \frac{\mathrm{d}P_1^i}{\mathrm{d}q_1} > \theta_1 \right\} \cdot P_2^i \left\{ \frac{\mathrm{d}P_2^i}{\mathrm{d}q_2} > \theta_2 \right\} \\ &= 1 - \prod_{j=1}^2 \left( 1 - P_j^i \left\{ \frac{\mathrm{d}P_j^i}{\mathrm{d}q_j} > \theta_j \right\} \right). \end{aligned}$$

For $\varepsilon > 0$, $3\varepsilon < \max_{j=1,2} \lambda_j(\infty)$

$$N(\mathcal{P}_j, \lambda_j(\infty) - \varepsilon) < \infty \text{ for } j = 1, 2$$

and by the *maximal code theorem*

$$N\big(\mathcal{P}_j, \lambda_j(\infty) - \varepsilon\big) > \theta_j \left( \frac{1}{N} \sum_{i=1}^{n} P_j^i \left\{ \frac{\mathrm{d}P_j^i}{\mathrm{d}q_j} > \theta_j \right\} - \big(1 + \varepsilon - \lambda_j(\infty)\big) \right)$$

for $j = 1, 2$ and *all* $\theta_j > 0$.

Hence, for a suitable $\theta_j^0$:

$$\frac{1}{N} \sum_{i=1}^{N} P_j^i \left\{ \frac{\mathrm{d}P_j^i}{\mathrm{d}q_j} > \theta_j^0 \right\} < 1 + 2\varepsilon - \lambda_j(\infty).$$

By Chebyshev's inequality, this implies

$$\frac{1}{N} \left| \left\{ i : P_j^i \left\{ \frac{\mathrm{d}P_j^i}{\mathrm{d}q_j} > \theta_j^0 \right\} > (1 + 2\varepsilon - \lambda_j(\infty))\delta \right\} \right| > 1 - \frac{\delta}{1+\delta} \text{ for } j = 1, 2$$

and hence also for $j = 1, 2$

$$\frac{1}{N} \left| \left\{ i : P_j^i \left\{ \frac{\mathrm{d}P_j^i}{\mathrm{d}q_j} > \theta_j^0 \right\} < \delta\big(1 + 2\varepsilon - \lambda_j(\infty)\big) \right\} \right| > 1 - \frac{2\delta}{1+\delta}.$$

Denote the set in brackets by $J$: $|J| > \frac{1-\delta}{1+\delta} N$.

For $i \in J$, this yields

$$P^i \left\{ \frac{\mathrm{d}P^i}{\mathrm{d}q} > \theta_1^0 \theta_2^0 \right\} \leq 1 - \prod_{j=1}^{2} \big(1 - (1 + 2\varepsilon - \lambda_j(\infty))\delta\big) = \gamma, \text{ say.}$$

The packing lemma yields

$$|J| < (1 - \lambda - \gamma)^{-1} \theta_1^0 \theta_2^0 \text{ for } \gamma < 1 - \lambda.$$

Notice that $\theta_1^0, \theta_2^0$ depend only on $N(\mathcal{P}_j, \lambda_j(\infty) - \varepsilon)$.

In order to achieve

$$\gamma = 1 - \prod_{j=1}^{2} \big(1 - (1 + 2\varepsilon - \lambda_j(\infty))\delta\big) < 1 - \lambda$$

or

$$\lambda < \prod_{j=1}^{2} \big(1 - (1 + 2\varepsilon - \lambda_j(\infty))\delta\big)$$

we choose $\delta_\varepsilon < 1$ such that

$$1 - (1 + 2\varepsilon - \lambda_j(\infty)) = \lambda_j(\infty) - 3\varepsilon > 0$$

and $\varepsilon_0$ such that

$$\lambda < (\lambda_1(\infty) - 3\varepsilon_0)(\lambda_2(\infty) - 3\varepsilon_0).$$

Then

$$N \leq \frac{1 + \delta_{\varepsilon_0}}{1 - \delta_{\varepsilon_0}} |J| \leq (1 - \lambda - \gamma)^{-1} \theta_1^0 \theta_2^0,$$

is a uniform bound. □

## 6.4 Bounds on $N(\mathcal{P}, \lambda)$ in Terms of Stochastic Inequalities

We study now the conditions between the growth of $N(\mathcal{P}, \cdot)$ and the information densities (resp. $\tilde{f}$).

**Definition 35** Let $M$ be a set and $\{p^i\}_{i=1,\dots n}$ be a finite subset of $M$. Let each $i$ be mapped to $a^i \in \mathbb{R}^+$ such that $\sum_{i=1}^n a^i = 1$. The PD on $(1, \dots, n) \times X$ defined by

$$\tilde{p}(i \times E) := a^i p^i(E)$$

is called the simultaneous distribution of $\{a^i\}_{i=1,\dots,n}$ and $\{p^i\}_{i=1,\dots,n}$ in respect to $M$.

The Radon–Nikodym density

$$\tilde{f} := \frac{\mathrm{d}\tilde{p}}{\mathrm{d}a \times \mathrm{d}q}$$

is called the simultaneous density of $\tilde{p}$.

We use only the maximal code theorem and the packing lemma. That is, the inequalities

$$N(\mathcal{P}, \lambda) > \sup_{\tilde{p}} \theta\big(\lambda - \tilde{p}\{\tilde{f} \leq \theta\}\big) \text{ for all } \theta, \quad 0 < \lambda < 1. \tag{6.4.1}$$

and

$$N \le (1 - \gamma - \lambda)^{-1}\theta, \tag{6.4.2}$$

if $\gamma + \lambda < 1$ and $N$ is the length of a $\lambda$-code $\{(p^i, D_i): \quad i = 1, \dots, N\}$ with

$$P_i\left\{\frac{dP_i}{dq} > \theta\right\} \le \gamma,\ i = 1, \dots, N$$

for some PD $q$ on $(\mathcal{Y}, \mathcal{B})$.

We derive now an upper bound on $N$, which uses the same quantities as (6.4.1). Choose $q := \frac{1}{N}\sum_{i=1}^N p^i$, then

$$\frac{1}{N}\sum_{i=1}^N p^i\left\{\frac{dp^i}{dq} > \theta\right\} = \tilde{p}\{\tilde{f} > \theta\}.$$

Clearly there exist $\left[\frac{N}{2}\right]$ indices $i$ such that

$$p^i\left\{\frac{dp^i}{dq} > \theta\right\} \le 2\tilde{p}\{\tilde{f} > \theta\}$$

and hence

$$N \le \theta \cdot 2\big(1 - \lambda - 2\tilde{p}\{\tilde{f} > \theta\}\big)^{-1}. \tag{6.4.3}$$

Finally,

$$N(\mathcal{P}, \lambda) \le \theta \cdot 2\left(1 - \lambda - 2\sup_{\tilde{p}} \tilde{p}\{\tilde{f} > \theta\}\right)^{-1}. \tag{6.4.4}$$

Using (6.4.1) and (6.4.4) one can derive characterizations for the *finiteness of* $N(\mathcal{P}, \lambda)$ in terms of stochastic inequalities for the information density.

**Lemma 67** *$N(\mathcal{P}, \lambda) < \infty\ \forall \lambda,\ (0 < \lambda < 1) \Leftrightarrow k > 0$ and a monotonically increasing function $m$ defined on $(0, \infty)$ satisfying $\lim_{r\to\infty} m(r) = \infty$ such that:*

$$m(r)\tilde{p}\{\tilde{f} > r\} < K \tag{6.4.5}$$

*for all $\tilde{p} \in \tilde{\mathcal{P}}$, $r > 0$.*

*Proof* First let $N(\mathcal{P}, \lambda) < \infty$ for all $\lambda$, $0 < \lambda < 1$. Set

$$g(\theta') := \begin{cases} N\left(\mathcal{P}, 1 - \frac{1}{\theta'}\right) & \text{for } \theta' > 1 \\ 1 & \text{for } 0 < \theta' \le 1 \end{cases}$$

yields with (6.4.1) and $\theta = g(\theta')\theta'$.

$$g(\theta') > \theta' g(\theta') \left(1 - \frac{1}{\theta'} - \tilde{p}\{\tilde{f} \leq \theta' g(\theta')\}\right) \text{ for } \theta' > 1$$

and hence

$$2 > \theta' \tilde{p}\{\tilde{f} > \theta' g(\theta')\}$$

for $\theta' > 1$ and also for $0 < \theta' \leq 1$.

Define now

$$m(r) := \sup\{\theta' : \theta' g(\theta') < r\} \text{ for } \quad r > 0.$$

Then $m$ satisfies all conditions of the lemma with $K = 2$.

Conversely, assume that (6.4.5) is satisfy. To any given $\lambda$, $0 < \lambda < 1$, choose $r$ so large that

$$\frac{K}{m(r)} \leq \frac{1-\lambda}{4}.$$

Let $\{(p^i, D_i) : i = 1, \dots, N\}$ be a $\lambda$-code of length $N < \infty$, with $q := \frac{1}{N}\sum_{i=1}^N p^i$ our hypothesis gives

$$\frac{1}{N}\sum_{i=1}^N p^i \left\{\frac{\mathrm{d}p^i}{\mathrm{d}q} > r\right\} < \frac{K}{m(r)} \leq \frac{1-\lambda}{4}$$

and hence by (6.4.3)

$$N \leq r \cdot 2\left(1 - \lambda - 2\frac{1-\lambda}{4}\right)^{-1} = 2r\left(\frac{1-\lambda}{2}\right)^{-1} = 4r(1-\lambda)^{-1}$$

and thus

$$N(\mathcal{P}, \lambda) \leq 4r(1-\lambda)^{-1} \text{ for all } \quad \lambda,\ 0 < \lambda < 1.$$ □

Next we derive estimates on the *speed* with which $N(\mathcal{P}, \lambda)$ tends to infinity if $\lambda$ tends to 1. Again the argument is based on maximal code theorem and packing lemma.

**Lemma 68** *Let $M$ be defined on $(0, \infty)$, positive and strictly monotonically increasing with $\lim_{r\to\infty} M(r) = \infty$, and let $g$ be a (strictly) positive function on $(0, \infty)$. Then*

$$M\left(\theta \cdot N\left(\mathcal{P}, 1 - \frac{1}{\theta}\right)\right) < g(\theta), \ \theta > 1 \tag{6.4.6}$$

*implies*

$$\theta \, \tilde{p}\{M(\tilde{f}) > g(\theta)\} < K, \ \tilde{p} \in \tilde{\mathcal{P}}, \ \theta > 0. \tag{6.4.7}$$

*for a suitable constant $K$. Then* (6.4.7) *implies*

$$g\left(\frac{4K}{1-\lambda}\right) > M\big((1-\lambda)N(\mathcal{P}, 1-\lambda)\big), \quad 0 < \lambda < 1. \tag{6.4.8}$$

*Proof* The function $M$ has the property that

$$M^{-1}\big(M(\theta)\big) = M\big(M^{-1}(\theta)\big) \ (\theta \geq \theta', \ \theta' \text{ suitable}).$$

Therefore and by assumption

$$\theta \, N\left(\mathcal{P}, 1 - \frac{1}{\theta}\right) < M^{-1}\big(g(\theta)\big)(\theta \geq \theta').$$

The maximal code theorem (MCT) gives

$$\frac{M^{-1}\big(g(\theta)\big)}{\theta} > M^{-1}\big(g(\theta)\big)\tilde{p}\left\{\tilde{f} > M^{-1}g(\theta)\right\} - \frac{1}{\theta}, \ \theta \geq \theta'.$$

Hence,

$$2 > \theta \, \tilde{p}\{\tilde{f} > M^{-1}\big(g(\theta)\big)\}, \quad \theta \geq \theta'.$$

For $\theta \leq \theta'$, the expression to the right is bounded and therefore $K$ exists.
Assume that (6.4.7) holds for $K > 0$.
Set for an $\lambda$-code $\{(p^i, D_i) : i = 1, \dots, N\}$ $q := \frac{1}{N}\sum_{i=1}^N P_i$.
Then

$$\frac{1}{N}\sum_{i=1}^{N} P^i \left\{M\left(\frac{\mathrm{d}p^i}{\mathrm{d}q}\right) > g(\theta)\right\} < \frac{K}{\theta}$$

or

$$P^i \left\{\frac{\mathrm{d}p^i}{\mathrm{d}q} > M^{-1}\big(g(\theta)\big)\right\} < \frac{2K}{\theta} = \gamma, \text{ say},$$

for at least $\left[\frac{N}{2}\right]$ of the $i$'s.

The packing lemma yields

$$N < 2(1-\lambda-\gamma)^{-1}M^{-1}\big(g(\theta)\big)$$

and if we choose $\theta = \frac{4K}{1-\lambda}$ and thus $\gamma = \frac{1-\lambda}{2}$,

$$N < \frac{1}{1-\lambda}M^{-1}\big(g(\theta)\big) \text{ and a fortiori}$$

$$M\big(N(\mathcal{P}, 1-\lambda)\cdot(1-\lambda)\big) < g\left(\frac{4K}{1-\lambda}\right).$$

□

**Corollary 30 (to Lemma** 68) *With $M$ as in Lemma* 68 *and the additional supposition*

$$\frac{M(\theta)}{\log^k(\theta)} \to 0 \ (\theta \to \infty) \qquad (6.4.9)$$

*for a natural number $K$ we have*

$$M\big(N(\mathcal{P}, 1-\lambda)\big) = O(1/(1-\lambda)) \ (\lambda \to 1) \qquad (6.4.10)$$

*iff there exists a constant $K > 0$ such that*

$$\theta\tilde{p}\big\{M(\tilde{f}) > \theta\big\} < K \text{ for all } \ \theta > 0, \tilde{p} \in \tilde{\mathcal{P}}.$$

$$M\big(N(\mathcal{P}, 1-\lambda)\big) = o(1/(1-\lambda)) \ (\lambda \to 1) \qquad (6.4.11)$$

*iff* $\theta\tilde{p}\left\{(M(\tilde{f}) > 0\right\}) \to 0 \ (\theta \to \infty)$ *uniformly in* $\tilde{p} \in \tilde{\mathcal{P}}$.

*Proof* The supposition (6.4.9) implies that

$$M\left(\frac{1}{1-\lambda}N(\mathcal{P}, 1-\lambda)\right) = O(1/(1-\lambda))(\lambda \to 1)$$

(resp. $= o(1/(1-\lambda))$) iff

$$M\big((1-\lambda)N(\mathcal{P}, 1-\lambda)\big) = O(1/1-\lambda)$$

(resp. $= o(1/1-\lambda)$).

Apply now the Lemma 68 with a suitable $g(1/(1-\lambda))$ instead of $O(1/(1-\lambda))$ resp. $o(1/(1-\lambda))$.

## 6.5 Lecture on Coding Theorem, Weak and Strong Converses for Stationary Channels

The results of this lecture are straightforward generalizations of the corresponding results for finite alphabet channels. With respect to strong converses, the situation is different as will be seen later.

The capacity for $(\mathcal{Y}, \mathcal{B}, \mathcal{K})$ is now given by

$$C(\mathcal{K}) = \sup_{\tilde{p} \in \tilde{\mathcal{P}}} \int \mathrm{d}\tilde{p} \log \tilde{f}. \tag{6.5.1}$$

The following lemma states some basic properties of $C(\mathcal{K})$.

**Lemma 69** (a) *For any two probability distributions $P_1$ and $P_2$ defined on $(\mathcal{Y}, \mathcal{B})$ with $P_1 \ll P_2$*

$$\int \mathrm{d}P_1 \log \frac{\mathrm{d}P_1}{\mathrm{d}P_2} \geq 0 \tag{6.5.2}$$

*with equality iff $P_1 = P_2$.*

$$\int \mathrm{d}P_1 \log^+ \frac{\mathrm{d}P_1}{\mathrm{d}P_2} \leq \int \mathrm{d}P_1 \log \frac{\mathrm{d}P_1}{\mathrm{d}P_2} + \frac{1}{e} \tag{6.5.3}$$

(b) *For any $(\mathcal{Y}, \mathcal{B}, \mathcal{K})$*

$$C(\mathcal{K}) \geq 0 \text{ with equality iff } |\mathcal{K}| = 1.$$

*Also $\int \mathrm{d}\tilde{p} \log^+ \tilde{f} \leq C(\mathcal{K}) + \frac{1}{e}$, $\tilde{p} \in \tilde{\mathcal{P}}$.*

(c) *For $\mathcal{K}^n = \mathcal{K}_1 \times \cdots \times \mathcal{K}_n$*

$$C(\mathcal{K}^n) = \sum_{t=1}^{n} C(\mathcal{K}_t).$$

(d) *Suppose that there exists a PD $\mu$ on $(\mathcal{Y}, \mathcal{B})$ such that $P \ll \mu$ for all $P \in \mathcal{K}$ and $\frac{\mathrm{d}P}{\mathrm{d}\mu} \leq K$ ($\mu$ almost everywhere) for some constant $K$, then*

$$C(\mathcal{K}) \leq \log K.$$

*In particular*

$$C\big(\{p^1 \ldots p^r\}\big) \leq \log r.$$

*Proof* For (a) see [7]

Apply now the statements in (a) to

$$\int \mathrm{d}\tilde{p} \log \tilde{f} = \int \mathrm{d}\tilde{p} \log \frac{\mathrm{d}\tilde{p}}{\mathrm{d}p \times \mathrm{d}q}.$$

If $|\mathcal{K}| > 1$, then there exists a $\tilde{p} \in \tilde{\mathcal{P}}$ with $\tilde{p} \neq p \times q$. This proves (b).
Since $\tilde{\mathcal{P}}^n \supset \tilde{\mathcal{P}}_1 \times \cdots \times \tilde{\mathcal{P}}_n$ obviously

$$\begin{aligned} C(\mathcal{K}^n) &\geq \sup_{\tilde{p}_1 \times \cdots \times \tilde{p}_n \in \tilde{\mathcal{P}}_1 \times \cdots \times \tilde{\mathcal{P}}_n} \int \mathrm{d}\tilde{p}_1 \times \cdots \times \mathrm{d}\tilde{p}_n \log \tilde{f}_1 \ldots \tilde{f}_n \\ &= \sum_{t=1}^{n} \sup_{\tilde{p}_t \in \tilde{\mathcal{P}}_t} \int \mathrm{d}\tilde{p}_t \log \tilde{f}_t = \sum_{t=1}^{n} C(\mathcal{K}_t). \end{aligned}$$

It remains to show that

$$C(\mathcal{K}^n) \leq \sum_{t=1}^{n} C(\mathcal{K}_t).$$

Let $\tilde{p} \in \tilde{\mathcal{P}}^n$, $\tilde{p}(i, E) = p(i) P^i(E)$ and let $P_t^i$ denote the $t$th component of $P^i$, then

$$\begin{aligned} \int \mathrm{d}\tilde{p} \log \tilde{f} &= \sum\nolimits_i p(i) \int \mathrm{d}P^i \log \frac{\mathrm{d}P^i}{\sum_j p(j) \mathrm{d}P^j} \\ &= \sum\nolimits_i p(i) \int \mathrm{d}P^i \log \frac{\mathrm{d}P^i}{\left(\sum_j p(j) \mathrm{d}P_1^j\right) \times \cdots \times \left(\sum_j p(j) \mathrm{d}P_n^j\right)} \\ &\quad - \sum\nolimits_i p(i) \int \mathrm{d}P^i \log \frac{\sum_j p(j) \mathrm{d}P^j}{\left(\sum_j p(j) \mathrm{d}P_1^j\right)}, \end{aligned}$$

where the last term is $\geq 0$ by (a).
Hence,

$$\int \mathrm{d}\tilde{p} \log \tilde{f} \leq \sum_{t=1}^{n} \left( \sum_i p(i) \int \mathrm{d}P_t^i \log \frac{\mathrm{d}P_t^i}{\left(\sum_j p(j) \mathrm{d}P_t^j\right)} \right) \leq \sum_{t=1}^{n} C(\mathcal{K}_t).$$

Finally, to see (d), write as above

$$\int \mathrm{d}\tilde{p} \log \tilde{f} = \sum\nolimits_i p(i) \int \mathrm{d}P^j \log \frac{\mathrm{d}P^i}{\sum_j p(j) \mathrm{d}P^j}$$

$$= \sum_i p(i) \int \mathrm{d}P^i \log \frac{\mathrm{d}P^i}{\mathrm{d}\mu} - \sum_i p(i) \int \mathrm{d}P^i \log \frac{\sum_j p(j)\mathrm{d}P^i}{\mathrm{d}\mu}$$

$$\leq \sum_i p(i) \int \mathrm{d}P^i \log \frac{\mathrm{d}P^i}{\mathrm{d}\mu} \leq \sum_i p(i) \log K = \log K.$$

For $\mathcal{K} = \{P^1, \dots, P^r\}$ choose $\mu := \frac{1}{r}\sum_{j=1}^r P^j$, then

$$\frac{\mathrm{d}P^i}{\mathrm{d}\mu} \leq r \text{ ($\mu$ almost everywhere)}.$$

□

**Theorem 131** (Coding theorem and weak converse)

(a) *Coding theorem*

(1) *For any $R < C(\mathcal{K}_1)$, $0 < \lambda < 1$:*

$$\underline{\lim}_{n\to\infty} \frac{1}{n} \log N(\mathcal{K}^n, \lambda) > R.$$

(2) *If $C(\mathcal{K}_1) = \infty$ then there exists a G-function $g$ ($g : \mathbb{R}^+ \to \mathbb{R}^+$, $g(o) = o$, $g(r) > o$ for $r > o$, $\frac{g(r)}{r} \to \infty$ $(r \to \infty)$) such that for all $\lambda$, $0 < \lambda < 1$ there exists a $t_o(\lambda)$:*

$$\log N(\mathcal{K}^n, \lambda) > g(t) \ \big(t > t_0(\lambda)\big).$$

(b) *Weak converse*

$$\log N(\mathcal{K}^n, \lambda) < \frac{1}{\gamma} C(\mathcal{K}_1) \cdot n + \frac{1}{e\gamma} - \log(1 - \lambda - \gamma) \ \ \textit{for all} \ \ \gamma, \ \gamma < 1 - \lambda.$$

*Remark* If $\mathcal{K}_1$ is relatively weakly compact and $C(\mathcal{K}_1) = \infty$, then always $N(\mathcal{K}^n, \lambda) < \infty$, but $N(\mathcal{K}^n, \lambda)$ grows faster than exponential.

*Proof* Given $R < C(\mathcal{K}_1)$. Choose $\tilde{p}_1 \in \tilde{\mathcal{P}}_1$ such that

$$\int \mathrm{d}\tilde{p}_1 \log \tilde{f}_1 > R.$$

For every $t > 1$ let $\tilde{p}_t \in \tilde{\mathcal{P}}_t$ be a copy of $\tilde{p}_1$. By the weak law of large numbers

$$\tilde{p}_1 \times \dots \times \tilde{p}_n\{\tilde{f}_1 \dots \tilde{f}_n > \exp\{Rn\}\} = \tilde{p}_1 \times \dots \times \tilde{p}_n \left\{ \frac{1}{n} \sum_{t=1}^n \log \tilde{f}_t > R \right\} \to 1 \ (n \to \infty) \tag{6.5.4}$$

and hence by the Corollary to the maximal code theorem of the discrete memoryless channel (see [7])

$$N(\mathcal{K}^n, \lambda) > \exp(Rn)\frac{\lambda}{2}$$

for large $n$, which yields (1)).

To see (2) first notice that by (1) there exists in the present case a monotonically increasing function $m(r)$ defined on $\mathbb{R}^+$ with $m(r) \to \infty$ $(r \to \infty)$ such that

$$\log N(\mathcal{K}^n, \lambda) > n\, m(n)\ (n > n(\lambda)).$$

Define $g(t) := \int_0^t m(r)dr$ then $g$ is $G$-function with $g(t) \leq t\, m(t)$.

Let $\{(P^i, E_i), i = 1, 2, \dots, N\}$ be a $\lambda$-code of length $N$ for $\mathcal{K}^n$ the weak converse is readily proved by using the packing lemma (§2) with $q := \frac{1}{N}\sum_{i=1}^{N} P^i$ and

$$\theta_i := \frac{1}{\gamma}\int \mathrm{d}P^i \log^+ \frac{\mathrm{d}P^i}{\mathrm{d}q}.$$

Then

$$\int \mathrm{d}P^i \left\{\frac{\mathrm{d}P^i}{\mathrm{d}q} \geq \mathrm{e}^{\theta_i}\right\} = \int \mathrm{d}P^i \left\{\log^+ \frac{\mathrm{d}P^i}{\mathrm{d}q} \geq \theta_i\right\} \leq \gamma, \text{ for } \ i = 1, 2, \dots, N$$

and by the packing lemma

$$N < (1-\lambda-\gamma)^{-1}\mathrm{e}^{\frac{1}{N}\sum\theta_i} \tag{6.5.5}$$

$$\leq (1-\lambda-\gamma)^{-1}\mathrm{e}^{\frac{1}{\gamma}\left(\frac{1}{N}\sum_{i=1}^{N}\int \mathrm{d}P^i \log \frac{\mathrm{d}P^i}{\mathrm{d}q}+\frac{1}{e}\right)} \text{ for } \gamma < 1-\lambda \tag{6.5.6}$$

and hence

$$\log N < -\log(1-\lambda-\gamma] + \frac{1}{\gamma}nC(\mathcal{K}_1) + \frac{1}{e\gamma} \text{ for all } \gamma < 1-\lambda, \tag{6.5.7}$$

□

Clearly this is the weak converse, because with $\lambda \to 0$ we can choose $\gamma$ arbitrarily close to 1. The theorem implies that

$$\inf_{\lambda} \overline{\lim}_{n\to\infty} \frac{1}{n}\log N(\mathcal{K}^n, \lambda) = \inf_{\lambda} \underline{\lim}_{n\to\infty} \frac{1}{n}\log N(\mathcal{K}^n, \lambda) = C(\mathcal{K}_1).$$

*Remark* The converse proof literally applies to the nonstationary case with $\sum_{t=1}^{n} C(\mathcal{K}_t)$ as capacity function of course. The direct part *did use stationarity* in applying the weak law of large numbers (WLLN).

## 6.6 Sharper Estimates for Channels with Moment Conditions

The coding theory for memoryless channels (stationary as well as nonstationary) easily extends to the infinite alphabet case, if the variances of the information functions are uniformly bounded, that is,

$$D^2(\mathcal{K}_t) = \sup_{\tilde{p}_t \in \tilde{\mathcal{P}}_t} \int \mathrm{d}\tilde{p}_t (\log \tilde{f}_t - \int \mathrm{d}\tilde{p} \log \tilde{f})^2 \qquad (6.6.1)$$

and $C(\mathcal{K}_t) < K$ for $t = 1, 2, \dots$

In the following, all integrals $\int \mathrm{d}\tilde{p} \log^2 \tilde{f} = \int \mathrm{d}p \times \mathrm{d}q \tilde{f} \log^2 \tilde{f}$, $\tilde{p} \in \tilde{\mathcal{P}}$, are finite, because $\tilde{f}$ is $p \times q$ almost everywhere bounded by a constant and $y \log^2 y$ is bounded in $[0, 1]$ by $4e^{-2}$.

**Theorem 132** *Suppose that $C(\mathcal{K}_t) < \infty$ for $t = 1, 2, \dots$, then*

$$\log N(\mathcal{K}^n, \lambda) \geq C(\mathcal{K}^n) - \left[\frac{1}{\lambda - \delta} \sum_{t=1}^{n} D^2(\mathcal{K}_t)\right]^{\frac{1}{2}} + \log \delta \qquad (6.6.2)$$

*for $0 < \lambda < 1$, $\delta < \lambda$ and*

$$\log N(\mathcal{K}^n, \lambda) \leq C(\mathcal{K}^n) + \left[\frac{1}{(1-\lambda)\delta} \sum_{t=1}^{n} D^2(\mathcal{K}_t)\right]^{\frac{1}{2}} - \log(1-\delta)(1-\lambda). \quad (6.6.3)$$

*If the variances are uniformly bounded then this implies*

$$|\log N(\mathcal{K}^n, \lambda) - C(\mathcal{K}^n)| = O(\sqrt{n}),$$

*in particular a coding theorem and a strong converse.*

*Proof* Write $\tilde{p}^n = \tilde{p}_1 \times \dots \times \tilde{p}_n$.

We use the maximal code theorem. Chebyshev's inequality yields

$$\tilde{p}^n \left\{ \log \tilde{f}^n > \int \mathrm{d}\tilde{p}^n \log \tilde{f}^n - \left[\frac{1}{\lambda - \delta} \mathrm{Var}(\log \tilde{f}^n)\right]^{\frac{1}{2}} \right\} > 1 - \lambda + \delta.$$

Hence,

$$\log N(\mathcal{K}^n, \lambda) \geq \sup_{\tilde{p}^n \in \tilde{\mathcal{P}}_1 \times \dots \times \tilde{\mathcal{P}}_n} \int \mathrm{d}\tilde{p}^n \log \tilde{f}^n - \left[\frac{1}{\lambda - \delta} \mathrm{Var}(\log \tilde{f}^n)\right]^{\frac{1}{2}} + \log \delta$$

$$\geq C(\mathcal{K}^n, \lambda) - \sup_{\tilde{p}^n \in \tilde{\mathcal{P}}_1 \times \cdots \times \tilde{\mathcal{P}}_n} \left[ \frac{1}{\lambda - \delta} \mathrm{Var}(\log \tilde{f}^n) \right]^{\frac{1}{2}} + \log \delta.$$

Since

$$\mathrm{Var}(\log \tilde{f}^n) = \sum_{t=1}^{n} \int \mathrm{d}\tilde{p}_t (\log \tilde{f}_t - \int \mathrm{d}\tilde{p}_t \log \tilde{f}_t)^2 \leq \sum_{t=1}^{n} D^2(\mathcal{K}_t) \text{ we get (6.6.2)}.$$

Let $\{(P^i, E_i), i = 1, 2, \dots, N\}$ be a $\lambda$-code of length $N$ for $\mathcal{K}^n$ to show (6.6.3) we use the packing lemma with

$$q : q_1 \times \cdots \times q_n := \left( \frac{1}{N} \sum_{i=1}^{N} P_1^i \right) \times \cdots \times \left( \frac{1}{N} \sum_{i=1}^{N} P_n^i \right)$$

and

$$\theta_i = \int \mathrm{d}P^i \log \frac{\mathrm{d}P^i}{\mathrm{d}q} + \left[ \frac{1}{(1-\lambda)\delta} \int \mathrm{d}P^i \left( \log \frac{\mathrm{d}P^i}{\mathrm{d}q} \int \mathrm{d}P^i \log \frac{\mathrm{d}P^i}{\mathrm{d}q} \right)^2 \right]^{\frac{1}{2}}.$$

From Chebyshev we get for $i = 1, 2, \dots, N$

$$P^i \left\{ \log \frac{\mathrm{d}P^i}{\mathrm{d}q} > \theta_i \right\} < (1-\lambda)\delta = \gamma, \text{ say},$$

and hence

$$N \leq (1 - \lambda - \gamma)^{-1} \mathrm{e}^{\frac{1}{N} \sum_{i=1}^{N} \theta_i} = \left[ (1-\delta)(1-\lambda) \right]^{-1} \mathrm{e}^{\frac{1}{N} \sum_{i=1}^{N} \theta_i}.$$

The evaluation of $\frac{1}{N} \sum_{i=1}^{N} \theta_i$ is exactly as in the finite case.

In the case of finite input but infinite output alphabets (the so-called semi-continuous case, see [20] for the result) the variances are always bounded by the size of the input alphabet. More generally the following holds.

**Lemma 70** *If there exists a probability measure $\mu$ on $(\mathcal{Y}, \mathcal{B})$ and a constant $K$ such that $\frac{\mathrm{d}p}{\mathrm{d}\mu} \leq K$ ($\mu$ almost everywhere) for all $p \in \mathcal{K}$, then*

$$D^2(\mathcal{K}) \leq \left( \frac{1}{e} + \log K \right)^2 + 7e^{-2}.$$

*In particular*

$$D^2(\{p^1, \dots, p^r\}) \leq \left( \frac{1}{e} + \log r \right)^2 + 7e^{-2}$$

(*Kemperman* [17]).

The previous theorem becomes trivial already (and does not imply coding theorem and converse) if $D^2(\mathcal{K}_t) = \infty$ or if $D^2(\mathcal{K}_t)$ goes faster than $o(t)$ to infinity as $t \to \infty$.

Under weakened moment conditions, some of those cases still can be handled with the use of

**Lemma 71** (Khinchin) *Let $f_1, \dots, f_n$ be independent RV's defined on a probability space $(\mathcal{Y}, \mathcal{B}, \mu)$ and let*

$$\int_{\mathcal{Y}} \mathrm{d}\mu f_t = 0 \, for \, t = 1, \dots, n$$

*then*

$$\int_{\mathcal{Y}} \mathrm{d}\mu \left| \sum_{t=1}^{n} f_t \right|^{1+\alpha} \leq 2^{1+\alpha} \sum_{t=1}^{n} \int_{\mathcal{Y}} \mathrm{d}\mu |f_t|^{1+\alpha} (0 \leq \alpha \leq 1).$$

Using in addition the generalized Chebyshev inequality, the proof of the Theorem 132 carries over verbatim and yields

**Theorem 133** (*) *Suppose that $C(\mathcal{K}_t) < \infty$ for $t = 1, 2, \dots,$ then*

$$\log N(\mathcal{K}^n, \lambda) \geq C(\mathcal{K}^n) - \left[ \frac{1}{\lambda - \delta} 4 \sum_{t=1}^{n} \sup_{\tilde{p}_t \in \tilde{\mathcal{P}}_t} \int \mathrm{d}\tilde{p}_t \left| \log \tilde{f}_t - \int \mathrm{d}\tilde{p}_t \log \tilde{f}_t \right|^{1+\alpha} \right]^{\frac{1}{1+\alpha}} + \log \delta \tag{6.6.4}$$

*for $0 < \lambda < 1, \delta < \lambda$ and $\alpha > 0$*

$$\log N(\mathcal{K}^n, \lambda) \leq C(\mathcal{K}^n) + [\dots\dots]^{\frac{1}{1+\alpha}}. \tag{6.6.5}$$

A further generalization is obtained by not just considering moments, that is functions $g(r) = r^{1+\alpha}$, but by allowing more general $G$-functions, specified below, instead. Fortunately, there is an extension of Khinchin's Lemma to this case. The extension of Theorem 132 is standard.

**Lemma 72** *Let $f_1, \dots, f_n$ be independent RV's defined on a probability space $(\mathcal{Y}, \mathcal{B}, \mu)$ and let $\int_{\mathcal{Y}} \mathrm{d}\mu f_t = 0$ for $t = 1, 2, \dots, n$, then for any G-function $g$ for which $g(\sqrt{r})$ is concave on $\mathbb{R}^+$:*

$$\int_{\mathcal{Y}} \mathrm{d}\mu g \left( \left| \sum_{t=1}^{n} f_t \right| \right) \leq 4 \sum_{t=1}^{n} \int_{\mathcal{Y}} \mathrm{d}\mu g(|f_t|).$$

*Proof* (1) We consider first the case where $f_t$ is symmetrically distributed:

$$\mu\{f_t < s\} = \mu\{-f_t < s\} \text{ for all } s \in \mathbb{R} \text{ for } t = 1, \dots, n.$$

Let $r_1, \dots, r_n$ be functions defined on $[0, 1]$ with values in $\{+1, -1\}$ and which are orthogonal:

$$\int_{[0,1]} \mathrm{d}x r_i(x) r_j(x) = 0 \quad (i \neq j).$$

(For instance choose any $n$ Rademacher functions.)

Then for any fixed $x \in [0, 1]$ obviously $\sum_{t=1}^n f_t$ and $\sum_{t=1}^n f_t r_t(x)$ have the same distribution functions. Therefore,

$$\begin{aligned}
&\int_{\mathcal{Y}} \mathrm{d}\mu g\left(\left|\sum_{t=1}^n f_t\right|\right) \\
&= \int_{\mathcal{Y}} \mathrm{d}\mu \left[\int_{[0,1]} \mathrm{d}x\, g\left(\left|\sum_{t=1}^n f_t(y) r_t(x)\right|\right)\right] \\
&\leq \int_{\mathcal{Y}} \mathrm{d}\mu \left[\int_{[0,1]} \mathrm{d}x\, g\left(\sqrt{\left|\sum_{t=1}^n f_t(y) r_t(x)\right|^2}\right)\right] \\
&\leq \int_{\mathcal{Y}} \mathrm{d}\mu g\left(\sqrt{\left[\int_{[0,1]} \mathrm{d}x \left(\sum_{t=1}^n f_t(y) r_t(x)\right)^2\right]}\right) \\
&= \int_{\mathcal{Y}} \mathrm{d}\mu g\left(\sqrt{\sum_{t=1}^n |f_t(y)|^2}\right).
\end{aligned}$$

The inequality is justified by Jensen's inequality applied to the concave function $g(\sqrt{\ })$. Moreover, from $g^{-1}(\theta) = \int_0^\theta \mathrm{d}r\, m(r)$ and

$$\int_0^{\theta_1+\theta_2} \mathrm{d}r\, m(r) = \int_0^{\theta_1} \mathrm{d}r\, m(r) + \int_{\theta_1}^{\theta_1+\theta_2} \mathrm{d}r\, m(r) \leq \int_0^{\theta_1} \mathrm{d}r\, m(r) + \int_0^{\theta_2} \mathrm{d}r\, m(r)$$

since $m(r)$ is monotonically decreasing, it follows that

$$g^{-1}(\theta_1 + \theta_2) \leq g^{-1}(\theta_1) + g^{-1}(\theta_2).$$

Hence

$$\int_{\mathcal{Y}} \mathrm{d}\mu g\left(\sqrt{\sum_{t=1}^{n} |f_t(y)|^2}\right) \leq \sum_{t=1}^{n} \int_{\mathcal{Y}} \mathrm{d}\mu g(|f_t|).$$

Thus, we have proved for symmetrically distributed independent random variables

$$\int_{\mathcal{Y}} \mathrm{d}\mu g\left(\left|\sum_{t=1}^{n} f_t\right|\right) \leq \sum_{t=1}^{n} \int_{\mathcal{Y}} \mathrm{d}\mu g(|f_t|).$$

(2) For the general case, we use in addition to the given data another probability space $(\mathcal{Y}', \mathcal{B}', \mu')$ and a set of independent RV's $f_{n+1}, \dots, f_{n+n}$ defined on $(\mathcal{Y}', \mathcal{B}', \mu')$ with $\mu\{f_t < s\} = \mu'\{-f_{n+1} < s\}$.

We put everything on one probability space by setting

$$\tilde{f}_t(y, y') := f_t(y),\ y' \in \mathcal{Y}',\ t = 1, \dots, n$$
$$\tilde{f}_{n+t}(y, y') := f_{n+t}(y'),\ y \in \mathcal{Y},\ t = 1, \dots, n.$$

Since all RV's have expected value 0, the Jensen inequality for conditional expectations gives ($g$ is convex!)

$$\begin{aligned}\int_{\mathcal{Y}} \mathrm{d}\mu g\left(\left|\sum_{t=1}^{n} f_t\right|\right) &\leq \int_{\mathcal{Y}\times\mathcal{Y}'} \mathrm{d}\mu \times \mathrm{d}\mu' g\left(\left|\sum_{t=1}^{n} (\tilde{f}_t + \tilde{f}_{n+1})\right|\right) \\ &\leq \sum_{t=1}^{n} \int_{\mathcal{Y}\times\mathcal{Y}'} \mathrm{d}\mu \times \mathrm{d}\mu' g(|\tilde{f}_t + \tilde{f}_{n+t}|)\end{aligned}$$

because $\tilde{f}_t + \tilde{f}_{n+t}$ is symmetrically distributed. Again by convexity of $g$:

$$\begin{aligned}\int_{\mathcal{Y}\times\mathcal{Y}'} \mathrm{d}\mu \times \mathrm{d}\mu' g(|\tilde{f}_t \times \tilde{f}_{n+t}|) &\leq \int_{\mathcal{Y}\times\mathcal{Y}'} \mathrm{d}\mu \times \mathrm{d}\mu' g\left(\left|\frac{2|\tilde{f}_t| + 2|\tilde{f}_{n+t}|}{2}\right|\right) \\ &\leq \int_{\mathcal{Y}\times\mathcal{Y}'} \mathrm{d}\mu \times \mathrm{d}\mu' \frac{1}{2}\left[g(2|\tilde{f}_t|) + g(2(\tilde{f}_{n+t}))\right] \\ &= \int_{\mathcal{Y}} \mathrm{d}\mu(2|f_t|).\end{aligned}$$

Furthermore,

$$g(2|f_t|) = g\left(\sqrt{4|f_t|^2}\right) \leq 4g\left(\sqrt{|f_t|^2}\right) = 4g(|f_t|)$$

and hence

$$\int_{\mathcal{Y}} \mathrm{d}\mu g\left(\left|\sum_{t=1}^{n} f_t\right|\right) \leq 4\sum_{t=1}^{n}\int_{\mathcal{Y}} \mathrm{d}\mu g(|f_t|).$$

□

**Theorem 134** *Suppose that $C(\mathcal{K}_t) < \infty$ for $t = 1, 2, \ldots$, then for any $G$-function $g$ with $g(\sqrt{\cdot})$ concave on $\mathbb{R}^+$:*

$$\log N(\mathcal{K}^n, \lambda) \geq C(\mathcal{K}^n) - g^{-1}\left(\frac{1}{\lambda - \delta}4\sum_{t=1}^{n}\left[\sup_{\tilde{p}_t \in \tilde{\mathcal{P}}_t} \int \mathrm{d}\tilde{p}_t g\left(|\log \tilde{f}_t - \int \tilde{p}_t \log \tilde{f}_t|\right)\right]\right) + \log \delta$$

*for $0 < \lambda < 1$, $\delta < \lambda$ and*

$$\log N(\mathcal{K}^n, \lambda) \leq C(\mathcal{K}^n) + g^{-1}\left(\frac{1}{\lambda - \delta}16\sum_{t=1}^{n}\left[\sup_{\tilde{p}_t \in \tilde{\mathcal{P}}_t} \int \tilde{p}_t g(|\log \tilde{f}_t|)\right]\right) - \log(\delta(1-\lambda)),$$

*for $0 < \lambda < 1$ and $0 < \delta < 1$.*

The factor 16 rather than 4 comes from the fact that one has to use for the converse part

$$\begin{aligned}
&\frac{1}{N}\sum_{i=1}^{N}\int \mathrm{d}P_t^i g\left(\left|\log\frac{\mathrm{d}P_t^i}{\mathrm{d}q_t} - \int \mathrm{d}p_t^i \log\frac{\mathrm{d}P_t^i}{\mathrm{d}q_t}\right|\right) \\
&\leq \frac{1}{N}\sum_{i=1}^{N}\int \mathrm{d}P_t^i \frac{1}{2}\left[g\left(2\left|\log\frac{\mathrm{d}P_t^i}{\mathrm{d}q_t}\right|\right) + g\left(2\left|\int \mathrm{d}P_t^i \log\frac{\mathrm{d}P_t^i}{\mathrm{d}q_t}\right|\right)\right] \\
&\leq \frac{1}{N}\sum_{i=1}^{N}\int \mathrm{d}P_t^i q\left(2\left|\log\frac{\mathrm{d}P_t^i}{\mathrm{d}q_t}\right|\right) \\
&\leq 4\frac{1}{N}\sum_{i=1}^{N}\int \mathrm{d}P_t^i g\left(\left|\log\frac{\mathrm{d}P_t^i}{\mathrm{d}q_t}\right|\right).
\end{aligned}$$

Theorem 134 gives a coding theorem and a strong converse, if for $G$-function $g$ with $g(\sqrt{\cdot})$ concave

$$\sup_t \sup_{\tilde{p}_t \in \tilde{\mathcal{P}}_t} \int d\tilde{p}_t \, g(|\log \tilde{f}_t|) < \infty. \tag{6.6.6}$$

In the next paragraph, we discuss this condition and related conditions as well. Those results described are probably only of interest to the infinite alphabet channels experts.

## 6.7 A Necessary and Sufficient Condition for the Strong Converse in the Stationary Case

Since coding theorem and weak converse always hold in the stationary case and the strong converse holds under certain moment conditions (see §1), it is natural to look for exact conditions for the strong converse to hold. We saw in Corollary 30 to Lemma 68 that the following two conditions are equivalent

$$C(\mathcal{K}) < \infty \quad \text{and} \log N(\mathcal{K}, \lambda) = o\left(\frac{1}{1-\lambda}\right) (\lambda \to 1) \tag{6.7.1}$$

and

$$C(\mathcal{K}) < \infty \quad \text{and} \sup_{\tilde{p} \in \tilde{\mathcal{P}}} \theta \tilde{p}\{\log \tilde{f} > \theta\} \to 0 \; (\theta \to 0). \tag{6.7.2}$$

Since maximal code length can be estimated from both sides by such stochastic inequalities, it is natural to look for a condition in terms of such inequalities. In fact, we shall prove the following

**Theorem 135** *For a stationary memoryless infinite alphabet channel with* $C(\mathcal{K}) < \infty$:

$$\overline{\lim}_{n \to \infty} \frac{1}{n} \log N(\mathcal{K}^n, \lambda) \leq C(\mathcal{K}) \quad \textit{for all } \lambda, 0 < \lambda < 1,$$

*iff*

$$\log N(\mathcal{K}, \lambda) = o\left(\frac{1}{1-\lambda}\right) (\lambda \to 1) \left[ \textit{or} \log N\left(\mathcal{K}, 1 - \frac{1}{n}\right) = o(n) \; (n \to \infty)\right].$$

*Proof* We show first that condition (6.7.1) is necessary by proving that its negation implies

$$\sup_\lambda \overline{\lim}_{n \to \infty} \frac{1}{n} \log N(\mathcal{K}^n, \lambda) = \infty.$$

Suppose therefore that there exists a $b > 0$ and a sequence $\{n_j\}_{j=1}^{\infty}$ of natural numbers such that

$$\log N\left(\mathcal{K}, 1 - \frac{1}{n_j}\right) \geq b\, n_j \text{ for } j = 1, 2, \dots .$$

By Lemma 65 we can also assume that $\mathcal{K}$ is convex. Fix $n = n_j$ and let $\{(\overline{P}_1^i, D_i)\}_{i=1}^N$ be a $(1 - \frac{1}{n})$-code for $\mathcal{K}_1$ of length $N > e^{bn}$. Define as usual

$$q_1 := \frac{1}{N} \sum_{i=1}^{N} \overline{P}_1^i$$

and consider the modified code $\{(P_1^i, D_i)\}_{i=1}^N$, where $P_1^i = \frac{1}{2}(\overline{P}_1^i + q_1)$.

This is a $(1 - \frac{1}{2n})$-code for $\mathcal{K}_1$ of length $N$ and

$$\frac{1}{N}\left|\left\{i : q_1(\mathcal{D}_i) > n\frac{1}{N}\left(\sum_{i=1}^{N} q_1(D_1)\right)\right\}\right| < \frac{1}{n}.$$

Also,

$$\frac{1}{N}\left|\left\{i : \frac{1}{q_1(D_i)} > \frac{N}{n}\right\}\right| > 1 - \frac{1}{n}$$

and

$$\frac{1}{N}\left|\left\{i : \frac{P_1^i(D_i)}{q_1(D_i)} > \frac{e^{bn}}{n}\frac{1}{2n}\right\}\right| > 1 - \frac{1}{n}.$$

We consider now $\mathcal{B}' = \mathcal{B}(\{D_i\})$ which denotes the finite $\sigma$-algebra generated by the decoding sets. The input–output distribution $\tilde{p}_1$, which is generated by $\left(\frac{1}{N}, \dots, \frac{1}{N}\right)$ and $\{\mathcal{P}_1^i\}$, is now defined on the product $\sigma$-algebra $\mathcal{B}(\{i\} \times D_\ell)$. Since (by definition of the modified code) $2\tilde{f}_1 \geq 1$ ($\tilde{p}_1$—almost everywhere) and by the analog construction also $2\tilde{f}_t \geq 1$, $t = 1, \dots, n$, we obtain:

$$\begin{aligned}
&\left(\prod_{t=1}^n \tilde{p}_t\right)\left\{\log\cdot\left(2^n \prod_{t=1}^n \tilde{f}_t\right) > K \log\left(\frac{1}{2n^2}e^{bn}\right)\right\} \\
&\geq \left(\prod_{t=1}^n {}^*\tilde{p}_t\right)\left(\left\{x^n = (x_1, \dots, x_n) : \tilde{f}_t(x^n) > \frac{1}{2n^2}e^{bn} \text{ for at least } K \text{ indices } t\right\}\right) \\
&> \sum_{j=K}^n \binom{n}{j}\left(1 - \frac{1}{n}\right)^j \left(\frac{1}{n}\right)^{n-j} .
\end{aligned}$$

Keeping $K$ fixed, for $n(= n_j)$ sufficiently large

$$\left(\prod_{t=1}^{n} \tilde{p}_t\right)\left(\left\{\log \prod_{t=1}^{n} \tilde{f}_t > K\, bn + K \log\left(\frac{1}{2n^2}\right) - n \log 2\right\}\right) \geq 1 - \delta > 0.$$

From the maximal code theorem, we get

$$N(\mathcal{K}^n, \lambda) > \exp\left\{K \cdot bn + K \log\left(\frac{1}{2n^2}\right) - n \log 2\right\}\left(1 - \delta - (1 - \lambda)\right)$$

and therefore: to every $K' > 0$ there exists an $\lambda(K')$ such that

$$\overline{\lim}_{n\to\infty} \frac{1}{n} \log N\left(\mathcal{K}^n, \lambda(K')\right) > K'.$$

This proves the necessity of condition (6.7.1).
We prove now the converse.

1. In a first step we show that

$$\log N\left(\mathcal{K}, 1 - \frac{1}{n}\right) = o(n) \quad (n \to \infty) \tag{6.7.3}$$

implies

$$\sum_{t=1}^{n} \sup_{\tilde{p}_t \in \tilde{\mathcal{P}}_t} \tilde{p}_t\{\log \tilde{f}_t > o_n(1)n\} \to 0 \ (n \to \infty) \text{ for a suitable } o_n(1). \tag{6.7.4}$$

This implies the existence of a function $h(n) = o(1)$ such that

$$\log N\left(\mathcal{K}, 1 - \frac{h(n)}{n}\right) \leq h(n)n. \tag{6.7.5}$$

Let $h'(n) = o(1)$ be another function with $\frac{h'(n)}{h(n)} \to \infty$ $(n \to \infty)$ and $h'(n) > \frac{1}{n}$.
From the MCT and the previous inequality:

$$\begin{aligned} h(n)n &\geq \log N\left(\mathcal{K}_t, 1 - \frac{h(n)}{n}\right) \\ &> h'(n)n + \log\left[\tilde{p}_t\{\log \tilde{f}_t > h'(n) \cdot n\} - \frac{h(n)}{n}\right]^+. \end{aligned}$$

Therefore for all $t = 1, 2, \dots$ and all $\tilde{p}_t \in \tilde{\mathcal{P}}_t$:

$$\tilde{p}_t\{\log \tilde{f}_t > h'(n) \cdot n\} < \mathrm{e}^{(h(n)-h'(n))n} + \frac{h(n)}{n} = \frac{1}{n} o_n(1)$$

(uniformly for all $t$ and all $\tilde{p}_t$), which implies (6.7.5).

2. The proof of the converse is again essentially based on the packing lemma. Since there are no moment conditions, an additional truncation argument is used. Set $r(n) = o_n(1)n$. For any sequence of $\lambda$-codes $\left(\{(P^i, D_i)\}_{i=1}^{N(\mathcal{K}^n,\lambda)}\right)_{n=1}^{\infty}$ and Fano*-output sources

$$q^n := q_1^n \times \dots \times q_n^n := \left(\frac{1}{N}\sum_{i=1}^{N} P_1^i\right) \times \dots \times \left(\frac{1}{N}\sum_{i=1}^{N} P_n^i\right)$$

we define

$$E_i := {}_1E_i \times \dots \times {}_nE_i = \left\{\log \frac{\mathrm{d}P_1^i}{\mathrm{d}q_1} < r(n)\right\} \times \dots \times \left\{\log \frac{\mathrm{d}P_n^i}{\mathrm{d}q_n} < r(n)\right\}.$$

Since by (6.7.5)

$$\sum_{t=1}^{n} \frac{1}{N} \sum_{t=1}^{N} P_t^i \left\{\log \frac{\mathrm{d}P_t^i}{\mathrm{d}q_t} > r(n)\right\} \to 0 \ (t \to 0)$$

there exists a function $o_n(1)$ such that

$$\frac{1}{N} \left| \left\{ i : \sum_{t=1}^{n} P_t^i \left\{\log \frac{\mathrm{d}P_t^i}{\mathrm{d}q_t} > r(n)\right\} < o_n(1) \right\} \right| \to 1.$$

Denoting the set in brackets by $U(n)$, then for any $\delta > 0$ there exists a $t_0(\delta)$ such that for $n \geq n_o(\delta)$

$$P^i(E_i^c) \leq \sum_{t=1}^{n} P_t^i \left\{\log \frac{\mathrm{d}P_t^i}{\mathrm{d}q_t} > r(n)\right\} < \delta \text{ for all } i \in U(n).$$

Now we operate only on the $E_i$'s, where the information densities are bounded uniformly.

For any $K > 1$ we have

$$P^i \left\{ E_i \cap \left\{ \frac{\mathrm{d}P^i}{\mathrm{d}q^n} > \left(\frac{2}{1-\lambda}\right)^K \left( \int_{E_i} \mathrm{d}P^i \left(\frac{\mathrm{d}P^i}{\mathrm{d}q}\right)^{1/K} \right) \right\} \right\} < \frac{1-\lambda}{2},$$

by Chebyshev and—by enlarging the terms to the right and removing $E_i$—

$$
\begin{aligned}
P^i & \left\{ \log \frac{\mathrm{d}P^i}{\mathrm{d}q^n} > K \log \frac{2}{1-\lambda} + K \sum_{t=1}^{n} \log \left( \int_{{}_tE_i} \mathrm{d}P_t^i \left( \frac{\mathrm{d}P_t^i}{\mathrm{d}q_t} \right) \right)^{1/K} \right\} \\
& < \frac{1-\lambda}{2} + P^i(E_i^c) \\
& \leq \frac{1-\lambda}{2} + \delta
\end{aligned}
$$

for $i \in U(n)$.

This is now exactly the condition for the packing lemma with $\gamma = \frac{1-\lambda}{2} + \delta$ and

$$
\theta_i = K \log \frac{2}{1-\lambda} + K \sum_{t=1}^{n} \log \left( \int_{{}_tE_i} \mathrm{d}P_t^i \left( \frac{\mathrm{d}P_t^i}{\mathrm{d}q_t} \right)^{1/K} \right),
$$

and therefore for $\delta < \frac{1-\lambda}{2}$

$$
|U(n)| \leq \left( \frac{1-\lambda}{2} - \delta \right)^{-1} \exp \left( \frac{1}{|U(n)|} \sum_{i \in U(n)} \theta_i \right).
$$

Finally, we choose now $K = K(n) = o_n(1) \cdot n$ and such that $\frac{K(n)}{r(n)} \to \infty \ (t \to \infty)$. This, definition of $E_i$ and by an elementary property of the log-function

$$
\begin{aligned}
\int_{{}_tE_i} \mathrm{d}P_t^i \left( \frac{\mathrm{d}P_t^i}{\mathrm{d}q_t} \right)^{1/K} & \leq \left( 1 + \frac{1}{K} \int_{{}_tE_i} \mathrm{d}P_t^i \, K \left[ \left( \frac{\mathrm{d}P_t^i}{\mathrm{d}q_t} \right)^{1/K} - 1 \right] \right) \\
& \leq 1 + \frac{1}{K} \int_{{}_tE_i} \mathrm{d}P_t^i \left( \log \frac{\mathrm{d}P_t^i}{\mathrm{d}q_t} \right) \left(1 + o_n(1)\right).
\end{aligned}
$$

Since

$$
\log \left( 1 + \frac{1}{K} \int_{{}_tE_i} \mathrm{d}P_t^i \left( \log \frac{\mathrm{d}P_t^i}{\mathrm{d}q_t} \right) \right) \left(1 + o_n(1)\right) = \frac{1}{K} \int_{{}_tE_i} \mathrm{d}P_t^i \log \frac{\mathrm{d}P_t^i}{\mathrm{d}q_t} \left(1 + o_n(1)\right),
$$

we get

$$|U(n)| \leq \left(\frac{1-\lambda}{2} - \delta\right)^{-1}$$

$$\exp\left\{-K \log \frac{2}{1-\lambda} - \sum_{t=1}^{n} \frac{1}{|U(n)|} \sum_{i \in U(n)} \int_{{}_t E_i} \mathrm{d}P_t^i \log \frac{\mathrm{d}P_t^i}{\mathrm{d}q_t}(1 + o_n(1))\right\}.$$

Recalling that $\frac{N(\mathcal{K}^n,\lambda)}{|U(n)|} \to 1 (n \to \infty)$ the desired result follows.

**Example** (Stationary channel without strong converse (for more details see [7]))
Let $\mu$ be the Lebesgue measure on [0, 1] and let $P^i$ $(i = 1, 2, \ldots)$ be such that

$$\frac{\mathrm{d}P^i}{\mathrm{d}\mu}(x) = 2\,(\text{Rademacher})\,, \mathcal{K} = \{P^i\}.$$

Then

$$\frac{1}{2^n} N\left(\mathcal{K}^n, \frac{1}{2^n}\right) \geq 1 \text{ for } n = 1, 2, \ldots$$

and therefore $N(\mathcal{K}^n, 1-\lambda) \neq o\left(\frac{1}{1-\lambda}\right)$.

However,

$$C(\mathcal{K}) \leq \sup_P \sum_{i=1}^{a} p(i) \int \mathrm{d}P^i \log \frac{\mathrm{d}P^i}{\mathrm{d}\mu},$$

since $\int \mathrm{d}q \log \frac{\mathrm{d}\mu}{\mathrm{d}q} \leq 0$.

But $\int \mathrm{d}P^i \log \frac{\mathrm{d}P^i}{\mathrm{d}\mu} \leq 1$ and hence $C(\mathcal{K}) \leq 1$.

The set $\mathcal{K}$ is *not* relatively compact in total variation norm. But also under the hypothesis of norm compactness examples with $C(\mathcal{K}) < \infty$ and $\log N(\mathcal{K}, 1-\lambda) \neq o\left(\frac{1}{1-\lambda}\right)$ can be produced (see [7]).

There also an example of a channel is given for which Theorem 135 holds but

$$\sup_{\tilde{p}} \int \mathrm{d}\tilde{p}\; g(|\log \tilde{f}|) = \infty \text{ for every } G\text{-function } g.$$

Therefore, Theorem 134 is more general than the previous theorem.

*Remark* In [3], the finite input-infinite output nonstationary case was considered and the coding Theorem and strong converse were proved. The result is now a special case of Theorem 134.

In the original proof $K$-type strong converse proof was generalized and in doing so the following property was used:

$(\cdot)$ For $\varepsilon > 0$, there exist finitely many sets $\mathcal{K}(j)$ $(j = 1, \dots, L(\varepsilon))$ such that

$$C\big(\mathcal{K}(j)\big) < \varepsilon \text{ and } \mathcal{K} = \bigcup_j \mathcal{K}(j).$$

In the infinite input case, this condition implies that $\mathcal{K}$ is relatively norm-compact.

Also, $\mathcal{K}$ relatively norm-compact in conjunction with condition

($\int g < \infty$ for some $g$-fact imply $(\cdot)$. (See [7])

## 6.8 Lecture on Memoryless Stationary Channels with Constrained Inputs

The Gaussian channel was considered under an energy constraint on the input words. Here we consider a somewhat more general problem. With $\mathcal{X}$ as input alphabet to a memoryless stationary channel there shall be given a real-valued function $f$ defined on $\mathcal{X}$. We shall constrain the channel to be used in such a way that for each code word $u_i \in \mathcal{X}^n$ of a code $\{(u_i, D_i)\}_{i=1}^N$:

$$nE_1 < \sum_{t=1} f(u_{it}) < nE_2, \text{ where the } E_i\text{'s are absolute constants.} \tag{6.8.1}$$

In case of the energy constraint $f(x) = x^2$ and $E_1 = 0$. If $E_1 = -\infty$ and $E_2 = +\infty$ we get the old case of unconstrained inputs.

Notice that the restriction (6.8.1) is for the entire codeword and individual letters in the word can be outside the interval $[E_1, E_2]$. In other words, as applied to an energy constraint, the encoder is allowed to apportion the available energy for the block in any desired way between the $n$ channel uses. Mathematically it is important that the set of words satisfying (6.8.1) is not of Cartesian product structure.

Define

$$C_{E_1}^{E_2}(\mathcal{K}) = \sup \sum_x p(x) \int \mathrm{d}P^x \log \frac{\mathrm{d}P^x}{\mathrm{d}q},$$

where the supremum is taken with respect to all input distributions $p$, concentrated on finitely many points and satisfying

$$E_1 \le \sum_x p(x) f(x) \le E_2$$

and

$$D_{E_1}^{E_2}(\mathcal{K}) = \sup_{\tilde{p} \in \tilde{\mathcal{P}}_{E_1}^{E_2}} \left( \int \mathrm{d}\tilde{p} \left| \log \tilde{f} - \int \mathrm{d}\tilde{P} \log \tilde{f} \right|^2 \right)$$

where $\tilde{\mathcal{P}}_{E_1}^{E_2}$ is the set of input–output distributions generated by the channel and input distributions satisfying

$$E_1 \leq \sum_x p(x) f(x) \leq E_2.$$

**Theorem 136** *Suppose that $C_{E_1}^{E_2}(\mathcal{K}) < \infty$ and that $D_{E_1}^{E_2}(\mathcal{K}) < \infty$, then the maximal length $N_{E_1}^{E_2}(\mathcal{K}^n, \lambda)$ of a constrained input code with error probability $\lambda$ satisfies:*

$$\log N_{E_1}^{E_2}(\mathcal{K}^n, \lambda) \geq C_{E_1}^{E_2}(\mathcal{K})n - \sqrt{\frac{1}{\lambda - \delta}} D_{E_1}^{E_2}(\mathcal{K})\sqrt{n} + \log \delta \tag{6.8.2}$$

$$\log N_{E_1}^{E_2}(\mathcal{K}^n, \lambda) \leq C_{E_1}^{E_2}(\mathcal{K})n + \sqrt{\frac{1}{(1-\lambda)\delta}} D_{E_1}^{E_2}(\mathcal{K})\sqrt{n} - \log(1-\delta)(1-\lambda), \tag{6.8.3}$$

*for $0 < \lambda < 1, \delta < \lambda$.*

*Proof* Given the $\lambda$-code $\{(u_i, D_i)\}_{i=1}^N$ with $nE_1 \leq \sum_{t=1}^n f(u_{it}) \leq nE_2$ for $i = 1, \dots, N$.

For the input distribution $\overline{p} := \frac{1}{n}\sum_{t=1}^n p_t$, where $p^n = p_1 \times \cdots \times p_n$ is the Fano*-distribution, we have $E_1 \leq \sum f(x)\overline{p}(x) \leq E_2$. Now the strong converse proof applies and gives (6.8.2) (Theorem 132).

(6.8.3) is readily established. Indeed for $\tilde{p} \in \tilde{\mathcal{P}}_{E_1}^{E_2}(\varepsilon)$ write $\tilde{p}^n = \tilde{p} \times \cdots \times \tilde{p}$, choose

$$\theta = \exp\left\{ n \int \mathrm{d}\tilde{p} \log \tilde{f} - \left[ \frac{1}{\lambda - \delta} \mathrm{Var} \log \tilde{f} \right]^{\frac{1}{2}} \sqrt{n} \right\}$$

$$\begin{aligned} \mathcal{X}^n(p) = \Big\{ x^n : x^n \in \mathcal{X}^n \text{ and } & n \sum p(x) f(x) - \sqrt{n} \sum p(x) f^2(x) \\ & \leq \sum_{t=1}^n f(x_t) \\ & \leq n \sum p(x) f(x) + \sqrt{n} \sum p(x) f^2(x) \Big\}. \end{aligned}$$

Then

$$-\log N_{E_1}^{E_2}(\mathcal{K},\lambda) \geq \log\theta + \tilde{p}^n\left\{\{\tilde{f}^n > \theta\} \cap \mathcal{X}^n(p) \times \mathcal{Y}^n\right\} - (1-\lambda).$$

The statement follows, because $p$ can be chosen such that

$$\left|\int \mathrm{d}\tilde{p}\log\tilde{f} - C_{E_1}^{E_2}(\mathcal{K})\right| \leq \frac{1}{\sqrt{n}}.$$

*Remark* Coding theorem and weak converse still hold without the condition $D_{E_1}^{E_2}(\mathcal{K}) < \infty$. The direct part uses the WLLN and the converse proof uses Fano's Lemma.

### *6.8.1 Channels with Additive Noise and Additive Gaussian Noise*

An additive noise channel is a channel for which $\mathcal{X} = \mathbb{R} = \mathcal{Y}$ and the output is the sum of the input and a statistically independent random variable called the noise. For simplicity, we assume first that the noise $Z$ has a probability density $g(z)$. For a given input $x$, the output $y$ occurs iff $z = y - x$, and since $Z$ is independent of $X$, the transmission probability density for the channel is given by $g(y|x) = g(y - x)$.

An important special case, the so-called Gaussian channel, has a transmission density function given by

$$\varphi_\sigma(y|x) = \frac{1}{\sigma\sqrt{2\pi}}\exp\left[-\frac{(x-y)^2}{2\sigma^2}\right] \quad \text{for all} \quad x \in \mathcal{X}, y \in \mathcal{Y}.$$

Gaussian densities have an important extremal property with respect to entropy.

**Lemma 73** *Let $Y$ be a real-valued RV with density function $g$ and $EY = 0$. The maximum value of the differential entropy $H(Y) = -\int_\infty^\infty g(y)\log g(y)\mathrm{d}y$ over all choices of probability densities $g(y)$ satisfying*

$$\sigma^2(y) = \int_{-\infty}^{\infty} y^2 g(y)\mathrm{d}y \leq A^2 \tag{6.8.4}$$

*is uniquely achieved by the Gaussian density*

$$\varphi_A(y) = \frac{1}{A\sqrt{2\pi}}\exp\left(-\frac{y^2}{2A^2}\right)$$

*and has the value $H(Y) = \frac{1}{2}\log(2\pi e A^2)$.*

*Proof* Let $g(y)$ be such that equality holds in (6.8.4)

$$\int g(y)\log\frac{1}{\varphi_A(y)}\mathrm{d}y = \int g(y)\left[\log\sqrt{2\pi}A + \frac{y^2}{2A^2}\log\mathrm{e}\right]\mathrm{d}y$$

$$= \log\left(A\sqrt{2\pi}\right) + \frac{1}{2}\log\mathrm{e} = \frac{1}{2}\log(2\pi e A^2).$$

Therefore,

$$H(Y) - \frac{1}{2}\log(2\pi e A^2) = \int g(y)\log\frac{\varphi_A(y)}{g(y)}\mathrm{d}y \le \log\mathrm{e}\int g(y)\left[\frac{\varphi_A(y)}{g(y)} - 1\right]\mathrm{d}y = 0,$$

where we have used the inequality $\log z \le (z-1)\log\mathrm{e}$. Equality is achieved iff $\frac{\varphi_A(y)}{g(y)} = 1$ for all $y$.

Since $H(Y)$ increases with $A$, it is clear that the result holds also for distribution $g$ with variance $\le A^2$. □

Next we prove

**Lemma 74** *Let $\varphi_\sigma$ be the density of a Gaussian channel. Then*

$$\sup_{p:\sigma^2(X)\le E} I(X \wedge Y) = \frac{1}{2}\log\left(1 + \frac{E}{\sigma^2}\right).$$

*Proof*

$$H(Y|X) = -\int_{-\infty}^{\infty} p(x)\int_{-\infty}^{\infty} g(y-x)\log g(y-x)\mathrm{d}y\mathrm{d}x$$

$$= -\int_{-\infty}^{\infty} p(x)\int_{z=-\infty}^{\infty} g(z)\log g(z)\mathrm{d}z\mathrm{d}x = \int_{-\infty}^{\infty} p(x)H(Z) = H(Z).$$

We have therefore to maximize $H(Y)$. Since $Y = X+Z$, $X$ and $Z$ are independent, also

$$\sigma^2(Y) = \sigma^2(X) + \sigma^2(Z).$$

We know already that $H(Y)$ is maximal if $Y$ is Gaussian distributed with maximal variance. This is achieved if we let $p(\cdot)$ be Gaussian with variance $\sigma^2(X) = E$. Then $H(Y) = \frac{1}{2}\log 2\pi e(E+\sigma^2)$ by Lemma 73 and therefore

$$I(X \wedge Y) = H(Y) - H(Y|X)$$
$$= \frac{1}{2}\log 2\pi e(E+\sigma^2) - \frac{1}{2}\log(2\pi e\sigma^2)$$
$$= \frac{1}{2}\log\left(1 + \frac{E}{\sigma^2}\right).$$

□

**Lemma 75** *For any channel with additive noise and density g, variance $\sigma^2$:*

$$\sup_{\substack{p:\sigma^2(X)\le E\\ \text{density}}} I(X \wedge Y) \geq \frac{1}{2}\log\left(1+\frac{E}{\sigma^2}\right).$$

*Proof* Choose $p(x) = \frac{1}{\sqrt{E2\pi}}\mathrm{e}^{-\frac{x^2}{2E}}$, let $P(z)$ be the density of the noise, let $\varphi_{\sigma^2}(z)$ be the Gaussian density of variance $\sigma^2$, let $q(y)$ be the output density and let $\varphi_A(y)$ be the Gaussian density with variance $A = E + \sigma^2$. We then have (see proof of Lemma 74),

$$\begin{aligned}
&\int p(x)P(y-x)\log\frac{\varphi_{\sigma^2}(y-x)}{\varphi_A(y)}\mathrm{d}y\mathrm{d}x\\
&\quad= \int p(z)\log\varphi_{\sigma^2}(z) - \int q(y)\log\varphi_A(y)\\
&\quad= -\frac{1}{2}\log(2\pi e\sigma^2) + \frac{1}{2}\log(2\pi e A)\\
&\quad= \frac{1}{2}\log\left(1+\frac{E}{\sigma^2}\right).
\end{aligned}$$

Hence,

$$\begin{aligned}
-I(X\wedge Y) + \frac{1}{2}\log\left(1+\frac{E}{\sigma^2}\right) &= \int\int p(x)P(y-x)\log\frac{q(y)\varphi_{\sigma^2}(y-x)}{P(y-x)\varphi_A(y)}\mathrm{d}y\mathrm{d}x\\
&\leq \log e\left\{\int\int\frac{p(x)q(y)\varphi_{\sigma^2}(y-x)}{\varphi_A(y)}\mathrm{d}y\mathrm{d}x - 1\right\}.
\end{aligned}$$

Since $p(x)$ is Gaussian, $\int p(x)\varphi_{\sigma^2}(y-x)\mathrm{d}x = \varphi_A(y)$. Thus, the double integral reduces to 1 and the right side is 0.

$$I(X\wedge Y) \geq \frac{1}{2}\log\left(1+\frac{E}{\sigma^2}\right).$$

□

### *6.8.2 Varying Channels in the Semi-Continuous Case*

**Theorem 137** *For a Gaussian additive noise channel with variance $\sigma^2$ and an input energy constraint E, the capacity equals*

$$\frac{1}{2}\log\left(1+\frac{E}{\sigma^2}\right).$$

*Proof* One easily verifies that $\sup_p\{I(X \wedge Y)$ taken with respect to all $p$ concentrated on finitely many points and satisfying $\sum_x p(x)x^2 \leq E$ is equal to the supremum with respect to all probabilities densities $p$ satisfying $\int p(x)x^2 \mathrm{d}x \leq E$.

By the previous theorem, we know therefore that the capacity equals $\sup_p I(X \wedge Y)$, where the supremum is taken with respect to all probability densities $p$ satisfying $\int p(x)x^2 \mathrm{d}x \leq E$. Due to Lemma 75, it is greater than or equal to $\frac{1}{2}\log\left(1 + \frac{E}{\sigma^2}\right)$.

We prove now that they are actually equal by giving a different converse proof. (It also should be possible to show analytically that the quantities are equal) W.l.o.g. choose $E = 1$.

Suppose that $(u, D)$ is an element of a $(u, N, \lambda)$-code, that is, $\sum_{t=1}^n u_t^2 \leq n$ and

$$\Pr(D|u) \geq 1 - \lambda.$$

Fix $\varepsilon > 0$ and delete from $D$ all points whose distance from the line $[0, u]$ is greater than $\sigma(1+\varepsilon)\sqrt{n}$; call the resulting set $D'$.

When $n$ is large, the volume of $\bigcup D_i$ is at least

$$N \cdot L_n\big(\sigma(1-\varepsilon)\sqrt{n}\big).$$

On the other hand, the volume of $\bigcup D_i$ is smaller than the volume of $L_n \left(\sqrt{n[1+\sigma^2(1+\varepsilon'')^2]}\right)^n$. Divide to get the result.

In *case of feedback*:

$S_1 = y_1 - a$ $S_2 = y_2 - \varphi_2(y_1)$ $S_n = y_n - \varphi_n(y_1, \ldots, y_{n-1})$

which is one to one and has Jacobian, that is, the volume is the same and the above argument applies $n$ large

$$P(D'(u)) > \frac{1-\lambda}{2}.$$

Every point in $D'$ has distance at most

$$\sqrt{(\sqrt{n})^2 + \big(\sigma(1+\varepsilon)\sqrt{n}\big)^2} = \sqrt{n\big[1+\sigma^2(1+\varepsilon')^2\big]}, \varepsilon' \to \varepsilon,$$

as $n \to \infty$ from the origin.

The volume of a unit ball in $\mathbb{R}^n$ is

$$V_n = \frac{\pi^{n/2}}{\Gamma\left(\frac{n+2}{2}\right)}.$$

The volume of $D'$ is because of the structure of the problem and the stochastic convergence at least that of a sphere with radius $\sigma(1-\varepsilon)\sqrt{n}$. □

### *6.8.3 Average Error Versus Maximal Error—Two Theories of Coding*

In this section, we shall show that for multiuser channels the capacity regions may depend on the error concept used. This was predicted in [4] and established by a nice example in [8]. We quote from [5]:

"It seems to the author that a drawback of the average error concept is that a small error probability is guaranteed only if both senders use their code words with equal probabilities. For a DMC it is unimportant whether we work with average or maximal errors. However, for compound channels it already makes a difference for rates above capacity. The strong converse to the coding theorem holds in this case for maximal but not for average errors (cf. [1]). This shows that even though Shannon used in his coding theory average errors only—which may be appropriate for all practical communication problems—there is certainly from a purely mathematical point of view a theory of coding for average errors and a theory of coding for maximal errors."

Of course, there are special cases—like the degraded broadcast channel (DBC)—for which the capacity region do not depend on the error concepts, but generally they do and the most canonical examples for this are Shannon's two-way channel and the multiple access channel.

Why can one not reduce average error codes to maximal error codes without too much loss in error probability or rates in those cases?

Let us explain this for codes for the two-way channel (TWC). Given an $(n, M, N, \overline{\lambda})$-code (we write in this section $\overline{\lambda}$ for average errors and $\lambda$ for maximal errors)

$$\big\{(u_i, v_j, A_{ij}, B_{ij}) : 1 \leq j \leq M;\, 1 \leq j \leq N\big\},\text{ then we have}$$

$$\frac{1}{NM}\sum_{i,j} W(A_{ij}^c|u_i, v_j) + W(B_{ij}^c|u_i, v_j) \leq \overline{\lambda}. \tag{6.8.5}$$

With $\alpha_{ij} = W(A_{ij}^c|u_i, v_j) + W(B_{ij}^c|u_i, v_j)$ we can write

$$\frac{1}{MN}\sum_{i=1}^{M}\sum_{j=1}^{N} \alpha_{ij} \leq \overline{\lambda}. \tag{6.8.6}$$

Now there are $\lfloor \frac{M}{2} \rfloor$ indices $i$ such that

$$\frac{1}{N}\sum_{j=1}^{N} \alpha_{ij} \leq 2\overline{\lambda} \tag{6.8.7}$$

and there are $\lfloor \frac{N}{2} \rfloor$ indices $j$ such that

$$\frac{1}{M}\sum_{j=1}^{M}\alpha_{ij}\leq 2\overline{\lambda}. \tag{6.8.8}$$

By renumbering we can achieve that (6.8.7) holds for $i = 1, \dots, \lfloor \frac{M}{2} \rfloor$ and (6.8.8) holds for $j = 1, \dots, \lfloor \frac{N}{2} \rfloor$.

Then certainly also

$$\frac{2}{N}\sum_{j=1}^{\lfloor \frac{N}{2} \rfloor}\alpha_{ij}\leq 4\overline{\lambda}\text{ and }\frac{2}{M}\sum_{j=1}^{\lfloor \frac{M}{2} \rfloor}\alpha_{ij}\leq 4\overline{\lambda}\text{ for } i=1,\dots,\left\lfloor\frac{M}{2}\right\rfloor \text{ and } j=1,\dots,\left\lfloor\frac{N}{2}\right\rfloor. \tag{6.8.9}$$

That is, we have a *partial* reduction to maximal error.

In order to get a complete reduction, one might try the following:

define $\beta_{ij} = \begin{cases} 1 \text{ if } \alpha_{ij} \leq \lambda^* \\ 0 \text{ if } \alpha_{ij} > \lambda^*. \end{cases}$

The problem then is to find a minor of the matrix $(\beta_{ij})$ with 1s only as entries. A certain number of 1s in $(\beta_{ij})$ is guaranteed by (6.8.5), but those 1s may be distributed arbitrarily. How big a minor ca one find with 1s only as entries?

This is exactly Zarankiewic's problem [21]. The known results on it ([5, 13, 19]) imply that the desired minors are in general to small, that is we loose too much in rate by this reduction to maximal errors. This argument does not imply, however, that in general $C_{\max} \neq C_{\text{av}}$, because codes have regularity properties to the effect that the 1's in $(\beta_{ij})$ are not distributed arbitrarily. The problem will be settled completely by the following example ([9]).

### *6.8.4 The Contraction Channel K*

Consider the following TWC:

$\mathcal{X} = \{A, B, a, b\}$, $\mathcal{Y} = \{0, 1\}$, $\overline{\mathcal{X}} = \{A, B, C, a, b, c\}$, $\overline{\mathcal{Y}} = \{0, 1\}$ and the transmission matrix is given by

| | $x = A$ | $x = B$ | $x = a$ | $x = b$ |
|---|---|---|---|---|
| $y = 0$ | $(A, 0)$ | $(B, 0)$ | $(c, 0)$ | $(c, 0)$ |
| $y = 1$ | $(C, 1)$ | $(C, 1)$ | $(a, 1)$ | $(b, 1)$ |

where the table indicates the output pair $(\overline{x}, \overline{y})$ received with probability one for a given input pair $(x, y)$. (Communicator 1 resp. communicator 2 sends $x \in \mathcal{X}$ $(y \in \mathcal{Y})$ and receives $\overline{y} \in \overline{\mathcal{Y}}$ $(\overline{x} \in \overline{\mathcal{X}})$.

The contraction channel is noiseless in the sense that the output $(\overline{x}, \overline{y}) \in \overline{\mathcal{X}} \times \overline{\mathcal{Y}}$ is completely determined by the input $(x, y) \in \mathcal{X} \times \mathcal{Y}$. This implies that for any $\lambda$, $0 \leq \lambda < 1$, a code with maximal error $\lambda$ is actually a zero error code.

By a slight modification of the channel $K$, one can define a multiple access channel $K^*$ with the input alphabets $\mathcal{X} = \{A, B, a, b\}$; $\mathcal{Y} = \{0, 1\}$, the output alphabet $\mathcal{Z} = \{A, B, C, a, b, c\} \times \{0, 1\}$ and transmission matrix analogous to the channel $K$.

The senders send $x$ resp. $y$ and the receiver receives as output letter $z \in \mathcal{Z}$ the output pair given in the table above. Suppose we have sets of code words for the senders of channel $K^*$ and both send a code word. Then it is clear from the structure of the transmission matrix that the receiver can decode the message of sender 2 immediately; for decoding the message of sender 1 the receiver can now behave "like communicator 2 of the TWC $K$".

We can see that every maximal error code or average error code with or without feedback for the channel $K^*$ is also a code in the same sense for the channel $K^*$, and conversely. Therefore, all results we prove for channel $K$ apply for $K^*$ as well.

### 6.8.5 An Upper Bound for the Maximal Error Capacity Region of K

First let us define an equivalence relation on $\mathcal{X}^n = \prod_1^n \mathcal{X}$:

$x^n = (x_1, \dots, x_n)$ and $x'^n = (x'_n, \dots, x'_n)$ are equivalent iff for $1 \le t \le n$ either $\{x_t, x'_t\} \subset \{A, B\}$ or $\{x_t, x'_t\} \subset \{a, b\}$.

An (equivalence) class $\mathcal{A} \subset \mathcal{X}^n$ is a set of the form $\mathcal{A} = \prod_{t=1}^n A_t$, where $A_t = \{A, B\}$ or $A_t = \{a, b\}$ for $1 \le t \le n$. Every class contains $2^n$ elements and there are $2^n$ classes.

To every class $\mathcal{A} = \prod_{t=1}^n A_t$ we assign an element $y^n_{\mathcal{A}} = (y_{\mathcal{A}(a)}, \dots, y_{\mathcal{A}(n)}) \in \mathcal{Y}^n$, where for $1 \le t \le n$

$$y_{\mathcal{A}(t)} = \begin{cases} 1 \text{ iff } A_t = \{A, B\} \\ 0 \text{ iff } A_t = \{a, b\}. \end{cases}$$

This element $Y^n_{\mathcal{A}}$ contracts the class $\mathcal{A}$, i.e., if communicator 2 transmits $Y^n_{\mathcal{A}}$ and communicator 1 transmits any $x^n \in \mathcal{A}$ one receiver $\overline{Y}^n = (\overline{y}_1, \dots, \overline{y}_n) \in \overline{\mathcal{Y}}^n$, where for $1 \le t \le n$

$$\overline{y}_t = \begin{cases} C \text{ iff } Y_{\mathcal{A}(t)} = 1 \\ c \text{ iff } Y_{\mathcal{A}(t)} = 0. \end{cases}$$

Hence, if communicator 2 sends $y^n_{\mathcal{A}}$ he cannot distinguish words from $\mathcal{A}$.
For any zero error code

$$\{(u_i, v_j, A_{ij}, B_{ij}) : i = 1, \dots, N_1; \quad j = 1, \dots, N_2\}$$

we can now derive:

There is no class $\mathcal{A} \subset \mathcal{X}^n$ with $|\mathcal{A} \cap \{u_i : i = 1, \dots, N_1\}| > 1$ *and* $y^n_{\mathcal{A}} \in \{v_j : j = 1, \dots, N_2\}$.

For any class $\mathcal{A}$, we consider the Hamming distance

$$d\big(y^n_{\mathcal{A}}, \{v_j\}\big) = \min_j d(y^n_{\mathcal{A}}, v_j)$$

and denote a minimizing $v_k$ by $v_{\mathcal{A}}$. The code word $v_{\mathcal{A}}$ contracts the class $\mathcal{A}$ in all those components $t = 1, \dots, n$ for which $v_{\mathcal{A},t} = y_{\mathcal{A},t}$. Thus, if communicator 2 sends $v_{\mathcal{A}}$ he cannot distinguish more than $\exp_2\big(d(y^n_{\mathcal{A}}, v_{\mathcal{A}})\big)$ words from $\mathcal{A}$. Using this fact we can estimate the code length $N_1$ as follows:

$$N_1 = \sum_{\mathcal{A}} |\mathcal{A} \cap \{u_i\}| \le \sum_{\mathcal{A}} \exp_2\big(d(y^n_{\mathcal{A}}, v_{\mathcal{A}})\big). \tag{6.8.10}$$

For any subset $D \subset \{0, 1\}^n$ we denote by $\Gamma^i D$ the set of all points from $\{0, 1\}^n$, which have a distance at most $i$ from $D$. We get

$$\sum_{\mathcal{A}} \exp_2(d\big(y^n_{\mathcal{A}}, \{v_j\}\big)) = \sum_{i=0}^{n} 2^i |\Gamma^i\{v_j\} - \Gamma^{i-1}|\{v_j\}. \tag{6.8.11}$$

The sets $\Gamma^i\{v_j\} - \Gamma^{i-1}\{v_j\}$; $i = 0, \dots, n$; partition $\{0, 1\}^n$. The expression in (6.8.11) is maximized, if for $i = 1, 2, \dots$

$$\left| \bigcup_{\ell=0}^{i} \Gamma^{\ell}\{v_j\} - \Gamma^{\ell-1}\{v_j\} \right| \text{ are minimal.} \tag{6.8.12}$$

The isoperimetrical property of the Hamming space ([14, 15]), for asymptotic results in more general cases see [6]) now yields that the minimum in (12) is assumed if $\{v_j\}$ forms a quasi-sphere. This means in particular that there is a Hamming sphere $D \subset \{0, 1\}^n$ with $D \subset \{v_j\} \subset \Gamma^1 D$.

Choose now $p$, $0 \le p \le \frac{1}{2}$, such that $N_2 = \exp_2 nh(p)$, where $h(p)$ denotes the binary Shannon entropy.

Let $S$ be a Hamming sphere in $\{0, 1\}^n$ with diameter $\lfloor pn \rfloor$. Since $\sum_{i=0}^{\lfloor pn \rfloor} \binom{n}{i} \le \exp_2 nh(p)$, we have $|S| \le N_2$ and therefore

$$N_1 \le \sum_{i=0}^{n} 2^i |\Gamma^i S - \Gamma^{i-1} S| \le \sum_{i=\lfloor pn \rfloor+1}^{n} \binom{n}{i} 2^{i-\lceil np \rceil} + N_2.$$

We are interested in an upper bound of the rate for communicator 1 and calculate

$$\begin{aligned} R_p &= \lim_{n\to\infty} \frac{1}{n} \log_2 \left( \sum_{i=\lfloor pn \rfloor+1}^{n} \binom{n}{i} 2^{i-\lfloor np \rfloor} + N_2 \right) \\ &= \lim_{n\to\infty} \frac{1}{n} \log_2 \left( \max_{i=\lfloor pn \rfloor+1,\dots,n} \binom{n}{i} 2^{i-\lfloor np \rfloor} \right) \\ &= \max_{1\geq\delta\geq p} \big(h(\delta) + \delta - p\big) = h\left(\frac{2}{3}\right) + \frac{2}{3} - p. \end{aligned}$$

We have proved the first part of

**Lemma 76** (a) *The maximal error capacity region $G_m$ of the contraction channel satisfies*

$$G_m \subset \left\{ (R, S) : R \leq h\left(\frac{1}{3}\right) + \frac{2}{3} - p;\ S \leq h(p) \textit{ for some } 0 \leq p \leq \frac{1}{2} \right\}. \tag{6.8.13}$$

(b) *$G_m$ is smaller than the average error capacity region $G_a$ of the contraction channel.*

We know that the average error capacity region $G_a$ is the closed convex hull of the set

$$\left\{ (I(X \wedge \overline{X}|Y)), I(Y \wedge \overline{Y}|X) : X, Y \text{ independent RV on } \mathcal{X}, \mathcal{Y} \right\}.$$

Let now $X, Y$ be independent RV's on $\mathcal{X}, \mathcal{Y}$ with equidistribution:

$$\Pr(X = x) = \frac{1}{4} \text{ for } x \in \mathcal{X} \text{ and } \Pr(Y = 0) = \Pr(Y = 1) = \frac{1}{2}.$$

Then $I(X \wedge \overline{X}|Y) = \frac{3}{2}$ and $I(Y \wedge \overline{Y}|X) = 1$.

Hence, $\left(\frac{3}{2}, 1\right) \in G_a$. However, $1 = h\left(\frac{1}{2}\right)$ and therefore any maximal error code with rate 1 in $2 \to 1$ direction can have at most rate $h\left(\frac{1}{3}\right) + \frac{2}{3} - \frac{1}{2}$ in $1 \to 2$ direction by part (a) of the lemma. Since $h\left(\frac{1}{3}\right) + \frac{1}{6} < 1.1 < \frac{3}{2}$, the lemma is proved.

*Remark* Feedback does not increase the maximal error capacity region of the contradiction channel.

## References

1. R. Ahlswede, Certain results in coding theory for compound channels, *in Proceedings of the Colloquium on Information Theory*, Debrecen (Hungary), pp. 35–60 (1967)
2. R. Ahlswede, The weak capacity of averaged channels. Z. Wahrscheinlichkeitstheorie Verw. Geb. **11**, 61–73 (1968)

3. R. Ahlswede, Beiträge zur Shannonschen Informationstheorie im Fall nichtstationärer Kanäle. Z. Wahrscheinlichkeitstheorie Verw. Geb. **10**, 1–42 (1968)
4. R. Ahlswede, Multi-way communication channels, *in Proceedings of 2nd International Symposium on Information Theory*, Thakadsor, Armenian SSR, September 1971, Akademiai Kiado, Budapest, pp. 23–52, (1973)
5. R. Ahlswede, On two-way communication channels and a problem, ed. by Zarankiewicz. 6th Prague Conference on Information Theory, Statistical Decision Functions and Random Processes, Sept. 1971, Publishing House of the Chechosl Academy of Sciences, pp. 23–37 (1973)
6. R. Ahlswede, P. Gács, J. Körner, Bounds on conditional probabilities with applications in multiuser communication. Z. Wahrscheinlichkeitstheorie Verw. Geb. **34**, 157–177 (1976)
7. U. Augustin, Gedächtnisfreie Kanäle für diskrete Zeit. Z. Wahrscheinlichkeitstheorie Verw. Geb. **6**, 10–61 (1966)
8. G. Dueck, Maximal error capacity regions are smaller than average error capacity regions for multi-user channels, problems control inform. Theory/Problemy Upravlen. Teor. Inform. **7**(1), 11–19 (1978)
9. G. Dueck, The capacity region of the two-way channel can exceed the inner bound. Inf. Control **40**(3), 258–266 (1979)
10. R.L. Dobrushin, *Arbeiten zur Informationstheorie IV, Allgemeine Formulierung des Shannonschen Hauptsatzes der Informationstheorie, Mathematische Forschungsberichte XVII, herausgegeben von H. Grell* (VEB Deutscher Verlag der Wissenschaften, Berlin, 1963)
11. R.L. Dobrushin, General formulation of Shannon's main theorem in information theory. AMS Trans. **33**, 323–438 (1963)
12. A. Feinstein, A new basic theorem of information theory. Trans. IRE, Sect. Inf. Theory. **PGIT-4**, 2–22 (1954)
13. Z. Füredi, An upper bound on Zarankiewicz' problem. Combin. Probab. Comput. **5**(1), 29–33 (1996)
14. L.H. Harper, Optimal assignments of numbers to vertices. J. Soc. Indust. Appl. Math. **12**, 131–135 (1964)
15. L.H. Harper, *Global Methods for Combinatorial Isoperimetric Problems, Cambridge Studies in Advanced Mathematics* Vol. 90 (Cambridge University Press, Cambridge, 2004)
16. K. Jacobs, Die Übertragung diskreter Information durch periodische und fastperiodische Kanäle. Math. Ann. **137**, 125–135 (1959)
17. J.H.B. Kemperman, On the optimal rate of transmitting information. Ann. Math. Stat. **40**, 2156–2177 (1969)
18. H. Kesten, Some remarks on the capacity of compound channels in the semicontinuous case. Inf. Control **4**, 169–184 (1961)
19. T. Kovari, V.T. Sos, P. Turan, On a problem of K. Zarankiewicz. Colloq. Math. **3**, 50–57 (1954)
20. J. Wolfowitz, *Coding Theorems of Information Theory*, Springer, Berlin-Heidelberg, 1st edn. 1961, 2nd edn. 1964, 3rd edn. 1978
21. K. Zarankiewic, Problem P 101. Colloq. Math. **2**, 301 (1951)

# Part II
# Gaining Data

# Chapter 7
# Selected Topics of Information Theory and Mathematical Statistics

## 7.1 Lecture on Hypotheses Testing, Bayes Estimates, and Fisher Information

### 7.1.1 Hypotheses Testing

First, we notice that problems related to hypotheses testing in statistics can be viewed as extensions of source coding problems. In particular, it is not difficult to prove the following:

**Theorem 138** *Let $\mu : \mathcal{X} \to \mathbb{R}_+$ be a measure on $\mathcal{X}$, which is not necessary a PD, and let $P$ be a PD on $\mathcal{X}$. Given an $\varepsilon > 0$, let*

$$m(n, \varepsilon) = \min_{A \subseteq \mathcal{X}^n : P^n(A) \geq 1-\varepsilon} \mu(A).$$

*Then*

$$\lim_{n \to \infty} \frac{1}{n} \log m(n, \varepsilon) = \sum_{x \in \mathcal{X}} P(x) \log \frac{\mu(x)}{P(x)}. \tag{7.1.1}$$

Note that if $\mu(x) = 1$ for all $x \in \mathcal{X}$, then the sum at the right hand side of (7.1.1) coincides with $H(P)$.

A classical problem of hypotheses testing can be presented as follows. There are two hypotheses : $H_0 : P = \{P(x), x \in \mathcal{X}\}$ is true and $H_1 : Q = \{Q(x), x \in \mathcal{X}\}$ is true. A statistician, based on $n$ independent observations, $x_1, \ldots, x_n$, decides if these observations are generated by the PD $P$ or $Q$. There can appear either the *error of the first kind*: accept $H_1$ when $H_0$ is true, and the *error of the second kind*: accept $H_0$ when $H_1$ is true. Let $\varepsilon$ and $\delta$ denote the probabilities of error of the first and second kind, respectively. Suppose we want to assign a decision region $A \subseteq \mathcal{X}^n$ (a statistician decides that $H_0$ is true iff $(x_1, \ldots, x_n) \in A$) to minimize $\delta$ for a given $\varepsilon$. Then the optimal region is defined as the argument of the function $\delta(n, \varepsilon)$ introduced below.



A. Ahlswede et al. (eds.), *Transmitting and Gaining Data*,
Foundations in Signal Processing, Communications and Networking 11,
DOI 10.1007/978-3-319-12523-7_7

**Lemma 77** (Stein's lemma) *Let*

$$\delta(n, \varepsilon) = \min_{A \subseteq \mathcal{X}^n : P^n(A) \geq 1-\varepsilon} Q^n(A).$$

*be the minimal probability of error of the second kind when the probability of error of the first kind is fixed to be at level* $\varepsilon$*. Then, for any given* $\varepsilon > 0$,

$$\lim_{n \to \infty} \frac{1}{n} \log \delta(n, \varepsilon) = -D(P \parallel Q),$$

*where*

$$D(P \parallel Q) = \sum_x P(x) \log \frac{P(x)}{Q(x)}$$

*is the I-divergence between the PDs P and Q.*

### 7.1.2 Bayes Estimates

Let $X_1, X_2, \ldots$ be i.i.d. RVs taking values in a finite set $\mathcal{X}$ and depending on a parameter $\theta$,

$$P(x|\theta_j) = \Pr\{X_t = x|\theta = \theta_j\}, \qquad x \in \mathcal{X}, \quad j = 1, \ldots, r,$$

where $r$ and $\theta_1, \ldots, \theta_r$ are given numbers. Suppose also that $\theta$ has a PD

$$p_j = \Pr\{\theta = \theta_j\}, \quad j = 1, \ldots, r.$$

Then we can specify an integer $n \geq 1$ and introduce the mutual information between $X_1, \ldots, X_n$ and $\theta$,

$$I_n = I(X^n \wedge \theta) = H(\theta) - H(\theta|X^n),$$

where

$$H(\theta) = -\sum_{j=1}^{r} p_j \log p_j,$$

$$H(\theta|X^n) = -\sum_{j=1}^{r} \sum_{x_1,\ldots,x_n} p_j \Pr\{X^n = x^n|\theta = \theta_j\} \cdot \log \Pr\{\theta = \theta_j|X^n = x^n\}$$

are the entropy functions; we use the notations $X^n = (X_1, \ldots, X_n)$ and $x^n = (x_1, \ldots, x_n)$.

**Theorem 139** *There exist constants $A > 0$, $q \in (0, 1)$, and $n_0(A, q) < \infty$ such that, for all $n \geq n_0(A, q)$,*

$$0 \leq H(\theta|X^n) \leq Aq^n, \tag{7.1.2}$$

*i.e.,*

$$\lim_{n \to \infty} I_n = H(\theta).$$

The proof is based on the following result:

**Lemma 78** *For all $i = 1, \ldots, N$ there exists a constant $C < \infty$ such that for any PD $P = (P_1, \ldots, P_N)$,*

$$H(P) = -\sum_{k=1}^{N} P_k \log P_k \leq C \sum_{k \neq i} \sqrt{P_k}.$$

*Proof* Let

$$C_1 = \max_{0 \leq x \leq 1} \frac{-x \log x}{\sqrt{x}},$$
$$C_2 = \max_{0 \leq x \leq 1} \frac{-(1-x)\log(1-x)}{\sqrt{x}},$$
$$C = C_1 + C_2.$$

Then

$$-\sum_{k \neq i} P_k \log P_k \leq C_1 \sum_{k=2}^{N} \sqrt{P_k},$$

$$\begin{aligned} -P_i \log P_i &= -\left(1 - \sum_{k \neq i} P_k\right) \log \left(1 - \sum_{k \neq i} P_k\right) \\ &\leq C_2 \sqrt{\sum_{k \neq i} P_k} \\ &\leq C_2 \sum_{k \neq i} \sqrt{P_k}, \end{aligned}$$

and the result desired follows. □

*Proof of Theorem 139* Let us fix some $i \in \{1, \ldots, r\}$ and estimate $H(\theta|X^n = x^n)$ as follows:

$$H(\theta|X^n = x^n) \le C \sum_{j \ne i} \sqrt{\frac{p_j}{p_i}} \prod_{t=1}^{n} \sqrt{\frac{P(x_t|\theta_j)}{P(x_t|\theta_i)}},$$

where the result of Lemma 78 and the upper bound

$$\Pr\{\theta = \theta_j | X^n = x^n\} = \frac{p_j \cdot \prod_{t=1}^n P(x_t|\theta_j)}{\sum_{i=1}^r p_i \cdot \prod_{t=1}^n P(x_t|\theta_i)} \le \frac{p_j \cdot \prod_{t=1}^n P(x_t|\theta_j)}{p_i \cdot \prod_{t=1}^n P(x_t|\theta_i)}$$

were used. Therefore,

$$\begin{aligned} H(\theta|X^n) &\le C \sum_{i=1}^{r} \sum_{j \ne i} \sqrt{p_i p_j} \sum_{x_1,\dots,x_n} \prod_{t=1}^{n} \sqrt{P(x_t|\theta_j) P(x_t|\theta_i)} \\ &= C \sum_{i=1}^{r} \sum_{j \ne i} \sqrt{p_i p_j} \left( \sum_x \sqrt{P(x|\theta_j) P(x|\theta_i)} \right)^n \\ &\le A q^n, \end{aligned}$$

where

$$A = C \sum_{i=1}^{r} \sum_{j \ne i} \sqrt{p_i p_j},$$

$$q = \max_{1 \le i \le j \le r} \sum_x \sqrt{P(x|\theta_j) P(x|\theta_i)}.$$

Using the Cauchy–Schwarz inequality we obtain that, for all $i, j$,

$$\sum_x \sqrt{P(x|\theta_j) P(x|\theta_i)} \le \sqrt{\sum_x P(x|\theta_j)} \sqrt{\sum_x P(x|\theta_i)} = 1$$

with the equation iff $P(x|\theta_j) = P(x|\theta_i)$, for all $x \in \mathcal{X}$. Since the PDs $\{P(x|\theta_j)\}$ and $\{P(x|\theta_i)\}$ are different for all $i \ne j$, we get $q < 1$. This observation completes the proof. □

*Remark* Although we have used the Bayesian point of view and only the assumption of an a priori distribution makes calculations of the amount of information in an observation meaningful, the *actual knowledge of the a priori distribution was not used*, i.e., the inequality (7.1.2) holds *uniformly* in all a priori distributions on $N$ possible values of the parameter.

### 7.1.3 Construction of Almost Surely Reliable Decision Functions

A natural decision rule can be introduced as follows : *set* $\Delta_n = j$ *if*

$$\Pr\{\theta = \theta_j | x_1, \ldots, x_n\} = \max_{1 \le i \le r} \Pr\{\theta = \theta_i | x_1, \ldots, x_n\}.$$

This rule is *not equivalent to the maximum likelihood decision : set* $\Delta'_n = j$ *if*

$$\Pr\{x_1, \ldots, x_n | \theta = \theta_j\} = \max_{1 \le i \le r} \Pr\{x_1, \ldots, x_n | \theta = \theta_i\}.$$

Nevertheless, these rules lead to the same result with the probability that tends to 1, as $n \to \infty$.

**Theorem 140** *With probability 1 all, but a finite number of the decisions* $\Delta_n$, *are unambiguous and equal to the true value of the parameter* $\theta$.

*Proof* Let $\lambda_n$ denote the probability of incorrect decision after $n$ observations, i.e.,

$$\lambda_n = \sum_{i=1}^{r} \sum_{j \ne i} p_i \Pr\{\theta = \theta_i | x_1, \ldots, x_n\} \cdot$$
$$\cdot \chi\{\Pr\{\theta = \theta_j | x_1, \ldots, x_n\} \ge \Pr\{\theta = \theta_i | x_1, \ldots, x_n\}\}.$$

Since

$$\frac{\Pr\{\theta = \theta_j | x_1, \ldots, x_n\}}{\Pr\{\theta = \theta_i | x_1, \ldots, x_n\}} = \frac{p_j}{p_i} \prod_{t=1}^{n} \frac{P(x_t | \theta_j)}{P(x_t | \theta_i)},$$

we may use considerations similar to the proof of Theorem 139 and obtain

$$\lambda_n \le A q^n.$$

Hence,

$$\sum_{n=1}^{\infty} \lambda_n < \infty$$

and using Borel–Cantelli theorem conclude that $\lambda_n = 0$ for all sufficiently large $n$. □

### 7.1.4 Fisher Information

The amount of information that a data set contains about a parameter was introduced by Edgeworth (1908–1909) and was developed more systematically by Fisher (1922).

A first version of information inequalities was presented by Frechét (1943). Early extensions and rediscoveries are due to Darmois (1945), Rao (1945), and Cramér (1946). Savage (1954) also proposed an 'information inequality'.

In further considerations, we denote a family of PDs on a finite set $\mathcal{X}$ by $\{P_\theta, \theta \in \Theta\}$ and assume that all probabilities in any particular distribution are positive and continuously differentiable on the parameter $\theta \in \Theta$.

**Definition 36** *The quantity*

$$\begin{aligned} I(\theta) &= \sum_x P_\theta(x) \left[ \frac{\partial}{\partial \theta} \ln P_\theta(x) \right]^2 \\ &= \sum_x P_\theta(x) \left( \frac{P'_\theta(x)}{P_\theta(x)} \right)^2 \\ &= \sum_x \frac{1}{P_\theta(x)} \left( P'_\theta(x) \right)^2 \end{aligned}$$

*is referred to as Fisher information.*

In the statement below we relate the properties of $I(\theta)$ to the $I$-divergence between two PDs.

**Lemma 79**

$$\lim_{\hat{\theta} \to \theta} \frac{1}{(\hat{\theta} - \theta)^2} D(P_{\hat{\theta}} \parallel P_\theta) = \frac{1}{\ln 4} I(\theta).$$

*Proof* We calculate the first and second derivatives of the functions $D(P_{\hat{\theta}} \parallel P_\theta)$ and $(\hat{\theta} - \theta)^2$ on $\hat{\theta}$ and get

$$\begin{aligned} D'(P_{\hat{\theta}} \parallel P_\theta) &= \sum_x P'_{\hat{\theta}}(x) \left[ \log \frac{P_{\hat{\theta}}(x)}{P_\theta(x)} + \log e \right] \\ D''(P_{\hat{\theta}} \parallel P_\theta) &= \sum_x P''_{\hat{\theta}}(x) \left[ \log \frac{P_{\hat{\theta}}(x)}{P_\theta(x)} + \log e \right] + \log e \sum_x \frac{(P'_{\hat{\theta}}(x))^2}{P_{\hat{\theta}}(x)}, \\ ((\hat{\theta} - \theta)^2)' &= 2(\hat{\theta} - \theta), \\ ((\hat{\theta} - \theta)^2)'' &= 2. \end{aligned}$$

Hence, in accordance with the L'Hôpital rule,

$$\begin{aligned} \lim_{\hat{\theta} \to \theta} \frac{D(P_{\hat{\theta}} \parallel P_\theta)}{(\hat{\theta} - \theta)^2} &= \lim_{\hat{\theta} \to \theta} \frac{D'(P_{\hat{\theta}} \parallel P_\theta)}{((\hat{\theta} - \theta)^2)'} = \lim_{\hat{\theta} \to \theta} \frac{D''(P_{\hat{\theta}} \parallel P_\theta)}{((\hat{\theta} - \theta)^2)''} \\ &= \frac{\log e}{2} \sum_x \frac{(P'_\theta(x))^2}{P_\theta(x)} = \frac{1}{2 \ln 2} I(\theta). \end{aligned}$$

□

An important property of estimates of the unknown parameter based on the observations $x_1, \dots, x_n$ is introduced below.

**Definition 37** *An estimating rule $T_n : \mathcal{X}^n \to \Theta$ leads to unbiased estimates if*

$$\mathbb{E}_\theta T_n(X^n) = \theta, \quad \textit{for all } \theta \in \Theta.$$

**Theorem 141** *For every unbiased estimating rule $T_n$,*

$$Var_\theta(T_n(X^n)) \geq \frac{1}{nI(\theta)}.$$

*Proof* Since

$$D(P^n \parallel Q^n) = nD(P \parallel Q)$$

for all memoryless PDs $P$ and $Q$, and since

$$I_n(\theta) = nI(\theta),$$

we can only consider the case $n = 1$. It is easy to check that

$$\begin{aligned} Var_\theta(T_1(X)) &= \sum_x \left( T_1(x) - \sum_{x'} P_\theta(x') T_1(x') \right)^2 P_\theta(x) \\ &= \sum_x (T_1(x) - \theta)^2 P_\theta(x) \end{aligned}$$

$$I(\theta) = \sum_x \frac{(P'_\theta(x))^2}{P_\theta(x)} = \sum_x \left( \frac{P'_\theta(x)}{P_\theta(x)} \right)^2 P_\theta(x).$$

Thus, using the Cauchy–Schwarz inequality we obtain

$$\begin{aligned} Var_\theta(T_1(X)) \cdot I(\theta) &\geq \left( \sum_x (T_1(x) - \theta) \frac{P'_\theta(x)}{P_\theta(x)} P_\theta(x) \right)^2 \\ &= \left( \sum_x (T_1(x) - \theta) P'_\theta(x) \right)^2 \\ &= \left( \sum_x T_1(x) P'_\theta(x) \right)^2 \\ &= \left( \frac{\partial}{\partial \theta} \sum_x T_1(x) P_\theta(x) \right)^2 \end{aligned}$$

$$= \left(\frac{\partial}{\partial\theta}\theta\right)^2$$
$$= 1.$$
□

## 7.2 Lecture on Information-Theoretical Approach to Expert Systems

### *7.2.1 Basic Properties of the Relative Entropy*

We present the results of Csiszár [2], who established some geometric properties of PDs and the relative entropy between two PDs that shows how much a PD $P$ differs from a PD $Q$. This quantity plays a similar role as squared Euclidean distance in metric spaces.

The relative entropy or Kullback–Leibler information [4] between two PDs $P$ and $Q$ defined on a measurable space can be introduced as a function

$$D(P \parallel Q) = \int \log p_Q \, \mathrm{d}P = \int p_Q \log p_Q \, \mathrm{d}Q$$

if $P$ is absolutely continuous with respect to $Q$ (we will write $P \ll Q$ ) and $D(P \parallel Q) = \infty$, otherwise. Hereafter, $p_Q$ denotes the corresponding density. If $R$ is any PD with $P \ll R$ and $Q \ll R$ then an equivalent definition can be given as

$$D(P \parallel Q) = \int p_R \log \frac{p_R}{q_R} \, \mathrm{d}R.$$

Note that $D(P \parallel Q) = 0$ if and only if $P = Q$. Otherwise, $D(P \parallel Q) > 0$. Note also that the following identity holds:

$$D(P \parallel R) - D(P \parallel Q) = \int p_R \log q_R \, \mathrm{d}R = \int \log q_R \, \mathrm{d}P \qquad (7.2.1)$$

Let us define by

$$S_\rho(Q) = \{P : \ D(P \parallel Q) \le \rho\}$$

the "ball" with the center $Q$ and radius $\rho$.

**Theorem 142** *Let $D(P_0 \parallel Q) = \rho$ for a given $0 < \rho < \infty$ and let $p_0 = dP_0/dQ$. Then, for every $P$ with $D(P \parallel Q) < \infty$,*

$$\int \log p_0(x) P(\mathrm{d}x)$$

*exists and*

(i) *if*

$$\int \log p_0 P(\mathrm{d}x) \geq \rho$$

*and* $P \neq P_0$, *then*

$$D(P \parallel Q) \geq D(P \parallel P_0) + D(P_0 \parallel Q) > \rho;$$

(ii) *if*

$$\int \log p_0 P(\mathrm{d}x) < \rho$$

*then*

$$D(P_\varepsilon \parallel Q) < \rho$$

*for sufficiently small* $\varepsilon > 0$, *where*

$$P_\varepsilon = \varepsilon P + (1 - \varepsilon) P_0.$$

*Proof* Let $P \ll Q$ be arbitrarily chosen. Then integration of the identity

$$p_Q(x) \log p_0(x) = p_Q(x) \log p_Q(x) - p_Q(x) \log \frac{p_Q(x)}{p_0(x)}$$

gives

$$\int \log p_0(x) P(\mathrm{d}x) = D(P \parallel Q) - D(P \parallel P_0)$$

and (i) follows.
To prove (ii) we denote

$$p_\varepsilon(x) = \varepsilon p_Q(x) + (1 - \varepsilon) p_0(x).$$

Since $-p_\varepsilon(x) \log p_\varepsilon(x)$ is a concave function of $\varepsilon$, the difference quotient

$$\frac{1}{\varepsilon} \left[ -p_\varepsilon(x) \log p_\varepsilon(x) + p_0(x) \log p_0(x) \right]$$

is nonincreasing as $\varepsilon \to 0$. If $D(P \parallel Q) < \infty$, then this difference quotient is bounded from below by an integrable function

$$-p_Q(x) \log p_Q(x) + p_0(x) \log p_0(x)$$

and

$$\begin{aligned}
&\lim_{\varepsilon\to 0}\frac{1}{\varepsilon}\left[D(P_0 \parallel Q) - D(P_\varepsilon \parallel Q)\right]\\
&\quad= \lim_{\varepsilon\to 0}\int \frac{1}{\varepsilon}\left[-p_\varepsilon(x)\log p_\varepsilon(x) + p_0(x)\log p_0(x)\right] Q(\mathrm{d}x)\\
&\quad= \int \frac{\mathrm{d}}{\mathrm{d}\varepsilon}\left[-p_\varepsilon(x)\log p_\varepsilon(x)\right]_{\varepsilon=0} Q(\mathrm{d}x)\\
&\quad= \int (-p_Q(x) + p_0(x))(\log p_0(x) + 1) Q(\mathrm{d}x)\\
&\quad= -\int \log p_0(x) P(\mathrm{d}x) + D(P_0 \parallel Q).
\end{aligned}$$

The last expression is positive if $\int \log p_0(x)P(\mathrm{d}x) < \rho$. □

Theorem 142 has the following geometric interpretation. The set

$$\left\{P : \int \log p_0(x)P(\mathrm{d}x) = \rho\right\}$$

can be viewed as a *tangential plane* of $S_\rho(Q)$ in the point $P_0$ :

- the "half space",

$$\left\{P : \int \log p_0(x)P(\mathrm{d}x) \geq \rho\right\}$$

  has exactly $P_0$ in common with $S_\rho(Q)$;
- for every element $P$ in a complementary "half space",

$$\left\{P : \int \log p_0(x)P(\mathrm{d}x) < \rho\right\}$$

  (with $D(P \parallel Q) < \infty$ ) the line connecting $P$ and $P_0$ necessarily intersects $S_\rho(Q)$;
- every $P$ in the tangential plane satisfies the equation

$$D(P \parallel Q) = D(P \parallel P_0) + D(P_0 \parallel Q).$$

**Lemma 80** *For any PDs $P$, $Q$, $R$, and $\alpha \in (0, 1)$,*

$$\begin{aligned}
\alpha D(P \parallel Q) + (1-\alpha)D(R \parallel Q) = {} & D(\alpha P + (1-\alpha)R \parallel Q)\\
&+ \alpha D(P \parallel \alpha P + (1-\alpha)R)\\
&+ (1-\alpha)D(R \parallel \alpha P + (1-\alpha)R).
\end{aligned} \tag{7.2.2}$$

*Furthermore, if* $D(\alpha P + (1-\alpha)R \parallel Q) < \infty$, *then also* $D(P \parallel Q) < \infty$ *and* $D(R \parallel Q) < \infty$.

*Proof* We may write

$$D(P \parallel Q) = \int p_Q \log p_Q \, \mathrm{d}Q,$$

$$D(R \parallel Q) = \int r_Q \log r_Q \, \mathrm{d}Q,$$

$$D(\alpha P + (1-\alpha)R \parallel Q) = \int (\alpha p_Q + (1-\alpha) r_Q) \log(\alpha p_Q + (1-\alpha) r_Q) \, \mathrm{d}Q.$$

However,

$$D(P \parallel \alpha P + (1-\alpha)R) = \int p_Q \log \frac{p_Q}{\alpha p_Q + (1-\alpha) r_Q} \, \mathrm{d}Q,$$

$$D(R \parallel \alpha P + (1-\alpha)R) = \int r_Q \log \frac{r_Q}{\alpha p_Q + (1-\alpha) r_Q} \, \mathrm{d}Q,$$

and Eq. (7.2.2) can be checked by inspection. Besides, the last two terms of (7.2.2) are bounded by $-\log \alpha$ and $-\log(1-\alpha)$ which gives the claim desired. □

Identity (7.2.2) has an interesting geometric interpretation since it is similar to the Parallelogram identity in Hilbert spaces. We illustrate this point in Fig.7.1 for $\alpha = 1/2$.

To investigate some further properties of the space of PDs that are similar to the properties of the Hilbert space let us introduce the following:

**Definition 38** *A PD* $P_0 \in \Pi$ *is called projection of* $Q$ *on a convex set* $\Pi$ *consisting of PDs defined on a measurable space if*

$$D(P_0 \parallel Q) = \inf_{P \in \Pi} D(P \parallel Q).$$

Note that Theorem 142 and identity (7.2.1) imply

**Theorem 143** *Necessary and sufficient conditions for a PD* $P_0$ *of being projection of* $Q$ *on* $\Pi$ *is*

$$\int \log p_0(x) P(\mathrm{d}x) \geq D(P_0 \parallel Q), \quad \textit{for all } P \in \Pi. \tag{7.2.3}$$

*In this case*

$$D(P \parallel Q) \geq D(P \parallel P_0) + D(P_0 \parallel Q), \quad \textit{for all } P \in \Pi. \tag{7.2.4}$$

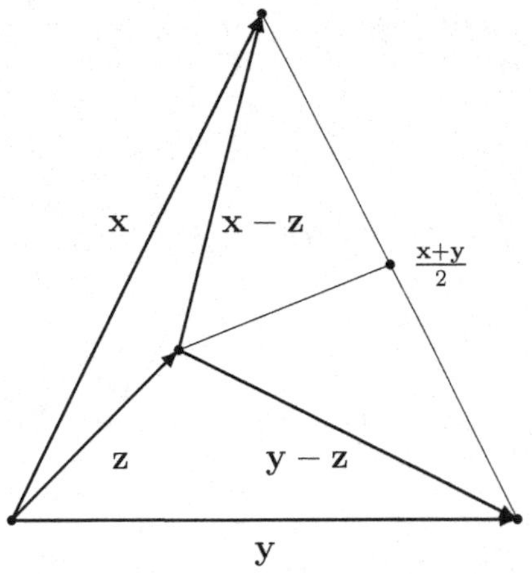


$$||\mathbf{x}-\mathbf{z}||^2+||\mathbf{y}-\mathbf{z}||^2=2||\tfrac{\mathbf{x}+\mathbf{y}}{2}-\mathbf{z}||^2+||\mathbf{x}-\tfrac{\mathbf{x}+\mathbf{y}}{2}||^2+||\mathbf{y}-\tfrac{\mathbf{x}+\mathbf{y}}{2}||^2$$

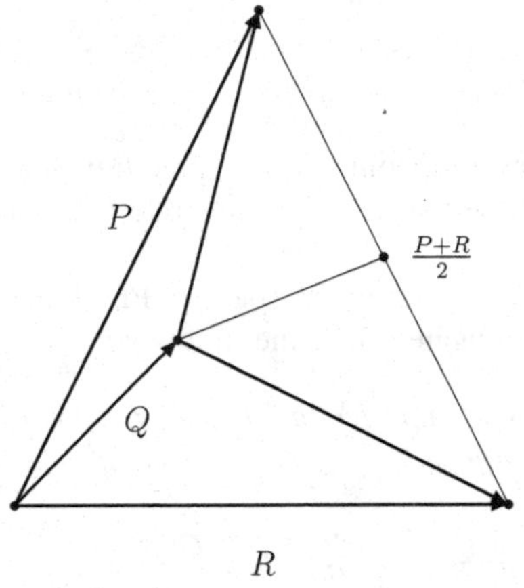


$$D(P \parallel Q)+D(R \parallel Q)=2D(\tfrac{P+R}{2} \parallel Q)+D(P \parallel \tfrac{P+R}{2})+D(R \parallel \tfrac{P+R}{2})$$

**Fig. 7.1** Geometric illustration of the analogy between parallelogram identities for the Hilbert space and space of PDs

*and if, for some $P$ there is equation in (7.2.3), then there is also equation in (7.2.4). These equations take place if $P_0$ is an algebraic inner point, i.e., for any $P' \in \Pi$ there exists an $R \in \Pi$ and $\alpha \in (0, 1)$ such that $\alpha P' + (1-\alpha)R = P$ and $D(P \parallel Q) < \infty$.*

**Theorem 144** *If the convex set $\Pi$ is closed in the norm of total variation, then there is a unique projection $P_0$ of $Q$ on $\Pi$.*

*Proof* Let us choose a sequence $P_n \in \Pi, n = 1, 2, \ldots$ with

$$D(P_n \parallel Q) \to \inf_{P \in \Pi} D(P \parallel Q), \quad \text{as } n \to \infty.$$

By convexity of $\Pi$ we have $(P_n + P_m)/2 \in \Pi$ and

$$\begin{aligned}\frac{1}{2}D(P_n \parallel Q) + \frac{1}{2}D(P_m \parallel Q) = D\left(\frac{P_n + P_m}{2} \parallel Q\right)\\ + D\left(P_n \parallel \frac{P_n + P_m}{2}\right)\\ + D\left(P_m \parallel \frac{P_n + P_m}{2}\right)\end{aligned}$$

Using Pinsker's inequality,

$$|P - Q| \le \sqrt{2D(P \parallel Q)}$$

we obtain that

$$|P_n - P_m| \le \left|P_n - \frac{P_n + P_m}{2}\right| + \left|P_m - \frac{P_n + P_m}{2}\right|$$

converges to 0 as $n, m \to \infty$ and, consequently, $P_n$ converges in variation to some PD $Q$, i.e.,

$$|P_n - Q| = \int |p_{nR} - q_R| dR \to 0, \quad \text{as } n \to \infty.$$

Thus,

$$\begin{aligned}\inf_{P \in \Pi} D(P \parallel Q) &= \lim_{n \to \infty} D(P_n \parallel Q)\\ &\ge D(P_0 \parallel Q)\\ &\ge \inf_{P \in \Pi} D(P \parallel Q)\end{aligned}$$

and

$$D(P_0 \parallel Q) = \inf_{P \in \Pi} D(P \parallel Q).$$

□

In applications to statistics it is important to have a description for a class of PD $P_0$ and $Q$ such that (7.2.4) is satisfied with the equation. Note that the set $\Pi$ consisting of PDs is a linear space : if $P, R \in \Pi$ and $\alpha \in [0, 1]$, then

$$\alpha P + (1 - \alpha)R \in \Pi.$$

We want to specify a subset $\Pi^*$ of $\Pi$ consisting of PDs with the following property: $\Pi^*$ is a linear subspace of $\Pi$ and if a PD $P_1$ is a projection of $Q$ on $\Pi$ and $P_2$ is a projection of $P_1$ on $\Pi^*$ then $P_2$ is a projection of $Q$ on $\Pi^*$. Meaning further extensions, we give the following:

**Definition 39** *A linear space $\Pi^* \subseteq \Pi$ is D-linear if for any $Q \in \Pi$ one can assign a $P_0 \in \Pi^*$ in such a way that* (7.2.4) *is satisfied with the equation, i.e.,*

$$D(P \parallel Q) = D(P \parallel P_0) + D(P_0 \parallel Q), \ \textit{for all} \ \ P \in \Pi^*. \tag{7.2.5}$$

**Theorem 145** *Let $\Pi_1 \subseteq \Pi$ be a D-linear subspace of $\Pi$ and let $\Pi_2 \subseteq \Pi_1$ be a convex linear subspace of $\Pi_1$. Then if $Q$ has a projection $P_i$ on $\Pi_i, i = 1, 2$, then $\Pi_2$ is a projection of $P_1$ on $\Pi_2$.*

*Proof* By assumption, $P_1$ satisfies (7.2.5) and $P_2$ satisfies (7.2.4). Thus,

$$\begin{aligned} D(P \parallel Q) &= D(P \parallel P_1) + D(P_1 \parallel Q) \\ &\geq D(P \parallel P_2) + D(P_2 \parallel Q), \quad \text{for all } P \in \Pi_2. \end{aligned} \tag{7.2.6}$$

Since $P_2 \in \Pi_2$,

$$D(P_2 \parallel Q) = D(P_2 \parallel P_1) + D(P_1 \parallel Q)$$

and using this equation in (7.2.6) we obtain

$$D(P \parallel Q) \geq D(P \parallel P_2) + D(P_2 \parallel P_1) + D(P_1 \parallel Q)$$

Hence,

$$D(P \parallel P_1) \geq D(P \parallel P_2) + D(P_2 \parallel P_1), \quad \text{for all } P \in \Pi_2.$$

On the other hand, (7.2.4) yields

$$\int \log \frac{\mathrm{d}P_2}{\mathrm{d}P_1} \mathrm{d}P = D(P \parallel P_1) + D(P \parallel P_2) \geq D(P_2 \parallel P_1).$$ □

### *7.2.2 Minimizing the Relative Entropy Under Linear Constraints*

A general formulation of a result known as minimum discrimination theorem is as follows [4]:

**Theorem 146** *For any not necessary convex set of PDs $\Pi^*$, if there exists a $Q \in \Pi^*$ with R-density $c \exp g(x)$ where*

$$\int g \, \mathrm{d}P_1 = \int g \, \mathrm{d}P_2 < \infty$$

.

*for any* $P_1, P_2 \in \Pi^*$, *then*

$$D(Q \parallel R) = \min_{P \in Pi^*} D(P \parallel R)$$

*and*

$$D(P \parallel R) = D(P \parallel Q) + D(Q \parallel R), \quad \textit{for all } P \in \Pi^*. \tag{7.2.7}$$

Note that this result immediately follows from (7.2.1). Two special cases are of main importance

(A) $\Pi^*$, is defined by constraints of the form $\int f_i \, \mathrm{d}P = a_i, i = 1, \ldots, k$. Then, if a $Q \in \Pi^*$ with

$$q_R(x) = c \exp\left\{\sum_{i=1}^{k} t_i f_i(x)\right\} \tag{7.2.8}$$

exists, then it is the $D$-projection of $R$ on $\Pi^*$ and (7.2.7) holds.

(B) Let $\Pi^*$ consist of PDs with given marginals $P_1$ and $P_2$ defined on a measurable spaces $X_1$ and $X_2$ and let the space be a Cartesian product of $X_1$ and $X_2$. If a $Q \in \Pi^*$ with

$$q_R(x_1, x_2) = a(x_1)b(x_2), \quad \log a \in L_1(P_1), \quad \log b \in L_1(P_2), \tag{7.2.9}$$

then $Q$ is a projection of $R$ on $\Pi^*$ and (7.2.7) holds.

**Theorem 147** *Let* $\{f_\gamma, \gamma \in \Gamma\}$ *be an arbitrary set of real-valued measurable functions on* $X$ *and* $\{a_\gamma, \gamma \in \Gamma\}$ *be real constants. Let* $\Pi^*$ *be the set of all those PDs* $P$ *for which the integrals* $\int f_\gamma dP$ *exist and equal* $a_\gamma$ *for all* $\gamma \in \Gamma$. *Then, if a PD* $R$ *has a projection* $Q$ *on* $\Pi^*$, *its* $R$ *density is of form*

$$q_R(x) = \begin{cases} c \exp\{g(x)\}, & \textit{if } x \notin N, \\ 0, & \textit{if } x \in N, \end{cases} \tag{7.2.10}$$

*where* $N$ *has* $P(N) = 0$ *for every* $P \in \Pi^* \cap S_\infty(R)$ *and* $g$ *belongs to the closed subspace of* $L_1(Q)$ *spanned by the* $f_\gamma$*s. On the other hand, if* $Q \in \Pi^*$ *has* $R$*-density of form* (7.2.10) *where* $g$ *belongs to the linear space spanned by the* $f_\gamma$*'s, then* $Q$ *is the* $D$*-projection of* $R$ *on* $\Pi^*$ *and* (7.2.7) *holds.*

**Corollary 31** *In case* (A) *or* (B) *above, a* $Q \in \Pi^*$ *is the projection of* $R$ *on* $\Pi^*$ *iff* $q_R$ *is of form* (7.2.8) *or* (7.2.9), *respectively; except possibly for a set* $N$ *where* $q_R$ *vanishes and* $P(N) = 0$ *for every* $P \in \Pi^* \cap S_\infty(R)$; *in both cases, the identity* (7.2.7) *holds. If, in particular, some* $P \in \Pi^*$ *with* $D(P \parallel R) < \infty$ *is measure-theoretically equivalent to* $R$ *then* (7.2.8) *or* (7.2.9) *is necessary and sufficient for* $Q$ *to be the projection of* $R$ *on* $\Pi^*$.

*Proof* If $Q$ is the $D$-projection of $R$ on $\Pi^*$ then for $N = \{x : q_R(x) = 0\}$ necessarily $P(N) = 0$ for each $P \in \Pi^* \cap S_\infty(R)$.

Let $\tilde{\Pi}^*$ be the set of PDs $P \in \Pi^*$ with $p_Q(x) \leq 2$. If $P \in \tilde{\Pi}^*$, there is a $\tilde{P} \in \tilde{\Pi}^*$ with $\tilde{p}_Q(x) = 2 - p_Q(x)$, and with it $Q = (P + \tilde{P})/2$; thus $Q$ is an algebraic inner point of $\tilde{\Pi}^*$. Applying Theorem 143 we obtain

$$\int \log q_R \, \mathrm{d}P = D(Q \parallel R),$$

i.e.,

$$\int \log q_R (p_Q - 1) \, \mathrm{d}Q = 0, \quad \text{for all } P \in \tilde{\Pi}^*.$$

However, for any measurable function $h$ with $|h(x)| \leq 1$ such that

$$\int h \, \mathrm{d}Q = 0, \quad \int f_\gamma h \, \mathrm{d}Q = 0, \quad \text{for all } \gamma \in \Gamma, \tag{7.2.11}$$

there exists a $P \in \tilde{\Pi}^*$ with $p_Q = 1 + h$. Thus, (7.2.11) gives

$$\int q_R h \, \mathrm{d}Q = 0 \tag{7.2.12}$$

for all $h$ and therefore also for all $h \in L_\infty(Q)$ satisfying (7.2.12). Hence, $\log q_R$ belongs to the closed subspace of $L_1(Q)$ spanned by 1 and the $f_\gamma$s. This proves the first assertion of Theorem 147.

Let us prove the second part. Suppose that $q_R$ is of stated form. Since $g$ is a finite linear combination of $f_\gamma$s, the integral $\int g \, \mathrm{d}P$ is constant for $P \in \Pi^*$ and

$$\int \log q_R \, \mathrm{d}P = \log c + \int g \, \mathrm{d}P = D(Q \parallel P). \tag{7.2.13}$$

However, for $P \in \Pi^*$ both $D(P \parallel R)$ and $D(P \parallel Q)$ are finite. Hence (7.2.7) follows from (7.2.1) and (7.2.13).

To prove Corollary 31, observe that the case (B) does fit into the considered model taking $f_\gamma$s the $P_i$-integrable functions $f_i(x)$, $i = 1, 2$. Theorem 147 clearly gives a necessary and sufficient condition on $q_R$and guarantees the validity of (7.2.7) for the projection $Q$ if the linear space spanned by the $f_\gamma$s is closed in $L_1(P)$ for each $P \in \Pi^*$. However, the later hypothesis is fulfilled in both cases (A) and (B), completing the proof. □

Theorem 147 and Corollary 31 leaves the question of existence of the projections open. If $\Pi^*$ is variation-closed, as in the case of bounded $f_\gamma$s or in the case (B), Theorem 144 guarantees the existence provided that $\Pi^* \neq \emptyset$ and $D(P \parallel R) < \infty$ for some $P \in \Pi^*$. As a consequence we obtain

**Corollary 32** *To given PDs $P_i$ on $X_i, i = 1, 2$ and $R$ on $X_1 \times X_2$, there exists a PD $Q$ on the product space with marginals $P_1$ and $P_2$ and with $R$-density of form $a(x_1)b(x_2)$, $\log a \in L_1(P_1)$, $\log b \in L_1(P_1)$ iff there exists any $P$ measure-theoretical equivalent to $R$ which has the prescribed marginals and satisfies $D(P \parallel R) < \infty$.*

Specializing Corollary 32 to finite sets we get

**Corollary 33** *Let $A$ be $m \times n$ matrix with nonnegative elements. For the existence of positive diagonal matrices $D_1$ and $D_2$ such that the row and column sums of $D_1 A D_2$ be given positive numbers, it is necessary and sufficient that some $B$ with nonnegative elements and with given row and column sums has the same zero entries as $A$.*

Finally, we consider the problem of existence of the projections in the case (A) and first prove the following auxiliary result.

**Lemma 81** *For any measurable function $f(x)$ for which $e^{tf(x)}$ is $Q$-integrable if $|t|$ is sufficiently small, $D(P_n \parallel Q) \to 0$ implies $\int f dP_n \to \int f dQ$.*

*Proof* Let $p_n$ denote the $Q$-density of $P_n$; it surely exists if $D(P_n \parallel Q) < \infty$. If $D(P_n \parallel Q) \to 0$, then

$$|P_n - Q| = \int |p_n - 1| \, \mathrm{d}Q \to 0.$$

Thus, it suffices to show that for any $\varepsilon > 0$ there exists $K$ such that

$$\limsup_{n\to\infty} \int_{X \setminus A_k} |f| \, \mathrm{d}P_n = \limsup_{n\to\infty} \int_{X \setminus A_k} |f| p_n \, \mathrm{d}Q < \varepsilon.$$

However, $D(P_n \parallel Q) \to 0$, implies

$$\lim_{n\to\infty} \int_A p_n \log p_n \, \mathrm{d}Q = 0 \tag{7.2.14}$$

for every $A$. Choosing $t > 0$ and $K$ to satisfy

$$\int_{X \setminus A_k} e^{t|f|} \, \mathrm{d}Q < \varepsilon t,$$

(7.2.14) follows from the inequality $ab < a \log a + e^b$, substituting $a = p_n(x)$ and $b = t|f(x)|$. □

**Theorem 148** *Let $\Pi^*(a_1, \ldots, a_k)$ be the set of PDs satisfying*

$$\int f_i \,\mathrm{d}P = a_i, \quad i = 1, \ldots, k$$

*and let $A_R$ be the set of points $(a_1, \ldots, a_k) \in E_k$ for which $\Pi^*(a_1, \ldots, a_k)$ contains some $P$ with $D(P \parallel R) < \infty$. Then, supposing that $A_R$ is open, the projection of $R$ on $\Pi^*(a_1, \ldots, a_k)$ exists for each inner point $(a_1, \ldots, a_k)$ of $A_R$, and its $R$-density is of form* (7.2.8).

*Proof* Because of convexity of $D(P \parallel R)$ on $P$, the set $A_R$ is a convex set and

$$F(a_1, \ldots, a_k) = \inf_{P \in \Pi^*(a_1, \ldots, a_k)} D(P \parallel Q). \tag{7.2.15}$$

is a finite valued convex function on $A_R$. Hence, if $(a_1, \ldots, a_k)$ is an inner point of $A_R$, there exists $(t_1, \ldots, t_k)$ such that

$$F(b_1, \ldots, b_k) \geq F(a_1, \ldots, a_k) + \sum_{i=1}^{k} t_i (b_i - a_i) \tag{7.2.16}$$

for all $(b_1, \ldots, b_k) \in A_R$. First, we show that $(t_1, \ldots, t_k) \in A_R$. Let $P_n \in \Pi^*(a_1, \ldots, a_k)$ and $D(P_n \parallel R) \to F(a_1, \ldots, a_k)$. Then $P_n$ converges in variation to some $Q$ by the proof of Theorem 144. Let $f_i^{(n)}(x) = f_i(x)$ if $f_i(x) \leq K_n$ and $f_i^{(n)}(x) = 0$ otherwise, where $K \to \infty$, and let $Q_n$ be the PD with $R$-density

$$q_{nR}(x) = c_n \exp\left\{\sum_{i=1}^{k} t_i f_i(x)\right\}.$$

Then

$$D(Q_n \parallel R) = \int \log q_{nR} \,\mathrm{d}P_n + \sum_{i=1}^{k} t_i \left(\int f_i^{(n)} \,\mathrm{d}Q_n - \int f_i^{(n)} \,\mathrm{d}P_n\right). \tag{7.2.17}$$

Since $(0, \ldots, 0) \in A_R$ and $A_R$ is open, the $f_i$s are $R$-integrable and thus $Q_n$-integrable and, choosing $K_n$ properly, we obtain $D(P_n \parallel Q_n) \to 0$. Setting $b_i = \int f_i \,\mathrm{d}Q$, similarly to (7.2.17) we have

$$D(Q \parallel R) = \int \log q_R \,\mathrm{d}P_n + \sum_{i=1}^{k} t_i (b_i - a_i) \tag{7.2.18}$$

and $D(P_n \parallel Q) \to 0$. Using the assumption that $A_R$ is open set, Lemma 81 gives

$$\int f_i \, dQ = \lim_{n\to\infty} \int f_i \, dP_n = a_i, \quad i = 1, \ldots, k.$$

This observation completes the proof. □

### 7.2.3 Axiomatic Results on Inference for Inverse Problems

Axiomatic approach to inverse problems was proposed by Shore and Johnson [5] and developed by Csiszár [1, 3]. We present some of these results, including an axiomatic characterization of the method minimizing an $L^p$-distance, where $1 < p < \infty$.

Rather often the values of some linear functionals $R_i, i = 1, \ldots, k$, applied to a function $f$ are known and we want to recover $f$ based on these observations. This problem is called inverse problem. Since the observations typically do not determine $f$ uniquely, a solution can be presented using a *selection rule* to pick one element of the set of feasible functions. If this selection depends also on a "prior" guess of the unknown $f$, the selected function is considered an abstract projection of the guess into the feasible set, and we may speak about *projection rule*. After some notations and definitions we will introduce various reasonable postulates that allow us to characterize these rules.

The real line and the positive half-line will be denoted by $R$ and $R_+$, respectively. The vectors in $R^n$ whose components are all zero or all one are denoted by $\mathbf{0}$ and $\mathbf{1}$. The set of $n$-dimensional vectors with positive components of sum 1 is denoted by

$$\Delta_n = \left\{ \mathbf{v} : \mathbf{v} \in R^n_+, \ \mathbf{1}^T \mathbf{v} = 1 \right\} \tag{7.2.19}$$

We will consider three cases in parallel, namely the basic set $S$ of all potentially permissible vectors is either of $R^n$, $R^n_+$ or $\Delta_n$ where $n \geq 3$ or, if $S = \Delta_n$, $n \geq 5$. According to the three cases, $V$ will denote $R^n$, $R^n_+$ or the open interval (0, 1), respectively. Unless stated otherwise, $u$, $v$, and $w$ will always denote the elements of $V$, and $\mathbf{u}, \mathbf{v}, \mathbf{w}$ are vectors in $V^n$. Further, $\mathcal{L}$ denotes the family of non-void subsets of $S$ defined by linear constraints. Thus $L \in \mathcal{L}$ iff

$$L = \{ \mathbf{v} : \mathbf{A}\mathbf{v} = \mathbf{b} = 1 \} \neq \emptyset \tag{7.2.20}$$

for some $k \times n$ matrix $\mathbf{A}$ and some $\mathbf{b} \in R^k$; in the case $S = \Delta_n$ it is assumed that $\mathbf{A}\mathbf{v} = \mathbf{b}$ implies $\mathbf{1}^T\mathbf{v} = 1$.

A *selection rule* with basic set $S$ is a mapping $\pi : \mathcal{L} \to S$ such that $\pi(L) = L$ for every $L \in \mathcal{L}$. A *projection rule* is a family of selection rules $\pi(\cdot|\mathbf{u})$, $\mathbf{u} \in S$, such that $\mathbf{u} \in L$ implies $\pi(L|\mathbf{u}) = \mathbf{u}$.

A selection rule is generated by a function $F(\mathbf{v})$, $\mathbf{v} \in S$, if for every $L \in \mathcal{L}$, $\pi(L)$ is a unique element of $L$ where $F(\mathbf{v})$ is minimized subject to $\mathbf{v} \in L$. A projection rule is generated by a function $F(\mathbf{v}|\mathbf{u})$, $\mathbf{u}, \mathbf{v} \in S$, if its component selection rules are generated by the functions $F(\cdot|\mathbf{u})$.

If a projection rule is generated by some function, it is also generated by a *measure of distance* on $S$, i.e., by a function with the property $F(\mathbf{v}|\mathbf{u}) \geq 0$, with equation iff $\mathbf{v} = \mathbf{u}$.

For any set of indices $J = \{j_1, \ldots, j_k\} \subseteq \{1, \ldots, n\}$ and any vector $\mathbf{a} \in R^n$, we denote by $\mathbf{a}_J$ the vector in $R^k$ defined by

$$\mathbf{a}_J = (a_{j_1}, \ldots, a_{j_k})^T. \tag{7.2.21}$$

For a selection rule $\pi$ we write $\pi_J(L)$ instead of $(\pi(L))_J$.

The basic axioms for selection rules are the following :

(1) (consistency) : if $L' \subset L$ and $\pi(L) \in L'$ then $\pi(L') = \pi(L)$;
(2) (distinctness) : if $L \neq L'$ are both $(n-1)$- or $(n-2)$-dimensional if $S = \Delta_n$, then $\pi(L') \neq \pi(L)$ unless both $L$ and $L'$ contain $\mathbf{v}^0 = \pi(S)$;
(3) (continuity) : the restriction of $\pi$ to any subclass of $\mathcal{L}$ consisting of sets of equal dimension is continuous;
(4) (locality) : if $L \in \mathcal{L}$ is defined by a matrix $\mathbf{A}$ in (7.2.21) such that for some $I \subseteq \{1, \ldots, k\}$ and $J \subseteq \{1, \ldots, n\}$ we have $a_{ij} = 0$ whenever $(i, j) \in (I \times J^c) \cup (I^c \times J)$, then $\pi_J(L)$ depends on $\mathbf{A}$ and $\mathbf{b}$ through $a_{ij}, i \in I, j \in J$, and $\mathbf{b}_I$ only.

For projection rules, these postulates are required to hold for all component selection rules $\pi(\cdot|\mathbf{u})$, $\mathbf{u} \in S$ (notice that in (2), $\pi(S|\mathbf{u}) = \mathbf{u}$ by definition), and in (4) it is additionally required that $\pi(L|\mathbf{u})$ depends on $\mathbf{u}$ through $\mathbf{u}_J$ only.

The key result of the paper by Csiszár [3] is the theorem formulated below. In that theorem, the term *standard n-tuple* with zero at $\mathbf{v}^0$ means an $n$-tuple of functions $(f_1, \ldots, f_n)$ defined on $V$ such that

(i) each $f_i$ is continuously differentiable and

$$f_i(v_i^0) = f_i'(v_i^0) = 0, \quad i = 1, \ldots, n;$$

(ii) in the cases $S = R_+^n$ or $\Delta_n$, $f_i'(v) \to -\infty$ as $v \to 0$;
(iii) $F(\mathbf{v}) = \sum_{i=1}^n f_i(v_i)$ is nonnegative and strictly quasi-convex on $S$, i.e., for any $\mathbf{v}$ and $\mathbf{v}'$ in $S$

$$F(\alpha\mathbf{v} + (1-\alpha)\mathbf{v}') < \max\{F(\mathbf{v}), F(\mathbf{v}')\}, \quad 0 < \alpha < 1. \tag{7.2.22}$$

**Theorem 149** (a) *If a selection rule* $\pi : \mathcal{L} \to S$ *satisfies the basic axioms* (1)–(4) *then it is generated by a function*

$$F(\mathbf{v}) < \sum_{i=1}^n f_i(v_i), \tag{7.2.23}$$

*where* $(f_1, \ldots, f_n)$ *is a standard n-tuple with zero at* $\mathbf{v}^0 = \pi(S)$*. Conversely, if* $(f_1, \ldots, f_n)$ *is a standard n-tuple with zero at* $\mathbf{v}^0 = \pi(S)$ *then* (7.2.23) *generates a selection rule with* $\pi(S) = \mathbf{v}^0$ *that satisfies the basic axioms.*

(b) *If a projection rule satisfies the basic axioms then it is generated by a measure of distance*

$$F(\mathbf{v}|\mathbf{u}) = \sum_{I=1}^{n} f_i(v_i|u_i), \qquad (7.2.24)$$

*where the functions* $f_1(\cdot|u_1), \ldots, f_n(\cdot|u_n)$ *form a standard n-tuple with zero at* $\mathbf{u}$*. Conversely, any such measure of distance generates a projection rule with satisfying the basic axioms.*

(c) *Two functions* $F$ *and* $\tilde{F}$ *in* (a) *and* (b) *generate the same selection or projection rule iff their terms* $f_i$ *and* $\tilde{f}_i$ *satisfy* $cf_i = \tilde{f}_i$, $i = 1, \ldots, n$, *for some constant* $c > 0$.

*Remark* If a selection rule $\Pi : \mathcal{L} \to S$ satisfies the basic axioms, its generating function (7.2.23) has also the following property :

(iv) $\mathrm{grad} F(\mathbf{v}) \neq \mathbf{0}$, and in the case $S = \Delta_n$ also $\mathrm{grad} F(\mathbf{v}) \neq \lambda \mathbf{1}$, for all $\mathbf{v} \in S$ with $\mathbf{v} \neq \mathbf{v}^0$.

On the other hand, for the converse assertion (a) in Theorem 149 this property is needed to check axiom (2). Thus, unless properties (i)–(iii) in the definition of a standard $n$-tuple already imply (iv), the latter has to be explicitly added to fill a minor gap in Theorem 149. It remains open whether (i)–(iii) imply (iv) if $S = \Delta_n$. In the cases $S = R^n$ and $S = R^n_+$, however, it is an immediate consequence of Theorem 152 that we present further.

The class of functions $F$ occurring in Theorem 149 can be restricted imposing some further postulates.

(5a) (scale invariance, for $S = R^n$ and $S = R^n_+$)

$$\Pi(\lambda L|\lambda \mathbf{u}) = \lambda \Pi(L|\mathbf{u}) \qquad (7.2.25)$$

for every $L \in \mathcal{L}$, $\lambda > 0$, $\mathbf{u} \in S$.

(5b) (translation invariance, for $S = R^n$)

$$\Pi(L + \mu \mathbf{1}|\mathbf{u} + \mu \mathbf{1}) = \Pi(L|\mathbf{u}) + \mu \mathbf{1} \qquad (7.2.26)$$

for every $L \in \mathcal{L}$, $\mu \in R$, $\mathbf{u} \in S$.

*Remark* Selection rules can be also scale invariant, i.e., satisfy $\Pi(\lambda L) = \lambda Pi(L)$ for every $L \in \mathcal{L}$ and $\lambda > 0$, but not only in the case $S = R^n$ (because $L = S$ yields $\Pi(S) = \mathbf{0}$). Translation invariance is not possible for selection rules.

It should be mentioned that in the case $S = R^n$, postulate (7.2.25) and its analog for selection rules could be imposed also in a stronger form, for $\lambda < 0$ as well. This stronger postulate, called *strong invariance postulate*, will be used in further considerations.

(6a) (subspace transitivity) for every $L' \subseteq L$ and $\mathbf{u} \in S$,

$$\Pi(L'|\mathbf{u}) = \Pi(L'|\Pi(L|\mathbf{u})); \tag{7.2.27}$$

(6b) (parallel transitivity) for every $L$ and $L'$ defined as in (7.2.20) with the same matrix $\mathbf{A}$, and for every $\mathbf{u} \in S$,

$$\Pi(L'|\mathbf{u}) = \Pi(L'|\Pi(L|\mathbf{u})). \tag{7.2.28}$$

**Theorem 150** *A projection rule satisfies the basic axioms* (1–4) *and the transitivity postulate* (7.2.27) *iff it is generated by*

$$F(\mathbf{v}|\mathbf{u}) = \Phi(\mathbf{v}) - \Phi(\mathbf{u}) - (\text{grad}\ \Phi(\mathbf{u}))^T(\mathbf{v} - \mathbf{u}), \tag{7.2.29}$$

*where*

$$\Phi(\mathbf{v}) = \sum_{i=1}^{n} \varphi_i(v_i),$$

*the functions $\varphi_i(v_i)$ defined on $V$ are continuously differentiable, $\Phi(\mathbf{v})$ is strictly convex in $S$, and in the cases $S = R^n_+$ or $\Delta_n$,*

$$\lim_{v \to 0} \varphi_i'(v) = -\infty.$$

*Further, subject to the basic axioms* (1–4), *the transitivity postulates* (7.2.27) *and* (7.2.28) *are equivalent.*

**Theorem 151** (a) *In the case $S = R^n$, a projection rule as in Theorem* 150 *is location and scale invariant iff it is generated by*

$$F(\mathbf{v}|\mathbf{u}) = \sum_{i=1}^{n} a_i(v_i - u_i)^2, \tag{7.2.30}$$

*for certain positive constants $a_1, \ldots, a_n$.*

(b) *In the case $S = R^n_+$, a projection rule as in Theorem 150 is scale invariant iff it is generated by*

$$F(\mathbf{v}|\mathbf{u}) = \sum_{i=1}^{n} a_i h_\alpha(v_i|u_i), \tag{7.2.31}$$

*where*

$$h_\alpha(v|u) = \begin{cases} v(\log v - \log u) - v + u, & \text{if } \alpha = 1, \\ \log u - \log v + v/u - 1, & \text{if } \alpha = 0, \\ (u^\alpha - v^\alpha)/\alpha + u^{\alpha-1}(v-u), & \text{otherwise} \end{cases} \tag{7.2.32}$$

*and* $a_1, \ldots, a_n$ *are positive constants.*

In the further considerations of this paragraph we first formulate the result saying that for selection rules $\Pi : \mathcal{L} \to S$ satisfying the basic axioms (1)–(4), the generating function (7.2.22) is necessarily convex if $S$ equals $R^n$ or $R^n_+$. Then we determine how the invariance postulates (7.2.25), (7.2.26) restrict the class of possible generating functions when—unlike in Theorem 151—a transitivity postulate is not imposed. Finally, we show that a simple additional axiom permits to uniquely extend any selection or projection rule, satisfying the basic axioms, from $\mathcal{L}$ to the class $\mathcal{C}$ of all closed convex subsets of $S$. The proofs of these results can be found in [1, 3].

**Theorem 152** *Let* $f_1, \ldots, f_n$ *be continuously differentiable functions on* $V = R$ *or* $R_+$ *having the properties* (i) *and* (iii) *in the definition of a standard n-tuple. Then each* $f_i$ *is convex and there can be at most one i for which* $f_i$ *is not strictly convex.*

A projection rule will be called *smooth* if for every $i \neq j$ and $t \in V$, the $i$th and $j$th components of $\Pi(L_{ij}(t)|\mathbf{u})$ depend continuously on $u_i$ and $u_j$.

**Theorem 153** (a) *A selection rule with basic set* $S = R^n$ *is scale invariant iff it is generated by*

$$F(\mathbf{v}) = \sum_{i=1}^{n} c_i(sign\ v_i)|v_i|^p, \quad p > 1, \tag{7.2.33}$$

*where* $c_i(sign\ v_i)$ *is a positive coefficient depending on i and* $sign\ v_i$. *Further, this selection rule is strongly scale invariant iff the coefficients do not depend on the sign of* $v_i$, *i.e., iff*

$$F(\mathbf{v}) = \sum_{i=1}^{n} a_i|v_i|^p, \quad p > 1. \tag{7.2.34}$$

(b) *A projection rule with basic set* $S = R^n_+$ *is smooth and scale invariant iff it is generated by*

$$F(\mathbf{v}|\mathbf{u}) = \sum_{i=1}^{n} c_i f_i^\alpha(v_i/u_i), \quad \alpha \in R, \tag{7.2.35}$$

*where* $(f_1, \ldots, f_n)$ *is a standard n-tuple with zero at* **1**.

(c) *A projection rule with basic set* $S = R^n$ *is scale invariant iff it is generated by*

$$F(\mathbf{v}|\mathbf{u}) = \sum_{i=1}^{n} f_i^{\alpha}(v_i|u_i), \tag{7.2.36}$$

$$f_i^{\alpha}(v|u) = \begin{cases} u^p f_i^+(v/u), & \textit{if } u > 0, \\ |u|^p f_i^-(v/u), & \textit{if } u < 0, \\ c_i(\text{sign } v_i)|v_i|^p, & \textit{if } u = 0, \end{cases}$$

*where* $(f_1^+, \ldots, f_n^+)$ *and* $(f_1^-, \ldots, f_n^-)$ *are standard n-tuples with zero at* $\mathbf{1}$, *the* $c_i(\text{sign } v_i)$ *are as in part (a), and* $p > 1$. *This projection rule is strongly scale invariant iff here* $f_i^+ = f_i^-$ *and the coefficients* $c_i(\text{sign } v_i)$ *do not depend on* sign $v_i$.

(d) *A projection rule with basic set* $S = R^n$ *is smooth and translation invariant iff it is generated by*

$$F(\mathbf{v}|\mathbf{u}) = \sum_{i=1}^{n} e^{\beta u_i} f_i(v_i - u_i), \quad \beta \in R, \tag{7.2.37}$$

*where* $(f_1, \ldots, f_n)$ *is standard n-tuples with zero at* $\mathbf{0}$.

(e) *A projection rule with basic set* $S = R^n$ *is both scale and translation invariant iff it is generated by*

$$F(\mathbf{v}|\mathbf{u}) = \sum_{i=1}^{n} c_i(\text{sign } (v_i - u_i))|v_i - u_i|^p, \quad p > 1. \tag{7.2.38}$$

*It is also strongly scale invariant iff here the coefficients do not depend on sign of* $v_i - u_i$, *i.e., iff*

$$F(\mathbf{v}|\mathbf{u}) = \sum_{i=1}^{n} a_i|v_i - u_i|^p, \quad p > 1. \tag{7.2.39}$$

*Remark* Assertions (a) and (e) provide axiomatic characterizations of the families of weighted $L^p$-norms and $L^p$-distances, respectively. We notice that the projection rules characterized in (c) are not necessarily smooth because the functions $f_i(v|u)$ in (7.2.36) may be discontinuous in $u$ at $u = 0$.

The domain $\mathcal{L}$ of selection and projection rules can be extended to the class $\mathcal{C}$ of all closed convex subsets of $S$ (when $S$ equals $R_+^n$ or $\Delta_n$, the sets $C \in \mathcal{C}$ are closed in the relative topology of $S$). Selection and projection rules with domain $\mathcal{C}$ are defined analogously to those with domain $\mathcal{L}$ considered so far; when needed to avoid ambiguity, we will speak about $\mathcal{C}$-and $\mathcal{L}$-selection (projection) rules, respectively. The need for considering the larger domain $\mathcal{C}$ arises in inverse problems where the available information consists in linear inequality constraints; then the feasible set consists of those $\mathbf{v} \in S$ that satisfy the given inequality constraints. More generally, nonlinear inequality constraints also often lead to convex feasible sets.

For $\mathcal{C}$-selection rules $\Pi : \mathcal{C} \to S$, we adopt the following modification of the consistency axiom (1):

1′ If $C' \subset C$ are in $\mathcal{C}$ and $\Pi(C) \in C'$ then $\Pi(C') = \Pi(C)$. In addition, if $\tilde{C}$ is determined by one linear inequality constraint, i.e.,

$$\tilde{C} = \begin{cases} \{\mathbf{v} : \mathbf{a}^T\mathbf{v} \geq b\}, & \text{if } S = R^n \text{ or } R^n_+, \\ \{\mathbf{v} : \mathbf{a}^T\mathbf{v} \geq b, \mathbf{1}^T\mathbf{v} = 1\}, & \text{if } S = \Delta^n \end{cases} \tag{7.2.40}$$

and $\tilde{C}$ does not contain $\mathbf{v}^0 = \Pi(S)$ then $\Pi(\tilde{C}) = \Pi(L)$, where $L \in \mathcal{L}$ is the boundary of $\tilde{C}$ defined by changing $\geq$ to $=$ in (7.2.40).

**Theorem 154** *Every $\mathcal{L}$-selection rule satisfying the basic axioms (1–4) can be uniquely extended to a $\mathcal{L}$-selection rule satisfying axiom (1′). This extension is still generated by the function $F(\mathbf{v})$ of Theorem* 149 (a), *i.e., $\Pi(C)$ for $C \in \mathcal{C}$ is that element of $C$ where $F(\mathbf{v})$ attains its minimum on $C$.*

Of course, a similar result holds for the extension of $\mathcal{L}$-projection rules to $\mathcal{C}$-projection rules, subject to the obvious analog of postulate (1′) for projection rules. Further, if an $\mathcal{L}$-projection rule is scale and/or translation invariant then so will be also the $\mathcal{C}$-projection rule obtained as its (unique) extension. A somewhat weaker assertion holds for the transitivity, as well. Namely, if an $\mathcal{L}$-projection rule satisfies the transitivity postulate (7.2.27) and the unique extension to $\mathcal{C}$ satisfying the analog of (1′), then for its scale and/or translation invariant the following hold: for any $L \in \mathcal{L}$ and $C \in \mathcal{C}$ with $C \subset L$, we have for every $\mathbf{u} \in S$,

$$\Pi(C|\mathbf{u}) = \Pi(C|\Pi(L|\mathbf{u})).$$

This is an immediate consequence of the fact that every measure of distance as in (7.2.29) has the "Pythagorean property": if $\Pi(L|\mathbf{u}) = \mathbf{v}$, $\mathbf{w} \in L$, $L \in \mathcal{L}$, then

$$F(\mathbf{v}|\mathbf{u}) + F(\mathbf{w}|\mathbf{u}) = F(\mathbf{w}|\mathbf{u}). \tag{7.2.41}$$

Notice that for $F(\mathbf{v}|\mathbf{u}) = ||\mathbf{v} - \mathbf{u}||^2$, (7.2.41) reduces to the Pythagorean theorem.

#### 7.2.3.1 Open Problems

The axiomatic theory of inference for inverse problems is still in its beginning stage, and there are so many open questions that it is hard to select only a few. Below we hint at some general directions for further research.

Undoubtedly the most important would be to extend the axiomatic approach to inverse problems involving errors, at least to the extent of covering the methods used in practice for this kind of inverse problems (more common than those without errors), and possibly to arrive at new methods, as well. Even if the possibility of errors is discounted, it may be still desirable to permit "solutions" that are not "data consistent", i.e., do not necessarily belong to the feasible set determined by the available constraints. At present, it is not clear how this situation could be treated axiomatically.

Within the framework considered before, there are at least three natural directions for further research. First, other choices of the basic set $S$ could be considered. Optimistically, one might try a "general" $S$ (say any convex subset of $R^n$), but already the natural modifications of our choices $S = R^n_+$ and $S = \Delta_n$ by adding the boundary to $S$ lead to substantial new mathematical problems.

Second, one should study selection (projection) rules whose domain is not the whole $\mathcal{L}$ (or $\mathcal{C}$) but perhaps a small subfamily thereof. If an inference method is "good" for a particular class of inverse problems where the possible feasible sets are of some restricted form (such as in $X$-tomography, for example) but it cannot be extended to a "good" selection (projection) rule with domain $\mathcal{L}$, it remains elusive in an axiomatic study dealing only with the latter. A problem whose positive solution would significantly enhance the power of the results is to show, for a possibly large class of subfamilies of $\mathcal{L}$, that "good" selection (projection) rules whose domain is such a subfamily of $\mathcal{L}$ must be restrictions of "good" selection (projection) rules with domain $\mathcal{L}$.

Third, even for the present choices of the basic set and domain, the basic axioms might be challenged. Since axiom (2) ("distinctness") is intuitively less compelling than the others, the consequences of dropping it (and perhaps introducing some other axiom) should be considered. It may be conjectured that Theorem 149 would still remain valid, except that the "generating function" were not necessarily differentiable. Another natural question is whether the axiom (4) ("locality") if dropped, could still a result like Theorem 149 be proved, with a "generating function" not necessarily of a sum form as in (7.2.23). Even if without axiom (2) or (4) a meaningful result could be obtained in general, this might become possible when imposing some other intuitively attractive postulates, such as invariance and/or transitivity.

## References

1. I. Csiszár, New axiomatic results on inference for inverse problems. Studia Scientarium Mathematicarum Hungarica **26**, 207–237 (1991)
2. I. Csiszár, *I*-Divergence geometry of probability distributions and minimization problems. Ann. Probab. **3**(1), 146–158 (1975)
3. I. Csiszár, Why least squares and maximum entropy? An axiomatic approach to inference for linear inverse problems. Ann. Stat. **19**(4), 2033–2066 (1991)
4. S. Kullback, *Information Theory and Statistics* (Wiley, New York, 1959)
5. J.E. Shore, R.W. Johnson, Axiomatic derivation of the principle of maximum entropy and principle of minimum of cross-entropy. IEEE Trans. Inf. Theory **26**, 26–37 (1980)

## Further Reading


6. P.J. Huber, V. Strassen, Minimax tests and the Neyman-Pearson lemma for capacities. Ann. Stat. **1**(2), 251–263 (1973)
7. E.L. Lehmann, *Testing Statistical Hypotheses* (Chapman & Hall, New York, 1959)
8. A. Wald, *Statistical Decision Functions* (Wiley, New York, 1950)

# Chapter 8
# $\beta$-Biased Estimators in Data Compression

## 8.1 Lecture on Introduction

In this chapter, we present the results of Krichevsky [6]. We consider the following estimation problem, which arises in the context of data compression, is discussed: For a given discrete memoryless source, we want to estimate the unknown underlying source probabilities by means of a former source output, assuming that the estimated probabilities are used to encode the letters of the source alphabet.

Rough underestimates of source probabilities may lead to long, possibly infinite, codewords. For example, assume the estimated source probability of an element $x_0 \in \mathcal{X}$ to be $\hat{p}(x_0) = 0$, where $\mathcal{X}$ denotes a finite alphabet. The noiseless coding theorem suggests to assign a codeword of length $\lceil -\log \hat{p}(x_0) \rceil$ to $x_0$. If the "true" probability of $x_0$ is strictly positive and $x_0$ happens to occur as the next source output, the encoding length of $x_0$ is infinite.

Gilbert proposed in [4] a class of estimators, the $\beta$-biased estimators, in order to avoid rough underestimates:

$$\left\{ \frac{\langle x^n \mid x \rangle + \beta}{n + \mathcal{B}\beta} : \beta \in \mathbb{R}^+ \right\}, \tag{8.1.1}$$

where $\langle x^n \mid x \rangle$ denotes the frequency of $x$ in the already received source sequence $x^n = (x_1, x_2, \ldots, x_n) \in \mathcal{X}^n$ for all $x \in \mathcal{X}$.

A theoretical motivation for the choice of this specific class of estimators will be given in Sect. 8.5, via Bayes' rule: every prior distribution on the set of discrete memoryless sources over the alphabet $\mathcal{X}$ yields an estimator, a prediction rule, but it is not clear which prior one should choose. For example, choosing the Dirichlet distribution with parameter $(\beta, \ldots, \beta)$ yields the $\beta$-biased estimator. In particular, the parameter $(1/2, \ldots, 1/2)$ is known as Jeffrey's rule. Universal encoding theory offers a natural solution to this problem: to every estimator $T$ we assign its redundancy



A. Ahlswede et al. (eds.), *Transmitting and Gaining Data*,
Foundations in Signal Processing, Communications and Networking 11,
DOI 10.1007/978-3-319-12523-7_8

$R^n(T)$. The redundancy is—according to Wald—a cost or loss function. It indicates how far from its minimum an optimal code, constructed by use of the estimator $T$, is.

The Occam Razor Principle states: do not multiply entities without necessity, or, in other words, always choose a simplest description. Rissanen called it the Minimum Description Length Principle (MDL). We shall refer to these principles in our problems.

A historical occurrence, concerning estimators and Bayes' rule in a background which is closely related to ours, is reported: Laplace used Bayes' rule to calculate the probability "that the sun will rise tomorrow, given that it has risen daily for 5000 years" [3].

We are concerned with two main problems. The first problem is to find an estimator of Gilbert's class (8.1.1) that is optimal, relative to the loss function $R^n$, i.e., a prediction rule which fulfills MDL and the Occam Razor Principle. The estimator to the parameter $\beta_0 = 0.50922\ldots$ turns out to be optimal.

The second problem is a question that arises naturally as a consequence of the first: how much better than the $\beta_0$-biased estimator is the "performance" of the best estimator if we do not restrict our considerations to Gilbert's class of estimators, i.e., we allow any estimator of the source probabilities? We can provide a good answer to this question: we find a lower bound for the "performance" of any estimator. This bound is insignificantly below the "performance" of the $\beta_0$-biased estimator. Thus, the advantage of general estimators seems to be negligible, especially, if we bear in mind that this lower bound does not suggest any way how to obtain an optimal general estimator.

In practice, people agreed on using the $\frac{1}{2}$-biased estimator. For example in [7] Willems, Shtarkov, and Tjalkens, use this parameter to estimate probabilities for sequences recursively. We want to supply the theoretical justification, showing that this choice is close to the optimum.

Gilbert [4] introduced a function $R^n_S$, a source-dependent redundancy that will be introduced in Sect. 8.4, in 1971 and found its asymptotic behavior. He concluded that in a real case, the 1-biased estimator is a safe choice. Cover [1] gave a Bayesian interpretation of [4] (cf. Sect. 8.5.6). Krichevsky's paper [6] may be considered as a continuation of [4], but the redundancy $R^n$ he deals with is more advanced than in Gilbert's case: instead of the redundancy for a single source, he considers the supremum of the redundancies over all possible sources.

The function $R^n$ was introduced by Krichevsky in 1975. He proved in [5] that if a word $x^n$ is followed by a lengthy word $x^m$ rather than by a single letter $x$, then Jeffreys' estimator (the $\frac{1}{2}$-biased) is asymptotically the best.

The chapter is organized as follows: Sect. 8.2 provides a general overview of source coding and data compression. Lecture 41 introduces criteria for the performance of estimators. In Sect. 8.4, we will derive a loss function, which enables us to define a "best" estimator, define the class of $\beta$-biased estimators, and examine these by the criteria of the previous lecture. In Sect. 8.5, we consider Bayes' rule and Bayesian estimators and apply these to a binary source in order to derive Laplace's estimator ($\beta = 1$) and review a paper of Cover [1], which generalizes the first part of this lecture. Approximations for the redundancy are derived in Sect. 8.6. Section 8.7 provides a preview of the main theorems and a sketch of the proofs, as well as an introduction to the lemmata. In Sect. 8.8, we state and prove Lemma 1 to 6 and Gilbert's Corollary, a direct consequence of Lemma 3. Gilbert's Corollary has been published for the first time in [4]. Section 8.9 deals with the main theorems and their proofs.

### *8.1.1 Notation*

By $p$ we denote a real number, $0 \leq p \leq 1, q \triangleq 1 - p$. We consider the (finite) alphabet $\mathcal{X}$. W.l.o.g. assume $\mathcal{X} = \{1, 2, \ldots, m\}$, $m \in \mathbb{N}$, $m \geq 2$, where $|\mathcal{X}| = m$ is the size of the alphabet $\mathcal{X}$.

Let $p_S = (p_S(1), \ldots, p_S(m))$ be a probability distribution on the alphabet $\mathcal{X}$. A source is a pair $(\mathcal{X}, p_S)$, consisting of an alphabet $\mathcal{X}$ and a probability distribution $p_S$ on $\mathcal{X}$. The sources we consider are always over the alphabet $\mathcal{X}$, hence, it is sufficient to identify the sources by their probability distribution. We denote the set of all discrete memoryless sources (DMS) over $\mathcal{X}$ by $\Sigma$.

$C$ stands for any positive constant.

We write $\lambda_S = np_S$. For $n \to \infty$ $\lambda$ is either bounded by a number $\lambda_0$ or goes to infinity, depending on the behavior of $p$ for $n$ tending to infinity. We use this notation similar to the notation one uses for approximating the Poisson distribution by binomial distributions, without the constraint that $\lim_{n\to\infty} np_S(x)$ is necessarily finite.

Consider a source $S = (p_S(1), \ldots, p_S(m))$. It is $\lambda_S(i) = np_S(i)$ and

$$\sum_{i=1}^{m} \lambda_S(i) = n.$$

Hence, for $n \to \infty$, there is at least one index $k_0$ with $\lambda_S(k_0) \xrightarrow{n\to\infty} \infty$.

## 8.2 Lecture on Models of Data Compression

### *8.2.1 Source Coding*

In Information Theory, one usually considers the following model of a communication system:

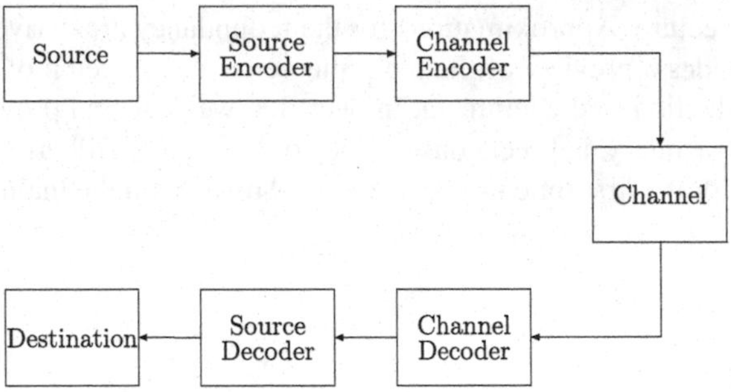


In the first step, the source encoder seeks for a shortest representation of the source output, this means the source encoder's subject is data compression. In the second step, the data has to be prepared for the transmission via the channel. The channel is considered to be noisy, the data may be disturbed. The channel encoder's task is to prepare the data for the transmission, i.e., to take care for the decodability of the transmitted, possibly disturbed data. Hence, the source encoder and the channel encoder have two contrary tasks: The source encoder removes redundancy in order to find a short representation, the channel encoder adds redundancy in order to secure correct decoding.

The channel decoder receives a sequence and "guesses" the source encoder's output. Finally, the source decoder translates the sequence to its original state.

In the following, we will concentrate on source coding, especially on Data Compression.

### 8.2.2 Data Compression

We consider a *Discrete Memoryless Source* (DMS). A DMS $S$ is a pair $S = (\mathcal{X}, p_S)$, consisting of a finite alphabet $\mathcal{X} = \{1, 2, \dots, m\}$ and a probability distribution $p_S$ on $\mathcal{X}$, furthermore are the letters of a sequence independent of each other, i.e., $p_S(x_1 x_2, \dots, x_n) = p_S(x_1) \cdot p_S(x_2) \cdot \ldots \cdot p_S(x_n)$ for all $x_1, x_2, \dots, x_n \in \mathcal{X}$. According to Sect. 8.1.1 we write $S = p_S$.

We denote the set of all finite sequences over the alphabet $\mathcal{Y}$ by

$$\mathcal{Y}^* \triangleq \bigcup_{i=0}^{\infty} \mathcal{Y}^i.$$

We adopt the source encoder's task: Find a (uniquely decipherable) code $c : \mathcal{X} \longrightarrow \mathcal{Y}^*$, which minimizes the expected code length

$$\bar{l}(c) \triangleq \sum_{x \in \mathcal{X}} p_S(x) \cdot l(x),$$

where $l(x)$ is the length of the codeword $c(x)$ for all $x \in \mathcal{X}$.

The Noiseless Coding Theorem states that for a given source $S = p_S$ it is

$$H(p_S) \leq \bar{l}(c)$$

for every code $c$ and that it is always possible to find a uniquely decipherable code $c_{\min} : \mathcal{X} \longrightarrow \mathcal{Y}^*$ with average length

$$H(p_S) \leq \bar{l}(c_{\min}) < H(p_S) + 1,$$

where $H(p_S)$ denotes the *entropy* of $p_S$, which is defined by

$$H(p_S) \triangleq -\sum_{x \in \mathcal{X}} p_S(x) \cdot \log_m p_S(x).$$

Hence, the minimal achievable average code length per letter for a source is approximately given by its entropy. The Huffman Coding Algorithm yields a prefix code which achieves this minimal average code length.

The Noiseless Coding Theorem, in combination with the Huffman Coding Algorithm, provides a good answer to the problem of data compression in the context of the DMS, if we assume, that the underlying probability distribution is known. In practice, this usually is not the case. In order to fix this deficit, one might get the idea to use the relative frequencies of the letters of $\mathcal{X}$ instead of the (not available) probability distribution. The drawback of this idea is that we have to wait until the source output is completely known, whereas in applications one usually demands, that the source output is encoded in "real time" or at least only with a short "delay."

Another idea to deal with sources, whose probabilities are unknown, is to use a universal code. This means that the code is asymptotically optimal for a whole set of sources, for example, the set of DMSs over $\mathcal{X}$. Surprisingly, there exist codes, which own these properties.

The following situation may be considered as a case in between complete knowledge of $p_S$ and almost none: the source $S$ has produced a word $x^n \in \mathcal{X}^n$. Hence, we cause $x^n$ to gain knowledge of $p_S$. This situation is called *Adaptive Encoding*. One may further distinguish between *Direct* and *Indirect* Adaptive Encoding. For Direct Adaptive Coding one uses the word $x^n$ itself, in the indirect case one estimates the probabilities by means of $x^n$.

*Example 1* Assume that we are given a computer text file. An often used code for computers is the ASCII code. The code length for every character is 8 bits. Hence every ASCII file, whose relative frequencies are not uniformly distributed, can be compressed more or less. We are looking for a code $c : \{0, 1\}^8 \longrightarrow \{0, 1\}^*$, which minimizes $\bar{l}(c)$:

1. Huffman Coding: As pointed out, in the case of Huffman Coding it is necessary to have knowledge of $p_S$. Because this is not the case, we use the relative frequencies of the letters instead of $p_S$. Consequently, we have to

   (a) generate a statistic of the relative frequencies
   (b) build the Huffman Code relative to the statistic
   (c) encode the file and write the encoded data into a new file, the compressed file.

   In order to be able to decode the compressed file, we have to include a "header," which contains either the used statistic or the codewords.
2. Universal Coding: We use a universal code and simply write the encoded data into a new file. A header is not necessary.
3. (Indirect) Adaptive Coding: In the beginning we are in the situation of Universal Coding, because there is no "source output" yet. Hence, we cannot derive any estimate.

   (a) encode the first letter, using a universal code, and write the generated data to a new file
   (b) use the already encoded letter(s) for an estimate of the underlying probabilities
   (c) build a code relative to the estimate, encode the next letter and write the encoded letter to the file
   (d) go to 3b, until the whole original file is encoded.

   Instead of using the whole encoded source output for an estimate, one may limit the data to the last $n$ letters. This is called the *Sliding Window* technique. The estimate of the probability distribution will be very rough in the beginning, but with increasing word length, it rapidly converges to the real distribution.

Therefore, the advantage of Universal and Adaptive Coding, in contrast to Huffman Coding, is their universality and the potentiality of immediate encoding. We do not have to wait for the end of the file and start counting letters. This means we can meet the demand of no delay, which is of no importance for the first example, but for the second. Another difference, in example 1, is the necessity to include a "header," which carries data for the reconstruction of the used code, for the Huffman compressed file. This is neither for Universal nor for Adaptive Coding necessary, but in the case of Adaptive Coding we pay with an impreciseness of the estimate.

*Example 2* Compression of a TV signal.
Assume we want to transmit a TV signal via a digital channel. Because of capacity constraints, we have to compress the signal. The signal consists of $s$ single pictures per second. For simplicity, we assume that one picture is a matrix, consisting of $m \cdot n$ black and white pixels. Hence, a compressing code can be considered as a code $c : \{b, w\}^{m \cdot n} \longrightarrow \{0, 1\}^*$, where $b$ stands for a black pixel and $w$ for a white, respectively.

In this example, it is obviously reasonable to demand a "real time" compression, a compression without or only little delay. This can only hardly be achieved by a Huffman Coding, because the data have to be encoded before they are completely known.

The following considerations will be made within the context of Indirect Adaptive Coding.

## 8.3 Lecture on Estimators

### *8.3.1 Estimators and Data Compression*

During the following considerations, we assume to act within the context of Indirect Adaptive Coding. We know we are dealing with a DMS $S = p_S$, but we have no direct knowledge about its probability distribution. The source $S$ has already produced a word $x^n \in \mathcal{X}^n$. Hence, we want to use $x^n$ to gain knowledge of $p_S$, i.e., we estimate $p_S$ by means of $x^n$.

An *estimator* for $p_S(x)$ and sample size $n$ is a random variable $T_x : \mathcal{X}^n \longrightarrow [0, 1]$ for all $x \in \mathcal{X}$, an *estimator* for $p_S$ and sample size $n$ is a random variable $T = (T_1, T_2, \ldots, T_m) : \mathcal{X}^n \longrightarrow [0, 1]^m$, whose component $T_x$ is an estimator for $p_S(x)$ and sample size $n$ for all $x \in \mathcal{X}$, and $\sum_{x \in \mathcal{X}} T_x = 1$. $T_x$ or $T$, respectively, applied to a word of $\mathcal{X}^n$ is called an *estimate*. The difference $T_x - p_S(x)$ is called the *error of the estimate*, the averaged error of the estimate $\mathbb{E}(T_x - p_S(x)) = \mathbb{E}(T_x) - p_S(x)$ is called the *bias* of $T_x$. $T_x$ is said to be *unbiased*, if the bias equals 0.

Let $\mathcal{G}_x(n)$ denote the set of all estimators for $p_S(x)$. A *loss function* is a mapping $L : \mathcal{G}_x(n) \longrightarrow \mathbb{R}$. It describes the "penalty" we have to "pay" by choosing a certain estimator. Therefore, a loss function is a criterion which we use to describe "the best" estimator: An estimator $T_{\text{opt}} \in \mathcal{G}_x$ is called *optimal* (relative to the loss function $L$) if $L$ assumes a global minimum in $T_{\text{opt}}$.

The first choice for an estimator of $p_S(x)$ one would take into consideration would probably be the so-called *naive estimator*

$$\frac{\langle x^n \mid x \rangle}{n},$$

where $\langle x^n \mid x \rangle$ denotes the frequency of $x$ in $x^n$. This estimator is unbiased and optimal relative to the loss function $\mathbb{E}[(T_x - \mathbb{E}(T_x))^2]$, which provides a good intuitive interpretation. But the choice of a loss function is arbitrary and there is no good reason for choosing this one. In the following, we will derive a different loss function, which is more suitable for our problem, i.e., a loss function which regards the background of Adaptive Coding.

Another argument against the naive estimator from a more practical point of view is the following: Assume that for a certain $x_0 \in \mathcal{X}$ it is $p_S(x_0) > 0$ and the source output $x^n \in \mathcal{X}^n$ is known, but $\langle x^n \mid x_0 \rangle = 0$. Hence, the naive estimator would yield 0 as the estimate for $p_S(x_0)$. If the next source output is $x_0$, we encode $x_0$ by a codeword of length $\lceil -\log_m \langle x^n \mid x_0 \rangle \rceil$, according to the Noiseless Coding

Theorem. Obviously this is no good choice. Consequently, it is preferable to avoid rough underestimates of the probabilities.

Laplace already examined a problem, which may be considered as a special case of the present; he considered a binary DMS and derived an estimator for the underlying probability distribution $(p, 1-p)$, $p \in [0, 1]$ fixed, from a Bayesian point of view. His estimate was

$$\frac{\langle x^n \mid x \rangle + 1}{n+2},$$

assuming that the source output $x^n \in \{0, 1\}^n$ is already known.

In Sect. 8.5.1, we will derive Laplace's estimator. Gilbert proposed in [4] a generalization of Laplace's estimate,

$$\hat{p}_S^{\beta}(x|x^n) \triangleq \frac{\langle x^n \mid x \rangle + \beta}{n + \beta m},$$

for $\beta > 0$ fixed, $x \in \mathcal{X}$ and $x^n \in \mathcal{X}^n$. We will refer to $\hat{p}_S^{\beta}(x|x^n)$ as the *$\beta$-biased estimator for $p_S(x)$* (given $x^n$).

Gilbert gives the following intuitive interpretation: "$\hat{p}_S^{\beta}(\cdot|x^n)$ is the estimate of $p_S$ one would obtain by the usual rule (the naive estimator), if $\beta$ extra letters of each kind were added to the sample."

Gilbert concluded that $\beta = 1$ is a "safe choice": "In a real case with unknown $p_S$ and large $n$, the value $\beta = 1$ is a safe choice. For, if some of the $p_S(x)$ are very small, Gilbert's Corollary (Sect. 8.8.1) shows that $\beta = 1$ is near the best value. If none of the $p_S(x)$ are very small, $\beta = 1$ may be far from the minimizing value; but there is high probability that each letter has many occurrences. Then even $\hat{p}(x) = \frac{\langle x^n|x\rangle}{n}$ (the estimate with $\beta = 0$) would have been good most of the time; taking $\beta = 1$ provides some insurance against the rare event $\langle x^n \mid x \rangle = 0$."

### 8.3.2 Criteria for Estimators

Different criteria for the "quality" of an estimator have been proposed:

1. *Unbiasedness*. The bias of an estimator has already been introduced in the previous paragraph.
2. *Efficiency*. Let $(X_\vartheta)_{\vartheta \in \Theta}$ be an indexed family of random variables with density $f(x; \vartheta)$ for all $\vartheta \in \Theta$. The *Fisher information* $J(\vartheta)$ is defined by

$$J(\vartheta) \triangleq \mathbb{E}_\vartheta\left(\left[\frac{\partial}{\partial \vartheta} \ln f(x; \vartheta)\right]^2\right). \quad (8.3.1)$$

The Fisher information of $n$ independent, identically distributed (i.i.d.) random variables $X_1, X_2, \ldots, X_n$ with density $f(x; \vartheta)$ turns out to be

$$J_n(\vartheta) = n \cdot J(\vartheta).$$

The *Cramér–Rao inequality* states that the variance of any *unbiased* estimator $T$ of the parameter $\vartheta$ is lower bounded by the reciprocal of the Fisher information:

$$\mathbb{V}(T) \geq \frac{1}{J(\vartheta)}. \tag{8.3.2}$$

Equation (8.3.2) induces the following definition: An unbiased estimator $T$ is said to be *efficient* if it meets the Cramér–Rao bound with equality, i.e., if $\mathbb{V}(T) = \frac{1}{J(\vartheta)}$. For a biased estimator $T$ with bias $b(\vartheta)$ the Cramér–Rao inequality (8.3.2) becomes

$$\mathbb{V}(T) \geq \frac{[1 + b'(\vartheta)]^2}{J(\vartheta)}, \tag{8.3.3}$$

where $b'(\vartheta)$ is the derivative of $b$ with respect to $\vartheta$.
"The Fisher information is therefore a measure of the amount of 'information' about $\vartheta$ that is present in the data. It gives a lower bound on the error in estimating $\vartheta$ from the data. However, it is possible that there does not exist an estimator meeting this lower bound."[2]

3. *Consistency*. An estimator $T_x : \mathcal{X}^n \to [0, 1]$ for $p_S(x)$ is called *consistent* if

$$\lim_{n \to \infty} p_S \left( \left| T_x(x^n) - p_S(x) \right| \geq \varepsilon \right) = 0$$

for every $\varepsilon > 0$ and $x \in \mathcal{X}$.
If $T_x$ is a consistent estimator for $p_S(x)$ and $n$ is large, then $T_x(x^n)$ is with high probability close to $p_S(x)$.

The bias of an estimator refers to its expected value, the "efficiency" to its variance. An unbiased estimator is worthless, if its variance is very great. Vice versa, a small variance does not guarantee a "good" estimator: Choosing $T$ constant yields an estimator with 0 variance, which is worthless in most cases.

Furthermore, there are criteria which may help to find a "minimal representation of the data without a loss of information":

1. *Sufficiency*. Let $(f_p)_p$ be a family of probability distributions on $\mathcal{X}^n$ indexed by $p$, $x^n$ a sample of a distribution from this family. A mapping $S : \mathcal{X}^n \to \mathbb{R}$ is called a *statistic*. A statistic $S$ is called *sufficient* to the family $(f_p)_p$, if the conditional distribution of $x^n$ given $S(x^n) = t$ does not involve $p$.
   Thus a sufficient statistic contains all the "information" of $p$.
2. *Minimal Sufficiency*. A statistic $S$ is a *minimal sufficient statistic* to the family $(f_p)_p$, if it is a function of every other sufficient statistic.

A minimal sufficient statistic maximally compresses the information about $p$ in the sample. Other sufficient statistics may contain additional irrelevant "information."

## 8.4 Lecture on a Suitable Loss Function and the Class of $\beta$-Biased Estimators

From now on, we will use the natural logarithm ln instead of the logarithm to base $m$ for all calculations. In particular, we introduce a "scaled entropy": for a source $S = p_S$ we define

$$H_S \triangleq -\sum_{x\in\mathcal{X}} p_S(x)\ln p_S(x).$$

The usage of the scaled entropy $H_S$ instead of the "correct" entropy $H(p_S)$ does not influence the results, it only scales the "redundancy" with a certain constant.

### 8.4.1 Derivation of the Loss Function

Let $S \triangleq p_S \in \Sigma$, $x^n \in \mathcal{X}^n$ be a word produced by $S$. By

$$\mathcal{G}(n) \triangleq \left\{ T = (T_1, T_2, \dots, T_m) : \mathcal{X}^n \longrightarrow [0,1]^m \,\middle|\, T_x \in \mathcal{G}_x(n), \sum_{x\in\mathcal{X}} T_x = 1 \right\}$$

we denote the set of all estimators for the probability distribution $p_S$.
Let $T \in \mathcal{G}(n)$. The application of $T$ to $x^n$ yields an estimate for $p_S$:

$$\hat{p}(x \mid x^n) \triangleq T_x(x^n) \quad \text{for all } x \in \mathcal{X}.$$

We can use this estimate to construct a code $c : \mathcal{X} \longrightarrow \mathcal{Y}^*$ with appertaining length function $l_S^T(\cdot \mid x^n)$ which is optimal relative to the estimate $\hat{p}(\cdot \mid x^n)$. Therefore it is $l_S^T(x \mid x^n) = -\ln \hat{p}(x \mid x^n)$, for all $x \in \mathcal{X}$. The average length per letter of the code $c$ is defined by

$$\bar{l}_S^T(x^n) \triangleq \sum_{x\in\mathcal{X}} p_S(x) \cdot l_S^T(x \mid x^n). \tag{8.4.1}$$

This average length depends on the word $x^n$, on which the estimate and consequently the code is based. In order to take all possible source outputs of length $n$ into account, we average (8.4.1) over all elements of $\mathcal{X}^n$:

$$\bar{L}_S^T(n) \triangleq \sum_{x^n\in\mathcal{X}^n} p_S(x^n) \cdot \bar{l}_S^T(x^n), \tag{8.4.2}$$

where $p_S(x^n) = \prod_{i=1}^{n} p_S(x_i)$ for all $x^n = (x_1, x_2, \dots, x_n) \in \mathcal{X}^n$ because $S$ is assumed to be memoryless. Nevertheless (8.4.2) still is an average length per letter!

From the Noiseless Coding Theorem follows

$$\overline{L}_S^T(n) \geq H_S \tag{8.4.3}$$

for every uniquely decipherable code. This inequality motivates to define the redundancy of an estimate on a source $S$ by

$$R_S^n(T) \triangleq \overline{L}_S^T(n) - H_S. \tag{8.4.4}$$

We also call $R_S^n : \mathcal{G}(n) \longrightarrow \mathbb{R}$ a loss function, a criterion for the performance of an estimate, but it is depending on the source $S$. A source-independent definition of redundancy is desirable, because we do not know $S$. Therefore, we define the (overall) redundancy of an estimate by[1]

$$R^n(T) \triangleq \sup_{S \in \Sigma} R_S^n(T) = \max_{S \in \Sigma} R_S^n(T). \tag{8.4.5}$$

### 8.4.2 Discussion of an Alternative Loss Function

By using (8.4.5), we choose the maximal redundancy of an estimate on a source to be our loss function. Instead, we also could use the *average* redundancy on a source:

$$\overline{R}^n(T) \triangleq \int_{\Sigma} R_S^n(T) d\mu(S),$$

where $\mu$ denotes a probability measure on $\Sigma$. From a practical point of view, one may favor $\overline{R}^n$, because $R^n$ does not take the probability distribution on $\Sigma$ into account. For example, it could happen that the source which achieves $R^n(T)$ is very unlikely to be chosen. On the other hand, it is hard to say which sources are likely to occur and which are not, because we have no knowledge of the distribution on $\Sigma$. But if

---

[1] We identify $\Sigma$ with the set

$$\left\{ x = (x_1, x_2, \dots, x_m) \in \mathbb{R}^m : x_i \geq 0 \text{ for } i \in \{1, \dots, m\} \text{ and } \sum_{i=1}^{m} x_i = 1 \right\}.$$

$\Sigma$ is compact in $\mathbb{R}^m$ and both $\overline{L}_S^T(n)$ and $H_S$ are continuous in $\Sigma$ relative to $S$. Therefore $R_S^n(T)$ is continuous and, understood as a function of $S$, assumes its maximum and minimum in $\Sigma$.

we want to calculate $\overline{R}^n$, it is necessary to define the probability measure $\mu$, and we are made to assume a Bayesian point of view:

$\mu$ establishes a prior distribution on $\Sigma$, i.e., that the source $S$ itself is chosen by a random process and $\mu$ must be defined explicitly, although it is absolutely not clear, which prior one should choose.

In this situation of no knowledge, one tends to assume a uniformly distributed prior. But this is no satisfying solution either. For the calculation of $R^n$, in contrast to $\overline{R}^n$, we only assume that every source $S \in \Sigma$ has a chance to be chosen. Hence, we also assume that a prior on $\Sigma$ exists, but it is not necessary to define its distribution on $\Sigma$.

Therefore, we choose $R^n$ as our loss function, because this decision enables us to avoid defining a prior on $\Sigma$.

### 8.4.3 Class of β-Biased Estimators

We define

$$\mathbb{E}_x(n) \triangleq \left\{ T_x^\beta = \frac{\langle \cdot \mid x \rangle + \beta}{n + \beta m} : \mathcal{X}^n \longrightarrow [0, 1] \;\middle|\; \beta \in \mathbb{R}^+ \right\}$$

to be the *set of all β-biased estimators for* $p_S(x)$ for all $x \in \mathcal{X}$, and

$$\mathbb{E}(n) \triangleq \left\{ (T_1^\beta, T_2^\beta, \dots, T_m^\beta) \mid T_x^\beta \in \mathbb{E}_x(n) \quad \text{for all } x \in \mathcal{X}, \beta \in \mathbb{R}^+ \right\}$$

to be the *set of all β-biased estimators for* $p_S$.

With these notations it is $\mathbb{E}_x(n) \subset \mathcal{G}_x(n)$ for all $x \in \mathcal{X}$ and $\mathbb{E}(n) \subset \mathcal{G}(n)$.

### 8.4.4 Application of β-Biased Estimators to Different Criteria

We apply the criteria given in Sect. 8.3.2 to an estimator $T_x^\beta \in \mathbb{E}_x(n)$:

1. *Bias.*[2]

$$\mathbb{E}(T_x^\beta) = \mathbb{E}\left( \frac{\langle \cdot \mid x \rangle + \beta}{n + \beta m} \right) = \sum_{x^n \in \mathcal{X}^n} p_S(x^n) \cdot \frac{\langle x^n \mid x \rangle + \beta}{n + \beta m}$$

[2] Instead of using the multinomial distribution for resolving the sum, the calculations show that it is sufficient to distinguish only between elements $x$ and (not $x$) and apply the binomial distribution.

$$
\begin{aligned}
&= \sum_{k=0}^{n} \sum_{x^n \in \mathcal{X}^n : \langle x^n | x \rangle = k} p_S(x^n) \cdot \frac{k+\beta}{n+\beta m} \\
&= \sum_{k=0}^{n} \sum_{(k_1,\dots,k_n): \sum k_y = n, k_x = k} \binom{n}{k_1, \dots, k_n} \prod_{y \in \mathcal{X}} p_S(y)^{k_y} \cdot \frac{k+\beta}{n+\beta m} \\
&= \sum_{k=0}^{n} \binom{n}{k} p_S(x)^k \\
&\quad \sum_{\sum k_y = n, k_x = k} \binom{n-k}{k_1, \dots, k_{x-1}, k_{x+1}, \dots, k_n} \prod_{y \neq x} p_S(y)^{k_y} \cdot \frac{k+\beta}{n+\beta m} \\
&= \sum_{k=0}^{n} \binom{n}{k} p_S(x)^k \left( \sum_{y \neq x} p_S(y) \right)^{n-k} \cdot \frac{k+\beta}{n+\beta m} \\
&= \sum_{k=0}^{n} \binom{n}{k} p_S(x)^k \, (1 - p_S(x))^{n-k} \cdot \frac{k+\beta}{n+\beta m} \\
&= \frac{\beta}{n+\beta m} + \frac{1}{n+\beta m} \cdot n p_S(x) \qquad (8.4.6) \\
&\xrightarrow{n \to \infty} \; p_S(x).
\end{aligned}
$$

Hence $T_x^\beta$ is biased by

$$\frac{\beta + n p_S(x)}{n + \beta m} - p_S(x), \tag{8.4.7}$$

but with $n$ tending to infinity, the bias tends to 0 for every $x \in \mathcal{X}$.

2. *Efficiency*. For brevity, we set $r \overset{\Delta}{=} p_S(x_0)$ for a fixed $x_0 \in \mathcal{X}$.
   The random variable

$$X_r \;\overset{\Delta}{=}\; \langle \cdot \mid x_0 \rangle : \mathcal{X}^n \to \{0, 1, \dots, n\} \quad \text{for } r \in [0, 1]$$

is binomially distributed to the parameters $n, r$. $(X_r)_{r \in [0,1]}$ is the family of random variables under consideration. It is $X_r = X_{\tilde{r}}$ for every $r, \tilde{r} \in [0, 1]$ but the images $X_r(p_S)$ differ with $r$.
In this discrete case, the Fisher information (8.3.1) is

$$
\begin{aligned}
J(r) &= \sum_{i=0}^{n} \left( \frac{\partial}{\partial r} \ln p_S(X_r = i) \right)^2 \cdot p_S(X_r = i) \\
&= \sum_{i=0}^{n} \left( \frac{\partial}{\partial r} \ln \left[ \binom{n}{i} r^i (1-r)^{n-i} \right] \right)^2 \cdot p_S(X_r = i)
\end{aligned}
$$

$$
\begin{aligned}
&= \sum_{i=0}^{n} \left( \frac{\partial}{\partial r} \left[ \ln \binom{n}{i} + i \ln r + (n-i) \ln(1-r) \right] \right)^2 \cdot p_S(X_r = i) \\
&= \sum_{i=0}^{n} \left( \frac{i}{r} - \frac{n-i}{1-r} \right)^2 \cdot p_S(X_r = i) \\
&= \sum_{i=0}^{n} \left( \frac{i-nr}{r(1-r)} \right)^2 \cdot p_S(X_r = i) \\
&= \frac{1}{r^2(1-r)^2} \cdot \mathbb{V}(X_r) = \frac{n}{r(1-r)}, \qquad (8.4.8)
\end{aligned}
$$

because $X_r$ is binomially distributed to the parameters $n, r$, and therefore $\mathbb{V}(X_r) = nr(1-r)$.
The derivative of the bias is

$$
\begin{aligned}
b'(r) &= \frac{\mathrm{d}}{\mathrm{d}r} \left( \frac{\beta}{n+\beta m} + \frac{1}{n+\beta m} \cdot nr - r \right) \\
&= \frac{n}{n+\beta m} - 1. \qquad (8.4.9)
\end{aligned}
$$

Combining (8.4.8) and (8.4.9), the Cramér–Rao inequality for the biased case (8.3.3) states

$$
\begin{aligned}
\mathbb{V}(T) &\geq \left( \frac{n}{n+\beta m} \right)^2 \cdot \frac{r(1-r)}{n} \\
&= \frac{nr(1-r)}{(n+\beta m)^2} \qquad (8.4.10)
\end{aligned}
$$

for every estimator $T$ for $r$.
Calculation of the variance of a $\beta$-biased estimator:
For a $\beta$-biased estimator $T_{x_0}^{\beta} \in \mathbb{E}_{x_0}(n)$ it is

$$
\begin{aligned}
\mathbb{E}((T_{x_0}^{\beta})^2) &= \sum_{x^n \in \mathcal{X}^n} p_S(x^n) \cdot \left( \frac{\langle x^n \mid x_0 \rangle + \beta}{n+\beta m} \right)^2 \\
&= \sum_{k=0}^{n} \binom{n}{k} r^k (1-r)^{n-k} \cdot \frac{k^2 + 2\beta k + \beta^2}{(n+\beta m)^2} \\
&\overset{(5)}{=} \frac{\beta^2 + 2\beta nr + n(n-1)r^2 + nr}{(n+\beta m)^2}. \qquad (8.4.11)
\end{aligned}
$$

Therefore the variance of $T_{x_0}^{\beta}$ is

$$\begin{aligned}\mathbb{V}(T_{x_0}^{\beta}) &= \mathbb{E}((T_{x_0}^{\beta})^2) - \mathbb{E}(T_{x_0}^{\beta})^2\\ &\overset{(11),(6)}{=} \frac{\beta^2 + 2\beta nr + n(n-1)r^2 + nr}{(n+\beta m)^2} - \left(\frac{\beta + nr}{n+\beta m}\right)^2\\ &= \frac{nr(1-r)}{(n+\beta m)^2}. \end{aligned} \tag{8.4.12}$$

Hence a $\beta$-biased estimator fulfills the Cramér–Rao inequality (8.4.10) with equality.

3. *Consistency*. Let $T_x^{\beta} \in \mathbb{E}_x(n)$, $\varepsilon > 0$. According to (8.4.12) it is

$$\lim_{n\to\infty} \mathbb{V}(T_x^{\beta}) = \lim_{n\to\infty} \frac{np_S(x)(1-p_S(x))}{(n+\beta m)^2} = 0$$

$$\overset{\text{Chebyshev}}{\Longrightarrow} \quad p_S\left(\left|T_x^{\beta} - \mathbb{E}(T_x^{\beta})\right| \geq \varepsilon\right) \leq \frac{1}{\varepsilon^2}\mathbb{V}(T_x^{\beta}) \xrightarrow{n\to\infty} 0. \tag{8.4.13}$$

Hence the estimator $T_x^{\beta}$ is consistent.

1. *Sufficiency*. We set $r \triangleq p_S(x_0)$ for a fixed $x_0 \in \mathcal{X}$ and define a random variable $B : \mathcal{X} \to \{0, 1\}$ by

$$B(x) = \begin{cases} 0, & x \neq x_0 \\ 1, & x = x_0 \end{cases} \quad \text{for } x \in \mathcal{X},$$

and $B^n : \mathcal{X}^n \to \{0, 1\}^n$ by

$$B(x^n) \triangleq (B(x_1), \dots, B(x_n)) \quad \text{for all } x^n = (x_1, \dots, x_n) \in \mathcal{X}^n.$$

Now it is

$$P_r(b^n) \triangleq p_S(B^n = b^n) = r^{\langle b^n | 1\rangle}(1-r)^{\langle b^n | 0\rangle} \qquad \text{for all } b^n \in \{0, 1\}^n.$$

We consider the random variable

$$X_r \triangleq \langle \cdot \mid 1\rangle : \{0, 1\}^n \to \{0, 1, \dots, n\}.$$

For every $r \in [0, 1]$, $t \in \{0, 1, \dots, n\}$, and $b^n \in \{0, 1\}^n$ with $\langle b^n \mid 1\rangle = t$ it is

$$P_r(b^n \mid X_r = t) = \frac{p_S(B^n = b^n)}{P_r(X_r = t)} = \frac{r^t(1-r)^{n-t}}{\binom{n}{t} r^t (1-r)^{n-t}} = \binom{n}{t}^{-1}.$$

This means

$$P_r(b^n \mid X_r = t) = \begin{cases} \binom{n}{t}^{-1}, & \text{if } \langle b^n \mid 1 \rangle = t \\ 0, & \text{otherwise} \end{cases} \tag{8.4.14}$$

for every $r \in [0, 1]$ and $b^n \in \{0, 1\}^n$.
Obviously (8.4.14) does not depend on $r$. Therefore, $X_r$ is a sufficient statistic relative to $(P_r)_{r \in [0,1]}$.

2. *Minimal Sufficiency*. The minimal sufficient estimator can be calculated by means of the likelihood function (cf. Sect. 8.5.3). $X_r$ turns out to be **a** minimal sufficient statistic.

## 8.5 Lecture on Derivation of β-Biased Estimators

### *8.5.1 Laplace's Estimator*

In this lecture, we will reproduce the derivation of Laplace's estimator. For the calculation we need Bayes' rule. Fundamentally being a reformulation of the definition of conditional probability, Bayes' rule provides a method to infer knowledge on the basis of observations. Although mathematically doubtlessly correct, Bayes' rule became ill reputed by some people's metaphysical applications. Laplace himself contributed to this development by using Bayes' rule to calculate the probability that the sun will rise tomorrow, provided that it has risen daily since the creation of the world.

Nowadays Bayes' rule splits the statisticians into two groups of believers: the Bayesians and the nonBayesians.

Assume we are given data, generated by an unknown distribution $F$ and we want to estimate a parameter $\theta$ which influences $F$. The Bayesian point of view is that $\theta$ itself is chosen according to a probability distribution $G_\theta$, called the prior or a priori distribution of $\theta$. By means of Bayes' rule, the prior distribution and the data are combined to "sharpen" the knowledge of $\theta$, leading to the so-called *posterior* or *a posteriori* distribution of $\theta$.

The nonBayesians doubt that the assumption of a parameter, which is chosen by a "stochastical process," is justified. Furthermore, in practice, one often does not know which prior distribution one should choose—provided one accepts the assumption of its existence. In this situation of no knowledge, one tends to choose the uniform distribution or tries to guess by means of subjective ideas about the parameter $\theta$. On the other hand, an argument for choosing the uniform distribution is the principle of maximal entropy.

### *8.5.2 Bayes' Rule*

Let $(\Omega, \mathcal{A}, P)$ be a probability space, $A, B \in \mathcal{A}$ two events. The *conditional probability* for the occurrence of the event $A$, provided that $B$ has already occurred and $P(B) > 0$ , is defined by

$$P(A \mid B) \triangleq \frac{P(A \cap B)}{P(B)}.$$

Consequently it is

$$P(A \cap B) = P(A \mid B) \cdot P(B) = P(B \mid A) \cdot P(A). \tag{8.5.1}$$

This yields

$$P(A \mid B) = \frac{P(B \mid A) \cdot P(A)}{P(B)}. \tag{8.5.2}$$

Let $A_1, \ldots, A_n \in \mathcal{A}$ be a partition of $\Omega$ (this means $\bigcup_{i=1}^n A_i = \Omega$ and $A_i \cap A_j = \emptyset$ for all $i \neq j$). Then it is

$$P(B) = P(B \cap \Omega) = P\left(B \cap \bigcup_{i=1}^n A_i\right) = P\left(\bigcup_{i=1}^n (B \cap A_i)\right) = \sum_{i=1}^n P(B \cap A_i) \tag{8.5.3}$$

Using (8.5.1) and (8.5.3), we can write (8.5.2) as

$$P(A_j \mid B) = \frac{P(B \mid A_j) \cdot P(A_j)}{\sum_{i=1}^n P(B \mid A_i) \cdot P(A_i)} \tag{8.5.4}$$

for all $j = 1, \ldots, n$, and obtained *Bayes' rule*. $P(A_j)$ is called the a priori or *prior probability* of $A_j$, $P(A_j \mid B)$ is called the *a posterior* or *posterior probability* for $A_j$ given $B$. For the application of Bayes' rule, it is obviously necessary to have knowledge of the a priori probability in order to obtain the desired a posterior probability. Unfortunately this knowledge is not always present, so one tends to assume uniformly distributed prior probabilities, "often used as synonymous for 'no advance knowledge' " [3], as we will also do for the calculation of (8.5.14). If this assumption is justified, has to be doubted.

### 8.5.3 *Generalization of Bayes' Rule to Continuous Parameters and* n *Observations*

We want to generalize Bayes' rule to a continuous parameter $p \in \mathbb{R}$, whose probability distribution is given by the density $\varphi$, and a discrete random variable $X$. Therefore, it is $\varphi \geq 0$ and $\int_{-\infty}^{\infty} \varphi(x)\mathrm{d}x = 1$.

From (8.5.4), we conclude that in this continuous case Bayes' rule can be written as

$$\varphi(p \mid X = x_0) = \frac{\varphi(p) \cdot P(X = x_0 \mid p)}{\int_{-\infty}^{\infty} \varphi(s) \cdot P(X = x_0 \mid s)\,\mathrm{d}s} \tag{8.5.5}$$

For a fixed $p \in \mathbb{R}$, the conditional probability for the observed $X = x_0$ given $p$ is also called the *Likelihood* for $x_0$:

$$L(p \mid X = x_0) \;\triangleq\; P(X = x_0 \mid p)$$

The denominator in (8.5.5) is a normalizing constant:

$$c' \;\triangleq\; \left( \int_{-\infty}^{\infty} \varphi(s) \cdot P(X = x_0 \mid s)\,\mathrm{d}s \right)^{-1}.$$

Hence we can rewrite (8.5.5) as

$$\varphi(p \mid X = x_0) = c' \cdot \varphi(p) \cdot L(p \mid X = x_0) \tag{8.5.6}$$

In this context, $\varphi$ serves as the prior and $\varphi(\cdot \mid X = x_0)$ as the posterior density of $p$.

The next generalization is to extend (8.5.6) to $n$ observations:

Let $X_1, \ldots, X_n$ be independent, identically distributed probability distributions. Because of the independence of $X_1, \ldots, X_n$, it is

$$P(X_1 = x_1, \ldots, X_n = x_n \mid p) = \prod_{i=1}^{n} P(X_i = x_i \mid p).$$

Now we can use (8.5.6) and obtain

$$\begin{aligned} &\varphi(p \mid X_1 = x_1, \ldots, X_n = x_n) \\ =\; &\frac{\varphi(p) \cdot \prod_{i=1}^{n} P(X_i = x_i \mid p)}{\int_{-\infty}^{\infty} \varphi(s) \prod_{i=1}^{n} P(X_i = x_i \mid s)\,\mathrm{d}s} && (8.5.7) \\ =\; &c \cdot \varphi(p) \cdot L(p \mid X_1 = x_1, \ldots, X_n = x_n), && (8.5.8) \end{aligned}$$

where $c \;\triangleq\; \left(\int_{-\infty}^{\infty} \varphi(s) \prod_{i=1}^{n} P(X_i = x_i \mid s)\,\mathrm{d}s\right)^{-1}$.

### 8.5.4 Bayesian Estimators

We received a conditional density for the parameter $p$ based upon the $n$ observations $X_1 = x_1, \dots, X_n = x_n$. Using this density, we want to obtain an estimate for $p$ conditioned on $X_1 = x_1, \dots, X_n = x_n$.

Again, we introduce a *loss function* $l : \mathbb{R}^2 \longrightarrow \mathbb{R}$. The idea is that the loss function $l$ assigns a "penalty" or "economic cost" $l(\hat{p}, p)$ to each guess $\hat{p}$ of the unknown $p$. Naturally, we are concerned to keep this "penalty" as small as possible. Therefore, we define the *Bayesian estimator* for $p$ by

$$\begin{aligned}&\hat{p}(X_1 = x_1, \dots, X_n = x_n)\\ &\quad \stackrel{\Delta}{=} \ \arg\min_{\hat{p}\in\mathbb{R}} \int_{-\infty}^{\infty} l(\hat{p}, s) \cdot \varphi(s \mid X_1 = x_1, \dots, X_n = x_n)\, \mathrm{d}s.\end{aligned}$$

In particular, choosing the squared error as our loss function

$$l(\hat{p}, p) \ \stackrel{\Delta}{=} \ (\hat{p} - p)^2, \tag{8.5.9}$$

yields

$$\begin{aligned}&\hat{p}(X_1 = x_1, \dots, X_n = x_n)\\ &= \arg\min_{\hat{p}\in\mathbb{R}} \int_{-\infty}^{\infty} (\hat{p} - s)^2 \cdot \varphi(s \mid X_1 = x_1, \dots, X_n = x_n)\, \mathrm{d}s\\ &= \arg\min_{\hat{p}\in\mathbb{R}} \Bigg[\hat{p}^2 - 2\hat{p} \underbrace{\int_{-\infty}^{\infty} s \cdot \varphi(s \mid X_1 = x_1, \dots, X_n = x_n)\, \mathrm{d}s}_{\stackrel{\Delta}{=}\ \overline{p}}\\ &\quad + \underbrace{\int_{-\infty}^{\infty} s^2 \cdot \varphi(s \mid X_1 = x_1, \dots, X_n = x_n)\, \mathrm{d}s}_{\stackrel{\Delta}{=}\ c}\Bigg]\\ &= \arg\min_{\hat{p}\in\mathbb{R}} \left[\hat{p}^2 - 2\overline{p}\hat{p} + c\right]\\ &= \arg\min_{\hat{p}\in\mathbb{R}} \left[\left(\hat{p} - \overline{p}\right)^2 - \overline{p}^2 + c\right]\\ &= \overline{p}.\end{aligned}$$

Therefore, if we choose the squared error as the loss function, the Bayesian estimator is the posterior mean:

$$\hat{p}(X_1 = x_1, \ldots, X_n = x_n) = \int_{-\infty}^{\infty} s \cdot \varphi(s \mid X_1 = x_1, \ldots, X_n = x_n)\, ds. \quad (8.5.10)$$

When authors write of the "Bayesian estimator" without further qualification, they usually refer to the posterior mean.

For the following calculations, we will choose (8.5.9) as our loss function.

### 8.5.5 Application of a Bayesian Estimator to a Binary Source

Consider a binary Discrete Memoryless Source $S = (\{0, 1\}, (1 - p, p))$ and denote the probability distribution $(1 - p, p)$ by $P_p$. We have no information of $p$, except $0 \leq p \leq 1$. Therefore, we assume every admissible value for $p$ to be equally probable and define

$$\varphi(p) \triangleq \begin{cases} 1, & 0 \leq p \leq 1 \\ 0, & \text{otherwise} \end{cases}$$

as its prior density. As mentioned before, there is no "good" justification for doing so.

Assume that a word $x^n = (x_1, \ldots, x_n) \in \{0, 1\}^n$ has already been produced by the independent and identically distributed random variables $X_1, \ldots, X_n$. It is

$$P_p(X_i = x_i) = \begin{cases} p, & x_i = 1 \\ 1 - p, & x_i = 0 \end{cases} \quad (8.5.11)$$

for all $i = 1, \ldots, n$, and

$$L(p \mid X_1 = x_1, \ldots X_n = x_n) = \prod_{i=1}^{n} P_p(X_i = x_i). \quad (8.5.12)$$

Define the random variable "number of successes"

$$S_n \triangleq \sum_{i=1}^{n} X_i$$

and assume $S_n(x^n) = \langle x^n \mid 1 \rangle = k$.
This yields according to (8.5.11) and (8.5.12)

$$L(p \mid S_n = k) = p^k \cdot (1 - p)^{n-k},$$

and by (8.5.8)

$$\varphi(p \mid S_n = k) = c \cdot p^k \cdot (1 - p)^{n-k} \tag{8.5.13}$$

for the posterior density of $p$.

Obviously, (8.5.13) is the density of the beta distribution to the parameters $k + 1$ and $n - k + 1$. (An introduction of the beta distribution is given in Sect. 8.10.2) Hence, we can calculate the normalizing constant $c$ by (8.10.1):

$$c = \frac{\Gamma(n+2)}{\Gamma(k+1) \cdot \Gamma(n-k+1)},$$

where $\Gamma$ denotes the gamma function.

In order to obtain the Bayesian estimator, we have to calculate $\hat{p}(S_n = k)$. We will do this by means of (8.5.10), (8.10.1), and the property of the gamma function

$$\Gamma(x+1) = x \cdot \Gamma(x)$$

for all $x \in \mathbb{R}^+$. We obtain

$$\begin{aligned}
\hat{p}(S_n = k) &\overset{(10)}{=} \int_0^1 t \cdot c \cdot t^k \cdot (1-t)^{n-k}\,\mathrm{d}t \\
&= c \int_0^1 t^{k+1} \cdot (1-t)^{n-k}\,\mathrm{d}t \\
&= \frac{\Gamma(n+2)}{\Gamma(k+1) \cdot \Gamma(n-k+1)} \cdot \frac{\Gamma(k+2) \cdot \Gamma(n-k+1)}{\Gamma(n+3)} \\
&= \frac{\Gamma(n+2)}{\Gamma(k+1)} \cdot \frac{\Gamma(k+2)}{\Gamma(n+3)} \\
&= \frac{k+1}{n+2}
\end{aligned} \tag{8.5.14}$$

This is the Bayesian estimate for the parameter $p$ of a binary DMS, first calculated by Laplace.

From our point of view, (8.5.14) is a 1-biased estimator for a binary source, a special case of Gilbert's proposal of $\beta$-biased estimators.

### 8.5.6 Admissibility of $\beta$-Biased Estimators

In the following, we want to review Cover's interpretation [1] of Gilbert's proposal of $\beta$-biased estimators [4]:

We consider a random variable $X$ with $p(x) \overset{\triangle}{=} p_S(X = x)$ for $x \in \mathcal{X}$. Let $p = (p(1), p(2), \ldots, p(\mathcal{B}))$, and $c = c(p)$ be a uniquely decipherable code for $X$, where $c$ consists of $\mathcal{B}$ codewords over a given alphabet with word lengths $l(1, c), l(2, c), \ldots, l(\mathcal{B}, c)$. Then the average code length is

$$\bar{l}(c, p) = \sum_{x \in \mathcal{X}} p(x) l(x, c).$$

The minimum average code length with respect to $p$ is defined by

$$\bar{l}_{\min}(p) \overset{\triangle}{=} \inf_c \bar{l}(c, p), \tag{8.5.15}$$

and due to the Noiseless Coding Theorem it is

$$H(p) \leq \bar{l}_{\min}(p) < H(p) + 1.$$

Let $c_{\min}(p)$ denote an encoding that achieves $\bar{l}_{\min}(p)$. It is known that Huffman encoding achieves $\bar{l}_{\min}(p)$.

Now let $p$ be a random variable drawn according to a density $g(p)$ on $\Sigma$. Then the expected code length for the encoding $c$ is

$$\begin{aligned}\int_\Sigma \bar{l}(c, p)\, \mathrm{d}g(p) &= \mathbb{E}\left(\bar{l}(c, p)\right) \\ &= \mathbb{E}\left(\sum_{x \in \mathcal{X}} p(x) l(x, c)\right) = \sum_{x \in \mathcal{X}} \mathbb{E}(p(x)) \cdot l(x, c).\end{aligned} \tag{8.5.16}$$

Define

$$\mathbb{E}(p) \overset{\triangle}{=} \left(\mathbb{E}(p(1)), \mathbb{E}(p(2)), \ldots, \mathbb{E}(p(\mathcal{B}))\right).$$

We infer from (8.5.15) and (8.5.16) that

$$\inf_c \mathbb{E}(\bar{l}(c, p)) = \inf_c \sum_{x \in \mathcal{X}} \mathbb{E}(p(x)) l(x, c) = \bar{l}_{\min}(\mathbb{E}(p)),$$

and $\bar{l}_{\min}(\mathbb{E}(p))$ is achieved by $c_{\min}(\mathbb{E}(p))$.

Suppose now that $n$ i.i.d. observations $X_1, X_2, \ldots, X_n$ are drawn according to a Dirichlet distribution on $\Sigma$ (an introduction of the Dirichlet distribution is provided in Sect. 8.10.3). Let $n_x$ equal the number of occurrences of $X_j = x$ in the $n = \sum_{x \in \mathcal{X}} n_x$ trials for all $x \in \mathcal{X}$.

By a realization of $X_1, X_2, \ldots, X_n$ and Bayes' rule, we obtain an "updated" Dirichlet distribution. We wish to show that $c_{\min}(\hat{p})$ minimizes the minimum expected average code length, where $\hat{p}$ represents a slightly generalized $\beta$-biased estimator:

Let $p$ be drawn according to a Dirichlet prior distribution with parameter $\lambda = (\lambda_1, \lambda_2, \ldots, \lambda_{\mathcal{B}})$, $\lambda_x \geq 1$ for all $x \in \mathcal{X}$, defined by the density function

$$g(p(1), p(2), \ldots, p(\mathcal{B})) = \frac{\Gamma(\lambda_1 + \lambda_2 + \cdots + \lambda_{\mathcal{B}})}{\Gamma(\lambda_1)\Gamma(\lambda_2), \ldots, \Gamma(\lambda_{\mathcal{B}})} \prod_{x \in \mathcal{X}} p(x)^{\lambda_x - 1} \tag{8.5.17}$$

According to (8.10.3) it is

$$\mathbb{E}_\lambda(p(x)) = \frac{\lambda_x}{\sum_{y \in \mathcal{X}} \lambda_y} \quad \text{for all } x \in \mathcal{X}, \tag{8.5.18}$$

where the expectation is taken relative to the Dirichlet distribution with the parameter $\lambda$.

Now we calculate the posterior distribution of the Dirichlet distribution via Bayes' rule (a slightly modified expression of (8.5.7)):

$$g(p \mid X_1 = x_1, \ldots, X_n = x_n) = \frac{g(p) \cdot \prod_{i=1}^n p(X_i = x_i \mid p)}{\int_\Sigma g(Q) \prod_{i=1}^n p(X_i = x_i \mid Q)\, dQ}. \tag{8.5.19}$$

Again we apply (8.10.3) for the calculation of the denominator of (8.5.19):

$$\int_\Sigma g(Q) \prod_{i=1}^n p(X_i = x_i \mid Q)\, dQ = C(\lambda) \cdot \frac{\Gamma(\lambda_1 + n_1), \ldots, \Gamma(\lambda_{\mathcal{B}} + n_{\mathcal{B}})}{\Gamma(\lambda_1 + n_1 + \cdots + \lambda_{\mathcal{B}} + n_{\mathcal{B}})}, \tag{8.5.20}$$

where

$$C(\lambda) = \frac{\Gamma(\lambda_1 + \cdots + \lambda_{\mathcal{B}})}{\Gamma(\lambda_1), \ldots, \Gamma(\lambda_{\mathcal{B}})}.$$

Inserting (8.5.20) into (8.5.19) yields

$$g(p \mid X_1 = x_1, \ldots, X_n = x_n) = \frac{\Gamma(\sum_{x \in \mathcal{X}} (\lambda_x + n_x))}{\prod_{x \in \mathcal{X}} \Gamma(\lambda_x + n_x)} \prod_{x \in \mathcal{X}} p(x)^{\lambda_x + n_x - 1} \tag{8.5.21}$$

The posterior distribution obviously is a Dirichlet distribution to the parameter $(\lambda_1 + n_1, \ldots, \lambda_{\mathcal{B}} + n_{\mathcal{B}})$. For brevity's sake, we will refer to this parameter as $\lambda + n$.

Furthermore, (8.5.18) implies

$$\begin{aligned} \mathbb{E}_{\lambda+n}(p(x)) &= \frac{\lambda_x + n_x}{\sum_{y \in \mathcal{X}} (\lambda_y + n_y)} = \frac{\lambda_x + n_x}{\sum_{y \in \mathcal{X}} \lambda_y + n} \\ &\stackrel{\Delta}{=} \hat{p}(x), \end{aligned}$$

for all $x \in \mathcal{X}$, where the expectation is taken relative to the posterior distribution (8.5.21). We define $\hat{p} \triangleq (\hat{p}(1), \dots, \hat{p}(\mathcal{B}))$.

Finally,

$$
\begin{aligned}
\inf_c \mathbb{E}_{\lambda+n}(\bar{l}(c, p)) &= \inf_c \mathbb{E}_{\lambda+n}\left(\sum_{x \in \mathcal{X}} p(x) l(x, c)\right) \\
&= \inf_c \left[\sum_{x \in \mathcal{X}} \mathbb{E}_{\lambda+n}(p(x)) l(x, c)\right] \\
&= \inf_c \left[\sum_{x \in \mathcal{X}} \hat{p}(x) l(x, c)\right] \\
&= \bar{l}_{\min}(\hat{p}). \qquad (8.5.22)
\end{aligned}
$$

Thus if $p$ is Dirichlet with parameter $\lambda$, and $X_1 = x_1, X_2 = x_2, \dots, X_n = x_n$ is observed, then by (8.5.22), (8.5.16) and the (8.5.16) following remark, $\mathbb{E}_{\lambda+n}\left(\bar{l}(c)\right)$ is minimized over all codes $c$, by the code $c_{\min}(\hat{p})$, which is the Huffman encoding with respect to $\hat{p}$.

In particular, this proves the optimality of 1-biased estimators for a uniformly distributed prior distribution, which is achieved by the Dirichlet distribution to the parameter $(1, 1, \dots, 1)$. We will use this result in the proof of Theorem 156.

Gilbert's calculations follow the same line as the calculations for Laplace's estimator in Sect. 8.5.1 in a more general context. They show that $\hat{p}(x)$ is a Bayesian estimator for every $x \in \mathcal{X}$.

Since $g(p)$ puts positive mass everywhere for any $\lambda = (\lambda_1, \lambda_2, \dots, \lambda_{\mathcal{B}})$ with $\lambda_x \geq 1$ for every $x \in \mathcal{X}$, it follows that $c_{\min}(\hat{p})$ is *admissible* in the sense that there exists no other encoding $\tilde{c}$ (based on $X_1 = x_1, X_2 = x_2, \dots, X_n = x_n$) such that

$$
\bar{l}(\tilde{c}, p) \leq \bar{l}_{\min}(\hat{p}) \quad \text{for all } p \in \Sigma.
$$

with strict inequality for a set of nonzero Lebesgue measure.

(Otherwise it is

$$
\begin{aligned}
\mathbb{E}_{\lambda+n}(\bar{l}(\tilde{c}, p)) &< \mathbb{E}_{\lambda+n}(\bar{l}_{\min}(\hat{p})) \\
&= \bar{l}_{\min}(\hat{p}) \\
&= \inf_c \mathbb{E}_{\lambda+n}(\bar{l}(c, p)).
\end{aligned}
$$

But this is obviously a contradiction.)

## 8.6 Lecture on Approximations for the Redundancy

In this lecture, we will introduce two functions $F$ and $\Phi$ that will be used for approximating the redundancy. For the derivation of $F$, we will simply use Taylor polynomials. Under certain conditions, $F$ can be approximated by $\Phi$.

We consider

$$\ln(n+m\beta) = \ln\left[n\left(1+\frac{m\beta}{n}\right)\right] = \ln n + \ln\left(1+\frac{m\beta}{n}\right)$$

and substitute $\ln(1+m\beta/n)$ by its Taylor polynomial at $x=1$:

$$\ln(n+m\beta) = \ln n + \frac{m\beta}{n} + O(1/n^2) \tag{8.6.1}$$

$$= \ln n + \frac{m\beta}{n} - \frac{m^2\beta^2}{2n^2} + O(1/n^3). \tag{8.6.2}$$

Furthermore it is, because of the nonnegativity of the third derivative of the logarithm,

$$\ln(n+m\beta) \geq \ln n + \frac{m\beta}{n} - \frac{m^2\beta^2}{2n^2}. \tag{8.6.3}$$

We rewrite the entropy $H_S$, using the substitution $p_S(x) = \lambda_S(x)/n$ for all $x \in \mathcal{X}$:

$$\begin{aligned} H_S &= -\sum_{x\in\mathcal{X}} p_S(x)\ln p_S(x) = -\sum_{x\in\mathcal{X}} \frac{\lambda_S(x)}{n}\ln\frac{\lambda_S(x)}{n} \\ &= -\frac{1}{n}\sum_{x\in\mathcal{X}} \lambda_S(x)\,[\ln\lambda_S(x) - \ln n] \\ &= \frac{\ln n}{n}\sum_{x\in\mathcal{X}}\lambda_S(x) - \frac{1}{n}\sum_{x\in\mathcal{X}}\lambda_S(x)\ln\lambda_S(x) \\ &= \ln n - \frac{1}{n}\sum_{x\in\mathcal{X}}\lambda_S(x)\ln\lambda_S(x). \end{aligned} \tag{8.6.4}$$

Using (8.6.2), the length function for a $\beta$-biased estimator $T^\beta \in \mathbb{E}(n)$ (8.4.2) changes to

$$\begin{aligned} \overline{L}_S^{T^\beta}(n) &= \sum_{x^n\in\mathcal{X}^n} p_S(x^n)\cdot \overline{l}_S^{T^\beta}(x^n) \\ &= \sum_{x^n\in\mathcal{X}^n} p_S(x^n)\sum_{x\in\mathcal{X}} p_S(x)\cdot l_S^{T^\beta}(x\mid x^n) \end{aligned}$$

$$
\begin{aligned}
&= -\sum_{x^n \in \mathcal{X}^n} p_S(x^n) \sum_{x \in \mathcal{X}} p_S(x) \ln \frac{\langle x^n \mid x \rangle + \beta}{n + m\beta} \\
&= \sum_{x \in \mathcal{X}} p_S(x) \sum_{x^n \in \mathcal{X}^n} p_S(x^n) \ln \frac{\langle x^n \mid x \rangle + \beta}{n + m\beta} \\
&= -\sum_{x \in \mathcal{X}} p_S(x) \sum_{k=0}^{n} b(k, n, p_S(x)) \cdot \ln \frac{k + \beta}{n + m\beta} \\
&= \sum_{x \in \mathcal{X}} p_S(x) \Bigg[ \ln n + \frac{m\beta}{n} - \frac{m^2\beta^2}{2n^2} + O(1/n^3) \\
&\quad - \sum_{k=0}^{n} b(k; n, p_S(x)) \ln(k + \beta) \Bigg] \\
&= \ln n + \frac{m\beta}{n} - \frac{m^2\beta^2}{2n^2} + O(1/n^3) \\
&\quad - \sum_{x \in \mathcal{X}} p_S(x) \sum_{k=0}^{n} b(k; n, p_S(x)) \ln(k + \beta) \\
&= \ln n - \frac{m^2\beta^2}{2n^2} + O(1/n^3) \\
&\quad + \frac{1}{n} \sum_{x \in \mathcal{X}} \left[ \beta - n p_S(x) \sum_{k=0}^{n} b(k; n, p_S(x)) \ln(k + \beta) \right].
\end{aligned} \tag{8.6.5}
$$

Using (8.6.4) and (8.6.5) it is

$$
\begin{aligned}
n R_S^n(T^\beta) &= n \left( \overline{L}_S^{T^\beta}(n) - H_S \right) \\
&= \sum_{x \in \mathcal{X}} \left[ \beta + \lambda_S(x) \ln \lambda_S(x) - \lambda_S(x) \sum_{k=0}^{n} b(k; n, p_S(x)) \ln(k + \beta) \right] \\
&\quad - \frac{m^2\beta^2}{2n} + O(1/n^2).
\end{aligned} \tag{8.6.6}
$$

Define

$$
F(n, \beta, p) \ \stackrel{\triangle}{=}\ \beta + \lambda \ln \lambda - \lambda \sum_{k=0}^{n} b(k; n, p) \ln(k + \beta).
$$

Now we can write (8.6.6) as

$$
n \cdot R_S^n(T^\beta) = \sum_{x \in \mathcal{X}} F(n, \beta, p_S(x)) - \frac{m^2\beta^2}{2n} + O(1/n^2). \tag{8.6.7}
$$

Using (8.6.1) instead of (8.6.2) yields, following the same line of calculations:

$$n \cdot R_S^n(T^\beta) = \sum_{x \in \mathcal{X}} F(n, \beta, p_S(x)) + O(1/n), \tag{8.6.8}$$

from the inequality (8.6.3) we derive

$$n \cdot R_S^n(T^\beta) \geq \sum_{x \in \mathcal{X}} F(n, \beta, p_S(x)) - \frac{m^2 \beta^2}{2n}. \tag{8.6.9}$$

We interpret $F$: asymptotically each letter $x \in \mathcal{X}$ contributes to the redundancy with about $F(n, \beta, p_S(x))/n$.

Some properties of $F$ will be used later:

1. $F(n, \beta, 0) = \beta$
2. For $F(, \beta, 1)$ it is

$$\begin{aligned} \lim_{n \to \infty} n \cdot F(n, \beta, 1) &= \lim_{n \to \infty} n\,(\beta + n \ln n - n \ln(n + \beta)) \\ &= \lim_{n \to \infty} n \left( \beta - \ln \left( 1 + \frac{\beta}{n} \right)^n \right) \end{aligned}$$

After several application of l'Hôpital's rule one finds that

$$\lim_{n \to \infty} n \cdot F(n, \beta, 1) = C,$$

i.e., $F(n, \beta, 1) = O(1/n)$.

3. Because of the convexity of $-\ln x$ and $\sum_{k=0}^{n} b(k; n, p) = 1$, we can apply Jensen's inequality:

$$\begin{aligned} F(n, \beta, p) &= \beta + \lambda \ln \lambda - \lambda \sum_{k=0}^{n} b(k; n, p) \cdot \ln(k + \beta) \\ &\geq \beta + \lambda \ln \lambda - \lambda \ln \left( \sum_{k=0}^{n} b(k; n, p)(k + \beta) \right) \\ &= \beta + \lambda \ln \lambda - \lambda \ln(\lambda + \beta) \\ &= \beta - \ln \left[ \left( 1 + \frac{\beta}{\lambda} \right)^{\lambda} \right] \\ &> \beta - \ln \left( e^{\beta} \right) \\ &= 0. \end{aligned}$$

For the last inequality we used, the fact that $(1 + x/n)^n$ approaches its limit $e^x$ from below for $n$ tending to infinity.
Therefore it is $F > 0$.

We define a function $\Phi$ by

$$\Phi(\beta, \lambda) \triangleq \beta + \lambda \ln \lambda - \lambda \sum_{k=0}^{\infty} \pi(k, \lambda) \ln(k + \beta).$$

where $\pi(k, \lambda)$ denotes the Poisson distribution to the parameters $k$ and $\lambda$, this means

$$\pi(k, \lambda) = \mathrm{e}^{-\lambda} \frac{\lambda^k}{k!}.$$

In the following lemmata, we will prove that for great $n$ and bounded $\lambda$, the function $\Phi$ can be used as an approximation for $F$.

## 8.7 Lecture on Outlook

### *8.7.1 Preview of the Main Theorems*

In the following we will prove two main theorems, Theorems 1 and 2: Theorem 1 deals with $\beta$-biased estimators. In Lemma 6, we derive a number $\beta_0 = 0.50922\ldots$.

**Theorem 155** *For $\beta \neq \beta_0$, $T^\beta, T^{\beta_0} \in \mathbb{E}(n)$ it is*

$$\liminf_{n \to \infty} n R^n(T^\beta) > \beta_0(m-1),$$

*and*

$$\lim_{n \to \infty} n R^n(T^{\beta_0}) = \beta_0(m-1).$$

This means that, for our choice of loss function, the $\beta_0$-biased estimator turns out to be the best among the $\beta$-biased estimators. Having found this solution for $\beta$-biased estimators, the next naturally arising question is: what is the best estimator if we do not restrict our considerations to $\beta$-biased ones? Theorem 2 provides a good answer to this question:

**Theorem 156** *For $T \in \mathcal{G}(n)$ it is*

$$\liminf_{n \to \infty} n R^n(T) \geq \frac{1}{2}(m-1).$$

This means, if we allow general estimators there may be some that perform better than the $\beta_0$-biased one. But since $\beta_0$ is close to 1/2, there was no big advantage in using those. In addition, Theorem 2 does not provide any idea how to find such an estimator.

*Conclusion*. The $\beta_0$-biased estimator is the best among $\beta$-biased ones and the effort to look for an optimal general estimator would only result in a minor improvement of the performance.

## 8.7.2 Sketch of the Proofs

1. Theorem 155
   (a) first statement: because of the definition of (overall) redundancy (8.4.5), it is $R^n(T^\beta) \geq R^n_S(T^\beta)$ for every DMS $S$ and every estimator $T^\beta \in \mathbb{E}(n)$. Bearing this fact in mind, we consider two cases separately, choosing for each a DMS with "worst case" behavior:
      (i) $\beta > \beta_0$: we choose a "worst case" source and use the asymptotic approximation $F$ for the redundancy
      (ii) $\beta < \beta_0$: again we choose a "worst case" source. Lemma 5 allows us to approximate $F$ by $\Phi$.
   This yields the stated inequality.
   (b) second statement: we use the asymptotic approximation to the redundancy $F$; it is for the estimator $T^{\beta_0} \in \mathbb{E}(n)$, large $n$ and $\varepsilon > 0$

$$n \cdot R^n_S(T^{\beta_0}) < \sum_{x \in \mathcal{X}} F(n, \beta_0, \lambda_x/n) + \varepsilon$$

   for any source $S = (\lambda_1/n, \ldots, \lambda_m/n)$. Lemma 83 yields an upper bound for $F$ if $\lambda$ is greater than a constant $\lambda_0$ which is specified there. For $\lambda \leq \lambda_0$ and large $n$, Lemma 86 states that we can approximate $F$ by $\Phi$. Hence, Lemma 83 and Lemma 86 complement one another. Application of these lemmata leads to an inequality. We also apply Lemma 87. Another inequality is derived by choosing a "worst case" source and making use of $F$. Combining these inequalities yields the second statement.

2. Theorem 156
   We infer from [1] (and Sect. 8.5.6, respectively) that for every estimator $T \in \mathcal{G}(n)$ and for the 1-biased estimator $T^1 \in \mathbb{E}(n)$ it is

$$R^n(T) \geq \int_\Sigma R^n_S(T^1)\,\mathrm{d}\mu(S), \tag{8.7.1}$$

   where $\mu$ denotes the uniformly distributed measure on $\Sigma$. We bound the right side of (8.7.1) by restricting the integration area to a subset $\Sigma_\delta$ which carries "almost all" probability, but contains only strictly positive sources. Now we can apply Lemma 84 and obtain Theorem 83

As pointed out in the sketches above, Lemmas 83, 84, 86, and 87 play an essential role for the Theorems 155 and 156. The following figure illustrates the relations between the lemmata and theorems (the arrows indicate which lemma contributes to which proof):

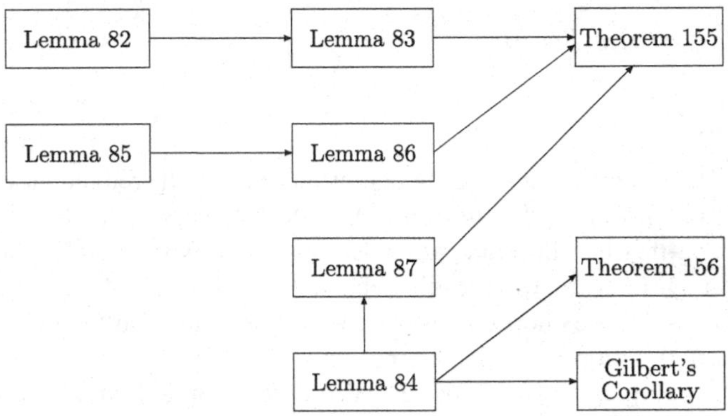


## 8.8 Lecture on the Mathematical Tools

**Lemma 82** *It is*

$$\sum_{k=0}^{\lambda-\delta} b(k; n, p) < C\lambda e^{-\frac{\delta^2}{\lambda}+\frac{(\delta+1)^2}{2(\lambda-\delta-1)}}$$

*for* $n > 2, 0 < p < 1, \lambda = np, 1 < \delta < \lambda$, *and a constant* $C$.

*Proof* For the binomial probabilities

$$b(k; n, p) = \binom{n}{k} p^k (1-p)^{n-k}$$

it is

$$\frac{b(k+1; n, p)}{b(k; n, p)} = \frac{(n-k)p}{(k+1)(1-p)},$$

and consequently

$$\frac{b(k+1; n, p)}{b(k; n, p)} < 1 \Leftrightarrow k+1 > (n+1)p.$$

This means $k \mapsto b(k; n, p)$ increases up to $k = \lfloor \lambda + p \rfloor$, and $b(\lfloor \lambda - \delta \rfloor; n, p) \geq b(k; n, p)$ for all $0 \leq k \leq \lfloor \lambda - \delta \rfloor$.

We define

$$r \overset{\Delta}{=} \lfloor \lambda - \delta \rfloor = \lambda - \delta_1 \quad \text{for a } \delta_1 \text{ with } \delta \le \delta_1 \le \delta + 1. \tag{8.8.1}$$

Now it is

$$\sum_{k=0}^{\lambda-\delta} b(k; n, p) \le \lambda b(r; n, p).$$

According to the Stirling formula it is

$$\binom{n}{r} \le C \sqrt{\frac{n}{r(n-r)}} \cdot \frac{n^n}{r^r (n-r)^{n-r}}. \tag{8.8.2}$$

We consider the function $y(r) = r(n-r)$. It is $y'(r) = n - 2r$, $y''(r) = -2$, this means $y(r)$ increases on the interval $1 \le r \le \frac{n}{2}$ and is symmetric with respect to $r = \frac{n}{2}$. Therefore it is

$$n - 1 \le r(n-r) \tag{8.8.3}$$

for $1 \le r \le n-1$ (for $r = 1$ and $r = n-1$ (8.8.3) holds with equality), and

$$\frac{n}{r(n-r)} \le \frac{n}{n-1} \le 2 \quad \text{for } n \ge 2. \tag{8.8.4}$$

Define the function $\nu(x) \overset{\Delta}{=} x \ln x$. Using (8.8.2), we derive an upper bound for

$$\begin{aligned} \ln b(r; n, p) &= \ln \left[ \binom{n}{r} p^r (1-p)^{n-r} \right] = \ln \binom{n}{r} + r \ln p + (n-r) \ln(1-p) \\ &\le \ln \left[ C \sqrt{\frac{n}{r(n-r)}} \cdot \frac{n^n}{r^r (n-r)^{n-r}} \right] + r \ln p + (n-r) \ln(1-p). \end{aligned}$$

It is

$$\begin{aligned} \ln \left[ C \sqrt{\frac{n}{r(n-r)}} \cdot \frac{n^n}{r^r (n-r)^{n-r}} \right] &= \ln C + \frac{1}{2} \ln \frac{n}{r(n-r)} + \ln \frac{n^n}{r^r (n-r)^{n-r}} \\ &\le C + n \ln n - r \ln r - (n-r) \ln(n-r), \end{aligned}$$

using (8.8.4) in the last step.
This yields

$$\begin{aligned} \ln b(r; n, p) &\le C + n \ln n - r \ln r - (n-r) \ln(n-r) \\ &\quad + r \ln p + (n-r) \ln(1-p) \\ &= C + n \ln n - (\lambda - \delta_1) \ln(\lambda - \delta_1) + (\lambda - \delta_1) \ln p \\ &\quad - (n - \lambda + \delta_1) \ln(n - \lambda + \delta_1) + (n - \lambda + \delta_1) \ln(1-p) \end{aligned}$$

$$\begin{aligned}
&= C + n\ln n - \nu(\lambda-\delta_1) - \nu(n-\lambda+\delta_1) + \lambda\ln\lambda - \lambda\ln n \\
&\quad - \delta_1\ln\lambda + \delta_1\ln n + (n-\lambda)\ln(n-\lambda) - (n-\lambda)\ln n \\
&\quad + \delta_1\ln(n-\lambda) - \delta_1\ln n \\
&= C - \nu(\lambda-\delta_1) - \nu(n-\lambda+\delta_1) + \nu(\lambda) \\
&\quad - \delta_1\ln\lambda + \nu(n-\lambda) + \delta_1\ln(n-\lambda). \qquad (8.8.5)
\end{aligned}$$

For $\nu(x) = x\ln x$ it is

$$\nu'(x) = \ln x + 1, \quad v''(x) = \frac{1}{x} > 0, \quad v'''(x) = -\frac{1}{x^2} < 0 \quad \text{for } x > 0.$$

We consider the Taylor polynomial of degree 2 of $\nu(x)$ at $x = \lambda - \delta_1$. Because of the nonpositivity of $\nu'''(x)$ it is

$$\nu(\lambda) \leq \nu(\lambda-\delta_1) + \delta_1\ln(\lambda-\delta_1) + \delta_1 + \frac{\delta_1^2}{2(\lambda-\delta_1)}. \qquad (8.8.6)$$

For the Taylor polynomial of degree 1 of $\nu(x)$ at $x = n - \lambda$ and because of the positivity of $\nu''(x)$ it is

$$\nu(x) \geq \nu(n-\lambda) + [\ln(n-\lambda) + 1](x - n + \lambda),$$

and in particular

$$\nu(n-\lambda+\delta_1) \geq \nu(n-\lambda) + \delta_1\ln(n-\lambda) + \delta_1. \qquad (8.8.7)$$

Applying (8.8.6) and (8.8.7) to (8.8.5) yields

$$\begin{aligned}
\ln b(r;n,p) &\leq C - \nu(\lambda-\delta_1) - \nu(n-\lambda+\delta_1) + \nu(\lambda) - \delta_1\ln\lambda \\
&\quad + \nu(n-\lambda) + \delta_1\ln(n-\lambda) \\
&\leq C - \nu(\lambda-\delta_1) - \nu(n-\lambda) - \delta_1\ln(n-\lambda) - \delta_1 + \nu(\lambda-\delta_1) \\
&\quad + \delta_1\ln(\lambda-\delta_1) + \delta_1 + \frac{\delta_1^2}{2(\lambda-\delta_1)} - \delta_1\ln\lambda \\
&\quad + \nu(n-\lambda) + \delta_1\ln(n-\lambda) \\
&= C + \delta_1\ln\left(1 - \frac{\delta_1}{\lambda}\right) + \frac{\delta_1^2}{2(\lambda-\delta_1)}. \qquad (8.8.8)
\end{aligned}$$

Using (8.8.2) and (8.8.8), and the inequality $\ln(1-x) \leq -x$ yields

$$\ln\sum_{k=0}^{\lambda-\delta} b(k;n,p) \leq \ln[\lambda b(r;n,p)] = \ln\lambda + \ln b(r;n,p)$$

$$\begin{aligned}
&\le \ln\lambda + C + \delta_1 \ln\left(1 - \frac{\delta_1}{\lambda}\right) + \frac{\delta_1^2}{2(\lambda - \delta_1)} \\
&\le \ln\lambda + C - \frac{\delta_1^2}{\lambda} + \frac{\delta_1^2}{2(\lambda - \delta_1)} \\
&\overset{(1)}{\le} \ln\lambda + C - \frac{\delta^2}{\lambda} + \frac{(\delta+1)^2}{2(\lambda - \delta - 1)},
\end{aligned}$$

and consequently

$$\sum_{k=0}^{\lambda-\delta} b(k; n, p) \le C\lambda e^{-\frac{\delta^2}{\lambda} + \frac{(\delta+1)^2}{2(\lambda-\delta-1)}}. \qquad \square$$

**Lemma 83** *The inequality*

$$\limsup_{\lambda\to\infty} \frac{F(n, \beta, \lambda/n)}{1 - \lambda/n} < \frac{1}{2}$$

*holds uniformly over n, this means, for every $\varepsilon > 0$ there exists a $\lambda_0$ such that for all $\lambda > \lambda_0$, all n, $p = \lambda/n$ it is*

$$F(n, \beta, p) < \frac{1-p}{2} + \varepsilon.$$

*Proof* We consider the Taylor polynomial of degree 3 of $\ln(k + \beta)$ at the point $\lambda$:

$$\ln(k+\beta) = \ln\lambda + \frac{1}{\lambda}(k + \beta - \lambda) - \frac{1}{2\lambda^2}(k + \beta - \lambda)^2 + \frac{1}{3\lambda^3}(k + \beta - \lambda)^3 + R(k), \tag{8.8.9}$$

where $R(k)$ is the remainder. In the Lagrange form it is

$$R(k) = -\frac{(k + \beta - \lambda)^4}{4\xi(k)^4} \quad \text{for a } \xi(k) \in (\lambda, k + \beta)$$

(or $\xi(k) \in (k + \beta, \lambda)$, respectively).

We use (8.8.9) to approximate $F(n, \beta, p)$:

$$\begin{aligned}
F(n, \beta, p) &= \beta + \lambda\ln\lambda - \lambda\sum_{k=0}^{n} b(k; n, p)\cdot\ln(k + \beta) \\
&= \beta + \lambda\ln\lambda - \lambda\sum_{k=0}^{n} b(k; n, p)\Bigg[\ln\lambda + \frac{k - \lambda + \beta}{\lambda} - \frac{(k - \lambda + \beta)^2}{2\lambda^2} \\
&\qquad + \frac{(k - \lambda + \beta)^3}{3\lambda^3} + R(k)\Bigg]
\end{aligned}$$

$$= \beta - \lambda \sum_{k=0}^{n} b(k; n, p) \Bigg[ \frac{\beta}{\lambda} - \left( \frac{1}{2\lambda^2}(k-\lambda)^2 + \frac{\beta}{\lambda^2}(k-\lambda) + \frac{\beta^2}{2\lambda^2} \right) + \left( \frac{1}{3\lambda^3}(k-\lambda)^3 + \frac{\beta}{\lambda^3}(k-\lambda)^2 + \frac{\beta^2}{\lambda^3}(k-\lambda) + \frac{\beta^3}{3\lambda^3} \right) + \frac{1}{\lambda}(k-\lambda) + R(k) \Bigg].$$

Now we will make use of the first three central moments of the binomial distribution, these are

$$\mu_1 = 0, \quad \mu_2 = \lambda q, \quad \mu_3 = \lambda q(q-p).$$

(Cf. appendix (8.10.4)).

This yields

$$\begin{aligned} F(n, \beta, p) &= \beta - \mu_1 - \beta + \frac{1}{2\lambda}\mu_2 + \frac{\beta}{\lambda}\mu_1 + \frac{\beta^2}{2\lambda} - \frac{1}{3\lambda^2}\mu_3 \\ &\quad - \frac{\beta}{\lambda^2}\mu_2 - \frac{\beta^2}{\lambda^2}\mu_1 - \frac{\beta^3}{3\lambda^2} - \lambda \sum_{k=0}^{n} b(k; n, p) \cdot R(k) \\ &= \left( \frac{1}{2\lambda} - \frac{\beta}{\lambda^2} \right) \mu_2 - \frac{1}{3\lambda^2}\mu_3 \\ &\quad + \frac{\beta^2}{2\lambda} - \frac{\beta^3}{3\lambda^2} - \lambda \sum_{k=0}^{n} b(k; n, p) \cdot R(k) \\ &= \left( \frac{1}{2\lambda} - \frac{\beta}{\lambda^2} \right) \lambda q - \frac{1}{3\lambda^2}\lambda q(q-p) \\ &\quad + \frac{\beta^2}{2\lambda} - \frac{\beta^3}{3\lambda^2} - \lambda \sum_{k=0}^{n} b(k; n, p) \cdot R(k) \\ &= \frac{q}{2} - \frac{\beta q}{\lambda} - \frac{1}{3\lambda}q(q-p) + \frac{\beta^2}{2\lambda} - \frac{\beta^3}{3\lambda^2} - \lambda \sum_{k=0}^{n} b(k; n, p) \cdot R(k) \\ &= \frac{q}{2} + \frac{1}{\lambda}\left[ \frac{\beta^2}{2} - \frac{1}{3}(q-p)q - \beta q \right] - \frac{\beta^3}{3\lambda^2} \\ &\quad - \lambda \sum_{k=0}^{n} b(k; n, p) \cdot R(k). \end{aligned} \tag{8.8.10}$$

Choose a number $\delta$ with $1 < \delta < \lambda$. There are two cases:

1. $0 \leq k \leq \lambda - \delta$: for great $\lambda$ it is $\xi(k) > \beta$ and consequently

$$R(k) = -\frac{(k-\lambda+\beta)^4}{4\xi(k)^4} \geq -\frac{(k-\lambda+\beta)^4}{4\beta^4}.$$

Using this inequality and $k-\lambda \le -\delta < \lambda$, it is

$$-\lambda\sum_{k=0}^{\lambda-\delta} b(k;n,p)\cdot R(k) \le \lambda\sum_{k=0}^{\lambda-\delta} b(k;n,p)\frac{(k-\lambda+\beta)^4}{4\beta^4} < \frac{\lambda(\lambda+\beta)^4}{4\beta^4}\sum_{k=0}^{\lambda-\delta} b(k;n,p). \qquad (8.8.11)$$

Now we choose $\delta \stackrel{\triangle}{=} \lambda^{\frac{3}{4}}$, let $\lambda$ tend to infinity, and apply Lemma 82. It is

$$-\frac{\delta^2}{\lambda}+\frac{(\delta+1)^2}{2(\lambda-\delta-1)} = -\lambda^{\frac{1}{2}}+\frac{\lambda^{\frac{6}{4}}+2\lambda^{\frac{3}{4}}+1}{2\lambda-2\lambda^{\frac{3}{4}}-2} = -\lambda^{\frac{1}{2}}\underbrace{\left(1-\frac{\lambda^{\frac{1}{4}}+2\lambda^{-\frac{1}{2}}+\lambda^{-\frac{5}{4}}}{2\lambda^{\frac{1}{4}}-2-2\lambda^{-\frac{3}{4}}}\right)}_{\stackrel{\lambda\to\infty}{\longrightarrow} C}.$$

This yields for great $\lambda$, according to Lemma 82,

$$\sum_{k=0}^{\lambda-\delta} b(k;n,p) < C\lambda e^{-C\sqrt{\lambda}}.$$

$\lambda(\lambda+\beta)^4$ is a polynomial in $\lambda$, $\beta$ a positive constant. The exponential function $e^x$ grows faster than any polynomial. Thus, for every $\varepsilon > 0$ and sufficiently great $\lambda$ it is

$$\left|-\lambda\sum_{k=0}^{\lambda-\delta} b(k;n,p)\cdot R(k)\right| < \frac{\varepsilon}{4}.$$

2. $\lambda-\delta \le k \le n$: for $\xi(k) \in (\lambda, k+\beta)$ it is obviously $\xi(k) > \lambda > \lambda-\lambda^{\frac{3}{4}}$. For $\xi(k) \in (k+\beta, \lambda)$ and $\delta = \lambda^{\frac{3}{4}}$ it is

$$k+\beta \ge \lambda-\delta+\beta = \lambda-\lambda^{\frac{3}{4}}+\beta > \lambda-\lambda^{\frac{3}{4}} = \lambda^{\frac{3}{4}}\left(\lambda^{\frac{1}{4}}-1\right).$$

This means

$$\xi(k) > \lambda^{\frac{3}{4}}\left(\lambda^{\frac{1}{4}}-1\right)$$

and furthermore

$$R(k) = -\frac{(k-\lambda+\beta)^4}{4\xi(k)^4} \ge -\frac{(k-\lambda+\beta)^4}{4\left(\lambda^{\frac{3}{4}}\left(\lambda^{\frac{1}{4}}-1\right)\right)^4}.$$

With this inequality, we infer

$$\begin{aligned}
&-\lambda \sum_{k=\lambda-\delta}^{n} b(k;n,p)R(k)\\
&\leq \frac{1}{4\lambda^2\left(\lambda^{\frac{1}{4}}-1\right)^4}\sum_{k=0}^{n} b(k;n,p)(k-\lambda+\beta)^4\\
&= \frac{1}{4\lambda^2\left(\lambda^{\frac{1}{4}}-1\right)^4}\sum_{k=0}^{n} b(k;n,p)[(k-\lambda)^4+4(k-\lambda)^3\beta+6(k-\lambda)^2\beta^2\\
&\quad +4(k-\lambda)\beta^3+\beta^4]\\
&= \frac{1}{4\lambda^2\left(\lambda^{\frac{1}{4}}-1\right)^4}[\mu_4+4\beta\mu_3+6\beta^2\mu_2+4\beta^3\mu_1+\beta^4]\\
&= \frac{1}{4\lambda^2\left(\lambda^{\frac{1}{4}}-1\right)^4}[3\lambda^2q^2+\lambda q(1-6pq)+4\beta\lambda q(q-p)+6\beta^2\lambda q+\beta^4]\\
&= \frac{3q^2}{4\left(\lambda^{\frac{1}{4}}-1\right)^4}+\frac{q(1-6pq)}{4\lambda\left(\lambda^{\frac{1}{4}}-1\right)^4}+\frac{\beta q(q-p)}{\lambda\left(\lambda^{\frac{1}{4}}-1\right)^4}+\frac{3\beta^2 q}{2\lambda\left(\lambda^{\frac{1}{4}}-1\right)^4}\\
&\quad +\frac{\beta^4}{4\lambda^2\left(\lambda^{\frac{1}{4}}-1\right)^4}\\
&< \frac{\varepsilon}{4}
\end{aligned}$$

for sufficiently great $\lambda$ and every $p, q, \beta$. For this calculation, we used the first four central moments of the binomial distribution (cf. (8.10.4)), the fourth is

$$\mu_4 = 3\lambda^2q^2+\lambda q(1-6pq).$$

Combining 1. and 2, we find that (8.8.10) simplifies to

$$\begin{aligned}
F(n,\beta,p) &= \frac{q}{2}+\frac{C}{\lambda}-\frac{C}{\lambda^2}-\lambda\sum_{k=0}^{n} b(k;n,p)\cdot R(k)\\
&< \frac{q}{2}+\frac{C}{\lambda}-\frac{C}{\lambda^2}+\frac{\varepsilon}{2}\\
&< \frac{q}{2}+\varepsilon
\end{aligned}$$

for sufficiently great $\lambda$. □

For a source $S = (p_S(1), p_S(2), \ldots, p_S(m))$, $p_S(x)$ strictly positive for all $x \in \mathcal{X}$, we define the number

$$\sigma_S \triangleq \sum_{x\in\mathcal{X}} \frac{1}{p_S(x)}.$$

**Lemma 84** *For a $\beta$-biased estimator $T^\beta \in \mathbb{E}(n)$ and $n$ tending to infinity it is*

$$n \cdot R_S^n(T^\beta) = \frac{1}{n}\left[\sigma_S\left(\frac{\beta^2}{2} - \beta + \frac{5}{12}\right) - \frac{m}{2} + m\beta + \frac{1}{12} - \frac{m^2\beta^2}{2}\right] + \frac{m-1}{2} + O(1/n^2). \quad (8.8.12)$$

*Furthermore, the inequality*

$$n \cdot R_S^n(T^\beta) \geq \frac{m-1}{2} - \frac{1}{n}\left[\frac{\sigma_S}{3} + \frac{2}{3} + \sigma_S\beta + \frac{m^2\beta^2}{2}\right] - \frac{\beta^3}{3n^2}\sum_{x\in\mathcal{X}} \frac{1}{p_S(x)^2}$$

*holds.*

*Proof* Again we consider the Taylor expansion of $\ln(k+\beta)$ (cf. (8.8.9)), but this time we choose the Peano form for the remainder

$$R(k) = -\frac{(k-\lambda+\beta)^4}{4\lambda^4} + o(1/\lambda^4).$$

This yields (with $\mu_4 = 3\lambda^2 q^2 + O(\lambda)$, $\mu_i = O(\lambda)$ for $i = 2, 3$)

$$\begin{aligned}
\sum_{k=0}^{n} b(k; n, p) \cdot R(k) &= \sum_{k=0}^{n} b(k; n, p)\left[-\frac{(k-\lambda+\beta)^4}{4\lambda^4} + o(1/\lambda^4)\right] \\
&= \sum_{k=0}^{n} b(k; n, p)\Bigg[-\frac{(k-\lambda)^4 + 4(k-\lambda)^3\beta}{4\lambda^4} \\
&\quad - \frac{6(k-\lambda)^2\beta^2 + 4(k-\lambda)\beta^3 + \beta^4}{4\lambda^4} + o(1/\lambda^4)\Bigg] \\
&= -\frac{1}{4\lambda^4}[\mu_4 + 4\beta\mu_3 + 6\beta^2\mu_2 + 4\beta^3\mu_1 + \beta^4] + o(1/\lambda^4) \\
&= -\frac{1}{4\lambda^4}[3\lambda^2 q^2 + O(\lambda)] + o(1/\lambda^4) \\
&= -\frac{3q^2}{4\lambda^2} + O(1/\lambda^3)
\end{aligned} \quad (8.8.13)$$

Obviously it is

$$\sum_{x\in\mathcal{X}} q(x) = \sum_{x\in\mathcal{X}}(1-p(x)) = m-1,$$

$$\sum_{x\in\mathcal{X}} \frac{q(x)}{p(x)} = \sum_{x\in\mathcal{X}} \frac{1-p(x)}{p(x)} = \sigma - m,$$

$$\sum_{x\in\mathcal{X}} \frac{q(x)^2}{p(x)} = \sum_{x\in\mathcal{X}} \frac{1-2p(x)+p(x)^2}{p(x)} = \sigma - 2m + 1,$$

and

$$\sum_{x\in\mathcal{X}} \frac{q(x)}{p(x)}(q(x)-p(x)) = \sum_{x\in\mathcal{X}} \frac{1-p(x)}{p(x)}(1-2p(x)) = \sigma - 3m + 2$$

for every source $(p(1), p(2), \dots, p(m)), q(x) = 1 - p(x)$ for all $x \in \mathcal{X}, \sigma = \sum_{x\in\mathcal{X}} \frac{1}{p(x)}$.

Taking these equalities into account, we plug (8.8.10) and (8.8.13) into (8.6.7):

$$\begin{aligned}
n \cdot R_S^n(T^\beta) &= \sum_{x\in\mathcal{X}} F(n,\beta,p_S(x)) - \frac{m^2\beta^2}{2n} + O(1/n^2) \\
&= \sum_{x\in\mathcal{X}} \left[\frac{1}{np_S(x)}\left[\frac{\beta^2}{2} - \frac{1}{3}(q_S(x)-p_S(x))q_S(x) - \beta q_S(x)\right]\right. \\
&\quad \left. - \frac{\beta^3}{3n^2 p_S(x)^2} + \frac{3q_S(x)^2}{4np_S(x)} + \frac{q_S(x)}{2} + O(1/n^2)\right] \\
&\quad - \frac{m^2\beta^2}{2n} + O(1/n^2) \\
&= \frac{m-1}{2} + \frac{\beta^2}{2n}\sigma_S - \frac{1}{3n}(\sigma_S - 3m + 2) - \frac{\beta}{n}(\sigma_S - m) \\
&\quad - \frac{\beta^3}{3n^2}\sum_{x\in\mathcal{X}} \frac{1}{p_S(x)^2} + \frac{3}{4n}(\sigma_S - 2m + 1) - \frac{m^2\beta^2}{2n} + O(1/n^2) \\
&= \frac{m-1}{2} + \frac{1}{n}\left[\sigma_S\left(\frac{\beta^2}{2} - \frac{1}{3} - \beta + \frac{3}{4}\right) + m - \frac{2}{3} + m\beta - \frac{3}{2}m\right. \\
&\quad \left. + \frac{3}{4} - \frac{m^2\beta^2}{2}\right] - \frac{\beta^3}{3n^2}\sum_{x\in\mathcal{X}} \frac{1}{p_S(x)^2} + O(1/n^2) \\
&= \frac{m-1}{2} + \frac{1}{n}\left[\sigma_S\left(\frac{\beta^2}{2} - \beta + \frac{5}{12}\right) - \frac{m}{2} + m\beta + \frac{1}{12} - \frac{m^2\beta^2}{2}\right] \\
&\quad + O(1/n^2).
\end{aligned}$$

This proves the first statement of Lemma 84.
From (8.8.10) we infer

$$F(n, \beta, p) \geq \frac{q}{2} + \frac{1}{\lambda}\left[\frac{\beta^2}{2} - \frac{1}{3}(q-p)q - \beta q\right] - \frac{\beta^3}{3\lambda^2}$$

since $R(k)$ in (8.8.10) is negative. Using this inequality in (8.6.9) yields

$$\begin{aligned}
n \cdot R_S^n(T^\beta) &\geq \sum_{x\in\mathcal{X}} F(n, \beta, p_S(x)) - \frac{m^2\beta^2}{2n} \\
&\geq \sum_{x\in\mathcal{X}} \left[\frac{1}{np_S(x)}\left(\frac{\beta^2}{2} - \frac{1}{3}(q_S(x) - p_S(x))q_S(x) - \beta q_S(x)\right)\right. \\
&\quad \left. - \frac{\beta^3}{3n^2 p_S(x)^2} + \frac{q_S(x)}{2}\right] - \frac{m^2\beta^2}{2n} \\
&= \frac{m-1}{2} + \frac{\beta^2}{2n}\sigma_S - \frac{1}{3n}(\sigma_S - 3m + 2) - \frac{\beta}{n}(\sigma_S - m) \\
&\quad - \frac{\beta^3}{3n^2}\sum_{x\in\mathcal{X}} \frac{1}{p_S(x)^2} - \frac{m^2\beta^2}{2n} \\
&= \frac{m-1}{2} + \frac{1}{n}\left[\sigma_S\left(\frac{\beta^2}{2} - \frac{1}{3} - \beta\right) + m - \frac{2}{3} + m\beta - \frac{m^2\beta^2}{2}\right] \\
&\quad - \frac{\beta^3}{3n^2}\sum_{x\in\mathcal{X}} \frac{1}{p_S(x)^2} \\
&\geq \frac{m-1}{2} + \frac{1}{n}\left[\sigma_S\left(-\frac{1}{3} - \beta\right) - \frac{2}{3} - \frac{m^2\beta^2}{2}\right] - \frac{\beta^3}{3n^2}\sum_{x\in\mathcal{X}} \frac{1}{p_S(x)^2} \\
&= \frac{m-1}{2} - \frac{1}{n}\left[\frac{\sigma_S}{3} + \sigma_S\beta + \frac{2}{3} + \frac{m^2\beta^2}{2}\right] - \frac{\beta^3}{3n^2}\sum_{x\in\mathcal{X}} \frac{1}{p_S(x)^2}.
\end{aligned}$$

This proves the second statement of Lemma 84. □

**Lemma 85** *For every $\lambda_0 > 0$, $\lambda \leq \lambda_0$, $m \in \mathbb{N}$, $\varepsilon > 0$ there exists an $n_0 \in \mathbb{N}$ with*

$$\lambda \sum_{k=0}^{m} |b(k; n, p) - \pi(k, \lambda)| \cdot \ln(k + \beta) < \varepsilon$$

*for all $n > n_0$, where $p = \lambda/n$ and $\pi(k, \lambda)$ denotes the Poisson distribution to the parameters $k$ and $\lambda$.*

*Proof* For $\lambda = np$, the Poisson distribution is an approximation to the binomial distribution. According to Feller [3], the approximation error can be estimated by

$$\pi(k,\lambda)e^{\frac{k\lambda}{n}} > b(k;n,p) > \pi(k,\lambda)e^{-\frac{k^2}{n-k}-\frac{\lambda^2}{n-\lambda}}.$$

Hence it is

$$\pi(k,\lambda)\left[e^{\frac{k\lambda}{n}}-1\right] > b(k;n,p)-\pi(k,\lambda) > \pi(k,\lambda)\left[e^{-\frac{k^2}{n-k}-\frac{\lambda^2}{n-\lambda}}-1\right]. \quad (8.8.14)$$

The left and right bound in (8.8.14) tend to 0 as $n$ tends to infinity.

From (8.8.14), we conclude that for every $\varepsilon > 0$ and sufficiently great $n$ it is

$$\lambda\sum_{k=0}^{m}|b(k;n,p)-\pi(k,\lambda)|\cdot\ln(k+\beta) < \varepsilon,$$

because $\lambda \le \lambda_0$, the sum is finite, $|b(k;n,p)-\pi(k,\lambda)|$ is arbitrarily small (according to (8.8.14)), and $\ln(k+\beta)$ is bounded by $\ln(m+\beta)$.

This proves Lemma 85. □

**Lemma 86** *For every $\lambda_0 > 0$ and $\varepsilon > 0$ there exists an $n_0 \in \mathbb{N}$ with*

$$|\Phi(\beta,\lambda)-F(n,\beta,p)| < \varepsilon$$

*for all $n \ge n_0$, $p = \lambda/n$, $\lambda \le \lambda_0$.*

*Proof* Fix $\lambda_0$. For every $n \in \mathbb{N}$, $\lambda > 0$, $m \in \mathbb{N}$ with $n > m > n\lambda$ it is

$$\begin{aligned}|\Phi(\beta,\lambda)-F(n,\beta,p)| &= \lambda\left|\sum_{k=0}^{\infty}\pi(k,\lambda)\ln(k+\beta)-\sum_{k=0}^{n}b(k;n,p)\ln(k+\beta)\right|\\ &\le \lambda\left|\sum_{k=0}^{m}(\pi(k,\lambda)-b(k;n,p))\ln(k+\beta)\right|\\ &\quad+\lambda\left[\sum_{k=m+1}^{n}b(k;n,p)\ln(k+\beta)\right.\\ &\quad\left.+\sum_{k=m+1}^{\infty}\pi(k,\lambda)\ln(k+\beta)\right]. \end{aligned} \quad (8.8.15)$$

For the analysis of the right side of (8.8.15), we will first consider its second sum: it is

$$\sum_{k=m+1}^{n}\lambda b(k;n,p)\ln(k+\beta) < \sum_{k=m+1}^{n}\lambda\ln(k+\beta)\sum_{l=k}^{n}b(l;n,p). \quad (8.8.16)$$

Let $B_n$ be a binomial distributed random variable, this means there is a probability distribution $P$ on $\mathbb{N}_0$ with $P(B_n = k) = b(k; n, p)$ for every $k \in \mathbb{N}_0$. Then it is by the Chebyshev inequality

$$\begin{aligned}\sum_{l=k}^{n} b(l; n, p) &= P(B_n \geq k) = P(B_n - \lambda \geq k - \lambda)\\ &\leq P(|B_n - \lambda| \geq k - \lambda)\\ &\leq \frac{\lambda(1-\lambda/n)}{(k-\lambda)^2}.\end{aligned}$$

Applying this inequality to (8.8.16) yields

$$\begin{aligned}\lambda \sum_{k=m+1}^{n} b(k; n, p)\ln(k+\beta) &< \sum_{k=m+1}^{n} \lambda \ln(k+\beta)\frac{\lambda(1-\lambda/n)}{(k-\lambda)^2}\\ &< \lambda^2 \sum_{k=m+1}^{\infty} \frac{\ln(k+\beta)}{(k-\lambda)^2}.\end{aligned} \tag{8.8.17}$$

The series in (8.8.17) is in $[0, \lambda_0]$ dominated by the series $\lambda_0^2 \sum_{k=m+1}^{\infty} \frac{\ln(k+\beta)}{(k-\lambda_0)^2}$, which is convergent:
For $\varepsilon > 0$ there exists an $x' \in \mathbb{R}$ such that

$$\frac{\ln(x+\beta)}{(x-\lambda_0)^{\frac{1}{2}}} < \varepsilon$$

for every $x > x'$.

We infer

$$\begin{aligned}\int_{m+1}^{\infty} \frac{\ln(x+\beta)}{(x-\lambda_0)^2}\,\mathrm{d}x &= \int_{m+1}^{\infty} \frac{\ln(x+\beta)}{(x-\lambda_0)^{\frac{1}{2}}} \cdot \frac{1}{(x-\lambda_0)^{\frac{3}{2}}}\,\mathrm{d}x\\ &< \int_{m+1}^{x'} \frac{\ln(x+\beta)}{(x-\lambda_0)^{\frac{1}{2}}} \cdot \frac{1}{(x-\lambda_0)^{\frac{3}{2}}}\,\mathrm{d}x + \varepsilon \int_{x'}^{\infty} (x-\lambda_0)^{-\frac{3}{2}}\,\mathrm{d}x\\ &< \infty.\end{aligned}$$

This induces the convergence of $\lambda_0^2 \sum_{k=m+1}^{\infty} \frac{\ln(k+\beta)}{(k-\lambda_0)^2}$.

Now we can apply the Weierstrass criterion and receive the uniform convergence of $\lambda^2 \sum_{k=m+1}^{\infty} \frac{\ln(k+\beta)}{(k-\lambda)^2}$ on $[0, \lambda_0]$. This means, for every $\varepsilon > 0$ there is an $m_1$ with

$$\lambda \sum_{k=m_1+1}^{n} b(k; n, p) \ln(k + \beta) < \frac{\varepsilon}{3} \tag{8.8.18}$$

for all $\lambda \le \lambda_0$, all $n$ and $p = \lambda/n$.

We consider the third sum in (8.8.15): let $X$ be a random variable which is Poisson distributed to the parameter $\lambda$, this means there exists a probability distribution $P$ on $\mathbb{N}_0$ such that $P(X = k) = \pi(k, \lambda)$ for all $k \in \mathbb{N}_0$. Again we apply the Chebyshev inequality:

$$\sum_{l=k}^{\infty} \pi(l, \lambda) = P(X \ge k) = P(X - \lambda \ge k - \lambda) \le P(|X - \lambda| \ge k - \lambda)$$
$$\le \frac{\lambda}{(k-\lambda)^2}.$$

This yields

$$\lambda \sum_{k=m+1}^{\infty} \pi(k, \lambda) \ln(k+\beta) \le \sum_{k=m+1}^{\infty} \lambda \ln(k+\beta) \cdot \sum_{l=k}^{\infty} \pi(l, \lambda) \le \sum_{k=m+1}^{\infty} \frac{\lambda^2 \ln(k+\beta)}{(k-\lambda)^2}.$$

This bound is the same series as in the calculations for the second sum (8.8.17). Hence we can proceed the same line of calculations and receive for every $\varepsilon > 0$ an $m_2$ such that it is

$$\lambda \sum_{k=m_2+1}^{\infty} \pi(k, \lambda) \ln(k + \beta) < \frac{\varepsilon}{3} \tag{8.8.19}$$

for all $\lambda \le \lambda_0$.

Let $\lambda_0$ and $\varepsilon > 0$ be given, choose $m = \max(m_1, m_2)$. Then (8.8.18) and (8.8.19) hold for all $n$, $\lambda \le \lambda_0$, $p = \lambda/n$. Furthermore, Lemma 85 provides an $n_0$ such that the first sum in (8.8.17) is less than $\varepsilon/3$. This proves Lemma 86. □

**Lemma 87** *There exist two numbers* $\beta_0 \triangleq 0.50922\ldots$ *and* $\lambda' \triangleq 5.22543\ldots$ *such that*

$$\max_{\lambda \ge 0} \Phi(\beta_0, \lambda) = \beta_0, \qquad \Phi(\beta_0, 0) = \Phi(\beta_0, \lambda') = \beta_0. \tag{8.8.20}$$

*For* $\beta < \beta_0$ *it is*

$$\Phi(\beta, \lambda') > \Phi(\beta_0, \lambda') = \beta_0.$$

*Proof* By numerical methods, including the usage of the package "Mathematica," Krichevsky found the two numbers $\beta_0$ and $\lambda'$ which fulfill the first statement.

For

$$\Phi(\beta, \lambda) = \beta + \lambda \ln \lambda - \lambda \sum_{k=0}^{\infty} \pi(k, \lambda) \ln(k + \beta)$$

it is

$$\frac{\partial}{\partial \beta}\Phi(\beta,\lambda) = 1 - \lambda \sum_{k=0}^{\infty} \frac{\pi(k,\lambda)}{k+\beta}$$
$$= 1 - \mathrm{e}^{-\lambda}\frac{\lambda}{\beta} - \mathrm{e}^{-\lambda}\sum_{k=1}^{\infty}\frac{\lambda^k}{k!}\cdot\frac{\lambda}{k+\beta}.$$

For $0 < \beta < 1$ it is furthermore

$$\frac{\partial}{\partial \beta}\Phi(\beta,\lambda) < 1 - \mathrm{e}^{-\lambda}\frac{\lambda}{\beta} - \mathrm{e}^{-\lambda}\sum_{k=1}^{\infty}\frac{\lambda^{k+1}}{(k+1)!}$$
$$= 1 - \mathrm{e}^{-\lambda}\frac{\lambda}{\beta} - \mathrm{e}^{-\lambda}(e^{\lambda} - 1 - \lambda)$$
$$= \mathrm{e}^{-\lambda}\left(1 + \lambda - \frac{\lambda}{\beta}\right) \overset{!}{<} 0$$
$$\Longrightarrow \beta < \frac{\lambda}{1+\lambda}.$$

The second derivative is

$$\frac{\partial^2}{\partial^2 \beta}\Phi(\beta,\lambda) = \lambda \sum_{k=0}^{\infty}\frac{\pi(k,\lambda)}{(k+\beta)^2}.$$

Hence, the first derivative with respect to $\beta$ is strictly negative for every $\lambda > 0$ and $0 < \beta < \frac{\lambda}{1+\lambda}$, the second derivative is strictly positive for every $\lambda > 0$.
We conclude that for every $\lambda > 0$ there is at most one null of the first derivative. For $\lambda' = 5.22543\ldots$ Krichevsky calculated via "Mathematica" this null to be $\beta' \triangleq 0.9780\ldots$. Therefore $\Phi(\beta, \lambda')$ decreases monotonically with respect to $\beta$ for $\beta < \beta'$. In particular, it is

$$\Phi(\beta,\lambda') > \Phi(\beta_0,\lambda') = \beta_0$$

for $\beta < \beta_0 < \beta'$. □

For a further analysis of the function $\Phi$ fix $\beta$: for small $\lambda$ the function $\Phi(\beta,\lambda)$ behaves like $\beta + \lambda \ln \lambda$. We ca use Lemma 84 to derive the asymptotic behavior of $\Phi$:

for $m = 2$, $S = (p_S(1), p_S(2))$, where $\lambda_S(1) \xrightarrow{n\to\infty} \infty$, $\lambda \triangleq \lim_{n\to\infty} \lambda_S(2) < \infty$ and $T^\beta \in \mathbb{E}(n)$ it is according to (8.6.8)

$$n \cdot R_S^n(T^\beta) = F(n, \beta, p_S(1)) + F(n, \beta, p_S(2)) + O(1/n).$$

Applying Lemma 83 and 86, which treat unbounded and bounded $\lambda$, respectively, it is for every $\varepsilon > 0$ and sufficiently great $n$:

$$\begin{aligned} n \cdot R_S^n(T^\beta) &= \frac{1 - p_S(1)}{2} + \varepsilon + \Phi(\beta, \lambda) + \varepsilon + O(1/n) \\ &= \Phi(\beta, \lambda) + O(1/n). \end{aligned} \tag{8.8.21}$$

The last equality holds for sufficiently small $\varepsilon > 0$ and because of

$$\begin{aligned} &\lim_{n\to\infty} n \cdot \frac{1 - p_S(1)}{2} = \lim_{n\to\infty} n \cdot \frac{p_S(2)}{2} = \frac{\lambda}{2} = C \\ \Rightarrow\ &\frac{1 - p_S(1)}{2} = O(1/n). \end{aligned}$$

On the other hand, we can apply Lemma 84: for sufficiently great $n$ it is

$$n \cdot R_S^n(T^\beta) = \frac{1}{n}\left[\sigma_S\left(\frac{\beta^2}{2} - \beta + \frac{5}{12}\right) - 1 + 2\beta + \frac{1}{12} - 2\beta^2\right] + \frac{1}{2} + O(1/n^2). \tag{8.8.22}$$

Thus, combining (8.8.21) and (8.8.22) yields

$$\begin{aligned} \Phi(\beta, \lambda) &= \frac{\sigma_S}{n}\left(\frac{\beta^2}{2} - \beta + \frac{5}{12}\right) + \frac{1}{2} + O(1/n) \\ &\xrightarrow{n\to\infty}{}^{1} \quad \frac{1}{2} + \frac{1}{\lambda}\left(\frac{\beta^2}{2} - \beta + \frac{5}{12}\right) \\ &\xrightarrow{\lambda\to\infty} \quad \frac{1}{2}. \end{aligned}$$

Hence, we find that for fixed $\beta$ and $\lambda$ tending to infinity, $\Phi$ approaches $1/2$. The sign of the term $\frac{\beta^2}{2} - \beta + \frac{5}{12}$ decides whether the approach is from above or below. In case of an approach from above, $\Phi(\beta, \lambda)$ assumes a local maximum.

Figure 8.1 shows a plot of the function $\Phi(\beta_0, \lambda)$. For $\beta_0$, the value $\Phi(\beta_0, 0)$ coincides with the value $\Phi(\beta_0, \lambda')$ that $\Phi(\beta_0, \lambda)$ assumes at its local maximum $\lambda'$.

As a further application besides Theorem 1 and 2, we give the following corollary, due to Gilbert [4], who obtained the result by different methods, as a consequence of Lemma 3.

### 8.8.1 Gilbert's Corollary

**Corollary 34** (Gilbert) *Let $p_S$ be a DMS, $p_S$ not uniformly distributed on $\mathcal{X}$ and strictly positive, $x^n \in \mathcal{X}$ a word, produced by $p_S$, $n$ tend to infinity.*

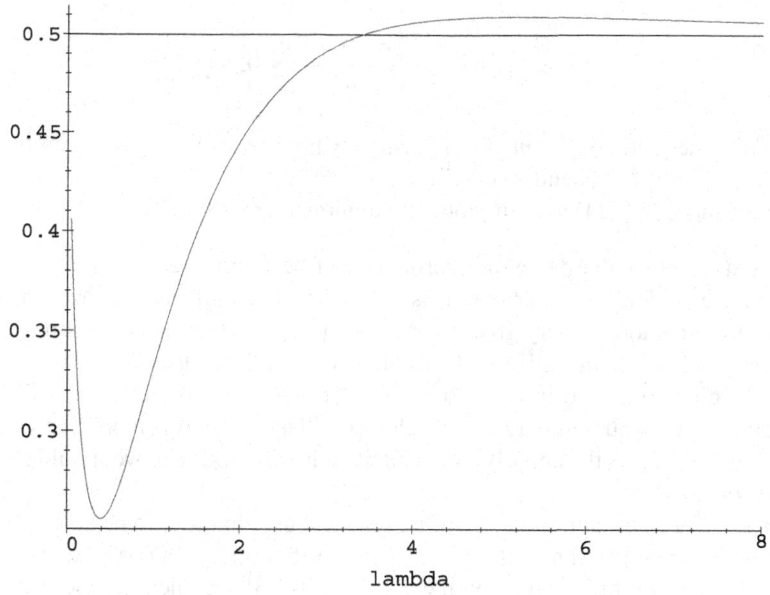


**Fig. 8.1** Plot of the function $\Phi(\beta_0, \lambda)$

*Then the best $\beta$-biased estimator $T^\beta \in \mathbb{E}(n)$, with respect to the source-dependent redundancy $R_S^n(T^\beta)$, is asymptotically given for the parameter*

$$\beta' = \frac{\sigma_S - m}{\sigma_S - m^2}, \qquad (8.8.23)$$

*where $\sigma_S = \sum_{x \in \mathcal{X}} 1/p_S(x)$.*

*This means that for $\beta \neq \beta'$ there is an $n_0$ such that $R_S^n(T^{\beta'}) < R_S^n(T^\beta)$ for $n > n_0$, $T^{\beta'}, T^\beta \in \mathbb{E}(n)$.*

*Proof* The redundancy caused by a $\beta$-biased estimator $T^\beta \in \mathbb{E}(n)$ for a source $S$ is asymptotically given by (8.8.12). We can write (8.8.12) as a polynomial in $\beta$:

$$R_S^n(T^\beta) = \frac{\sigma_S - m^2}{2n^2}\beta^2 + \frac{m - \sigma_S}{n^2}\beta + O(1/n). \qquad (8.8.24)$$

We examine $R_S^n(T^\beta)$ for local extrema: it is

$$\frac{\partial}{\partial \beta} R_S^n(T^\beta) = \frac{\sigma_S - m^2}{n^2}\beta + \frac{m - \sigma_S}{n^2} \stackrel{!}{=} 0$$

$$\Longrightarrow \quad \beta = \frac{\sigma_S - m}{\sigma_S - m^2} > 0$$

and

$$\frac{\partial^2}{\partial^2 \beta} R_S^n(T^\beta) = \frac{\sigma_S - m^2}{n^2} > 0.$$

We used the inequality $\sigma_S > m^2$. This inequality holds because $p_S$ is not uniformly distributed on $\mathcal{X}$ (cf. Appendix 8.10.5).

We conclude: (8.8.24) is asymptotically minimized by (8.8.23). □

The statement of the preceding corollary can be interpreted as follows: For a DMS, whose probability distribution is "close" to the uniform distribution, it is advantageous to choose a very great $\beta$, because this guarantees that our estimates for $p_S$ will stay "close" to the uniform distribution, even if the following source outputs favor certain elements. If, in contrast, $p_S$ is "located close to the edge of $\Sigma$," the probability is concentrated only on few elements. Therefore, it is more likely that a source output "reflects the underlying probability distribution" and we should choose $\beta$ relatively small.

Gilbert's Corollary appears to be of no practical use: In order to be able to apply the corollary, we need to know the probability distribution $p_S$. In this case, we could estimate by the optimal $\beta$ given by the corollary. But if we knew $p_S$ there was no need of estimating any more.

Gilbert argued in [4], referring to his corollary, that the 1-biased estimator is a good choice, being close to the suggested $\beta'$ (8.8.23) in "most cases."

## 8.9 Lecture on Main Results

*For $\beta \neq \beta_0 = 0.50922\ldots$ and $T^\beta, T^{\beta_0} \in \mathbb{E}(n)$ it is*

$$\liminf_{n\to\infty} n \cdot R^n(T^\beta) > \beta_0(m-1)$$

and

$$\lim_{n\to\infty} n \cdot R^n(T^{\beta_0}) = \beta_0(m-1).$$

*Proof* We consider two cases:

1. $\beta > \beta_0$: define the source $S^* \stackrel{\Delta}{=} (1, 0, 0, \ldots, 0)$. It is

$$\begin{aligned} nR^n(T^\beta) &= n \sup_{S\in\Sigma} R_S^n(T^\beta) \geq nR_{S^*}^n(T^\beta) \\ &\stackrel{(8)}{=} \sum_{x\in\mathcal{X}} F(n, \beta, p_{S^*}(x)) + O(1/n) \\ &= (m-1)\beta + O(1/n) \\ &> (m-1)\beta_0. \end{aligned}$$

The last inequality holds for sufficiently great $n$. We also used properties of $F$ (cf. Sect. 8.6).

2. $\beta < \beta_0$: we choose $\lambda_0 > \lambda' = 5.22543\ldots$ and $\varepsilon > 0$. By Lemma 86 there exists an $n_0$ with

$$F(n, \beta, p) > \Phi(\beta, \lambda) - \varepsilon \tag{8.9.1}$$

for all $n \geq n_0, \lambda \leq \lambda_0$, and $p = \lambda/n$.
For a fixed $n \geq n_0$ we consider the source

$$S' \stackrel{\Delta}{=} \left(\frac{\lambda'}{n}, \ldots, \frac{\lambda'}{n}, 1 - (m-1)\frac{\lambda'}{n}\right).$$

Now it is

$$\begin{aligned} nR^n(T^\beta) &= n \sup_{S \in \Sigma} R^n_S(T^\beta) \geq nR^n_{S'}(T^\beta) \\ &= \sum_{x \in \mathcal{X}} F(n, \beta, p_{S'}(x)) + O(1/n) \\ &= (m-1) \cdot F\left(n, \beta, \frac{\lambda'}{n}\right) \\ &\quad + F\left(n, \beta, 1 - (m-1)\frac{\lambda'}{n}\right) + O(1/n) \\ &\stackrel{F \geq 0}{\geq} (m-1) \cdot F\left(n, \beta, \frac{\lambda'}{n}\right) + O(1/n) \\ &\stackrel{(1)}{>} (m-1)[\Phi(\beta, \lambda') - \varepsilon] + O(1/n) \\ &= (m-1)\Phi(\beta, \lambda') - (m-1)\varepsilon + O(1/n) \\ &> (m-1)\beta_0 - (m-1)\varepsilon + O(1/n). \end{aligned}$$

For the last inequality we used Lemma 87.
Choosing $\varepsilon > 0$ sufficiently small and $n$ sufficiently large yields

$$n \cdot R^n(T^\beta) > (m-1)\beta_0.$$

This proves the first statement of Theorem 82.

For $\varepsilon_1 > 0$, Lemma 83 yields a $\lambda_0$ with

$$F\left(n, \beta_0, \frac{\lambda}{n}\right) < \frac{1}{2}\left(1 - \frac{\lambda}{n}\right) + \varepsilon_1 \tag{8.9.2}$$

for all $\lambda > \lambda_0$ and all $n \in \mathbb{N}$.

Lemma 86 provides an $n_0 \in \mathbb{N}$ with

$$\left|\Phi(\beta_0, \lambda) - F\left(n, \beta_0, \frac{\lambda}{n}\right)\right| = F\left(n, \beta_0, \frac{\lambda}{n}\right) - \Phi(\beta_0, \lambda) < \varepsilon_1 \tag{8.9.3}$$

for all $n \geq n_0$, $\lambda \leq \lambda_0$.

For $n > n_0$ choose an arbitrary DMS $S \triangleq \left(\frac{\lambda_S(1)}{n}, \dots, \frac{\lambda_S(m)}{n}\right)$, $\lambda_S(1), \dots, \lambda_S(m) \in \mathbb{R}$. By (8.6.8) it is

$$nR_S^n(T^{\beta_0}) = \sum_{x \in \mathcal{X}} F\left(n, \beta_0, \frac{\lambda_S(x)}{n}\right) + O(1/n).$$

Choose $n$ sufficiently large, such that $O(1/n) < \varepsilon_1$. Then it is

$$nR_S^n(T^{\beta_0}) < \sum_{x \in \mathcal{X}} F\left(n, \beta_0, \frac{\lambda_S(x)}{n}\right) + \varepsilon_1. \tag{8.9.4}$$

Some of the numbers $\lambda_S(1), \lambda_S(2), \dots, \lambda_S(m)$ are greater $\lambda_0$, some are not. W.l.o.g. assume $\lambda_S(i) > \lambda_0$ for $1 \leq i \leq k$ and

$$\lambda_S(i) \leq \lambda_0 \qquad \text{for } k+1 \leq i \leq m, \tag{8.9.5}$$

$k \in \{0, 1, \dots, m\}$.

Obviously it is $\sum_{x \in \mathcal{X}} \lambda_S(x) = n$. This means, for $n$ tending to infinity there exists at least one $\lambda_S(y)$, $y \in \{1, \dots, m\}$, that exceeds $\lambda_0$. Hence, it is $k \in \{1, \dots, m\}$ and we can continue refining (8.9.4) by means of (8.9.2) and (8.9.3): for $k < m$ it is

$$\begin{aligned} nR_S^n(T^{\beta_0}) &< \sum_{x \in \mathcal{X}} F\left(n, \beta_0, \frac{\lambda_S(x)}{n}\right) + \varepsilon_1 \\ &= \sum_{i=1}^{k} F\left(n, \beta_0, \frac{\lambda_S(i)}{n}\right) + \sum_{i=k+1}^{m} F\left(n, \beta_0, \frac{\lambda_S(i)}{n}\right) + \varepsilon_1 \\ &< \sum_{i=1}^{k} \left[\frac{1}{2}\left(1 - \frac{\lambda_S(i)}{n}\right) + \varepsilon_1\right] + \sum_{i=k+1}^{m} [\Phi(\beta_0, \lambda_S(i)) + \varepsilon_1] + \varepsilon_1. \end{aligned}$$

If $k = m$, "each second sum" equals 0.

By Lemma 87 we know that the maximum of $\Phi(\beta_0, \lambda)$ is $\beta_0$, therefore it is

$$nR_S^n(T^{\beta_0}) < \frac{1}{2}\sum_{i=1}^{k}\left(1 - \frac{\lambda_S(i)}{n}\right) + (m-k)\beta_0 + (m+1)\varepsilon_1. \tag{8.9.6}$$

We consider the sum in (8.9.6). It is

$$\begin{aligned}\sum_{i=1}^{k}\left(1-\frac{\lambda_S(i)}{n}\right) &= \sum_{i=1}^{k} 1 - \frac{1}{n}\sum_{i=1}^{k}\lambda_S(i) = k - \frac{1}{n}\left(n - \sum_{i=k+1}^{m}\lambda_S(i)\right)\\ &= k-1+\frac{1}{n}\sum_{i=k+1}^{m}\lambda_S(i)\\ &\overset{(5)}{\leq} k-1+\frac{m}{n}\lambda_0.\end{aligned}$$

Using this bound we infer

$$\begin{aligned}nR_S^n(T^{\beta_0}) &< \frac{1}{2}\left(k-1+\frac{m}{n}\lambda_0\right)+(m-k)\beta_0+(m+1)\varepsilon_1\\ &< \frac{k-1}{2}+\frac{m}{n}\lambda_0+(m-k)\beta_0+(m+1)\varepsilon_1\\ &= m\beta_0 - k\underbrace{\left(\beta_0-\frac{1}{2}\right)}_{>0}-\frac{1}{2}+\frac{m}{n}\lambda_0+(m+1)\varepsilon_1\\ &< m\beta_0-\left(\beta_0-\frac{1}{2}\right)-\frac{1}{2}+\frac{m}{n}\lambda_0+(m+1)\varepsilon_1\\ &= \beta_0(m-1)+(m+1)\varepsilon_1+\frac{m}{n}\lambda_0.\end{aligned}$$

This means: for every $\varepsilon > 0$ and every source $S \in \Sigma$ we find an $n_1$ with

$$nR_S^n(T^{\beta_0}) < \beta_0(m-1)+\varepsilon \tag{8.9.7}$$

for all $n > n_1$.

On the other hand, for the source $S^* = (1, 0, \ldots, 0)$ and $\varepsilon > 0$ it is

$$\begin{aligned}nR^n(T^{\beta_0}) \geq nR_{S^*}^n(T^{\beta_0}) &= \sum_{x\in\mathcal{X}} F(n,\beta,p_{S^*}(x))+O(1/n)\\ &\geq (m-1)\beta_0+O(1/n)\\ &> (m-1)\beta_0-\varepsilon \end{aligned}\tag{8.9.8}$$

for sufficiently great $n$.

Combining (8.9.7) and (8.9.8) yields the second statement of Theorem 155. □

*Let $T \in \mathcal{G}(n)$ be an estimator for $p_S$. Then it is*

$$\liminf_{n\to\infty} n\cdot R^n(T) \geq \frac{m-1}{2}.$$

*Proof* The average is always a lower bound for the supremum. Therefore it is

$$R^n(T) = \sup_{S \in \Sigma} R_S^n(T) \geq \int_{\Sigma} R_S^n(T)\,\mathrm{d}\mu(S), \tag{8.9.9}$$

where $\mu$ denotes the uniformly distributed measure on the set $\Sigma$ of all DMSs. Hence, $\mu$ is the Dirichlet distribution on $\Sigma$ to the parameter $(1, 1, \dots, 1)$. Cover showed in [1] that the right side of (8.9.9) is minimized by $T^1$, the 1-biased estimator (cf. Sect. 8.5.6). Therefore it is

$$R^n(T) \geq \int_{\Sigma} R_S^n(T^1)\,\mathrm{d}\mu(S).$$

For a $\delta > 0$ we define a set $\Sigma_\delta \subset \Sigma$ by

$$\Sigma_\delta \triangleq \{(p_1, p_2, \dots, p_m) \in \Sigma : p_i \geq \delta \text{ for } \; i = 1, \dots, m\},$$

this means $\Sigma_\delta$ is "$\Sigma$ except the edge."

For $\varepsilon > 0$, we choose $\delta > 0$ in such a way that

$$\int_{\Sigma_\delta} \mathrm{d}\mu(S) \geq 1 - \varepsilon.$$

Due to the Noiseless Coding Theorem, the redundancy $R_S^n$ is nonnegative. Therefore it is

$$R^n(T) \geq \int_{\Sigma_\delta} R_S^n(T^1)\,\mathrm{d}\mu(S).$$

For all $S \in \Sigma_\delta$, we can apply Lemma 84 because $\Sigma_\delta$ does not contain any source S, whose probability distribution is not strictly positive:

$$\begin{aligned} nR_S^n(T^1) &\geq \frac{m-1}{2} - \frac{1}{n}\left[\frac{\sigma_S}{3} + \frac{2}{3} + \beta\sigma_S + \frac{m^2\beta^2}{2}\right] - \frac{\beta^3}{3n^2}\sum_{x \in \mathcal{X}} \frac{1}{p_S(x)^2} \\ &\geq \frac{m-1}{2} - \frac{C}{n} - \frac{C}{n^2} \end{aligned}$$

for constants $C$ that depend on $m$ and $\delta$, but not on $n$.

($\sigma_S$ and $\sum_{x \in \mathcal{X}} \frac{1}{p_S(x)^2}$ are continuous functions on $\Sigma$. $\Sigma$ can be identified with a compact subset of $\mathbb{R}^m$. Consequently, both functions assume their minimum and maximum in $\Sigma$ and are bound by these on $\Sigma_\delta$, since $\Sigma_\delta$ is a subset of $\Sigma$. Therefore, we can choose the constants $C$ appropriately.)

Now it is

$$
\begin{aligned}
nR^n(T) &\geq \int_{\Sigma_\delta} nR^n_S(T^1)\,\mathrm{d}\mu(S) \geq \int_{\Sigma_\delta} \left[\frac{m-1}{2} - \frac{C}{n} - \frac{C}{n^2}\right] \mathrm{d}\mu(S) \\
&\geq \left[\frac{m-1}{2} - \frac{C}{n} - \frac{C}{n^2}\right](1-\varepsilon) \\
&\xrightarrow{n\to\infty} \frac{m-1}{2}(1-\varepsilon)
\end{aligned}
$$

for every $\varepsilon > 0$. This induces

$$\liminf_{n\to\infty} nR^n(T) = \frac{m-1}{2}. \qquad \square$$

## 8.10 Lecture on Distributions

### *8.10.1 The Poisson Distribution*

The Poisson distribution to the parameter $\lambda > 0$ is defined by

$$\pi(k,\lambda) \;\triangleq\; e^{-k} \cdot \frac{\lambda^k}{k!}, \quad k \in \mathbb{N}_0.$$

Poisson's Limit Theorem states that binomial distributions converge to the Poisson distribution:

If $np \xrightarrow{n\to\infty} \lambda > 0$, it is

$$\lim_{n\to\infty} b(k;n,p) = \pi(k,\lambda).$$

### *8.10.2 The Beta Distribution*

A beta distribution to the parameters $\kappa, \lambda > 0$ has the density function

$$\varphi_{\kappa,\lambda}(x) \;\triangleq\; \frac{\Gamma(\kappa+\lambda)}{\Gamma(\kappa)\cdot\Gamma(\lambda)} \cdot x^{\kappa-1}(1-x)^{\lambda-1}, \quad 0 \leq x \leq 1,$$

where $\Gamma$ denotes the gamma function,

$$\Gamma(x) \;\triangleq\; \int_0^\infty t^{x-1}e^{-t}\,\mathrm{d}t$$

for all $x > 0$. The equation

$$\int_0^1 x^{\alpha-1}(1-x)^{\beta-1}\,\mathrm{d}x = \frac{\Gamma(\alpha)\cdot\Gamma(\beta)}{\Gamma(\alpha+\beta)} \tag{8.10.1}$$

holds. Consequently it is

$$\int_0^1 \varphi_{\kappa,\lambda}(x)\,\mathrm{d}x = 1$$

and the average value of a beta distributed random variable is

$$\begin{aligned}\int_0^1 x\cdot\varphi_{\kappa,\lambda}(x)\,\mathrm{d}x &= \frac{\Gamma(\kappa+\lambda)}{\Gamma(\kappa)\cdot\Gamma(\lambda)}\cdot\frac{\Gamma(\kappa+1)\cdot\Gamma(\lambda)}{\Gamma(\kappa+\lambda+1)}\\ &= \frac{\Gamma(\kappa+\lambda)}{\Gamma(\kappa)\cdot\Gamma(\lambda)}\cdot\frac{\kappa\cdot\Gamma(\kappa)\cdot\Gamma(\lambda)}{(\kappa+\lambda)\cdot\Gamma(\kappa+\lambda)}\\ &= \frac{\kappa}{\kappa+\lambda}.\end{aligned}$$

### 8.10.3 The Dirichlet Distribution

The Dirichlet distribution may be considered as a generalization of the beta distribution.

A $(k-1)$-*dimensional Dirichlet distribution to the parameter* $\lambda = (\lambda_1, \lambda_2, \ldots, \lambda_k)$, $\lambda_j > 0$ *for all* $j$, has the density function

$$f_\lambda(x_1, x_2, \ldots, x_{k-1}) = C(\lambda)\cdot {x_1}^{\lambda_1-1}{x_2}^{\lambda_2-1}\cdot\ldots\cdot {x_{k-1}}^{\lambda_{k-1}-1}\left(1-\sum_{j=1}^{k-1}x_j\right)^{\lambda_k-1}$$

for $x_j \geq 0$ for all $j$, $\sum_{j=1}^{k-1} x_j \leq 1$, where

$$C(\lambda) = \frac{\Gamma(\lambda_1+\lambda_2+\cdots+\lambda_k)}{\Gamma(\lambda_1)\Gamma(\lambda_2)\cdot\ldots\cdot\Gamma(\lambda_k)}.$$

By applying the substitution rule, we verify

$$\int_0^{1-a} t^{k-1}(1-a-t)^{l-1}\,\mathrm{d}t = (1-a)^{k+l-1}\int_0^1 x^{k-1}(1-x)^{l-1}\,\mathrm{d}x \tag{8.10.2}$$

for all $0 \le a \le 1$, $k, l > 0$.

Successive application of (8.10.1) and (8.10.2) yields

$$\int_0^1 \int_0^{1-x_1} \cdots \int_0^{1-\sum_{j=1}^{k-2} x_j} x_1^{\lambda_1 - 1} x_2^{\lambda_2 - 1}, \dots, x_{k-1}^{\lambda_{k-1}-1} \left(1 - \sum_{j=1}^{k-1} x_j\right)^{\lambda_k - 1} \mathrm{d}x_{k-1}\, \mathrm{d}x_{k-2}, \dots, \mathrm{d}x_1$$

$$= \int_0^1 \int_0^{1-x_1} \dots x_1^{\lambda_1 - 1}, \dots, x_{k-2}^{\lambda_{k-2}-1} \int_0^{1-\sum_{j=1}^{k-2} x_j} x_{k-1}^{\lambda_{k-1}-1} \left(1 - \sum_{j=1}^{k-2} x_j - x_{k-1}\right)^{\lambda_k - 1} \mathrm{d}x_{k-1}\, \mathrm{d}x_{k-2}, \dots, \mathrm{d}x_1$$

$$\overset{(2)}{=} \int_0^1 \int_0^{1-x_1} \dots x_1^{\lambda_1 - 1}, \dots, x_{k-2}^{\lambda_{k-2}-1} \left(1 - \sum_{j=1}^{k-2} x_j\right)^{\lambda_{k-1}+\lambda_k - 1} \int_0^1 x^{\lambda_{k-1}-1}(1-x)^{\lambda_k - 1}\, \mathrm{d}x\, \mathrm{d}x_{k-2}, \dots, \mathrm{d}x_1$$

$$\overset{(1)}{=} \int_0^1 \int_0^{1-x_1} \dots x_1^{\lambda_1 - 1}, \dots, x_{k-2}^{\lambda_{k-2}-1} \left(1 - \sum_{j=1}^{k-2} x_j\right)^{\lambda_{k-1}+\lambda_k - 1} \frac{\Gamma(\lambda_{k-1})\Gamma(\lambda_k)}{\Gamma(\lambda_{k+1} + \lambda_k)} \mathrm{d}x_{k-2}, \dots, \mathrm{d}x_1$$

$$= \cdots$$

$$= \frac{\Gamma(\lambda_{k-1})\Gamma(\lambda_k)}{\Gamma(\lambda_{k-1} + \lambda_k)} \cdot \frac{\Gamma(\lambda_{k-2})\Gamma(\lambda_{k-1} + \lambda_k)}{\Gamma(\lambda_{k-2} + \lambda_{k-1} + \lambda_k)} \cdot \dots \cdot \frac{\Gamma(\lambda_1)\Gamma(\lambda_2 + \dots + \lambda_k)}{\Gamma(\lambda_1 + \dots + \lambda_k)}$$

$$= \frac{\Gamma(\lambda_1)\Gamma(\lambda_2) \cdot \dots \cdot \Gamma(\lambda_k)}{\Gamma(\lambda_1 + \lambda_2 + \dots + \lambda_k)}. \tag{8.10.3}$$

Thus

$$\int_\Sigma f_\lambda(x)\, \mathrm{d}x = 1.$$

This means: A $(k-1)$-dimensional Dirichlet distribution establishes a density on the $(k-1)$-dimensional simplex $\Sigma$ in $\mathbb{R}^k$.

## 8.10.4 The Central Moments

Let $(\Omega, \mathcal{A}, P)$ be a probability space. The $n$th central moment of a random variable $X : \Omega \longrightarrow \mathbb{R}$ is defined to be

$$\begin{aligned}\mu_n(X) &\stackrel{\triangle}{=} \mathbb{E}((X - \mathbb{E}(X))^n)\\ &= \mathbb{E}((X-\mu)^n),\end{aligned}$$

where $\mu \stackrel{\triangle}{=} \mathbb{E}(X)$. Taking advantage of the linearity of the expected value of a random variable, we easily obtain the following expressions for the first four central moments:

$$\begin{aligned}\mu_1(X) &= 0\\ \mu_2(X) &= \mathbb{E}(X^2) - \mu^2\\ \mu_3(X) &= \mathbb{E}(X^3) - 3\mu\mathbb{E}(X^2) + 2\mu^3\\ \mu_4(X) &= \mathbb{E}(X^4) - 4\mu\mathbb{E}(X^3) + 6\mu^2\mathbb{E}(X^2) - 3\mu^4. \end{aligned} \tag{8.10.4}$$

Let $X$ be a binomial distributed random variable to the parameters $n, p$; $q \stackrel{\triangle}{=} 1 - p$. This yields

$$\begin{aligned}\mu = \mathbb{E}(X) &= \sum_{k=1}^{n} k \cdot \binom{n}{k} p^k q^{n-k} = np \sum_{k=0}^{n-1} \binom{n-1}{k} p^k q^{n-1-k} = np,\\ \mathbb{E}(X^2) &= \sum_{k=0}^{n} k^2 \cdot \binom{n}{k} p^k q^{n-k} = \sum_{k=2}^{n} k(k-1) \cdot \binom{n}{k} p^k q^{n-k} + \mu\\ &= \sum_{k=2}^{n} \frac{n!}{(k-2)!(n-k)!} p^k q^{n-k} + \mu\\ &= n(n-1)p^2 + np. \end{aligned} \tag{8.10.5}$$

By the same "trick" it is

$$\begin{aligned}\mathbb{E}(X^3) &= n(n-1)(n-2)p^3 + 3(n(n-1)p^2 + np) - 2np\\ \mathbb{E}(X^4) &= n(n-1)(n-2)(n-3)p^4 + 6n(n-1)(n-2)p^3\\ &\quad + 7n(n-1)p^2 + np.\end{aligned}$$

Inserting these terms in (8.10.4) yields, after some calculations, the first four central moments for a binomial distributed random variable $X$ to the parameters $n, p$:

$$\mu_1 \stackrel{\triangle}{=} \mu_1(X) = 0$$

$$\mu_2 \stackrel{\Delta}{=} \mu_2(X) = npq$$

$$\mu_3 \stackrel{\Delta}{=} \mu_3(X) = npq(q-p)$$

$$\mu_4 \stackrel{\Delta}{=} \mu_4(X) = 3n^2p^2q^2 + npq(1-6pq).$$

### 8.10.5 Minimum of $\sigma_{\mathbf{p}}$

**Lemma** *For a probability distribution* $\mathbf{p} = (p_1, \dots, p_m)$ *on* $\mathcal{X}$, $p_i > 0$ *for all* $i$, $\sigma_{\mathbf{p}} = \sum_{x\in\mathcal{X}} \frac{1}{p_x}$ *is minimal if* $\mathbf{p}$ *is uniformly distributed on* $\mathcal{X}$ *and its minimum is* $m^2$.

*Proof* Define

$$M \stackrel{\Delta}{=} \left\{(p_1, \dots, p_{m-1}) \in \mathbb{R}^{m-1} : p_i > 0 \ \text{ for all } \ i \ \text{ and } \ \sum_{i=1}^{m-1} p_i < 1\right\}$$

and consider the function $f : M \to \mathbb{R}$,

$$f(p_1, \dots, p_{m-1}) \stackrel{\Delta}{=} \sum_{k=1}^{m-1} \frac{1}{p_k} + \frac{1}{1 - \sum_{k=1}^{m-1} p_k}.$$

It is

$$\frac{\partial}{\partial p_i} f(p_1, \dots, p_{m-1}) = -\frac{1}{p_i^2} + \frac{1}{\left(1 - \sum_{k=1}^{m-1} p_k\right)^2}.$$

f assumes a local extremum in $p$, iff $\operatorname{grad} f(p) = 0$. Here it is

$$\operatorname{grad} f(p) = 0 \Leftrightarrow \sum_{k\neq i} p_k + 2p_i = 1 \qquad \text{for all } i = 1, 2, \dots, n$$

and $p = (p_1, \dots, p_{m-1}) \in M$. The unique solution of these $n$ linear equations is

$$p_0 \stackrel{\Delta}{=} \left(\frac{1}{m}, \dots, \frac{1}{m}\right) \in M.$$

$f$ assumes a finite value $f(p)$ for every $p \in M$, but $f$ tends to infinity for every sequence $(p_k)_{k\in\mathbb{N}} \subset M$ that converges to "the edge" of $M$, $p_0$ is the only local extremum in $M$.
Consequently, $f$ assumes a global minimum in $p_0$ and this minimum is $f(p_0) = m^2$.
□

## References

1. T.M. Cover, Admissibility of Gilbert's encoding for unknown source probabilities. IEEE Trans. Inf. Theory **18**(1), 216–217 (1972)
2. T.M. Cover, J.A. Thomas, *Elements of Information Theory*, Wiley Series in Telecommunications (Wiley, New York, 1991)
3. W. Feller, *An Introduction to Probability Theory and Its Applications*, 3rd edn, vol. I (Wiley, New York, 1968)
4. E.N. Gilbert, Codes based on inaccurate source probabilities. IEEE Trans. Inf. Theory **3**, 304–314 (1971)
5. R. Krichevsky, *Universal Compression and Retrieval* (Kluwer Academic Publisher, Dordrecht, 1994)
6. R. Krichevsky, Laplace's law of succession and universal encoding. IEEE Trans. Inf. Theory **44**(1), 296–303 (1998)
7. F. Willems, Y. Shtarkov, T. Tjalkens, The context tree weighting method: basic properties. IEEE Trans. Inf. Theory **IT–41**, 653–664 (1995)

## Further Reading

8. G. Box, G. Tiao, *Bayesian Inference in Statistical Analysis* (Wiley, New York, 1992)
9. H. Cramér, *Mathematical Methods of Statistics* (Princeton University Press, Princeton, 1974)
10. R. Krichevsky, Laplace's Law of Succession and Universal Coding, Preprint, Institute of Mathematics of the Siberian Branch of the Russian Academy of Sciences, No. 21, Novosibirsk (1995)
11. K. Stange, *Bayes-Verfahren* (Springer, Berlin, 1977)

# About the Author

## In Memoriam Rudolf Ahlswede 1938–2010[1]

by Ning Cai, Imre Csiszar, Kingo Kobayashi, and Ulrich Tamm

Rudolf Ahlswede, a mathematician, one of the truly great personalities of Information Theory, passed away on December 18, 2010 in his house in Polle, Germany, due to a heart attack. He is survived by his son Alexander. His untimely death, when he was still very actively engaged in research and was full with new ideas, is an irrecoverable loss for the IT community.

Ahlswede was born on September 15, 1938 in Dielmissen, Germany. He studied Mathematics, Philosophy and Physics in Göttingen, Germany, taking courses, among others, of the great mathematicians Carl Ludwig Siegel and Kurt Reidemeister. His interest in Information Theory was aroused by his advisor Konrad Jacobs, of whom many students became leading scientists in Probability Theory and related fields.

In 1967 Ahlswede moved to the US and became Assistant Professor, later Full Professor at Ohio State University, Columbus. His cooperation during 1967–1971 with J. Wolfowitz, the renowned statistician and information theorist, contributed to his scientific development. Their joint works included two papers on arbitrarily varying channels (AVCs), a subject to which Ahlswede repeatedly returned later.

His first seminal result was, however, the coding theorem for the (discrete memoryless) multiple-access channel (MAC). Following the lead of Shannon's Two-Way Channel paper, this was one of the key results originating Multiuser Information Theory (others were those of T. Cover on broadcast channels and of D. Slepian and J. Wolf on separate coding of correlated sources), and it was soon followed by an extension to two-output MACs, requiring new ideas. Also afterwards, Ahlswede continued to be a major contributor to this research direction, in collaboration with J. Körner (visiting in Columbus in 1974) and later also with other members of the Information Theory group in Budapest, Hungary. In addition to producing joint

---

[1] This obituary first appeared in IEEE Information Theory Society Newsletter, Vol. 61, No. 1, 7–8 2011.



A. Ahlswede et al. (eds.), *Transmitting and Gaining Data*,
Foundations in Signal Processing, Communications and Networking 11,
DOI 10.1007/978-3-319-12523-7

papers enriching the field with new results and techniques, this collaboration also contributed to the Csiszár–Körner book where several ideas are acknowledged to be due to Ahlswede or have emerged in discussions with him.

In 1975 Ahlswede returned to Germany, accepting an offer from Universität Bielefeld, a newly established "research university" with low teaching obligations. He was Professor of Mathematics there until 2003, and Professor Emeritus from 2003 to 2010. For several years he devoted much effort to building up the Applied Mathematics Division, which at his initiative included Theoretical Computer Science, Combinatorics, Information Theory, and Statistical Physics. These administrative duties did not affect his research activity. He was able to develop a strong research group working with him, including visitors he attracted as a leading scientist, and good students he attracted as an excellent teacher. In the subsequent years Ahlswede was heading many highly fruitful research projects, several of them regularly extended even after his retirement which is quite exceptional in Germany. The large-scale interdisciplinary project "General Theory of Information Transfer" (Center of Interdisciplinary Research, 2001–2004) deserves special mentioning. It enabled him to pursue very productive joint research with many guests and to organize several conferences. An impressive collection of new scientific results obtained within this project was published in the book "General Theory of Information Transfer and Combinatorics" (Lecture Notes in Computer Science, Springer, 2006).

During his research career Ahlswede received numerous awards and honours. He was recipient of the Shannon Award of the IEEE IT Society in 2006, and previously twice of the Paper Award of the IT Society (see below). He was member of the European Academy of Sciences, recipient of the 1998/99 Humboldt-Japan Society Senior Scientist Award, and he received honorary doctorate of the Russian Academy of Sciences in 2001. He was also honored by a volume of 50 articles on the occasion of his 60th birthday (Numbers, Information and Complexity, Kluwer, 2000).

Ahlswede's research interests included also other fields of Applied and Pure Mathematics, such as Complexity Theory, Search Theory (his book "Search Problems" with I. Wegener is a classic), Combinatorics, and Number Theory. Many problems in these disciplines that aroused Ahlswede's interest had connections with Information Theory, and shedding light on the interplay of IT with other fields was an important goal for him. He was likely the first to deeply understand the combinatorial nature of many IT problems, and to use tools of Combinatorics to solve them.

In the tradition of giants as Shannon and Kolmogorov, Ahlswede was fascinated with Information Theory for its mathematical beauty rather than its practical value (of course, not underestimating the latter). In the same spirit, he was not less interested in problems of other fields which he found mathematically fascinating. This is not the right place to discuss his (substantial) results not related to IT. We just mention the celebrated Ahlswede–Daykin "Four Functions Theorem" having many applications in Statistical Physics and in Graph Theory, and the famous Ahlswede-Khachatrian "Complete Intersection Theorem". The latter provided the final solution of a problem of Paul Erdős, which had been very long-standing even though Erdős offered $500— for the solution (Ahlswede and Khachatrian collected). For more on this, and also on combinatorial results of information theoretic interest, see his book "Lectures on Advances in Combinatorics" with V. Blinovsky (Springer, 2008).

Even within strict sense Information Theory, Ahlswede's contributions are too wide-ranging for individual mentioning, they extend as far as the formerly exotic but now highly popular field of Quantum Information Theory. Still, many of his main results are one of the following two kinds.

On the one hand, Ahlswede found great satisfaction in solving hard mathematical problems. Apparently, this is why he returned again and again to AVCs, proving hard results on a variety of models. By his most famous AVC theorem, the (average error) capacity of an AVC either equals its random code capacity or zero. Remarkably, this needed no hard math at all, "only" a bright idea, the so-called elimination technique (a kind of derandomization). He was particularly proud of his solution of the AVC version of the Gelfand-Pinsker problem about channels with non-causal channel state information at the sender. To this, the elimination technique had to be combined with really hard math. Another famous hard problem he solved was the "zero excess rate" case of the Multiple Descriptions Problem (the general case is still unsolved).

On the other hand, Ahlswede was eager to look for brand new or at least little studied models, and was also pleased to join forces with coauthors suggesting work on such models. His most frequently cited result (with Cai, Li and Yeung), the Min-Cut-Max-Flow Theorem for communication networks with one source and any number of sinks, belongs to this category. So do also his joint results with Csiszár on hypothesis testing with communication constraints, and with Dueck on identification capacity, receiving the Best Paper Award of the IT Society in 1988 and 1990. Later on, Ahlswede has significantly broadened the scope of the theory of identification, for example to quantum channels (with Winter). Further, a two-part joint paper with Csiszár provides the first systematic study of the concept of common randomness, both secret and non-secret, relevant, among others, for secrecy problems and for identification capacity. The new kind of problems studied in these papers support Ahlswede's philosophical view that the real subject of information theory should be the broad field of "information transfer", which is currently unchartered and only some of its distinct areas (such as Shannon's theory of information transmission and the Ahlswede-Dueck theory of identification) are in view. Alas, Rudi is no longer with us, and extending information theory to cover such a wide scope of yet unknown dimensions will be the task of the new generation.

# Comments by Gerhard Kramer

Rudolf Ahlswede played a key role in the development of information theory during 1970–2010. He contributed to all areas of Shannon theory, including arbitrarily varying (AV) channels, identification, secrecy, and information networks. As a graduate student, I always enjoyed listening to his lectures at conferences and meetings, not only because of the technical content (for which I usually lacked the knowledge to understand the details) but because the lectures were entertaining and full of surprises. For example, the first Ahlswede lectures that I attended were at the IEEE International Symposium on Information Theory in Trondheim, Norway, in 1994. Professor Ahlswede had co-authored four papers:

(1) "Erasure, list and detection zero-error capacities for low noise and a relation to identification";
(2) "On interactive communication";
(3) "Localized random and arbitrary errors in the light of AV channel theory"; and
(4) "Identification via wiretap channels".

Two of these talks were scheduled in succession, giving him $2 \times 20$ min to speak. After the first talk, Professor Thomas Cover asked for a copy of the full paper. Ahlswede was obviously pleased by the request, he grabbed a copy that he had with him, he sprinted up the stairs with a speed that astonished the audience, and he proudly handed over the document. For the second talk, Ahlswede spent perhaps three minutes describing the problem, he looked up at the audience, a look of sorrow crossed his face, he stated "You have had enough", and he simply stopped. The audience was not displeased to have a longer coffee break. Such light moments were part of every Ahlswede lecture that I attended, and it demonstrated this brilliant mathematician's desire for interaction with his audience.

I began to study Ahlswede's work in 2002 or so, when I was interested in multiple-description coding. Later on, his foundational work on network coding directly influenced my work on networks with broadcast constraints. His papers were often difficult to read, but once one understood the proof, one marveled at the creativity and insight with which the ideas were presented.



A. Ahlswede et al. (eds.), *Transmitting and Gaining Data*,
Foundations in Signal Processing, Communications and Networking 11,
DOI 10.1007/978-3-319-12523-7

The final time I met Professor Ahlswede was in 2010 at the International ITG Conference on Source and Channel Coding in Siegen, Germany. He had grown older but was as passionate as always about research and community. During a memorable introduction of James Massey, he emphasized the importance of bringing together researchers from different disciplines to solve fundamental research problems. His introduction went on for more than 15 min, yet the session chair didn't dare to stop Rudolf Ahlswede. The Information Theory Society was very fortunate to have such a dedicated and intellectually provocative mathematician contributing to the foundations of our field.

Gerhard Kramer

# Notations

| | |
|---|---|
| $\mathcal{X} = \{1, 2, \ldots, m\}$ | finite alphabet |
| $\mathcal{B}$ | cardinality of $\mathcal{X}$ |
| $\mathcal{X}^n$ | set of all sequences of length $n$ over $\mathcal{X}$ |
| $\mathcal{X}^*$ | set of all finite sequences over $\mathcal{X}$ |
| $p_S$ | probability distribution of the source $S$ on $\mathcal{X}$ |
| $S = (\mathcal{X}, p_S)$ | discrete memoryless source $S$ |
| DMS | discrete memoryless source |
| $\Sigma$ | set of all sources over $\mathcal{X}$ |
| $\Sigma_\delta$ | set of all sources over $\mathcal{X}$ which are $\delta$-bounded from 0 probabilities |
| $H(p_S)$ | entropy of $p_S$ |
| $H_S$ | scaled entropy of $p_S$ |
| $\log_b$ | logarithm to base $b$ |
| log | logarithm to base $\mathcal{B}$ |
| ln | natural logarithm |
| $\langle x^n \mid x \rangle$ | frequency of $x$ in $x^n$ |
| $\mathbb{E}(X)$ | expected value of $X$ |
| $\mathbb{V}(X)$ | variance of $X$ |
| $\mathcal{G}(n)$ | set of all estimators for $p_S$ |
| $\mathcal{G}_x(n)$ | set of all estimators for $p_S(x)$ |
| $\mathbb{E}(n)$ | set of all $\beta$-biased estimators for $p_S$ |
| $\mathbb{E}_x(n)$ | set of all $\beta$-biased estimators for $p_S(x)$ |
| $\hat{p}_S(x \mid x^n)$ | estimated value for $p_S(x)$ |
| $l_S^T(x \mid x^n)$ | length of the codeword for $x$ |
| $\overline{l}_S^T(x^n)$ | average length of a code (depending on $x^n$) |
| $\overline{l}_S^T(n)$ | average length of a code |
| $R_S^n(T)$ | redundancy of $T$ on $S$ |
| $R^n(T)$ | maximal redundancy of $T$ on $\Sigma$ |
| $\overline{R}^n(T)$ | average redundancy of $T$ on $\Sigma$ |
| $\mu$ | probability measure on $\Sigma$ |



A. Ahlswede et al. (eds.), *Transmitting and Gaining Data*,
Foundations in Signal Processing, Communications and Networking 11,
DOI 10.1007/978-3-319-12523-7

| | |
|---|---|
| $\mathbb{N}$ | set of natural numbers $\{1, 2, 3, \ldots\}$ |
| $\mathbb{N}_0$ | $\mathbb{N} \cup \{0\}$ |
| $\mathbb{R}$ | set of real numbers |
| $\mathbb{R}^+$ | set of positive real numbers |
| $\lambda_S(x)$ | $np_S(x)$ |
| $\sigma_S$ | $\sum \frac{1}{p_S(x)}$ |
| $\varphi_{\kappa,\lambda}$ | beta distribution to the parameters $\kappa, \lambda > 0$ |
| $\Gamma$ | gamma function |
| $\mu_n(X)$ | $n$-th central moment of $X$ |
| $\lceil x \rceil$ | smallest integer larger or equal $x$ |
| $\mathcal{W}$ | a class of channels called matching channel |
| $\mathcal{W}_0$ | deterministic matching channel |
| $W$ | DMC associated with $\mathcal{W}_0$ |
| $\mathcal{X}_W(\cdot), \overline{Z}(n,p)_{V,c}(\cdot)$ | column supports of matrices |
| $\mathcal{Y}_W(\cdot), \overline{Z}(n,p)_{V,r}(\cdot)$ | row supports of matrices |
| MC | matching codes |
| MDC | matching zero-error detection codes (for deterministic channels) |
| MDCF | matching zero-error detection codes with feedback |
| $M^n_{\text{de}}(W)$ | largest size of zero-error detection codes for $W^n$ |
| $M^n_{\text{mde}}(W)$ | largest size of MDC for $W^n$ |
| $C(\mathcal{W})$ | capacity of matching channel $\mathcal{W}$ |
| $C(\mathcal{W}_0)$ | capacity of deterministic matching channels $\mathcal{W}_0$ |
| $C_f(\mathcal{W})$ | capacity of matching codes with feedback for $\mathcal{W}$ |
| $C(\geq \cdot, < \cdot)$ | various second order "identification capacities" for a DMC |
| $M^n_{--}(W)$ | the largest sizes |
| $M^n_{-+}(W)$ | of 4 kinds of |
| $M^n_{+-}(W)$ | pseudo-matching |
| $M^n_{++}(W)$ | 0-error detection codes for channel $W^n$ |
| $M^n_0(W)$ | the largest size of zero-error codes for $W^n$ |
| $\mathcal{G}_1 \otimes \mathcal{G}_2$ | product of graphs $\mathcal{G}_1$ and $\mathcal{G}_2$ |
| $\mathcal{G}^{\otimes n}$ | $n$th power of graph $\mathcal{G}$ |
| $d_{\mathcal{G}}(v)$ | degree of vertex $v$ in graph $\mathcal{G}$ |
| $\Gamma_{\mathcal{G}}(v)$ | vertices connected with $v$ |
| $\nu(\mathcal{G})$ | matching number of graph $\mathcal{G}$ |
| $\tau(\mathcal{G})$ | vertex covering number of $\mathcal{G}$ |
| $\gamma(\mathcal{G})$ | $\lim_{n\to\infty} \frac{1}{n} \log \nu(\mathcal{G}^{\otimes n})$ |
| $\mathcal{K}(\mathcal{G})$ | König-Hall pair of distributions |
| $T^n_P, T^n_Z, \overline{Z}(n,p)^n(P)$ | the set of $P$-typical sequences |
| $T^n_{V,\delta}(x^n)$ | $(V, \delta)$-generated sequences of $x^n$ |

# Author Index





A. Ahlswede et al. (eds.), *Transmitting and Gaining Data*,
Foundations in Signal Processing, Communications and Networking 11,
DOI 10.1007/978-3-319-12523-7

# Subject Index



 
A. Ahlswede et al. (eds.), *Transmitting and Gaining Data*,
Foundations in Signal Processing, Communications and Networking 11,
DOI 10.1007/978-3-319-12523-7

Printed in the United States
By Bookmasters